Introduction to Telecommunications

Voice, Data, and the Internet

Second Edition

Marion Cole
Samsung Telecommunications America, Inc.

Upper Saddle River, New Jersey
Columbus, Ohio

Library of Congress Cataloging-in-Publication Data

Cole, Marion.
  Introduction to telecommunications : voice, data, and the Internet / Marion Cole.
    p. cm.
  ISBN 0-13-060890-4
  1. Telecommunication.   I. Title.

  TK5101 .C63496 2002
  621.382--dc21

2001036798

Editor in Chief: Stephen Helba
Assistant Vice President and Publisher: Charles E. Stewart, Jr.
Production Editor: Alexandrina Benedicto Wolf
Production Coordination: Custom Editorial Productions, Inc.
Design Coordinator/Cover Designer: Diane Ernsberger
Cover Image: Stock Market
Production Manager: Matthew Ottenweller

This book was set in Palatino by Custom Editorial Productions, Inc. It was printed and bound by R.R. Donnelley & Sons Company. The cover was printed by Phoenix Color Corp.

Pearson Education Ltd., *London*
Pearson Education Australia Pty. Limited, *Sydney*
Pearson Education Singapore, Pte. Ltd.
Pearson Education North Asia Ltd., *Hong Kong*
Pearson Education Canada, Ltd., *Toronto*
Pearson Educación de Mexico, S.A. de C.V.
Pearson Education—Japan, *Tokyo*
Pearson Education Malaysia, Pte. Ltd.
Pearson Education, *Upper Saddle River, New Jersey*

10 9 8 7 6 5 4 3 2 1
ISBN: 0-13-060890-4

Dedicated to my loving wife Mary and my children
John, Christine, Kim, and Mark

In memory of Kenneth Michael Cole

Contents

CHAPTER 5   THE MEDIUM   153

Preface

The telecommunications industry has experienced tremendous changes during the last 25 years on both the legislative and technical fronts. This book was written to cover some of the most important changes that have occurred. Older technologies are covered briefly but with enough depth to show readers how the technologies have changed and what benefits the new developments offer. The sections on voice, data, and the Internet are written for nontechnical individuals and gives an excellent overview of existing technology to allow you to better manage voice and data networks.

This edition provides updated information on the various components that make up the PSTN, the packet (or public) data network (PDN), and the Internet. The public switched telephone network (PSTN), has evolved from a voice-only analog network to a digital network handling both voice and data. The on-ramp to the PSTN remains mostly an analog twisted copper wire pair. Technologies such as Integrated Services Digital Network (ISDN) and Asymmetric Digital Subscriber Line (ADSL) are being implemented to change this last remaining analog component to a digital-access medium. Many analog and digital technologies are discussed in depth, but the primary focus is on how the technology works, not how to repair systems or components. That level of detail can be found in an electronics engineering curriculum or in a manufacturing school.

Technicians must attend a manufacturer's school to learn a particular system and how to perform diagnostics on it. The technology behind a particular component in the PSTN is basically the same for all manufacturers of the component. Each manufacturer puts its own twist on a particular generic technology to turn it into a proprietary system with its brand name on it. This book will help you understand many telecommunications technologies in the generic sense. A technician with a basic understanding of technology will easily understand and gain much more from the manufacturer's training program.

Appendixes A–F round out the book by providing an in-depth treatment of basic electricity and other topics. Material on the basic principles of electricity in Appendixes A, B, and C is available to professors who find this information critical to their introductory classes. When telecommunications is taught as a class in an electronics engineering program, coverage of electrical principles as applied to telecommunications is essential, but becomes less essential for a class in a business administration or telecommunications management program. My course starts with legislation because I think it is important for telecommunications managers to understand how legislation shapes the industry and the services available. This approach also allows the class to start at a better pace than if we tried going right into basic electricity discussions.

OVERVIEW

The introductory chapters of *Introduction to Telecommunications: Voice, Data, and the Internet* deal with the evolution of telecommunications (Chapters 1 and 3) and with the relevant legislative history (Chapter 2). Chapter 4 discusses station equipment, Chapter 5 focuses on the twisted-pair local loop and other media, and Chapter 6 covers multiplexing. Chapter 7 treats analog and digital signals. Chapter 8 covers data communication, and Chapter 9 discusses ISDN and ADSL. Chapters 10 and 11 take up local area networks and wide area networks, respectively. Chapter 12 surveys Internet services, Chapter 13 is devoted to mobile telephones and personal communication systems, and Chapter 14 turns to management issues.

ACKNOWLEDGMENTS

First, I must thank Charles Stewart of Prentice Hall for his encouragement to write this book. I also want to thank Delia Uherec of Prentice Hall for her support and encouragement. In addition, I thank Elizabeth Judd and Gina Smith for their tremendous help in bringing the manuscript in line with standard publishing conventions. The book would have taken a very different form without their editorial expertise.

I drew on my knowledge of telecommunications technology and experience in managing telecommunications operations to write this book. My knowledge of telecommunications was gained from reading books, attending company and manufacturers' schools, and on-the-job training. What I know, I learned from other people. I thank all of my past teachers and coworkers.

A special thanks to the professors using this book and for the feedback provided. Your support, suggestions, and positive comments make the effort that went into writing this book rewarding. I appreciate the invaluable suggestions provided by the following reviewers: Maurice Fritz, Texas State Technical College; Bill Gann, Carl Albert State College, Oklahoma; John Johnson, Ayers State Technical College, Alabama; and M.T. Martinez, Mt. Sierra College, California.

Thanks to all who have adopted this book or offered suggestions. I hope this preface gives you some insight into how we arrived at the final product for this second edition.

1

Introduction to Telecommunications

KEY TERMS

Analog Signal
Automatic Call Distributor (ACD)
Binary Signal
Central Office
Codec
Computer Telephony Integrated
  (CTI)
Data Communication
Dense Wave Division
  Multiplexer (DWDM)
Digital Signal
Ethernet
Hypertext Markup Language
  (HTML)

Hypertext Transport Protocol
  (HTTP)
Interactive-Voice-Response (IVR)
Interexchange Carrier (IEC)
Internet
Keysystem
Line Circuit
Local Area Network (LAN)
Local Exchange Carrier (LEC)
Microwave Radio
Modem
Multiplexing
Network Interface Card (NIC)
Packet (or Public) Data Network
    (PDN)

Private Branch Exchange (PBX)
Public Switched Telephone
  Network (PSTN)
Radio Waves
Relay Center
RJ-11 Jack
RJ-45 Connectors
Signaling System 7 (SS7)
Tags
Telecommunications
UnPBX
Voice Communication
Voice-Over IP (VOIP)
Wiring Hub

OBJECTIVES

On completion of this chapter, you should:

1   Be able to explain what telecommunications is.

2   Have a basic understanding of the various networks used to transmit voice, video, and data signals from one location to another.

3   Know the types of media used to convey telecommunications signals between a sender and a receiver.

4   Understand the basics of the three major voice communication technologies (keysystem, private branch exchange, and automatic call distributor) available to a business enterprise and the type of business best served by each of these technologies.

5   Have a basic understanding of personal computer-based voice communication systems, referred to as computer telephony integrated (CTI) systems.

6 Have a basic understanding of what a local area network and a wide area network are and how a business uses them to meet data communication needs.

7 Have a basic understanding of Hypertext Markup Language and web browsers.

Telecommunications is communication over a distance. It encompasses all forms of communications and includes communication by voice, video, and data. Telecommunications signals carry the voice, video, or data that we wish to transmit from one point to another. People often separate the communication of voice and data into two categories, using the term *telecommunications* to denote the transmission of voice signals and *data communication* to refer to the transmission of data signals.

In today's *public switched telephone network (PSTN)*, voice signals emerge from a telephone as *analog signals*. The word *analog* is related to *analogous*. The electrical signals caused by a voice wave hitting the diaphragm of a telephone set's transmitter are analogous (similar) to the shape of the voice wave. All telephones connect to a local *central office,* which has an automated switching system that connects callers to their desired destination. The local central office contains a *line circuit* for every telephone connected to it. The line circuit in turn contains a device called a *codec* (coder/decoder) that converts analog voice signals received from the telephone into *digital signals.* These digital voice signals are then carried over the digital circuits that comprise the PSTN (see Figure 1-1).

Data is converted into digital codes when it is stored inside computers. The transmission of these digital codes results in a digital signal. Digital data can be

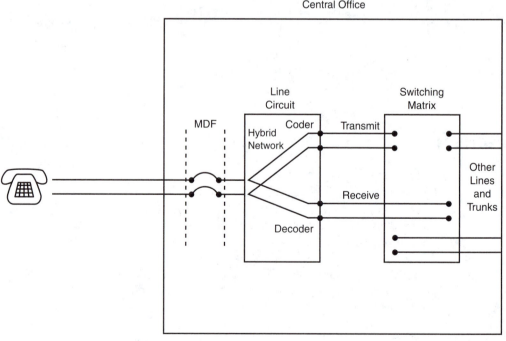

FIGURE 1-1 Telephone connected to central office.

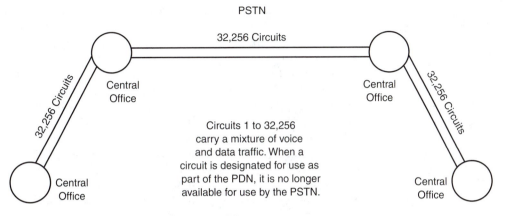

FIGURE 1-2 Packet (or public) data network.

transmitted over the PSTN by using *modems* or can be transmitted over a *packet data network* (also known as *public data network (PDN)*. The PDN is a wide area network and is discussed in Chapter 11. The PDN actually uses facilities in the PSTN that have been reserved for data transmission (see Figure 1-2). All circuits in the PSTN are digital circuits and can carry digitized voice, data, and video. It is appropriate today, therefore, to consider telecommunications as including voice, data, and video. It is also appropriate to break these components out into separate categories for study purposes. Thus, we can study *voice communication*, *data communication*, and video communication, but they all belong to a larger group or classification: telecommunications.

1.1 BEGINNING OF TELECOMMUNICATIONS

Telecommunications basically had its beginnings with the invention of the telegraph by Samuel F. B. Morse (1791–1872) in 1837. Morse formed a telegraph company based on his technology in 1845. The Western Union Telegraph Company was established in 1856 and within ten years had bought out its competitors, becoming the single largest telecommunications company in the world. Morse had developed a method of transmitting information by sending electricity over wire. By the early 1800s, scientists had developed ways to generate and transmit electricity. When electrical signals travel over wire, they gradually lose energy. Morse developed a device called a *repeater* that could be used to regenerate electrical signals. He also developed a code—known as *Morse code*—to represent each letter of the alphabet. The telecommunications industry had its first standard with the implementation of Morse code.

Morse code represents each letter of the alphabet as a combination of short and long signals. These signals can be visual or electrical. You can use light as a signal and turn the light on to represent a signal. By varying the length of time the light is turned on, it is possible to represent each part of the code. If the light is on for a short time, it represents a short signal, called a *dot*. If the light is on for a longer interval of time, it represents a long signal, called a *dash*. Thus, by turning a light on and off rapidly, with the duration of each on period being either short (dot) or long (dash), you could use the code to transmit letters of the alphabet. Morse used bursts of electricity to represent the dot and dash. A short burst of electricity represented a dot; a longer burst of electricity represented a dash.

Each letter of the alphabet has a code. The letter S is represented by three dots (dot, dot, dot). The letter O is represented by three dashes (dash, dash, dash). The emergency message "Save Our Ship" (SOS) was transmitted by sending dot, dot, dot, brief pause, dash, dash, dash, brief pause, dot, dot, dot. The pause between each character code was very brief (about 1 second). With Morse code in hand, a telegrapher could operate the telegraph key to send short or long bursts of electrical energy by controlling how long the key was held in a closed position.

The telegraph system contained many relay points. Electrical signals could only travel about 50 miles before they needed to be regenerated. Thus, the telegraph system needed a telegrapher every 50 miles to receive a message and repeat it to the next station down the line, or the line needed to have the automated repeater developed by Morse. It is interesting to note that one of the people working on the invention of the telephone also invented a repeater used by Western Union on its telegraph lines. That inventor was Elisha Gray, and his company was the Western Electric Company. In 1882 Gray sold Western Electric to the Bell Company.

As mentioned earlier, the telegraph system was responsible for the first commercial use of electricity to send signals. Since electricity travels best over wire, telegraph wires were strung on poles between towns. The signals being sent over the wires consisted of a two-state signal (short or long electrical pulse). A two-state signal is referred to as a *binary signal.* Most digital signals are composed of two-state (binary) signals. Digital signals are used to transfer voice, data, and video in today's telecommunications systems, but they were first associated with the transmission of data. For this reason, many people refer to the telegraph system, with its use of binary signals, as the first data communication system. Binary electrical signals were used to convey data (the sender's message) from one town to another.

The next device to use electricity for communication purposes was the "wireless telegraph" (radio) invented by Italian Guglielmo Marconi (1874–1937). Marconi discovered how to make electric energy radiate from a wire into air (electromagnetic radiation). The first radio was used as a wireless telegraph, but it was not long before voice signals were being carried by *radio waves.* In the 1940s, radio signals were used to transmit video, and television was born. Radio waves are used to carry many forms of information. *Microwave radio* towers dot our landscape. Until a few years ago, most long distance calls were carried by radio waves that used coaxial cable or microwave radio waves as their transmission media.

In the 1980s, scientists developed a way to transmit voice, data, and video using light signals. They also developed a way to transmit light signals over a narrow ribbon of glass (called a *glass fiber*). Sprint was able to take advantage of fiber optic technology when it built its long distance network. Because it was building a completely new network, it was not stuck with old technology that had to be gradually phased out, and it was the first *interexchange carrier (IEC)* with an all-digital network for long distance calls. Telecommunications signals can be transmitted as electrical signals over wire, or they can be transmitted using radio waves and light waves. The media used for telecommunications can be wire, air, and glass fiber.

1.2 INVENTION OF THE TELEPHONE

As noted, telecommunications began with the telegraph in 1837. In 1876, Elisha Gray and Alexander Graham Bell filed papers with the patent office for an invention called the *telephone*. Gray filed a disclosure notification, and Bell's father-in-law filed a patent application for Bell. The Supreme Court ruled in a split decision that Bell is to be recognized as the inventor of the telephone. Western Union made many improvements to the telegraph system over the years, but nothing could beat the direct person-to-person contact that the telephone offered.

1.3 RELAY CENTERS AND SPEED OF DELIVERY

The telegraph system served our country well in earlier periods. It was far superior to the U.S. Mail system, allowing people to receive messages quickly. Teletype messages could be delivered across the country in less than an hour while the U.S. Mail took a month to deliver messages. But when using a telegraph, the sender of a message had to wait hours for a response due to the delay involved in the messenger locating the recipient and giving them the message.

When telephone service was first established in the 1870s, the telephone company used the same approach to delivering messages that was used by the telegraph company. A customer would go to the telephone office and give the operator a message along with the location of the recipient. The telephone company would place a call to the telephone center in the next town. It would in turn relay the message to the telephone center in the next town and so on. Thus, telephone messages used the same relaying technique the telegraph company employed (see Figure 1-3).

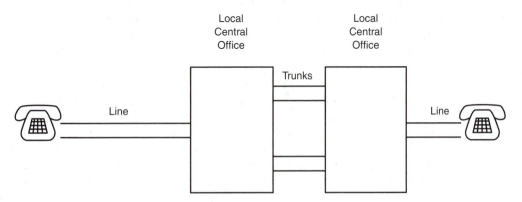

FIGURE 1-3 Connection of telephone offices.

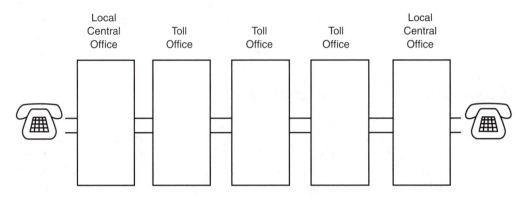

FIGURE 1-4 Long distance network.

The Bell Company established a toll network that made it possible to reduce the number of relays made to complete a long distance call (see Figure 1-4). This long distance network was managed by the AT&T Long Lines department. All long distance calls were handled by AT&T. All local telephone companies (both Bell and non-Bell) had connections to an AT&T toll center. The local telephone company today is referred to as a *local exchange carrier (LEC)*. When someone was making a long distance call, they would call their local operator and give them the telephone number and location of the person they wished to call. Their local operator would instruct them to wait near their telephone and would call them back when a connection was established.

When long distance telephone calls were placed, it would take up to an hour for the telephone company to get a connection through to a telephone near the called party. Remember that many people did not have telephones. The originating party knew where a telephone was located that would accept calls for the called party, or they would call the operator in the town where the person they wanted lived. Once someone brought the called party to the telephone, the originating local operator was notified. The operator would then ring the originator of the call and tell them to go ahead with their conversation to the called party.

The functions of operators have been automated. First the local operator functions were automated, then this was done with the long distance operator functions. Today, almost all telephone calls are processed by computer-controlled switching systems. Even collect calls and person-to-person calls have been automated by using *interactive-voice-response (IVR)* systems. These systems can recognize and carry out the instructions of voice commands. IVR is used to automate the functions operators previously performed. These automated operator systems let the caller provide information via voice responses to scripted questions. The responses provided give information to the computer that directs call switching.

Operators are needed to handle calls rejected by the automated system. A call will be rejected if the caller did not enter the correct information requested by IVR, did not enter any information when prompted by IVR, or selected an option that allows bypassing IVR and going to an operator. If a caller entered an invalid calling card number, the call would be sent to an operator. If a person feels more comfortable using an operator to place the call, they are given an option that will force the call out of automated operator and that will send the call to an operator. The automation of operator functions has drastically reduced the number of operators needed.

1.4 ANOTHER TYPE OF RELAY CENTER

A major telecommunications service that requires the use of operators is telecommunications for the hearing impaired. Each state government contracts with a major telephone company to provide telecommunications services for the hearing impaired via telephone offices called *relay centers.* Operators at the relay center are called *agents.* The relay agent will communicate via voice with a person that can hear the agent and will communicate with a hearing impaired person via a teletype device (see Figure 1-5).

Messages received over the voice connection from the speaking person are relayed by the agent to the hearing impaired person via the teletype connection. The agent's teletype connection is via a personal computer equipped with special terminal emulation programs that emulate a teletype. Teletypes use a different code from a PC for the transmission of data. The teletype connection does not use a regular modem; it uses a special purpose modem designed for teletype connections.

1.5 INCREASING SPEED OF TELECOMMUNICATIONS

Automating the switching systems and eliminating the need for operators to make connections have significantly increased the speed with which a telephone call can be made. The development of computer-controlled switching systems was the next step in speeding up the delivery of services. The next big step in improving the PSTN was the development of a signaling system to connect the CPUs in all computer-controlled switching systems together so they can inform each other when connections need to be established between them for a call. This system is called *Signaling System 7 (SS7)* (see Figure 1-6).

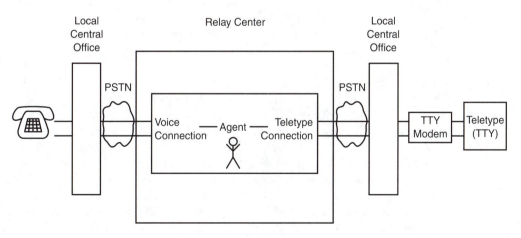

FIGURE 1-5 Relay center.

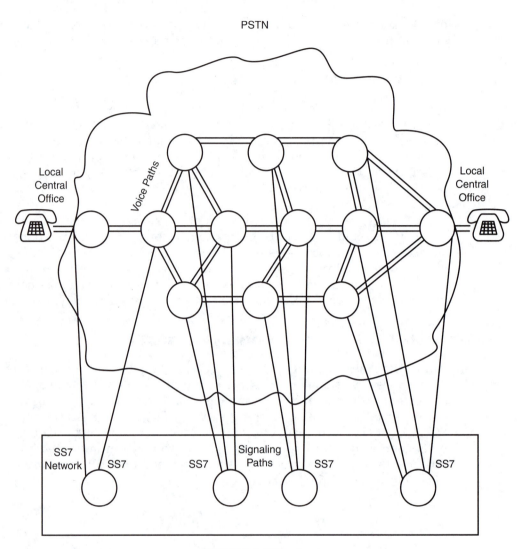

FIGURE 1-6 SS7 network.

SS7 contains many special purpose computers and databases connected together in a data network that stretches across the United States. All computer-controlled switching systems connect to the SS7 network. The major purpose of SS7 is to provide a path that allows information from one computer-controlled switching system to reach another one. If you have the caller ID option on your telephone line, your local switching system's computer requests the telephone number of the person calling you. It asks the computer at the originating switching system to send the telephone number of the caller to it. The request for the number and the transmission of the number are done over the SS7 network. This network is extremely fast and efficient; it establishes connections for calls in a few milliseconds.

1.6 INCREASING ABILITY TO HANDLE DATA COMMUNICATION

The PSTN has gradually evolved into an all-digital network between switching systems. Most telephones remain analog and require an interface at the local switching office to convert their analog voice signals into digital signals. Because the network connecting switching systems together is digital, it can carry voice, data, and video signals that have been converted into digital signals. The hardware devices that handle digital signals are designed into devices called *very large-scale integrated (VLSI)* circuit chips. These small, special purpose integrated circuit chips are designed using the same design techniques utilized to design the Intel Pentium chips found in personal computers.

These VLSI circuit chips are not large. They are small in both size and cost. *Very large* pertains to the large number of electronic components contained in miniaturized form within the chip. This keeps the size and cost of telecommunications equipment low. As chip technology and design have improved, the ability to send digital signals faster and faster has been achieved. The faster a signal can be sent, the more individual signals a transmission system can send. The use of fiber optic cables for transmission media has helped make enormously fast signal changes possible. Fiber optic facilities are used to carry voice, data, and video signals for the PSTN, the PDN, and the Internet.

Devices have been designed that can place many signals from many different conversations or data inputs into one very fast signal stream. These devices use a technique called *multiplexing* to place many signals over one transmission medium (multiplexing technology is discussed in Chapter 5). The multiplexing devices used to place many signals over glass fiber have developed to a point where they change the light signal used to transmit signals over fiber at a rate of almost 10 billion times a second. Just as astounding is the development of receivers that can detect these rapid changes in the light wave (see Figure 1-7).

Recent advances in fiber optic transmission have led to the use of 32 different lightwaves to carry signals over one fiber. This device—called a *Dense Wave Division*

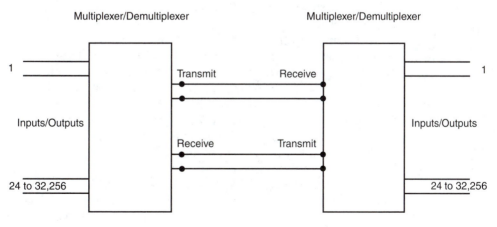

FIGURE 1-7 Multiplexer.

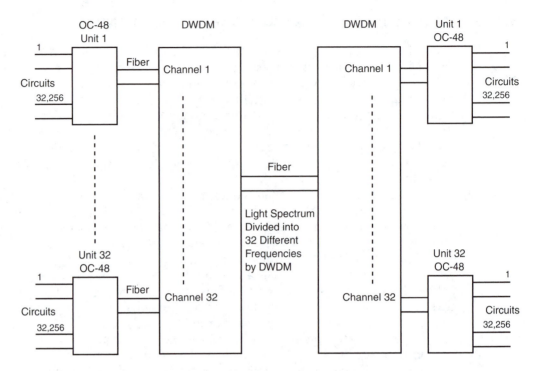

FIGURE 1-8 Multiplexing using DWDM and OC-48.

Multiplexer (DWDM)—allows for the connection of 32 different multiplexers to one fiber. Each multiplexer is sending signals at almost 10 billion signal changes per second. Thus, with DWDM, one fiber can carry 320 billion signal changes per second. One fiber cable can contain up to 140 fibers. Fiber optic technology requires two fibers for each multiplexer (one transmit and one receive). This means that one cable containing 70 pairs of fibers and using DWDM on each pair can send and receive signals that total over 22 trillion signals a second. This is a tremendous capacity (see Figure 1-8).

The telecommunications companies use their fiber cables to handle all types of telecommunications services. Some fibers are used to handle telephone calls, some are used to carry data or video, and some are used to provide facilities that carry the Internet traffic. These services do not require their own individual fiber. All four services can be commingled over one fiber if desired. As the speed of signals has increased, so also has the ability to carry more data and carry it faster. Companies that used satellites in the past to carry large volumes of data between their locations now employ services leased to them by telecommunications companies that utilize fiber cables to carry the data (see Figure 1-9).

As we move into the 21st century, we are placing more demands on the telecommunications systems and networks. Almost every large business has installed data communication networks to link personal computers, file servers, and mainframes together. The number of companies and people using the Internet continues to grow

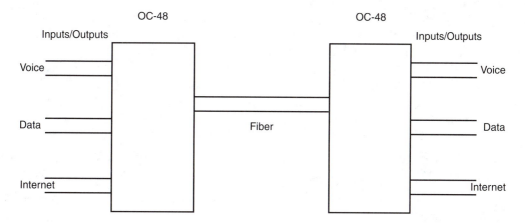

FIGURE 1-9 Fiber carrying all types of circuits.

exponentially. Businesses cannot survive in the future without sophisticated telecommunications networks that can handle their voice, data, and video communication needs. It will not be long before cars come equipped with personal communication systems that allow voice and data connections to your company's network. The car will contain a built-in PC that allows you to transmit voice and/or data to anyone.

The developments in cellular radio technology will accelerate, and soon you will no longer need to have your telephones tied via an umbilical cord (cable pair) to the local central office. The switching centers that connect to mobile telephones will be capable of directing all calls to your mobile phone and will indicate whether the call is for your business or residence telephone number. It will also provide caller ID. These technologies are so real that they may even be implemented between the time I write this book and when it is published.

1.7 TELECOMMUNICATIONS FOR VOICE APPLICATIONS IN BUSINESS

Businesses are relying more each day on telecommunications technology. In fact, large businesses have a telecommunications department that ensures the company is using the right mix of telecommunications technology. Because data communication comprises almost 80% of all telecommunications that companies use, they often assign responsibility for the telecommunications function to the director of information technology. This person is usually in the management information systems (MIS) department.

MIS is usually concerned with the data network that all its computers are attached to. In the past, many MIS departments paid little attention to the voice services requirements of a company, which led to splitting the telecommunications functions between a manager for the voice services applications and a manager for the data communication applications. It will not be long before voice and data are both delivered over the company's data network using digital signaling technology, so the need for separate managers will diminish. This service is called *Voice-Over IP*

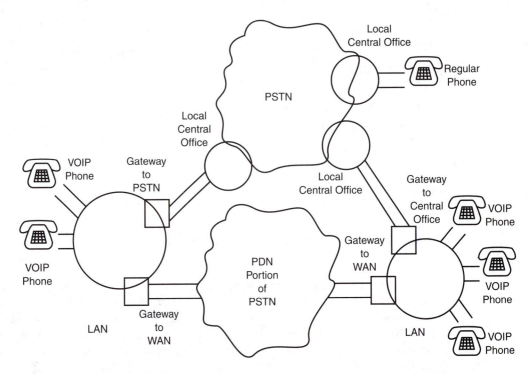

FIGURE 1-10 Voice-over IP.

(VOIP). As the quality of VOIP improves and as standards are developed for the technology, it will be used to replace analog telephones (see Figure 1-10).

Currently most large businesses use a private voice switching system called a *private branch exchange (PBX)* to handle the voice communication needs of their business. This switching system has all telephones of the business connected to it; it also has connections to the local central exchange that are commonly called *C.O. lines.* Employees can call one another via the PBX. They can call telephones on the PSTN by dialing 9. When they do this, the PBX switching system connects their telephone to one of the C.O. lines. The local central office supplies a dial tone and will look at subsequent digits dialed to determine where the caller should be connected (see Figure 1-11).

PBX systems are now being furnished via personal computers (PCs). In their expansion slots, these PC-based systems have special cards containing **RJ-11 jacks** (standard jacks that a telephone plugs into). The RJ-11 jacks provide a means of connecting to telephones or central exchange lines. These PC-based systems are called **UnPBXs** because when they are used you do not need the older-style PBX. This type of technology is catching on rapidly and will explode when voice-over IP becomes more prevalent (see Figure 1-12).

The use of a PC to serve as a PBX also has the benefit of allowing the computer to provide information about the call on its screen. Customer information can be kept in a database file on the PC, and the PC can use that database to get

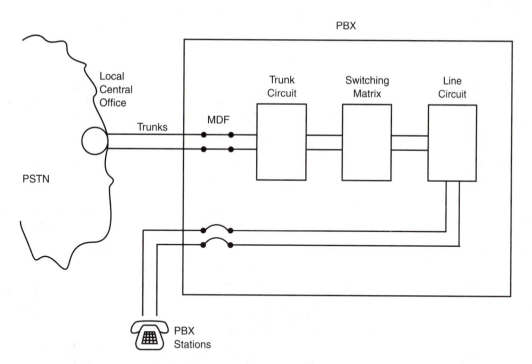

FIGURE 1-11 PBX switching system.

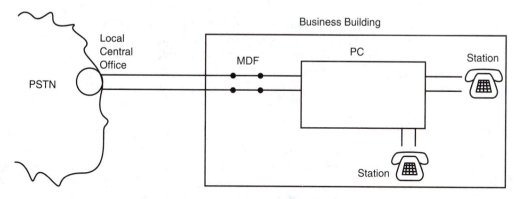

FIGURE 1-12 PC-based PBX or UnPBX.

information that helps with handling the call. Several software programs are available that enhance the PC's ability to interact with incoming calls according to information associated with a call.

Sometimes the linkage between the call and the database is established by a simple procedure, such as asking callers for their telephone numbers and keying them into the computer. The computer then uses this number to retrieve a customer's record in the database. Some systems will use caller ID to automatically

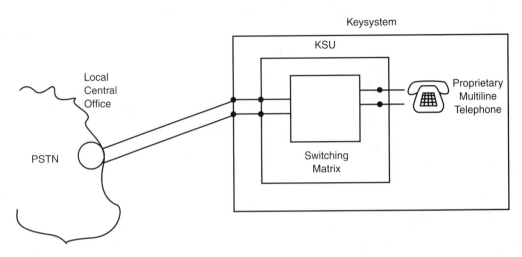

FIGURE 1-13 Keysystems.

identify a caller's telephone number. PC-based PBX systems are referred to as *computer telephony integrated (CTI)* systems.

Many small businesses do not need a PBX and will use a *keysystem* to connect their telephones to the local central exchange (keysystems are discussed in Chapter 4). PBXs are used when a company needs more than 100 telephones for its employees or needs more than 24 connections to the central exchange. Keysystems are used when a firm needs less than 24 telephones and less than 12 connections to the central exchange. Systems that handle requirements between those of a PBX and a keysystem are called *hybrid keysystems*. Hybrid, in telecommunications, generally refers to the combining of two different technologies or systems. In this case, a hybrid keysystem combines some of the features and functions found in a PBX with a keysystem. Today's telecommunications manager needs to understand how PBXs and keysystems can be used to satisfy the voice communication needs of a business (see Figure 1-13).

Hybrid keysystems and PBXs require the user to dial 9 to access a circuit to the local central exchange. These circuits are accessed on a regular keysystem by pressing a button on the telephone assigned to the circuit desired. A keysystem has a separate button to push for every circuit to the central office. This is why keysystems are usually limited to 24 central office circuits. It is difficult to get more than 24 buttons on the telephone. As discussed in Chapter 4, a keysystem is the perfect low-cost communication system for a small business, doctor's office, dentist's office, and so on.

Another technology used to handle voice services is an *automatic call distributor (ACD)*. Large call centers such as those used by the operator services of AT&T, Sprint, and MCI use an ACD to distribute incoming calls to their operators. Many companies use an ACD and set up a call center to handle customers' calls. The call center may be for an airline company to handle reservation and ticket requests, or it may be set up by a credit card company to handle customer service (see Figure 1-14).

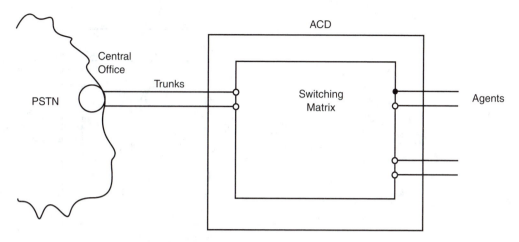

FIGURE 1-14 Automatic call distributor.

All of us are familiar with telemarketing firms that call us to sell the latest carpet cleaning special or award us a trip to the Caribbean. Large telemarketing companies also use an ACD to place outbound calls automatically. These ACDs contain predictive dialing software. It predicts when the next agent will be available to handle a call and starts placing the telephone call before the agent has finished the call they are currently working on. This makes the agents very productive. They do not have to spend time dialing calls and waiting for a customer to answer and do not waste time on unanswered calls or calls to busy numbers.

If an agent does not finish their current call within the time predicted, a call may be placed and answered by a customer before an agent is available to handle the call. Most customers will stay on the line after answering a call and say "Hello" several times in an attempt to find out who is calling them. By this time an agent is probably free and will be assigned to the call. If the agent does not become free within a few seconds, most customers think the call was a nuisance call and will hang up the phone. These predictive dialers will disconnect a call after three rings so that an answering machine does not answer them. They do not want to connect agents to answering machines. They will also place unanswered and busy signal calls back into the calling queue for a call later in the day.

An ACD designed to handle inbound calls automatically connects the caller to an operator or service agent. Outbound ACDs automatically dial telephone numbers, and when someone answers the call, the ACD automatically connects them with a sales agent. The use of an ACD improves productivity tremendously. The money a company saves via the productivity gains of its agents or operators has always justified the installation of an ACD.

Many inbound ACD installations also include the installation of a *local area network (LAN)* for data (see Figure 1-15). When calls are received by the ACD, the call center also receives information about the call from the PSTN via the SS7 network. This information is used to provide information to the operator or agent

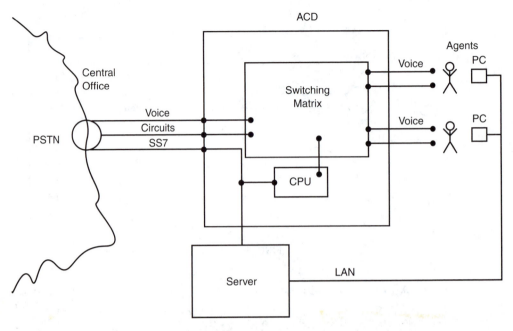

FIGURE 1-15 ACD and LAN.

about the caller. The information is presented on the monitor associated with the PC at the agent's position that was assigned to handle the call by the ACD. Information will also be presented on the screen instructing the agent on how the call should be answered and processed. Several PC-based ACDs are available today that combine these functions (CTI) into one system.

PBXs, ACDs, and keysystems use another technology that we have discussed. Some systems will be equipped with a form of IVR that automates the operator process. A business usually has a receptionist answer all calls; thus the receptionist also serves as the switchboard operator for the business. With automated operator, the incoming calls are answered by a recording that will ask the caller to enter the extension number of the party they wish to speak to. The IVR will recognize the digits dialed, and the PBX, ACD, or keysystem can use this information to connect to the called number (this feature will be discussed in more detail in Chapter 4).

1.8 TELECOMMUNICATIONS FOR DATA APPLICATIONS IN BUSINESS

A large business needs a special data communication network that ties all its computers together. Computers within one building will be connected together by a LAN. Computers located in different buildings will be on different LANs, and these LANs will be connected together by a *metropolitan area network (MAN)* or a *wide area network (WAN)*. These data networks are discussed in detail in Chapters 10 and 11 (see Figure 1-16).

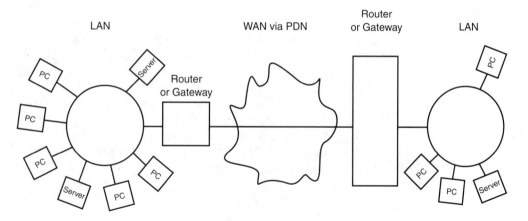

FIGURE 1-16 LAN to WAN.

Eighty percent of all LANs use a technology called **Ethernet** to control the access that PCs on the network have to the common communication medium used by the LAN. Today, PCs are connected to an Ethernet LAN via two twisted-pair copper wires. The four wires terminate in **RJ-45 connectors** on each end of the wires. One RJ-45 plug is inserted into a special **network interface card (NIC)** in turn inserted into an expansion slot of the PC. The other RJ-45 connector is inserted into a device called a **wiring hub**. The PC can be an individual PC at a worker's desk (workstation), or it can be a special PC called a server that shares programs and data with all workstations.

LANs require someone to make sure they are operating properly and to add new workstations, servers, and software applications to the LAN. The person who does these things is called a *LAN administrator* or *LAN manager*. LANs require a special software program to allow LAN administrators to do their job. This software is contained within a special *network operating system (NOS)* loaded onto workstations and servers. To properly manage a LAN, the LAN manager must completely understand the NOS, the hardware of the PC, the PC's operating system, and the hardware associated with a LAN (NICs and hubs). The LAN manager must also have a good understanding of the devices and WAN technology used to connect LANs together.

Two of the primary data applications in use today are e-mail and the **Internet**. The use of e-mail is especially prevalent. Everyone at work uses their PC and LAN to send messages to each other. The growth of the Internet allows people to have access to a tremendous volume of data. In fact, so much information is available that you can easily experience information overload. A person has to be careful about how they search for information and must use special software called *search engines* to find data. The Internet has not only given us the ability to locate information; it has provided a delivery system that allows us to send e-mail to anybody. (Chapter 12 is devoted entirely to the Internet.)

1.9 THE INTERNET

The Internet has grown tremendously over the past few years. It has had as much impact on our society as the telegraph, telephone, television, and PC. Access to the Internet is crucial to those who seek information; thus, the U.S. government has funded a program to establish an information superhighway to allow for more effective commerce over the Internet. This program is called the *National Information Infrastructure (NII)*. Voice over the Internet is a fast-growing segment of Internet traffic. As more and more business is conducted over the Internet, its highways must grow in their ability to handle the increased traffic. Fortunately, fiber optic and DWDM technologies can provide the capacity needed for the foreseeable future (these technologies will be covered in Chapter 5). For more on the Internet, refer to Chapter 12.

1.10 THE WORLD WIDE WEB

The World Wide Web (WWW) organizes the resources of the Internet, which consists of many computers linked together via high-speed data facilities. These computers contain files that are accessed by Internet surfers. The collections of files available on the Internet are usually accessed by a web browser. The web browser uses a language called HTML, which is sent over the Internet as HTTP.

Tim Berners-Lee developed *Hypertext Markup Language (HTML)* and *Hypertext Transport Protocol (HTTP)* while doing research at the European Particle Physics Laboratory (CERN) near Geneva, Switzerland, in 1991. *World Wide Web* was a term coined by Berners-Lee's colleagues to refer to the HTML and HTTP development project. HTML was based on Standard Generalized Markup Language (SGML). HTML is not a programming language; it is a markup language that is used to mark up documents. Special words (language) are used that will be recognized by the program displaying the text on a screen. These special words are called *tags*. HTTP is the protocol developed to transport HTML files across a network.

Mosaic and Netscape browsers are based on the HTML developed by Berners-Lee. HTML and HTTP standards were established by the WWW Internet Engineering Task Force as RFC 1866 (HTML version 2). Although Netscape and Windows NT browsers follow the standard, they have also developed proprietary tags that only work with their browsers. HTML tags control the format of text displayed on the screen. The tag indicates how the browser should paint the screen; it will paint the screen that way until another tag is encountered.

Many HTML tags require an opening tag and a closing tag. A typical opening tag is <HTML>. This tag indicates that the following document is an HTML document. The closing tag at the end of the document is </HTML>. Following the opening <HTML> tag are the <HEAD><TITLE> tags. The title of the document is placed after these tags and will serve as the title of the web page. After the title is typed, it is closed with the </TITLE></HEAD>tags. Next comes the <BODY> tag, which is followed by the text of the document. After all the text of the document is the closing </BODY> tag, immediately followed by the </HTML> tag.

There are numerous tags to help control the formatting of a web page on the screen of your PC (refer to Appendix E). Web pages are not difficult to write. In fact, two of Microsoft's applications (Word and PowerPoint) allow for saving the document as an HTML document. You can also get permission from most other web page authors to copy their page and edit it to fit your needs. You must first save their page to a file on your computer. When you open the saved HTML file that is now on your local hard drive with a Netscape or Explorer browser equipped with editing capabilities, you can edit the file.

After opening the saved HTML file with your browser, click on View in the menu bar and then click on Source. You will see the HTML document and can edit it to fit what you want displayed on your web page. If your version of Netscape or Explorer does not contain editing capabilities, you will find that you cannot make changes to the source document. Of course, you can always read a book on HTML and design a web page from scratch. You can use almost any text editor to create an HTML file.

Along with HTTP, web browsers and HTML are the tools that led to the explosive growth of the Internet. These tools allow the use of a *graphical user interface (GUI)* to control moving around on the Internet. Simply point your mouse and click. This ease of use has encouraged many people to join the Internet, but the downside of GUI is the increase in data that needs to be transmitted to display web pages. Some people include pictures in their web pages, which take a lot of time to transmit. If a user tries to surf the WWW using a 14.4-Kbps modem, it quickly becomes the World Wide Wait. The minimum speed needed for your modem is 28.8 Kbps.

Although the Internet has grown tremendously, many households still lack PCs and Internet access. Thus, this area is still ripe for continued growth. It is just as important to have a PC and Internet access as it was to have a telephone or television 50 years ago. Over time, the telephone and TV came to be seen as necessities; Internet access and a PC in every home will follow along the same lines. You will need access to the Internet to gather information that will help reinforce the topics covered in this book.

1.11 METRIC NUMBERING SYSTEM:

In the telecommunications and electrical industry, we use a metric numbering system. In Appendixes A, B, and C, we use metric prefixes to describe voltage, current and resistance. In Chapter 6 we use metric prefixes to describe speed or bandwidth of multiplexing systems. Metric prefixes derived from Greek, Latin, and French are used to designate quantities. The Greek prefixes used are:

$$K = kilo = 1000$$
$$M = mega = 1\ million$$
$$G = giga = 1\ billion$$
$$T = tera = 1\ trillion$$

Thus, 1.544 Mbps means the system is handling 1,544,000 bits per second. Latin and French prefixes are used for numbers less than 1:

$$d = \text{deci} = 1/10$$
$$m = \text{milli} = 1/1000$$
$$\mu = \text{micro} = 1/1,000,000$$
$$n = \text{nano} = 1/1,000,000,000$$
$$p = \text{pico} = 1/1,000,000,000,000$$

All numbers can be expressed as powers of 10. For example, 1000 (kilo) = 1×10^3 ($10 \times 10 \times 10$). Milli ($1/1000$) = $1/(10^3)$ and is also equal to 1×10^{-3}. The prefixes *kilo, mega,* and *giga* can be represented by 10 to a positive power: 10^{+3} = kilo; 10^{+6} = mega, and 10^{+9} = giga. The prefixes *milli, micro, nano,* and *pico* can be represented by 10 to a negative power: 10^{-3} = milli; 10^{-6} = micro; 10^{-9} = nano, and 10^{-12} = pico. Note that each of these prefixes can be stated as either a fraction or as 10 to a negative power—for example, milli = $1/1000$ or = 1×10^{-3}. Since 1000 = 10^{+3}, $1/1000$ can also be stated as $1/10^{+3}$. Therefore, milli can be stated as 10^{-3} or as $1/10^{+3}$. This is a basic mathematics theorem, which states that N^{-x} is equal to $1/N^{+x}$ and $1/N^{-x}$ is equal to N^{+x}. For instance, $1/10^{-3} = 1/0.001$ and $1/0.001 = 1000$ or 10^{+3}.

When calculating how long an event can be in order to do X amount in a given time frame, it is possible to change the sign of the power of 10 to its opposite sign to arrive at the calculation. If we wish to do 1.544 Mbps ($1.544 \times 10^{+6}$), then each bit must be no longer than $(1/1.544) \times 10^{-6}$. Further, $1/1.544 = 0.6476684$ and $(1/1.544) \times 10^{-6} = 0.6476684 \times 10^{-6} = 0.0000006476684$. This is equal to 647.6684×10^{-9}. Since 10^{-9} power $= 1/1,000,000,000 = $ nano, the time frame is 647.6684 ns for each bit. If each bit sent on the TDM medium is only 647.6684 ns long, the system can transmit 1,544,000 bps. Thus, $F = 1/T = 1/(647.6684 \times 10^{-9}) = 1.544$ Mbps. Note that the same answer is achieved by using the reciprocal power method: Since $1/10^{-9} = 10^{+9}$, $F = 1/T = (1/647.6684) \times 10^{+9}$. Moreover, $1/647.6684 = 0.001544$ and $0.001544 \times 10^{+9} = 1,544,000$.

1.12 SUMMARY

Telecommunications means to communicate over a distance using electrical signals, radio waves, or light waves. The telegraph was the first technological device that allowed us to communicate over vast distances within a short time frame. The telegraph system eventually evolved into the teletype network. The invention of the telephone allowed for two-way interactive communication and rapidly replaced the telegraph for the transmission of voice calls. The telegraph continued to be the choice of customers who wished to transmit data until the establishment of data communication networks in the early 1980s.

Advances in telecommunications technology have enabled the PSTN to serve as a medium for the transmission of voice, data, and video. The PSTN has gradually evolved into a digital network between switching systems. This digital highway has continued to expand its speed and capacity. The development of fiber optic technologies has provided a transport medium that can handle the faster speeds. This high-speed fiber optic network is also being used to transport Internet traffic.

The continual evolution of telecommunications technology, the deregulation of the telecommunications industry, the explosive growth of personal computing, the need for LANs and WANs to connect PCs together, the increasing needs of information technology, and the growth of the Internet have all contributed to an enormous growth in the job market for telecommunications managers and technicians. We are in the information age, and every business of any size needs a telecommunications manager.

Today's telecommunications manager will probably be managing both data communication networks and voice communication networks. The manager will need to decide whether a keysystem, PBX, or ACD will best serve their voice communication needs. They will need to decide whether to use CTI-based systems or stand-alone systems. They will be called on to select from a variety of LAN and WAN technologies. After finishing this book, you will be well on your way to understanding the breadth and depth of a telecommunications manager's functions.

Because telecommunications technologies continue to grow, the telecommunications manager must stay abreast of what's happening via reading books and trade journals, surfing the net, and attending trade shows and training seminars. Only by staying abreast of changes to the technologies can a manager make the best decision on the technology needed for a particular application. Even with a good understanding of technologies, the manager should seek the advice of technicians, vendors, and consultants before leaping into the implementation of a specific technology.

CHAPTER REVIEW QUESTIONS

1. What advantages does the telephone have over the telegraph?
2. What did Samuel F. B. Morse invent that made the telegraph possible?
3. What was the name of the company that manufactured repeaters for Western Union, and who was the primary stockholder of that company?
4. What does *telecommunications* encompass?
5. What types of signals are used in telecommunications?
6. What types of media are used for the signals mentioned in Question 5?
7. Since most calls are handled by automated switching systems, why do we still have operators?
8. What data network was developed to allow computer-controlled switching systems to communicate switching instructions to one another?
9. What system is used to transmit caller ID to the requester?
10. How does the speed of a signal relate to its capacity for transmitting data?
11. When is a keysystem preferred over a PBX?

12. What are PC-based PBXs called?

13. What does an ACD do?

14. What is a PBX?

15. What is a keysystem?

16. What is a hybrid keysystem?

17. What is a LAN used for?

18. What is a WAN used for?

19. What is a binary signal?

20. What medium is used to carry light waves?

PROJECTS AND CASE STUDIES:

Project #1

Research Interactive Voice Response (IVR) and write a brief paper that describes IVR and some typical applications. Discuss the advantages and disadvantages of IVR. List some companies that misuse IVR to keep callers waiting a long time before connecting the caller to a person. If you were the manager of telecommunications for this company, could you improve on the current situation?

Case Study #1

American Telecommunications is a small owner-operated business that provides telecommunication-consulting services to businesses and telephone companies. This company employs 20 consultants. Does this company need a web site? Does it need e-mail capabilities? What would you suggest? Would your answer be different if this company employed fewer consultants (2-5 consultants)? Could you design a web site for this company? Do some research and make a proposal on what the company needs in the way of hardware and software. Does this company need an Internet server and connection to the Internet or can they use an ISP to provide e-mail and a web site? Your professor may want to assign development of a web page for American Telecommunications. Appendix E will help you get started.

GLOSSARY

Analog Signal   An electrical signal that is analogous (similar) to a voice signal. The signal continuously varies in amplitude and frequency. An analog signal has an infinite number of values for voltage, current, and frequency.

Automatic Call Distributor (ACD)   A switching system used by telemarketing and/or customer service centers to connect incoming calls to agents. Calls are assigned to agents according to which agent has been least busy. ACDs can also serve as outbound systems, and with predictive dialing software, they can automatically dial telephone numbers and connect the called party to an agent.

Binary Signal   A signal that assumes one of two discrete states.

Central Office   A switching system. They are used in the PSTN to connect the calling party to the called party or to another switching system that can connect through the PSTN to the called party.

Codec   Coder/decoder is the hardware device that converts an analog signal into a digital signal and vice versa.

Computer Telephony Integrated (CTI)   The integration of personal computer technology and telecommunications technology to provide PBX and/or ACD services via a personal computer system.

Data Communication   The transfer of data over telecommunications facilities.

Dense Wave Division Multiplexer (DWDM)   A device that places many voice or data circuits over one glass fiber by splitting the infrared light spectrum into several different frequencies and then attaching a time division multiplexer to each of these frequencies.

Digital Signal   A signal that assumes one of several discrete steps. The signal may assume one of four discrete steps, as it does in the ISDN signal between the central office and customer, or it may be one of two distinct steps (binary signal), as is most often the case.

Ethernet   A protocol used on local area networks (LANs) to control a user's access to the LAN medium.

Hypertext Markup Language (HTML)   A structured use of special words called *tags* utilized in turn by a browser

as instructions on how to display a web page.

Hypertext Transport Protocol (HTTP)   The protocol developed to transport HTML files across a network.

Interactive-Voice-Response (IVR)   An automated system that responds to user input. The user input is often done via the touch-tone keypad of a telephone, but some systems accept voice input. On receiving input from a user, the IVR system will respond with information or will request additional input.

Interexchange Carrier (IEC)   A telecommunications company that handles long distance calls.

Internet   A network that connects internet service providers (ISPs) together. Some companies will serve as their own ISP and will connect their local network to the Internet via a router or server, but most people access the Internet via an ISP. Anyone who can access the Internet can access any computer attached to the Internet.

Keysystem   A telephone system used for small businesses. It allows several telephone lines from the local telephone office to be connected to it. It also allows for the connection of several telephones to the system, and any telephone can access all the lines to the telephone office.

Line Circuit   The hardware device in a local telephone office (switching exchange) that a telephone connects to. Every telephone must connect to its own line circuit if the phone is on a private line.

Local Area Network (LAN)   A network that exists in one building to connect the computers within that building to one another so they can share information.

Local Exchange Carrier (LEC)   The local telephone company.

Microwave Radio Utilizes signals in the gigahertz band of frequencies. These signals must take a line-of-sight path, and due to the earth's curvature, microwave relay stations must be placed about 30 miles apart.

Modem A modulator/demodulator; used to connect a PC to a regular telephone line.

Multiplexing A technique used on circuits between two central office exchanges so that many circuits can be placed over one transmission medium.

Network Interface Card (NIC) A printed circuitboard placed inside a PC that will allow the PC to connect to a LAN.

Packet (or Public) Data Network (PDN) A public network that connects computers together and transports data between them. It is composed of circuits leased from IECs and LECs.

Private Branch Exchange (PBX) A small switching system used by a business. It is a smaller version of the switching system used by a local telephone company. Since the owner of the business owns it, it is a private system. Telephones inside the business connect to the PBX.

Public Switched Telephone Network (PSTN) The network formed by connecting the switching systems of LECs and IECs together using the transmission media owned by the LECs and IECs. Every telephone connects to a local telephone exchange that is connected to the PSTN.

Radio Waves Electromagnetic waves. When these waves cut across a wire (antenna), they induce a voltage signal in the wire.

Relay Center A center that relays information or data from one customer to another. Today the term designates telecommunications centers that relay information between people who are hard of hearing and those who are not.

RJ-11 Jacks Regular telephone jacks. The jack that a telephone plugs into is an RJ-11 jack. These jacks contain four wires, though only two are used for a telephone.

RJ-45 Connectors Telephone jacks that contain eight wires. They are used for data devices.

Signaling System 7 (SS7) A network of switches and databases used to establish and control voice paths in the PSTN. SS7 is a data network that connects switching systems together and allows them to exchange information.

Tags HTML tags. These tags are used by a browser as instructions on how a web page should be displayed.

Telecommunications The transmission of voice, data, or video over a distance using electrical, electromagnetic, or light signals.

UnPBX A PBX system that is CTI-based, and thus there is no standalone CBX system. The system is within the PC that serves to replace a standalone CBX.

Voice Communication The transmission of voice signals over telecommunications facilities.

Voice-Over IP (VOIP) The transmission of voice using Internet protocol-based packets that contain digitized voice signals.

Wiring Hub A device used to connect computers to a LAN when the medium used by the LAN is twisted-pair copper wire.

2

Telecommunications Legislative History

KEY TERMS

Bell Operating Companies
  (BOCs)
Carterphone Decision of 1968
Common Carrier
Communications Act of 1934
Computer Inquiry
Computer Inquiry I
Computer Inquiry II
Computer Inquiry III
Customer-Provided Equipment
  (CPE)
Deregulation
Divestiture
Equal Access
E-Rate Fund

Feature Group
Federal Communications
  Commission (FCC)
Graham Act of 1921
Interexchange Carriers (IECs or
  IXCs)
Kingsbury Commitment of 1913
Local Access Transport Area
  (LATA)
Local Exchange Carrier (LEC)
MCI Decision of 1976
MCI Ruling of 1969
Modified Final Judgment (MFJ)
Monopoly
Network

Point of Presence (POP)
Preferred Interexchange Carrier
  (PIC)
Private-Line Services
Public Utilities Commission (PUC)
Regulated Monopoly
Revenue Sharing
Rural Electrification
  Administration (REA)
Specialized Common Carrier
  Decision of 1971
Station Equipment
Tariff
Telecommunications Reform Act
  of 1996

OBJECTIVES

On completion of this chapter, you should:

1 Be able to discuss how legislative activity has affected the telecommunications industry.
2 Be aware of the differences between the regulated monopoly and competitive environments.
3 Know the difference between an Independent telephone company and Bell Telephone Company.
4 Define the differences between an incumbent local exchange carrier, a competitive local exchange carrier, and an interexchange carrier.
5 Understand what a tariff is and with whom a tariff is filed.
6 Understand the universal service fund and educational rate fund concepts.
7 Understand the impact of the Carterphone and MCI decision rulings by the federal courts and how these lawsuits opened the industry to competition in station equipment and long distance service respectively.

8 Understand how the 1984 Modified Final Judgment and the 1996 Telecommunications Reform Act have impacted the industry and customers.

9 Be able to describe the PSTN and its evolution from a five-level network to a three-level network.

10 Understand what a local access transport area and a point of presence are.

Telecommunications is the communication of voice or data over long distances using the *public switched telephone network (PSTN)* or privately owned *networks.* Legislation governing the telecommunications industry has changed a great deal over the past 125 years; and in particular, many legislative and technological changes have affected the PSTN in the last two decades. Laws covering the telephone industry did not exist when the industry was in its infancy, but as the *Bell Operating Companies (BOCs)* grew, such laws were enacted to direct growth in the industry. Initial legislation led to a regulated and monopolistic industry. Over the past 20 years, legislation has been aimed at deregulation of the industry to encourage competition.

2.1 MONOPOLY SERVICE PROVIDERS

Initial legislation provided for a regulated monopoly because a monopoly service provider could gain "economies of scale" to provide services at a lower cost. This is the basic economic principle underlying any *monopoly.* Legislators initially thought that by regulating the monopoly service providers, they could ensure that the monopoly would provide low rates. But legislators have done an about-face. They now believe a competitive environment will lead to lower rates. Some people question the wisdom of this argument. It is well known that a monopoly provider can provide service at a lower cost than any other form of business. The problem with a monopoly is its tendency to overcharge for services. If the government agencies charged with regulating the monopoly do their job, the cost savings achieved by the monopoly should be passed on to customers.

2.2 REGULATION OF MONOPOLY SERVICE PROVIDERS

Regulated monopolies are allowed to charge a rate for their services that provides them with a certain return on their investment. How much a monopoly can earn, or the return on investment it is allowed to earn, is set by the government regulatory agencies. These agencies monitor rates and investments of the monopoly to ensure that rates are not inflated. If the agencies do not do their job well, the monopoly service provider will find ways to inflate rates. Since the rate of return on an investment for a regulated monopoly is set, the only way for a regulated monopoly to raise what it earns is to increase what it invests.

The amount of investment allowed is the area that must be controlled to control rates. Some legislators believe that monopoly service providers have little incentive to improve productivity and lower costs. The higher their costs, the more money they can earn. These legislators believe that rates are inflated because the monopoly service providers are not seeking to keep costs low. They believe costs will be lower in a competitive environment. This is tantamount to saying government regulators

[handwritten margin note: Federal: PUC FERC State: PUC, PSC]

have not performed their job properly. If in fact the regulators have not been able to properly regulate what the monopolies have been doing, competition may result in lower rates. As we have seen with the deregulation of other industries, an initial reduction in rates can be achieved by sacrificing customer service.

2.3 INDEPENDENT TELEPHONE COMPANIES—1893

Between 1870 and 1875, several individuals were working on devices to convert sound waves into electrical energy that could be sent over telegraph lines and then to convert the electrical energy back into sound waves at the other end. Alexander Graham Bell was one of the successful inventors. He filed a patent for the telephone in 1876 and formed the Bell Telephone Company in 1877. The patent on his device ran out in 1893, after which others could use his technology to form their own telephone company. This led to the establishment of other telephone companies called *Independents*. By 1894 there were about 90 Independents. In 1900 there were approximately 4000 Independents serving about 40% of all telephone customers; BOCs served the other 60%. Bell had created telephone companies in most large cities but did not consider rural communities profitable. Independent companies were formed all across America in cities as well as small towns. Some Independents even competed directly against Bell. The largest cities had several telephone companies competing against each other and refusing to interconnect services. This meant a person needed phone service and a phone from each company. Bell bought out the Independents occupying prime territory. In 1882, Bell bought the Western Electric Company and in 1885 incorporated as American Telephone and Telegraph Company (AT&T). AT&T purchased large amounts of Western Union stock between 1900 and 1913 and bought out major competitors in local telephone markets.

Consider: In 1877 Bell offers his Co. to Western Union for $100,000 and is

2.4 MANN-ELKINS ACT OF 1910   turned down!

In 1910 Congress had passed the Mann-Elkins Act as the first step toward regulating the telecommunications industry. The Interstate Commerce Commission (ICC) was placed in charge of enforcing the regulations of the Mann-Elkins Act.

2.5 KINGSBURY COMMITMENT OF 1913

The Department of Justice was concerned that AT&T was monopolizing the telecommunications industry and was contemplating using the Sherman Antitrust Act to bring charges against AT&T. The acquisitions made by AT&T had substantially lessened competition. In response to the concerns of the Justice Department, the vice president of AT&T, Nathan Kingsbury, made a commitment called the Kingsbury Commitment. The *Kingsbury Commitment of 1913* stated that AT&T would not buy any more Independents without Justice Department approval and would allow the connection of Independent companies to the AT&T network. This meant homes only needed one phone and one telephone company to provide service. The provider of their service could interconnect with other Independents and

AT&T to complete calls anywhere in America. The Kingsbury Commitment also stated that AT&T would sell its interest in Western Union. By entering into this agreement, AT&T successfully headed off action by the Justice Department and prevented sanctions by the Sherman Antitrust Act.

2.6 GRAHAM ACT OF 1921

Between 1913 and 1921 AT&T lobbied the government to exempt telecommunications from the Sherman Antitrust Act. In 1921, Congress passed the Graham Act, which made the telecommunications industry (and AT&T) exempt from the Sherman Antitrust Act. As a result, companies were able to consolidate and become monopoly service providers. This measure prevented duplication of service providers and encouraged AT&T to build a nationwide network. Congress believed that a monopoly service provider under government control could provide phone service at the lowest cost to the public, but the *Graham Act of 1921* did not provide the controls needed to ensure low cost for customers. Congress recognized the error of this legislation and passed the Communications Act of 1934 to regulate the interstate telecommunications provided by AT&T.

2.7 COMMUNICATIONS ACT OF 1934

The *Communications Act of 1934* was designed to regulate broadcast radio and interstate telecommunications; it also made it possible to pursue certain actions under the Sherman Antitrust Act. In 1910 Congress had passed the Mann-Elkins Act as an initial step toward regulation. Under this act, enforcement was under the control of the ICC, but the ICC was ineffective in its role as regulator of the telecommunications industry. The Communications Act of 1934 addressed this concern by establishing the *Federal Communications Commission (FCC)* to oversee interstate telecommunications regulation. The act has been modified over time but still regulates the telecommunications and broadcast industries. Congress left the regulation of telecommunications within a state (intrastate) to the jurisdiction of each state. States passed legislation and set up a *Public Utilities Commission (PUC)* in each state to govern intrastate telecommunications.

2.8 TARIFFS

Regulation required telephone companies to file a document detailing any new service and the proposed charge for the service. The document defining in detail any new service is called a *tariff.* For interstate services under the jurisdiction of the FCC, the tariff is filed with the FCC. For intrastate services, it is filed with the governing state's PUC. The tariff is a public document. Any company can view the tariff filed by a competitor and gain an understanding of the technology required as well as the costs and profit the competitor has anticipated. The regulation of rates by the government guaranteed that monopoly service providers were giving fair rates to customers. When AT&T was a monopoly service provider there was no competitor, and publication of tariffs served the public interest. With deregulation and competition, the filing of a tariff puts the filer at a competitive disadvantage. Therefore, the requirement to file tariffs has been relaxed as the industry has become deregulated. Local telephone services are still regulated, and tariffs must be filed with the PUCs. When the phone company wants to raise rates for local phone service, for example, a new tariff must be filed and approved by the PUC.

2.9 RURAL ELECTRIFICATION ACT OF 1936

During the first half of the 20th century, rural America was still being deprived of electricity and telephone service. It was more profitable to install equipment and wiring in large cities, where the density of homes per square mile was high and the cost of equipment could be spread over a large customer base. In rural areas, wire had to be run many miles just to serve a few homes. To encourage investment in these rural areas, the Rural Electrification Act was passed in 1936 and created the *Rural Electrification Administration (REA).* This act made low-cost, government-backed loans available to entrepreneurs willing to establish electric service in rural areas. In 1949 the act was extended to cover low-cost loans for the establishment of telephone service in rural areas. The REA has been renamed and is part of the Rural Utilities Services of the U.S. Department of Agriculture.

2.10 STATE LEGISLATION: PUBLIC UTILITIES COMMISSION

no they don't!

State legislation guaranteed telephone companies a positive rate of return on their investment. A company could install electric or phone service in a community and provide the PUC with documentation on the amount of money invested to provide that service. The PUC would depreciate the investment over a definite time frame (5 to 25 years depending on the asset), add yearly maintenance expenses, and determine the rates a phone company could charge customers to receive as much as a 7% return on its investment. The investment could be borrowed from the REA at rates as low as 1%. A company could make 6% a year profit by providing the entrepreneurial resources and borrowing needed capital from the REA. The REA loan rates today are between 5% and 10%. The higher cost of money requires the PUC to authorize higher rates of return to the telecommunications companies. The PUC also grants a franchised territory to each company. Only one company could provide local telephone service, and the provider of that service was regulated by the PUC. The regulations established by state legislatures eliminated multiple providers of phone service and the problems associated with multiple telephone companies in one town refusing to cooperate. Today, deregulation is beginning to impact the franchised territory principle, since in a regulated system, one company was given sole marketing rights in a specific geographic area—that is, only one telephone company existed in each geographic area (no competition). In 1996, Congress passed legislation leading to the authorization of more than one provider for local service in the same community. Today's deregulation opens the door to competition by allowing many service providers in any area.

2.11 AT&T CONSENT DECREE OF 1956

The Department of Justice (DOJ) sued AT&T in 1949 for violations of the Sherman Antitrust Act. The BOCs would only purchase equipment from Western Electric (both companies were part of AT&T). Absence of competition in the sales of equipment to Bell companies effectively prevented regulation of rates charged by these companies. Western Electric could charge high prices on sales of equipment to the BOCs. The BOCs included the cost of equipment in the tariffs filed for rate increases. High costs for equipment meant high rates for phone service. Western Electric was not a regulated company and could charge whatever price the market would pay. The BOCs were a captive market. They would pay whatever price Western Electric charged because they were guaranteed a return of the investment, plus a profit on the investment.

Independent telephone companies would buy their equipment from manufacturers offering the lowest price. The DOJ suspected that Western Electric was selling equipment to Independents at a much lower price than it sold it for to the BOCs. The DOJ thought Western Electric was undercutting the pricing of smaller equipment manufacturers. This type of anticompetitive practice could be very detrimental to the smaller equipment manufacturers. The DOJ was looking into allegations that AT&T was involved in the actions just described and determined that the results of its investigation warranted antitrust action.

To prevent the use of the Sherman Antitrust Act against it, AT&T settled the suit by reaching an agreement with the government. This agreement is known as the *Consent Decree of 1956* (or sometimes as the *final judgment of 1956*). AT&T was permitted to keep Western Electric but was ordered to limit its sales to the BOCs and to sell at competitive prices. The decree forced the Independents to buy switching equipment from companies such as Northern Telecom, Automatic Electric, Leich Electric, Stromberg Carlson, and Northern Electric. This ruling allowed Automatic Electric to become a major equipment provider to the Independents in the 1960s and 1970s. In the 1980s, Northern Telecom became a major switching systems manufacturer supplying computer-controlled switching systems to both Independents and the Regional Bell Operating Companies (RBOCs).

2.12 CARTERPHONE DECISION OF 1968 BY THE FEDERAL COMMUNICATIONS COMMISSION

In 1968, the FCC ruled that AT&T must allow customers to use a device called the *Carterphone.* This device would let ham radio operators attach a phone line to their radio system. The ham operator could use the ham radio to establish connections across the country with other ham radio operators. The ham radio operators would then use their phone line to call someone in their town and connect the called parties together with the Carterphone. Prior to the *Carterphone Decision of 1968*, AT&T would disconnect phone lines where it found a Carterphone connected. Carter Electronics Corporation of Dallas sued AT&T and won. This is the decision that eventually let customers buy their telephones and business phone systems from someone other than the phone company. These alternate equipment providers were called *interconnect companies.*

2.13 STATION EQUIPMENT AND CUSTOMER-PROVIDED EQUIPMENT DEREGULATION: 1981 COMPUTER INQUIRY II

The telephone belongs to a larger classification of equipment known as *station equipment.* Station equipment is the equipment located at the telephone customer's residence or business location and generally consists of a telephone, modem, or PC terminal with internal modem. Station equipment was the first classification of telephone equipment to be deregulated. When customers provided their own equipment, it was called *customer-provided equipment (CPE).* Now that telephone companies no longer provide station equipment as part of the services covered by the monthly service charge, all customers must provide their own station equipment and all station equipment is customer owned.

Deregulation has reached out to include more than station equipment. Another area to be deregulated in the early 1980s was private telecommunications systems for businesses (keysystems and private switching systems called *private branch exchanges* or *PBXs*). If a business is large enough to require a keysystem or PBX, it must own or lease these systems from an interconnect company. The local telephone company could no longer provide these systems as of 1981 due to the Computer Inquiry II legislation. The Telecommunications Reform Act of 1996 allows local telephone companies to reenter this market by striking down the provisions of Computer Inquiry II.

All BOCs and Independents were forced by the deregulation rulings of the FCC (Computer Inquiry II of 1981) to set up separate subsidiaries as interconnect companies to handle the marketing and sales of keysystems and PBXs to business customers. The federal government felt a separate subsidiary was necessary to prevent the local telephone companies from using the revenues from local telephone service charges to subsidize discounts on sales of CPE. The existing business communication and marketing organizations of the local telephone companies were spun off in 1982 as separate subsidiaries that could not receive funds from their parent companies. The new subsidiary could sell CPE. The meaning of CPE was changed from "customer-provided equipment" to "customer-premise equipment." CPE included all equipment at the customer's location that had to be owned by someone other than the telephone company, according to the 1981 Computer Inquiry II legislation.

In 1996, Congress passed legislation opening the way for competition in the local phone service markets. As part of this legislation, telephone companies are allowed to sell CPE and no longer need a separate subsidiary to handle CPE sales and service. Shortly after the passage of the Telecommunications Reform Act of 1996, many phone companies began merging the CPE subsidiary they spun off in 1982 back into the local phone company organization.

2.14 INTERCONNECT COMPANIES

AT&T fought deregulation of the telephone instrument on the grounds that interconnect companies would provide inferior equipment for attachment to the network and degrade its performance. The FCC initially tried to alleviate these concerns by ruling that an interconnect device would be required when using a piece of equipment not provided by the telephone company. The interconnect device was furnished by the local phone company and was placed between the CPE and phone company equipment. This added unnecessary cost and inconvenience for purchasers of interconnect equipment. The manufacturers of telephones sold their telephones to telephone companies and to retail stores. If you bought the phone from a retail store, you needed a protective coupler from the telephone company. If you had the telephone company supply that same manufacturer's phone as part of your service, no protective coupler was needed.

The FCC recognized the error of its earlier ruling on the need for protective devices and dropped the requirement for an interconnect device. In lieu of such a device, the FCC requires registration with it of all equipment intended for connection to the telephone network. Through the registration requirement, the FCC ensures that all telecommunications station equipment provided by manufacturers to interconnect companies, retail outlets, and local phone companies meets minimum established service standards.

2.15 MICROWAVE COMMUNICATIONS INC. 1969 RULING BY THE FEDERAL COMMUNICATIONS COMMISSION AND SPECIALIZED COMMON CARRIER RULING OF 1971

In 1969, the FCC ruled that AT&T could not prohibit customers from using local phone lines for access to Microwave Communications Inc. (MCI). The *MCI ruling of 1969* made it possible for businesses to use companies other than AT&T for private business communication networks. In the *Specialized Common Carrier Decision of 1971,* the FCC broadened the scope of the MCI decision. It allowed any qualified common carrier the right to carry *private-line services.* Southern Pacific Railroad was quick to join MCI in providing private-line networks for business customers. In 1975, MCI began providing long distance service to the general public. AT&T filed a complaint with the FCC asking it to stop MCI from providing regular long distance service. The FCC dutifully complied with the request and instructed MCI to cease providing this service. MCI sued, and the appeals court reversed the FCC ruling in 1976. As a result of the *MCI decision of 1976,* the FCC opened the long distance market to competition. The FCC ordered AT&T to provide access, via the BOCs, to MCI customers. In addition to providing private networks for businesses, MCI, Southern Pacific, and other companies could now provide long distance service to anyone.

2.16 DEREGULATION

Deregulation of the industry began in 1968. From the late 1960s until today, proponents of deregulation have steadfastly held that deregulation would provide consumers with more choice and stimulate competition within the telecommunications industry. Proponents stated this competition would lead to lower costs for telephone service. Opponents of deregulation have held that deregulation would lead to inferior service and higher costs. History has proved they were both wrong (but the opponents of deregulation were partially correct). Deregulation has not led to inferior service; in fact, service has improved. Long distance phone service was deregulated, though local service is still regulated. Deregulation of long distance service reduced long distance charges but led to increased local service rates.

Prior to 1984, AT&T shared long distance revenues with the local phone companies. With deregulation in 1984, the local phone companies lost this important source of revenue. To ensure that they still received a guaranteed rate of return on their investment, these companies have been permitted to increase local rates and charge customers for access to long distance companies. Thus, access charges were developed as a means of compensating the local phone companies for losing their share of toll revenue. The long distance service providers and all telephone users pay access charges, which replaced *revenue sharing*. Access charges account for the largest expense for a long distance service provider.

Local phone rates are also higher than they were with sharing of toll revenue. If long distance rates were a lot lower than they were prior to regulation, perhaps they would offset higher local service costs and access charges, but long distance rates have not fallen enough to offset the tremendous rise in local service rates. Business customers do have a significantly lower overall cost due to deregulation of long distance. Because of the volume of long distance calls they make, they can negotiate large discounts with the long distance service providers. The loss of long distance revenue to the local telephone companies had to be offset somehow. It has been offset by the higher rates paid by residential and business customers for local service.

In the past, business users complained that long distance and business charges to them were extraordinarily high and they were subsidizing residential service. It is true that toll revenue and business rates subsidized local residential phone service. This made it possible to keep local residential service rates low and almost everyone could afford phone service. It is also true that a business customer uses the phone system much more than a residential user and should pay more. Now that the residential customer is paying rates that subsidize business rates, no one is complaining. In this respect deregulation has been a success. Business customers have a far wider range of choices for services, and competition results in lower costs for business. But you would have a hard time convincing opponents of deregulation that the residential customer has benefited. In 1980, residential service was about $10 a month and the phone was provided as part of the service. Today residential service is approximately $30 a month and the customer must purchase the phone.

2.17 HIGHER LOCAL SERVICE RATES AS A RESULT OF DEREGULATION

The local telephone companies have started filing tariffs to increase local rates again. They are citing the Telecommunications Reform Act of 1996 as one of the reasons they need to increase local service rates. They are stating that under the new law they must lower the access charges they are making to the long distance carriers (*interexchange carriers*, known as *IECs* or *IXCs*). Losing the IEC revenue must be offset by charging higher local rates. The PUCs have held hearings on this issue and have raised local residential rates again to offset the loss of revenue from access charges. It remains to be seen if deregulation of the local service arena will create competition that will lower local service rates. The local telephone companies *(local exchange carriers* or *LECs)* have reduced operating expenses by making capital investments in electronic switching systems, computerized record systems, and new outside plant. These systems require less maintenance and have led to dramatic productivity improvements, which have allowed the companies to make drastic reductions in management personnel and the craft workforce. With deregulation, many people install their own CPE. This has reduced the workload of the LECs and has allowed the LECs to reduce personnel.

Although the LECs have dramatically reduced costs, these reductions have not been passed on to the consumer in the form of a rate reduction. The vast majority of all tariffs filed seek permission to increase rates; it is rare for a LEC to seek a reduction in rates. Because most PUCs approve most requests from LECs to raise rates, it appears as if the motto of many PUCs is: "If you file a tariff, we will approve it." The PUCs hold public hearings on the requested rate increase, but schedule them at times and places that limit attendance by the public. Although they seem to approve all rate hike requests, the PUCs have ordered rate reductions in some instances. If the PUCs had not ordered these rate reductions, it is highly improbable they would have been made voluntarily by the LECs. If in fact local rates are artificially high, competition in the local services arena will lead to lower rates. Conversely, if competition does lead to lower rates, we know that our PUCs were allowing the LECs to overcharge us. If deregulation does not lead to lower rates, deregulation of local service is a failure from the consumer's point of view. In a deregulated environment, there is not much need for regulators; the competitive market serves as its own regulation device. Thus, elimination of regulated monopolies should lead to a reduction of personnel in the PUCs. It remains to be seen if our government agencies reduce their workforce in response to a reduced workload.

2.18 COMPUTER INQUIRY I OF 1971, COMPUTER INQUIRY II OF 1981, AND COMPUTER INQUIRY III OF 1986

Several rulings by the FCC during the 1970s and 1980s were aimed at defining whether regulation should apply to computer and data services; each ruling was known as a *Computer Inquiry.* The first was *Computer Inquiry I* in 1971. This ruling was issued after a study of the data industry by the FCC. The study found that the computer industry should not be subject to regulation and was beyond the FCC's control. In 1981 the *Computer Inquiry II* ruling was issued. As mentioned earlier, this ruling deregulated CPE and required telephone companies to set up separate subsidiaries to handle CPE. The ruling also allowed the BOCs to enter the nonregulated data processing business. In 1986, the *Computer Inquiry III* ruling was issued detailing the extent to which AT&T and the Bell companies could compete in the nonregulated enhanced services arena.

2.19 DIVISION OF TOLL REVENUE PRIOR TO 1984

The major opponents of deregulation were AT&T and the telephone companies. AT&T was the only provider of long distance service, and its BOCs furnished toll access to the Independents. Through toll revenue sharing, the Independents received a portion of the toll revenue their customers paid. If the customers of a local Independent phone company generated $100,000 a month in toll billing, the Independent would pass this amount on to Bell. The Independent would submit documents to Bell outlining what percentage of calls were toll calls and what percentage of their total investment was apportioned to provide both incoming and outgoing toll call completion. This process was called *separations and settlements.* Most Independents received more than half the toll revenue they collected. Some Independents had separations and settlements departments that provided documentation, which resulted in the Independent getting more money back than it gave Bell. In the example above, the Independent would probably receive $106,000 for the $100,000 paid to Bell. Without the Independent companies, Bell customers would not be able to make long distance calls to the Independents' customers. Bell kept all toll revenue generated by its own customers. Paying the Independents more toll revenue than their customers generated was basically a way to pay the Independents something to complete the toll calls of Bell's customers to the customers of the Independents. Deregulation of the telecommunications industry led to abolishing the separations and settlements process. The revenue lost from the demise of this process was replaced with access charges.

2.20 MODIFIED FINAL JUDGMENT OF 1984: DIVESTITURE AND DEREGULATION

Deregulation of the telephone industry has occurred mostly as a result of court rulings. Carter Electronics had to file suit to end anticompetitive practices by AT&T. It took a lawsuit filed by MCI against AT&T to open the doors to competition in the long distance market. The FCC was not in the forefront of establishing a competitive environment. Allegations of anticompetitive practices against AT&T led to involvement of the Justice Department in deregulation of the industry. The DOJ began to pursue antitrust action against AT&T, filing an antitrust lawsuit against the company on November 20, 1974. AT&T fought the lawsuit for eight years and then settled out of court, in 1982, with another "compromise." AT&T had come to regret the 1956 compromise agreement because it limited AT&T's business ventures to telecommunications. AT&T wanted to get into the computer industry, but the 1956 agreement prevented it from entering the computer market. The 1974 lawsuit was settled on August 24, 1982, as a modification of the 1956 Consent Decree's Final Judgment and is known as the *Modified Final Judgment (MFJ).* The MFJ took effect on January 1, 1984, replacing the 1956 Consent Decree.

The 1984 MFJ addressed *divestiture* and deregulation. AT&T agreed to divest itself of the BOCs, but one wonders if this action really hurt AT&T. It appears that AT&T, not the DOJ, actually drafts these settlement agreements. The 1996 Telecommunications Reform Act allows AT&T to reenter the local telecommunications service market. AT&T will be using newer technology than it left behind with the BOCs. It has aggressively entered several prime markets and is competing against the LECs and other IECs in these local markets to provide customers with local telephone service. Remember that local service was also defined as toll calls within a *local access transport area (LATA).* All major IECs can compete against the LECs for this business and can use their existing networks to provide intra-LATA toll service.

2.21 RBOCS, LECS, IECS, AND COMMON CARRIERS

On divestiture, 21 BOCs were called *Baby Bells.* The Baby Bells merged and formed 7 Regional Bell Operating Companies. These RBOCs are Atlantic Bell, Ameritec, Bell South, NYNEX, Pacific Telesis, Southwestern Bell, and US West. These regional companies are made up of several of the original 21 BOCs. For example, Ameritec includes the former companies of Ohio Bell, Indiana Bell, Illinois Bell, Michigan Bell, and Wisconsin Bell (see Figure 2-1). These local BOCs and local Independent telephone companies are now called LECs. Long distance service providers (AT&T, MCI, SPRINT, and so on) are called IECs.

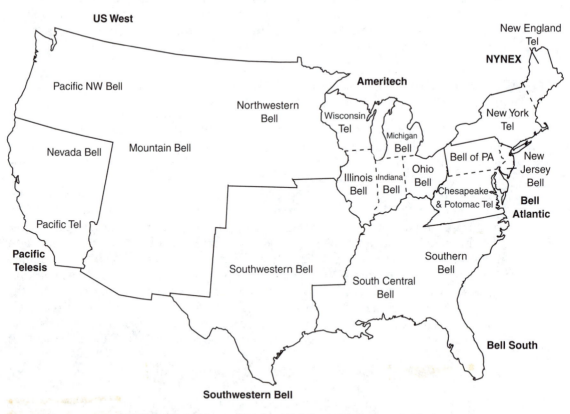

FIGURE 2-1 A diagram of Regional Bell Operating Company territories.

Companies offering telecommunications service as part of the PSTN are all referred to as **common carriers.** In addition to LECs and IECs, there are *satellite carriers* (such as RCA and American Satellite Company), *specialized common carriers* or *SCCs* (such as Metropolitan Fiber), and *value added network carriers* or *VANs* (such as GTE Telenet). Electronic Data Interexchange (EDI) networks are VANs. Some SCCs own their own networks; others use the networks of other carriers and act as resellers. SCCs such as Metropolitan Fiber offer business customers an alternative to using the LEC's local loop and central offices to access IECs. In large cities, Metropolitan Fiber has a cable network that rings the city and connects to IECs. A large business can buy IEC access from Metropolitan Fiber and bypass the LEC. Telenet is the oldest VAN in the United States and offers services such as electronic mail and access to services such as Prodigy and America Online (AOL) via Telenet's packet switched network. This network is made up of leased circuits from an IEC tied into packet assembler/disassembler (PADs) hardware owned by Telenet. Sprint also offers access to large, commercial Internet service providers (such as AOL and Prodigy) via its packet switched network called *SprintNet*. Common carriers provide service to the general public as part of the PSTN. They also provide services for private networks, which are owned and operated by companies primarily for their own use but may lease services to other companies. Private networks may be constructed by leasing circuits from common carriers to form part or all of the network.

2.22 EQUAL ACCESS TO LONG DISTANCE

In addition to forcing AT&T to divest itself of the BOCs, the MFJ contained language to deregulate the long distance market and required the BOCs to provide all IEC customers with **equal access** to their IECs. The MFJ ruling also required that local BOCs furnish space in their central offices for IECs to place their equipment. This space provides a point at which the LEC can connect customers to an IEC and is called the **point of presence (POP).** Since the 1976 ruling opening the long distance market to competition, AT&T had an advantage in marketing its service because AT&T customers only had to dial the digit 1 for access to AT&T long distance. Customers of MCI, Sprint, and other IECs had to dial a seven- or ten-digit phone number to access the IEC. After accessing the IEC, the customer had to dial a personal identification number (PIN) of four digits, then dial the area code and seven-digit phone number of the called party. AT&T customers dialed eleven digits to place a long distance call: 1 + area code + seven-digit phone number. Other IEC customers dialed 25 digits to place a call: 1 + 800 + seven digits + four-digit PIN + area code + seven-digit phone number.

When AT&T handled all long distance calls, it could be reached by dialing the digit 1. The digit 1 was interpreted by the translator of the central office switch as a call that needed connection to an AT&T trunk circuit, and the call was connected to an AT&T toll trunk. This is referred to as *trunk side access.* Much information can be passed over trunk signaling circuits such as the originating caller's telephone number for billing purposes. Equipment called *Automatic Number Identification (ANI)* was added to each line circuit. On long distance calls, ANI automatically provided the caller's telephone number to AT&T. When other IECs came into being, they could not use numeral 1 for access because it had already been assigned to AT&T.

Only one IEC could use 1 for access to toll and since AT&T had already been given it, AT&T customers had 1+ access. It was necessary to use telephone numbers to reach other IECs. The use of telephone numbers for access is called *line side access*. The originating callers' telephone numbers (ANI) could not be passed through a line circuit; thus, line side access required callers to identify themselves to the IEC billing machine through the use of PINs. Line side access is called *Feature Group A* access (one of several types of *Feature Group* access).

2.23 FEATURE GROUP B ACCESS

The BOCs devised a way to provide trunk side access to non-AT&T IECs. This access was called *Feature Group B*. Feature Group B used the code 950-10XX. The last two digits of the code identified a unique IEC. This cut down on the number of digits required to place a long distance call using an IEC other than AT&T. The switching system could identify IECs from the translation of the seven-digit 950-10XX number and route the call to a trunk circuit instead of a line circuit. With trunk side access, no PIN was needed because the caller's telephone number was available via ANI over the trunk circuit to the IEC. The Feature Group B access code was changed from 950-10XX to 10XXX access in 1987 and was changed to 10-10-XXX in 1998.

2.24 FEATURE GROUP D ACCESS

It was technically impossible for Bell to provide equal access in older electro-mechanical local exchanges. To provide all IECs with equal access via the digit 1 *(Feature Group D)* would require the replacement of these offices with computer-controlled switching systems. On dialing 1, a computer-controlled system can look at the originating line's database to determine its *preferred interexchange carrier (PIC)* and route the call to that IEC. The Justice Department recognized that time would be required to comply with this MFJ directive and gave the BOCs until 1987 to meet the provisions of equal access. True equal access where all IECs can be accessed by the digit 1 is called Feature Group D. A customer can inform the LEC of a PIC, and the line for that customer is PICed to that IEC by making an entry in the switching equipment's database. Feature Group D also allows a caller to override a PIC and access any IEC by dialing the Feature Group B access code (10XXX). The access code for Feature Group B was changed from the seven-digit 950-10XX to the five-digit 10XXX. To use AT&T, callers dial 10288. To use Sprint, 10333 should be dialed. Other IECs also have five-digit Feature Group B access codes of the form 10XXX.

Deregulation requires that the LECs assign IECs to pay stations in private business locations, based on the desires of the proprietor of the business where the pay phone is located. All pay phones located in public places must be assigned to an IEC by the public authorities or on an allocation basis by the LEC. The PIC used at pay stations might prevent people from using calling cards. If a pay station is picked to Sprint, it will not accept an AT&T calling card and vice versa. Feature Group B access allows callers to override the PIC. The 10XXX code allows pay phone users to override the PIC (the assigned IEC) and use the IEC of their choice. This allows them to use the phone card provided by their IEC.

2.25 UNITED TELECOMMUNICATIONS, INC. ADAPTS TO THE ERA OF DEREGULATION

Even though telephone companies originally opposed deregulation, some have benefited from it. As any company should, they have been quick to adjust to the new rules of the game. A prime example would be United Telecommunications, Inc. United was an aggressively growing local exchange company in the 1970s. With deregulation of the long distance market, it quickly seized the opportunity to enter this market. It purchased some railroad right of way to lay fiber optic cable. It formed alliances with Southern Pacific Railroad and later with General Telephone. It subsequently bought out both companies and is the sole owner of Sprint. The company that grew up as United Telecommunications and formed a subsidiary called *Sprint* has seen the child grow up and take the dominant role. United Telecommunications companies are now subsidiaries of Sprint.

The major portion of Sprint's revenues come from its deregulated long distance services, not from the regulated local exchange companies (although most profit is still generated by the LECs). Company stockholders have benefited from deregulation due to the visionary leadership at United Telecommunications. Where others could only see deregulation as a problem, they saw deregulation as an opportunity. They have aggressively pursued this opportunity and are now forming alliances with telecommunications companies overseas to become as successful in providing intercontinental long distance. Sprint has also formed a division to provide local phone service in RBOC territories. Sprint will use personal communication systems (PCS) technology as well as alliances with cable TV companies to provide local telephone service.

2.26 LOCAL EXCHANGE CARRIERS

Before the deregulation movement got underway, there were about 14,000 Independent (non-Bell) local telephone companies. These Independent companies and the local BOCs provide local telephone service and are LECs. The LECs—briefly discussed earlier—are the local phone companies. They connect our telephones to a central office, from which we are able to call anywhere in the world over the PSTN. When we pick up our telephone handset, we are automatically connected to switching equipment in the local central office. The switching equipment will receive dialed digits from our telephone and determine from those digits where the call should be switched to. Prior to deregulation, the LECs provided local service and AT&T was the sole provider of long distance service. Today there are several long distance service providers. As we have seen, companies offering long distance service are called IECs or IXCs.

Many of the Independent LECs have consolidated, and today there are about 1300 Independent LECs in addition to the 23 BOCs. The LEC is the carrier authorized to provide a physical connection from telephones to the telecommunications network. Recent legislation has approved competition for local telephone service in several areas of the United States. In these areas, the LEC must rent facilities to other competitors wishing to provide local exchange service using wire facilities. Some competitors will use PCS to provide local service. PCS employs signals similar to a mobile telephone and pager. PCS does not require a pair of wires to connect a phone to the central office switch. PCS providers will install their own switching systems and use radio waves to connect their switching system to their customers. The PCS switch will have trunks connecting it to the LEC switching center, and PCS customers will be able to call LEC customers.

2.27 LOCAL ACCESS TRANSPORT AREAS

The MFJ divided the country into 184 areas known as *LATAs*. A LATA is considered a local toll calling area, and the MFJ recognized toll calls within a LATA as calls that must be handled by an LEC and not an IEC. Many LATAs will serve the same basic geographic boundary covered by an area code; however, some LATAs serve more than one area code. LATA 524 has three POPs in Missouri (Kansas City, Moberly, and St. Joseph). The Kansas City, Missouri, POP serves customers making a toll call from both Kansas and Missouri. The area code for Kansas City, Kansas, is 913, and the area code for Kansas City, Missouri, is 816. Thus, this POP serves two area codes and also serves customers in two states (see Figures 2-2 and 2-3).

2.28 POINT OF PRESENCE

In most LATAs the majority of intercity calls within the LATA (intra-LATA) are handled by the BOC for that LATA. Of the original 184 LATAs, 156 were served by Bell LECs and 28 by Independent LECs. The MFJ required calls between LATAs to be served by an IEC. The MFJ authorized AT&T and other IECs to provide service on inter-LATA calls (calls that cross from one LATA to another) and forbid the offering of this inter-LATA traffic by the RBOCs (or any LEC). Recent legislation by Congress has opened intra-LATA service to competition from the IECs. The RBOCs are petitioning for the right to provide inter-LATA (long distance) services.

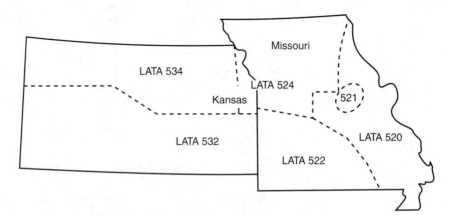

FIGURE 2-2 Kansas/Missouri LATAs.

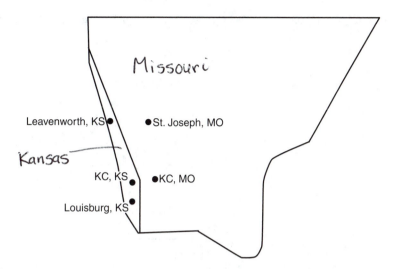

FIGURE 2-3 LATA 524.

Each LATA has established one or more access points—or POPs—for connection to the IECs for long distance services between LATAs. To comply with the 1984 provision, which mandated that the BOCs provide equal access to all IECs, the POPs were set up in the class 4 toll office and were called *equal access tandems*. Because class 5 electromechanical central offices have been replaced with computerized *stored program control (SPC)* central offices, it is possible to place the POP at the class 5 office if desired. For a long distance call, the LEC will connect a customer to the POP where the call will be connected to that customer's preferred IEC, and that IEC will handle the call. Telecommunications legislation of 1996 approved the entrance of RBOCs into the long distance market if they demonstrate compliance with the portion of the 1996 legislation providing for competition in the local service arena.

2.29 NETWORK HIERARCHY

When AT&T handled all long distance calls, it used a network hierarchy of five levels. The local exchange that telephones are wired to was designated as a class 5 office. There were about 20,000 class 5 offices across the United States. All calls originate at a class 5 office, and the class 5 office will connect to the network via a class 4 central office. When handling a long distance call, the local exchange would connect the call to a class 4 office when the caller dialed a 1 or 0. The class 4 office would serve as an access point to AT&T long distance. All class 5 offices within a small geographic area would connect to the class 4 office serving their area. The class 4 offices connected to a class 3 office. The United States was divided into different areas and an area code assigned to each area. Each area code would contain one class 3 office, which was known as a *primary center*. The class 3 office served all class 4 offices in the area code. Several primary centers (class 3 offices) would

Five-Level Public Switched Telephone Network

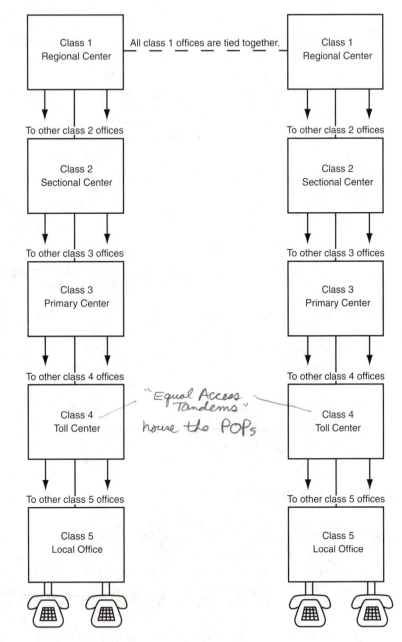

also called Exchange Levels

"Equal Access Tandems" house the POPs

FIGURE 2-4 All telephones connect to a class 5 local central office.

connect to a *sectional center* (class 2 office). Each sectional center, serving a geographic section of the country, would connect calls to a *regional center* (class 1 office). AT&T has 12 class 1 regional switching centers in the United States (see Figure 2-4).

If a long distance call was made between towns served by the same class 4 office, the caller's class 5 office would be connected to the class 4 office. The class 4 office would connect the call to the called class 5 office. Thus, the connection would be: originating telephone, originating class 5 office, class 4 toll office, terminating class 5 office, and terminating telephone. If the long distance call was from one region of the country to another, the connection would be: originating telephone, originating class 5 office, class 4 office, class 3 office, class 2 office, class 1 office serving the calling region, class 1 office in the region of the called number, class 2 office, the primary (class 3) office serving the area code for the called number, class 4 office, the class 5 office serving the called telephone, and the called telephone.

An analogy to airline travel may help clarify the different levels of switching and the trade-off that occurs between the number of transmission facilities and these levels of switching. The airline companies do not fly to every town in the country because the cost of having passenger terminals and routes to every town is prohibitive. If we wish to travel from a small town to a destination across the country, we will drive to a nearby airport and take a small commuter plane to a larger airport. At the larger airport, we catch a flight to a regional airport. Then we change planes and fly to the regional airport serving the region of our final destination. Here we board another commuter airline's plane for the final leg of our trip. For example, to fly from Dodge City, Kansas, to Pinehurst, North Carolina, we would need to change planes several times because no direct route exists between these two towns. We would drive from our small community (class 5 office) to an airport in Dodge City. Dodge City is the equivalent of our class 4 office. From Dodge City, we fly to Kansas City. Our transmission facility from Dodge City to Kansas City is a small airplane. There is no need to use a large airplane because the amount of traffic is small.

At Kansas City (our class 3 office), we board a larger plane (larger transmission facility) and fly to St. Louis (our class 2 office). Many passengers originating in various communities around the state of Kansas have joined us for our flight to St. Louis. At St. Louis, we leave some of our fellow passengers. Some have reached their final destination and will drive by car on the final leg of their journey to a local community in the St. Louis area. Others are going in a different direction and will board other planes heading for New York, Los Angeles, Denver, Dallas, and so on. We will board a plane to Chicago (our class 1 office), along with some passengers from Kansas City as well as other passengers who came from the St. Louis area or flew in from a different airport. The transmission facility is a large plane because a lot of people are flying to Chicago. At Chicago, we are switched again to another plane and board a plane heading for Atlanta (another class 1 office). In Atlanta, we board a plane bound for Charlotte, North Carolina (a class 2 switching center). In Charlotte, we switch to a plane flying into Raleigh (a class 3 office). In Raleigh, we are switched to a smaller plane flying to Fayetteville, North Carolina (a class 4 office), and we take a plane from Fayetteville to Pinehurst (our class 5 office and the end of our trip).

Of course, travelers were not happy having to change planes so many times, and as the amount of air travel grew, additional transmission facilities (additional routes and planes) were added between class 3 centers to flatten the levels of the network. In telecommunications, it does not matter to us how many times our call is being switched or how many switching offices are used to connect the call. The switching is so fast that the number of offices we use to connect the call makes no difference. The PSTN network was flattened to reduce cost. The development of fiber optic transmission facilities has provided our industry with a transmission medium capable of handling 516,096 different conversations over one pair of glass fibers. This technology has made it possible to flatten the PSTN. Instead of needing thousands of smaller transmission media to connect offices, only one fiber optic facility is required. IECs have flattened the network to reduce the number of switching stages needed, thereby reducing capital expenditures for replacement offices and lowering maintenance expenses. Fewer offices require fewer maintenance people.

2.30 SPRINT'S FLAT NETWORK

This hierarchy has been flattened with the introduction of computer-controlled electronic switching systems, high levels of multiplexing using technology that allows thousands of conversations to be carried by one device, and fiber optic transmission systems. When AT&T set up the five-level hierarchy, it had to use electromechanical switching systems for the central offices and wire cable as the medium between switching systems. These older technologies and low orders of multiplexing limited the volume of calls that could be handled by one switching center. When Sprint and MCI set up their long distance networks using electronic switching systems, high levels of multiplexing, microwave technology, and fiber optic cables, they were able to combine the functions of the regional and sectional offices within one office. Sprint has approximately 45 sectional offices. These sectional offices are interconnected using fiber optic cable, and each office connects to many POPs (class 4 toll tandem offices and some class 5 offices). POPs were required because it was cheaper for the Bell LECs to furnish the switching technology necessary to provide equal access at the class 4 office rather than replace all class 5 electromechanical switching offices. As noted, these class 4 offices are called equal access tandems.

As the class 5 electromechanical central exchanges were changed out to computerized switching systems, it became possible to provide equal access from that local office. Class 4 offices still serve as POPs to aggregate calls from small exchanges with low long distance calling volumes, but IECs also have direct connections to the larger class 5 computer-controlled central offices because these offices are capable of providing equal access. A caller using Sprint as the IEC of choice will connect from an originating class 5 office to the class 4 office serving as a POP (or directly from the local class 5 office to Sprint in some cases). The POP will connect to a Sprint sectional office. The originating caller's Sprint sectional office will determine which of the other 45 Sprint offices the call should be connected to, in order to connect to the POP serving the called number. The call will be connected to that office. This second Sprint office will then connect to the terminating POP. The terminating POP class 4 toll office connects to the terminating class 5 office for connection to the called telephone. In a small rural area, your IEC and the LEC use a POP in the class 4 toll exchange. In larger cities, the POP may be at the class 5 local exchange (see Figure 2-5 for the Sprint Network.)

2.31 TELECOMMUNICATIONS REFORM ACT OF 1996

On March 30, 1995, Senator Larry Lee Pressler of South Dakota introduced Senate Bill 652. This bill was amended many times over the next year and was signed into law on February 8, 1996, as Public Law 104-104. Telecommunications managers should be familiar with the *Telecommunications Reform Act of 1996* since it governs how we can conduct our business. (The complete text of the bill can be downloaded over the Internet from the Library of Congress at **http://thomas.loc.gov.**) This law provides for further deregulation of both the telecommunications industry and the cable television industry. When Senate Bill 652 was under consideration, there was little debate on the portions of the bill dealing with deregulation of telecommunications, but it received a tremendous amount of debate over its provisions to deregulate rates charged by cable television companies, to limit pornographic materials on the Internet, and to require manufacturers to include a programmable chip in television sets that parents could program to restrict what their children watch. *& "gave" established TV stations HDTV bandwidth for free!*

Vchip

The local BOCs and their parent RBOCs actively lobbied for support of the Telecommunications Reform Act of 1996 because it contained provisions that would remove restrictions placed on them by the 1984 MFJ. As applied to the telecommunications industry, Public Law 104-104 supersedes the 1984 MFJ three years from the date of becoming law. Therefore, this law effectively replaced the MFJ on February 8, 1999. The law removes the restrictions imposed on both AT&T and the RBOCs. It also eliminates restrictions on all IECs. They are allowed to carry intra-LATA calls and can now compete in the local telephone service markets. LECs must cooperate with IECs and Competitive Access Providers (CAPs), such as Metropolitan Fiber Systems, that wish to provide local exchange services. Even before the president signed this legislation, the FCC and PUCs were authorizing multiple local access providers in the same locality. Under the Telecommunications Reform Act of 1996, the BOCs are now allowed to provide long distance (inter-LATA) service, if they prove they have not hindered competition for local services in their local exchanges. AT&T is allowed to reenter the local services market and the RBOCs are allowed to manufacture telecommunications equipment.

One significant area this bill did not provide for what was "universal service." Several amendments were offered to make enhanced services such as last number redial, caller ID, and local area networks available in all local service areas, but these amendments could not gain support of the majority in either the Senate or the House. The bill asks that LECs strive to provide "universal service" at the same cost these services are provided at in large cities, but this bill does not mandate universal service. Therefore, many areas of rural America will not receive the enhanced services now available in larger cities. If the services are provided, they will cost more because the necessary investment cannot be spread over a large customer base.

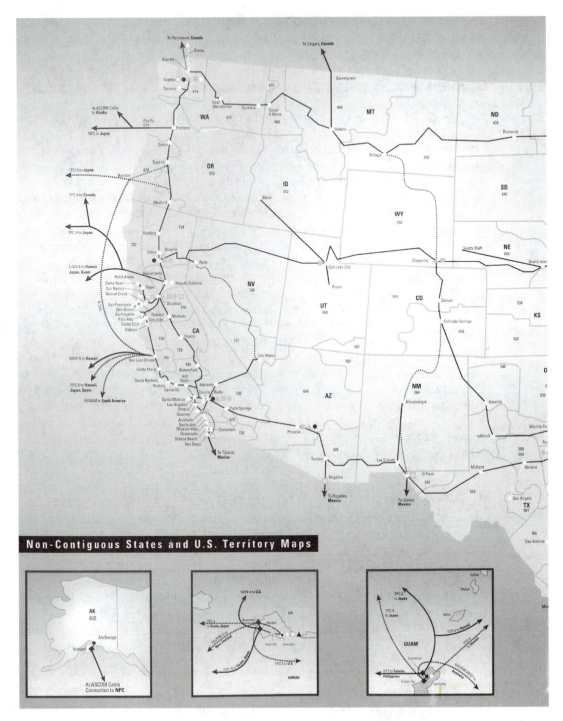

FIGURE 2-5 Sprint network. Used with permission from Sprint.

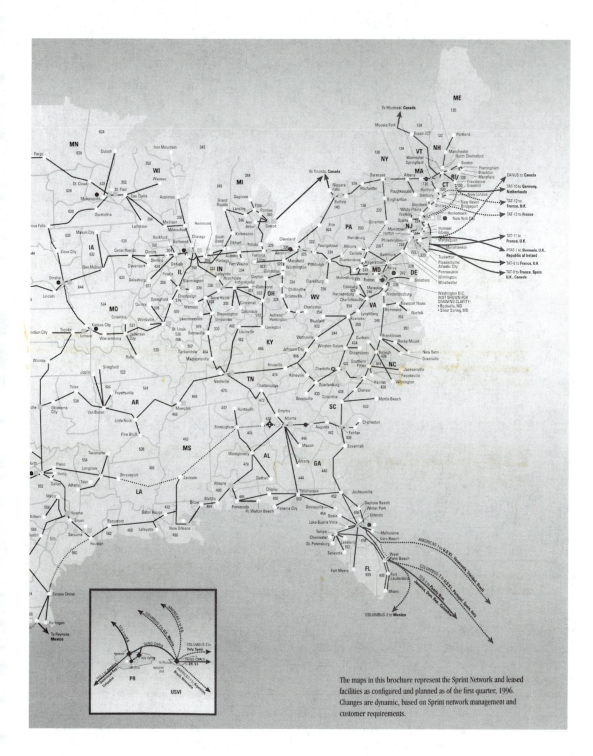

The maps in this brochure represent the Sprint Network and leased facilities as configured and planned as of the first quarter, 1996. Changes are dynamic, based on Sprint network management and customer requirements.

2.32 1997 AND 1998 EVENTS REGARDING THE 1996 TCOM REFORM ACT

Congress charged the FCC with administering the 1996 Telecommunications Act in a manner that would stimulate competition. Many people charge that the act is vague and has created a chaotic atmosphere. Still others charge that the FCC has overstepped its authority in setting rules for deregulation of local telecommunications services. The 1996 Telecommunications Reform Act has caused a great deal of commotion; nobody seems happy with the results thus far in deregulating local telecommunications services. This should have been anticipated. Anytime you make such sweeping changes in the way a business or industry can operate, it is bound to create a wave of point-counterpoint. The 1996 Telecommunications Act has become a political football and a livelihood for the many law firms filing lawsuits regarding the FCC's enforcement of the act.

Following the mandate of Congress to promote competition in the local markets, the FCC established a 14-point checklist that incumbent local exchange carriers (ILECs) were to meet before they could receive permission to provide interstate long distance service (which is controlled by the FCC). The FCC also set national pricing guidelines called *total element long-run incremental costs (TELRIC)* that ILECs were to use as a guide in pricing services to CLECs. In addition, the FCC established a Universal Service Fund to provide funding for enhanced services in rural areas. An educational fund was also provided to fund Internet services to schools and libraries. This fund is known as the ***E-rate fund***. Immediately, the FCC rulings were challenged in the courts.

2.33 REDUCED IEC ACCESS CHARGES AND E-RATE FUNDING

As a result of the 1996 Telecommunications Reform Act, the FCC also ordered the ILECs to reduce access charges to the IECs. Access charges are the single biggest expense that an IEC has. Since the FCC has helped IECs reduce their cost of providing service by lowering the access fees they pay to LECs, it thought the IECs would not object to paying the fees necessary to fund the E-rate fund. The E-rate fee was miniscule compared to what the FCC saved the IECs in access charges, and you would think the IECs would be happy with their large piece of the savings pie. Alas, such is not the case. IECs want the whole pie, and they proceeded to pass the E-rate fees on to their consumers.

The FCC said IECs could not do this. The IECs raised opposition to the E-rate fund via their lawyers, public relations departments, and lobbyists in Washington. The IECs claim that the FCC simply replaced one fee (access charges) with another fee (E-rate charge). Congress was quick to come to the aid of big business. After all, the IECs as a group contribute a lot of money to help members of Congress get elected.

The political nature of the E-rate fund is highlighted by the fact that Republicans refer to it as the *Gore tax* simply because, as vice-president, Al Gore was the main politician pushing to make the Internet an information superhighway. Democrats and Republicans continue to fight over this program. It has been made political, in part, because members of Congress do not want to be viewed as denying Internet access to underprivileged children. The majority of our legislators supported E-rate funding. I would think that most people reading this book know how important Internet access is to our children's future.

Opponents of E-rate funding also have good reason to oppose it. The fund was originally administered by a private organization. The major official at this company is a friend of Gore's. This company did not do a good job of administering the E-rate funds. Since the FCC took control of the funding, it has also done a poor job. Very little of the annual E-rate fund budget was dispensed in 1998, and most of what was dispensed was not used for Internet services. Most of the funds are being used to pay for local telephone service.

Due to pressure from opponents of E-rate funding, some Representatives have attacked the FCC rulings as well as Gore's involvement. Log on to the Internet and check the records for yourself. See if your representatives represent your view on this issue. Their votes, words, and actions provide a good picture of their stance on E-rate funding. This funding seems like a good idea, but it appears that the funds are either being misappropriated or not properly disbursed.

Here is what financier and political candidate Steve Forbes had to say in the August 10, 1998 issue of *Forbes:*

> Congress should repeal a nasty, not-so-little levy that it never intended to enact in the first place. Buried in the 1996 Telecommunications Act is a provision put in at the behest of Vice President Gore that the Federal Communications Commission has interpreted as giving it the right to impose a special tax on long distance telephone customers. The exaction is supposed to fund a program to connect schools and libraries to the Internet. By 2003 revenue from this so called Gore Tax could be twice that of the federal gas tax.

Why do *Forbes* and others refer to the E-rate fund as the *Gore tax?* Would opponents of E-rate funding rather launch political attacks than attack the issue?

Forbes goes on to state that legislators are outraged at this highhandedness because the FCC as an unelected group of bureaucrats has no authority to levy taxes. Only a legislature can impose taxes. True, but the FCC has the right to levy fees. The response from some is: "If it looks like a tax, walks like a tax, and quacks like a tax, it must be a tax." This sounds cute, but being cute does not make a fee a tax. If people do not want agencies to have the power to levy fees, they should encourage Congress to issue legislation making it illegal for all government agencies to levy fees.

Opponents of E-rate funding deride supporters of the funding, as Gore tax advocates. The response from the supporters of E-rate funding is: "If your argument is weak, don't debate the issue; instead, attack the opposition." Rich people in our society can afford to provide their children with Internet access. Why would the rich want to deprive disadvantaged children of one of the tools that will help them better their lives? *Forbes* is quick to state that 80% of our schools have Internet access, and there is no need for the E-rate fund. This may or may not be so, but the goal should be 100%, not 80%. The 20% of our schools without Internet access are the very schools that need access the most. Should Congress reduce funding that helps educate our children before reducing funding for the arts? Congress has to listen to many special interest groups that seek funds for their programs. Who is to say that E-rate funding is any more or less deserving than other programs?

The 1996 Telecommunications Reform Act included provisions for an E-rate fund at $2.125 million. In June 1998, that fund was cut to $1.25 million at the direction of Congress. They were responding to pressure from people such as Forbes, the IECs, and their lobbyists, who have attacked the FCC and the administration of the fund. Most of the attacks have been political in nature, but they do have a point. Administration of the E-rate fund needs to be improved. Politicians can justify spending millions on their pork belly projects, and on the impeachment or defense of a president, but they cannot justify spending $2 million on Internet access to help educate children. Do they need to rethink their priorities?

You now see just one reason why the 1996 act has caused a great deal of commotion. The E-rate issue is a political issue, but it should not be; it should stand or fall on its own merit. Unfortunately the politicization of the issue creates problems for advocates on both sides. Internet access for all children is an issue deserving of debate without resorting to political mudslinging. Hopefully, we can find a bipartisan solution to the E-rate issue and take steps to prepare our children for the new millennium in the most productive and economical way. Just as we strive to make a telephone in every home affordable, we must strive to make the Internet accessible everywhere.

2.34 FCC 14-POINT CHECKLIST AND PRICING GUIDELINES

Earlier, I mentioned that the FCC established a 14-point checklist that ILECs must comply with if they wish to provide inter-LATA long distance service. SBC Communications, Inc. (Southwestern Bell and others) filed for an appeal of this FCC ruling on July 2, 1997. On December 31, 1997, a U.S. District Court judge in Wichita Falls, Kansas, ruled that RBOCs do not have to comply with the 14-point provisions of the FCC before they can sell long distance service. That decision was appealed and on September 4, 1998, in a split decision, the U.S. Court of Appeals for the Fifth Circuit reversed the lower court decision. The appeals court restored the provision of the Telecommunications Act of 1996 that RBOCs must demonstrate they have opened up their local markets to competition before they will be allowed to handle long distance calling. This decision was appealed to the Supreme Court, which settled the issue in favor of the FCC in January 1999. The Supreme Court ruled that the FCC had the authority to set conditions that have to be met before allowing RBOCs to compete in the long distance markets.

Another area of contention is the national pricing guidelines called TELRIC that were established by the FCC. The FCC instructed the ILECs to use TELRIC as a guide in pricing services to CLECs. In response to a lawsuit filed by one of the RBOCs, the Eighth U.S. District Court of Appeals in St. Louis ruled in January 1998 that the FCC overstepped its authority; the court allowed RBOCs to set their own charges for services. This ruling was appealed and also settled by the Supreme Court in January 1999. The court ruled that the FCC has the authority to set pricing guidelines on the unbundling of local services.

In addition to charging CLECs for the use of their cable plant to serve customers, ILECs also charge for floor space in their central exchanges when CLECs wish to use their equipment to serve customers. The RBOCs and other ILECs are charging anywhere from $25,000 to $250,000 for the establishment of a 10-foot by 10-foot floor space in their central office for a CLEC to place its equipment. ILECs then charge anywhere from $10,000 to $15,000 a month for rent and utilities (electricity, heating, and air-conditioning).

2.35 CLEC VS. ILECS: WARFARE IN THE LOCAL MARKETS

It has been alleged that the ILECs drag their feet on applications from CLECs to secure floor space and the conversion of customers to their company. Several CLECs have been especially critical of the RBOCs for dragging out the application process by requiring the CLEC to follow a procedure that details each piece of equipment, how it will be used, and its conformity to performance and safety specifications established by Bellcore. In far too many instances, the ILECs simply state that no space is available for CLECs because it is all in use or planned for use. This seems implausible.

The ILECs have replaced older switching systems with new computer-controlled switching systems and have replaced T1 carrier systems with OC-48 multiplexer and fiber optic technologies. Conversion to these new technologies freed up a lot of floor space and also freed up a lot of space on the main distributing frame. ILECs used to open their central offices to tours for the public. They no longer do this. Is it possible they are afraid someone will see there is ample space in their offices for CLEC equipment when they state otherwise?

Three years after the 1996 act was signed into law, the new CLECs had less than 1% of local customers and less than 4% of the dollars spent by consumers for local service. If Internet service is excluded from the above numbers, the CLECs have almost no business. So far competition in the local services arena is almost nonexistent. Some people charge that the act has resulted in actions counter to its original purpose. They charge that the merger of existing competitors has in fact lessened competition. The RBOCs have stated that the only way they can compete effectively in the long distance market is to merge networks. So it appears that rather than stimulating competition, the 1996 act has created an atmosphere of buying potential competitors' businesses. As famous baseball player Yogi Berra once said, "It's déjà vu all over again." This buying of competitors by AT&T at the turn of the century is what led to the Kingsbury Commitment.

The established ILECs have all filed applications to provide long distance services. On April 20, 1998, Bell Atlantic received a prefiling statement from the New York PSC outlining steps it must take to receive approval for entry into the long distance market. Basically, Bell Atlantic must agree to provide unbundled services to competitors in its local markets. Potential competitors NYNEX and GTE are no longer competitors due to their merger with Bell Atlantic.

2.36 MERGER MANIA

SBC merged with Pacific Telesis in April 1997. With that merger, SBC's local exchange region expanded to include more than 32 million access lines in 7 states—Arkansas, California, Kansas, Missouri, Nevada, Texas, and Oklahoma. In January 1998, SBC announced its intention to merge with Southern New England Telecommunications. This merger was completed on October 26, 1998. SBC announced in May 1998 a proposed merger with Ameritech. The FCC began holding hearings on the proposed merger in December 1998. Among those who oppose this merger are Sprint and Customer Service Groups in California.

The merger of MCI and WorldCom was approved on September 15, 1998, after MCI agreed to sell its Internet backbone facilities to Cable and Wireless. WorldCom retained its Internet facilities (UUNET) and is the number one provider of Internet backbone services. Cable and Wireless is number two. Had MCI been allowed to keep its Internet facilities, MCI WorldCom would truly have had a monopoly position in Internet access services.

While the FCC approves megamergers, it has not approved the request of start-up IEC Quest Communications. Quest Communications has established a very high capacity, state-of-the-art, fiber optic network. It attempted to expand its long distance service customer base through reseller arrangements with US West and Ameritec. In November 1998, the FCC blocked the deals between Ameritec, US West, and Quest. The FCC felt this agreement would provide a mechanism whereby US West and Ameritec could enter the long distance market prior to meeting the FCC requirements of opening their local markets to competition. The FCC and Congress also blocked the merger of Sprint and MCI/WorldCom.

NYNEX was merged into Bell Atlantic in 1997, and then on August 3, 1998, Bell Atlantic and GTE agreed to merge. Among those who opposed this merger at the December 1998 FCC hearings were AT&T and Sprint. In January 1998, AT&T acquired Teleport Communications Group Inc. (TCG). On July 27, 1988, AT&T began to pursue TCI. Both companies will provide AT&T with cable TV and Internet networks that can be used to pursue local telephone and Internet access market opportunities. In 1997 AT&T acquired McCaw Cellular, and it will also use cellular radio to provide local services. In January 1999, AT&T reached an agreement with Time Warner Cable to sell local services via its cable TV facilities.

2.37 UNIVERSAL SERVICE FUND

Another area of major conflict with the 1996 act is the mandate to the FCC from Congress to establish a universal service fund. The FCC implemented an initial approach to funding in May 1997. SBC, Bell South, and GTE filed a joint brief with the U.S. District Court in San Antonio, Texas, on February 20, 1998, asking that the FCC order on universal service funding be overturned. The premise behind universal service is the desire to provide all Americans with basic telecommunications services no matter where they live.

The goal of universal service between 1940 and 1990 was to provide all Americans with affordable voice communication service. Prior to deregulation of the telecommunications industry, charging much more for business services and long distance calls than it costs to provide the service, funded universal service. Some people also indicate that regular residential rates in a city were overpriced to help support services to rural America. A portion of the higher rates went into a universal service fund that was distributed by the PUC to rural telephone companies.

The cost to provide service is higher in a rural area than in a city because rural customers are spread over a much larger geographic area. In a city, the cost of a central office switching system can be spread over a much larger base of customers, and the cable lengths are much shorter. The rates, which a PUC sets for local telephone service, vary depending on the number of customers a given caller can reach. In a city, a caller can reach many more customers and pays a higher rate than a rural customer pays.

A rural telephone company faces higher costs in providing telephone service and also has to charge lower rates than a telephone company in the city can charge. Do you see a problem here? If rural telephone customers paid according to the cost of service, their rates would be higher than the rates charged in a city. The only way to make telephone service affordable to all rural Americans was via the use of a universal service fund that was funded by higher business and toll rates. These higher rates subsidized residential rates.

Since business rates and toll rates were much higher than the cost to provide these services, they offered competitors in the market the highest profit margins and most lucrative business opportunities. There are two sides to every story. Universal service is a great idea, but if someone competes for your customers and they do not have to contribute to a universal service fund, they can "cherry pick" customers from you. If a LEC has to charge $2 more for a business line just to pay into a universal service fund, competitors can automatically enter the market for business customers with a $2 cost advantage.

Business customers are where the profit is. This is why CLECs could care less about getting residential customers. It will be some time before competition occurs in the residential market to the degree necessary to drive residential rates down. If universal service funding is dropped, the ILECs will be able to reduce business rates. This will allow the ILECs to be competitive with CLECs for business services. Loss of subsidization for residential rates means the residential rates must and will go up.

Just as it was with the deregulation of long distance services, the deregulation of local services means business customers end up with lower telephone bills and residential customers end up with higher bills. There is no such thing as a free lunch. The phone company must make a profit like any other business concern; total revenue must be a certain percentage above total cost. If a company gives a discount to one group of customers, products, or services, the charges made with respect to other customers, products, or services must be raised to offset the discount and thereby maintain the same profit margin.

The LECs price all services to generate a certain amount of total revenue. If total revenue is to remain the same when business rates go down, residential rates must go up. This is precisely what happened when the FCC ordered ILECs to reduce access fees to IECs. The ILECs filed tariffs with the PUCs requesting permission to raise residential and business line access charges to offset the loss of revenue caused by lower access charges to IECs. Thus, you and I end up having to pay higher telephone charges to enable the IECs to receive reduced rates from the LECs.

The 1996 Telecommunications Act did not require universal service but simply asked that telephone companies (IECs and LECs) strive to provide universal service at affordable prices. The FCC controls a universal service fund that is funded by a portion of the access charges paid by telephone users and IECs to the LECs. Debate now rages over whether this fund should be used to help keep the cost of telephone and Internet services to rural America low.

2.38 ENHANCED LOCAL SERVICES VIA DSL

On January 26, 1998, Bell Atlantic filed an application with the FCC to provide packet data services. This application was filed under Section 706 of the 1996 Telecommunications Act. Section 706 specifically directed the FCC to promote advanced data networks by removing regulatory restrictions that hinder deployment of broadband services. The FCC was directed by Congress to have a plan developed by August 1, 1998, to promote broadband services. On February 25, 1998, US West, Inc. filed a similar application, and on March 5, 1998, Ameritec Corporation also filed under Section 706 to provide broadband services. All three companies are asking for permission to build broadband networks that can be used for data and Internet access.

AT&T, MCI, and World Com oppose this separate action by Bell Atlantic to provide packet switched services before approval is received to provide long distance services. The IECs fear that the RBOCs will use the broadband networks for voice as well as data and circumvent the 14-point provision checklist requirements. In response to these concerns, on August 10, 1998, the FCC ruled that RBOCs can provide broadband services such as DSL but must establish a separate nonregulated subsidiary for these services.

The FCC included restrictions on the types of broadband services that the new deregulated subsidiary could offer. It could not provide long distance data transport service. It could not offer long distance service until receiving proper authority and approval. It must provide office space for competitors who wish to provide their own DSL line equipment. It must provide competitors with copper access lines to customers at wholesale rates. This ruling was scheduled to take effect March 1, 1999.

The RBOCs were not happy with the ruling and filed an appeal in U.S. District Court. On August 24, 1998, SBC filed with the U.S. Court of Appeals in St. Louis asking it to review the order requiring it to establish a separate subsidiary for DSL and other broadband services. Since most of the FCC rulings have been appealed all the way to the Supreme Court, this ruling will probably also end up there. Telecommunications managers must stay abreast of these and other legislative and judicial actions in order to know what actions they should take. One way to stay informed is to spend time on the Internet and visit the sites of Congress, the FCC, the RBOCs, and the IECs.

2.39 ISPS AND THE 1996 TELECOMMUNICATIONS ACT

On September 14, 1998, Bell South demanded access fees for voice-over IP. It stated that ISPs would have to start paying the same access fees as the IECs. Bell South cited the FCC, which said it was considering requiring IP telephony carriers to pay access charges. Bell Atlantic took a different approach with ISPs. It established an

IP-to-PSTN gateway and provided information for each call to the IP provider, which the ISP could use for billing its IP customer. Bell Atlantic charges less than IEC access charges and more than its regular IP telephony access charge.

On March 30, 1998, Congress came to the aid of ISPs and warned the FCC against treating ISPs as other carriers with respect to paying into the universal service rate fund and being regulated by the FCC. On August 20, 1998, the Eighth Circuit Court of Appeals in St. Louis ruled that ISPs are not to be charged access fees as if they were IECs because they are not IECs. Debate continues to rage over who, if anyone, should contribute to the E-rate and universal service funds. Since ISPs do not have to contribute to these funds and cannot be charged regular IEC access charges, their cost to provide long distance voice via IP is 2 or 3 cents a minute less than an IEC's cost.

2.40 SUMMARY

The telecommunications industry is a dynamic industry and is in a constant state of change. Legislation covering telecommunications and the technology available to provide various services are changing continually. As these changes occur, they create new market opportunities. The astute entrepreneur stays abreast of these changes and is quick to take advantage of new opportunities. The first to market reaps the greatest rewards, and you can only be first in a new market by keeping informed of changes in that market. When the telecommunications industry was a monopoly, a business owner had one source for all services, the LEC. The LEC had a marketing department, a service department, and a staff of consultants to assist customers with the design and provisioning of a telecommunications network for the company. With deregulation, businesses have a wider array of service providers to choose from, but they can no longer rely on the LEC to solve their problems. Businesses have had to form telecommunications departments and hire managers who understand telecommunications technology. Companies must be capable of designing, developing, implementing, and servicing their own telecommunications functions. This requirement has led to the creation of many telecommunications job opportunities in the private business sector.

Deregulation of the telecommunications industry began with the deregulation of station equipment on customers' premises. The event having the most impact on deregulation of station equipment was the Carterphone decision of 1968 by the FCC. This decision allowed customers to purchase telephones from independent manufacturers or retailers. These companies were referred to as interconnect companies. In 1981, the Computer Inquiry II decision of the FCC extended deregulation to cover keysystems and private branch exchange switching systems used by business customers. This decision basically stated that all equipment on a customer's premises must be owned by the customer (or an interconnect company). Computer Inquiry II also prevented the LECs from selling CPE; they were forced to establish separate companies to market CPE. The 1996 Telecommunications Reform Act abolished the need for a separate company to sell CPE, and the LECs have merged these separate subsidiaries back into the regular operations of the LEC.

Long distance telephone service was the second area of telecommunications to be deregulated. Deregulation of this service began with the 1969 FCC ruling often referred to as the MCI decision. This ruling allowed MCI to pursue its completion of a long distance network that could serve private-line networks of business customers. In 1971, the FCC broadened this decision to allow any common carrier the right to carry private-line services. This ruling is called the Specialized Common Carrier Decision of 1971. In 1975, MCI began offering long distance service to the general public. The FCC ordered MCI to cease providing service to the general public and to restrict its service to the private-line business. MCI sued and the appeals court reversed the FCC ruling. In 1976, the FCC opened the provisioning of long distance service to the general public to competition.

The Modified Final Judgment (MFJ) of 1984 addressed the deregulation of telecommunications and the breakup of the Bell System. AT&T was ordered to divest itself of the BOCs. AT&T was left with the long distance facilities and could not provide local service. The BOCs were left with the local service facilities and were restricted to providing local service. Local service was defined as calls that originated and terminated within a small geographic area called a LATA. The 1984 MFJ divided the country into 184 LATAs. According to this agreement, calls within a LATA must be carried by a LEC, even if they are toll calls, and calls between LATAs must be carried by an IEC. The 1996 Telecommunications Reform Act opened both areas to competition between the LECs, IECs, and other common carriers.

The 1984 MFJ also ordered the BOCs to provide equal access (1+ access) to all customers of an IEC by 1987. To meet this requirement, the BOCs had to replace electromechanical toll offices with computer-controlled switching systems. They also had to establish locations for their facilities to meet the facilities of the IEC. These locations are called POP locations. The IECs were allowed to rent space in the BOC exchanges for the termination of their long distance network. Customers could specify who they wanted as their long distance service provider. The BOCs enter this information into the switching system's database. When the customer makes an inter-LATA call, the BOC automatically connects the caller to the preferred IEC.

Telecommunications Events and Legislation

Year	Title	Explanation
1876	Patent issued for telephone to Alexander Graham Bell	Recognized Alexander Graham Bell as inventor of telephone
1877	Bell Telephone Company formed	Investors needed to expand
1882	Bell buys Western Electric	Bell needed a manufacturer
1885	Bell incorporates as AT&T	To provide protection to company officers
1893	Bell's patent expires	Independents formed
1910	Mann-Elkins Act	Regulated using ICC
1913	Kingsbury Commitment	To settle antitrust action by DOJ
1921	Graham Act	Exempted AT&T from antitrust law
1934	Communications Act of 1934	Established FCC and state PUCs

1949	REA Act amended	To provide low-cost loans to Independents
1949	DOJ brings antitrust action	AT&T charged under Sherman Act
1956	1956 consent decree	1949 lawsuit settled by DOJ/AT&T
1968	FCC Carterphone decision	AT&T must permit use of CPE
1969	MCI ruling by FCC	AT&T must allow MCI access
1971	Specialized common carrier decision	Any common carrier can provide private-line service
1971	Computer Inquiry I	Computer industry exempt from regulation
1974	DOJ files antitrust action	AT&T charged by DOJ
1975	MCI decision	Appeals court upheld MCI
1976	FCC orders AT&T to provide access to customers of SCCs	Any SCC can provide long distance service to the general public
1981	Computer Inquiry II	All CPE deregulated
1982	Modified Final Judgment (MFJ)	1974 lawsuit settled by consent
1984	MFJ takes effect	Replaced 1956 Final Judgment
1987	Equal access (1984 MFJ)	RBOCs must provide equal access
1996	1996 Telecommunications	Replaced 1984 MFJ Reform Act

CHAPTER REVIEW QUESTIONS

1. Who invented the telephone and when was it patented?
2. When was the first BOC established?
3. When was the first Independent telephone company formed?
4. What was the purpose of the Kingsbury Commitment?
5. When was the Graham act repealed?
6. In telecommunications, what is the purpose of a tariff?
7. Which government agency regulates interstate telecommunications?
8. Which government agency regulates intrastate telecommunications?
9. How did the government encourage companies to build telecommunications systems as part of the PSTN in low-profit, rural areas?
10. What are companies called that provide telecommunications services to the general public?
11. What was the outcome of the antitrust lawsuit brought by the Department of Justice against AT&T in 1949?
12. What FCC ruling led to deregulation of station equipment?

13. What name is given to equipment on a customer's premises not owned by the LEC?
14. What is the name given to privately held companies selling equipment and providing services in the CPE market?
15. Which FCC ruling led the way for deregulation of long distance service?
16. Has deregulation of the telecommunications industry led to degradation of performance in the PSTN and lower local service rates?
17. Which ruling resulted in the BOCs buying equipment not only from Western Electric but from other manufacturers as well?
18. The MFJ accomplished two specific and separate actions. What are these two broad categories of actions called?
19. What was the toll revenue sharing procedure between the Independents and the BOCs called? What ruling killed this procedure?
20. Was the legislative process leading to deregulation and divestiture brought about by the FCC?
21. What is the name given to common carriers providing local phone service and connection to the PSTN?
22. What is the name given to common carriers providing long distance service?
23. What is the difference between the PSTN and private networks?
24. What did the Carterphone decision of 1968 accomplish?
25. What are the major provisions of the 1984 MFJ?
26. What is a PIC?
27. How can the user of a pay phone (or any phone) override the PIC?
28. What is a LATA? How many LATAs are there? Do LATAs follow the same boundaries established by area codes? Can LATAs cross state boundaries?
29. What is a POP? Where is it located?
30. What technologies have led to flatter network hierarchies?
31. As Public Law 104-104 applies to telecommunications, what were the significant provisions of the Telecommunications Reform Act of 1996?

PROJECTS AND CASE STUDIES:

Project #1

In Chapter Two, Section 2.23, reference is made to a 14-point checklist. What are the 14 points? What was the purpose of the 14 points? You can search the web or go to **http://thomas.loc.gov** and look up Public Law 104-104 (104 session of Congress and the 104th law passed by that Congress). The full text of the law contains Section 271 Bell Operating Company Entry into Interlata Services. Bell companies must file under Section 271 for permission to offer long distance services that cross LATA boundaries.

Case Study #1

Jefferson Middle School would like to install 20 PCs on a LAN with ADSL access to an ISP for access to the Internet. What will E-rate funding cover? How do they apply? There are many sites with E-rate funding information. You might choose to start at **www.fcc.gov** to see what is at their site. The FCC site contains a link to Universal Services Administration Company—Schools and Libraries Division. This site offers a lot of information. I would also suggest searching at Alta Vista for E-rate.

Case Study #2

Joe Smith in a small midwestern town wants last number redial, caller ID, ADSL, and Internet access. Are these part of Universal Service? What is Universal Service? How is the Universal Service fund created and administered? Check the 1996 Telecommunications Reform Act (Public Law 104-104) as well as the FCC site to see what they have on Universal Service.

GLOSSARY

Bell Operating Companies (BOCs)   Often referred to after deregulation as the *Baby Bells.* These were the local Bell Telephone companies. Before the 1984 Modified Final Judgment, 23 BOCs existed as subsidiaries of AT&T.

Carterphone Decision of 1968   A ruling by the Federal Communications Commission forcing AT&T to allow the attachment of a Carterphone to telephone lines at the customer's residence. This ruling was the beginning of deregulation of station equipment.

Common Carrier   A company that offers telecommunications services to the general public as part of the public switched telephone network.

Communications Act of 1934   Legislation passed by Congress and signed into law by the president in 1934 to regulate interstate telecommunications and radio broadcasts. This law created the Federal Communications Commission to administer the act.

Computer Inquiry   Any of a series of rulings by the Federal Communications Commission. Issued Computer Inquiry I in 1971; stated that the Federal Communications Commission would not regulate computer services.

Computer Inquiry I   A 1971 ruling by the FCC stating that it would not regulate computer services and networks.

Computer Inquiry II   Issued in 1981; mandated that station equipment was to be deregulated and could not be provided by a local exchange carrier.

Computer Inquiry III   In 1986, detailed the extent to which AT&T and the Bell Operating Companies could compete in the nonregulated enhanced services arena.

Customer-Provided Equipment (CPE)   Also called *customer-premise equipment.* This is station equipment, on a customer's premises, that according to Computer Inquiry II must be provided by someone other than the local exchange carrier. The 1996 Telecommunications Reform Act overrides Computer Inquiry II and allows LECs to reenter the CPE market.

Deregulation   The change of the telecommunications industry from a regulated monopoly to a competitive nonregulated market. One of the major provisions of the 1984 Modified Final Judgment was to deregulate long distance services.

Divestiture   A major provision of the 1984 Modified Final Judgment. This provision forced AT&T to divest itself of the 23 Bell Operating Companies.

Equal Access   A key provision of the 1984 Modified Final Judgment. This provision forced the Bell Operating Companies to provide 1+ access to toll for all interexchange carriers.

E-rate Fund   Established by a provision of the 1996 Telecommunications Reform Act; intended to provide schools (K–12) with funds to buy computer network services and to pay for Internet access services. Financed by fees levied on common carriers by the FCC.

Feature Group   A series of services established by the Bell Operating Companies to provide access to interexchange carriers (IECs). Feature Group A gave customers access to IECs by having them dial a seven- or ten-digit telephone number plus a personal identification number. Feature Group B provided access via a seven-digit code of the form 950-10*XX*. This access code was later changed to 10*XXX*. Feature Group D is the 1+ access to IECs that was mandated by the 1984 Modified Final Judgment.

Federal Communications Commission (FCC)   The federal agency created by the Communications Act of 1934 to oversee enforcement of the act by regulating interstate telecommunications and broadcast communication.

Graham Act of 1921   Legislation passed by Congress and signed by the president in 1921 to exempt telecommunications from antitrust legislation.

Interexchange Carriers (IECs or IXCs)   Common carriers that provide long distance telephone service. Major IECs are AT&T, MCI, Sprint, LDI, and so on.

Kingsbury Commitment of 1913   An agreement signed by the vice president of AT&T in 1913. This agreement stated that AT&T would allow other telephone companies access to its network, would sell its Western Union stock, and would not buy any more Independent telephone companies without getting permission from the government.

Local Access Transport Area (LATA)   A key provision of the 1984 Modified Final Judgment (MFJ). This provision established 184 geographic regions that conform to the Standard Metropolitan Statistical Index used by marketing organizations. The LATA was defined as an area within which calls must be carried by a local exchange carrier (LEC) and could not be carried by an interexchange carrier (IEC or IXC). These calls within the LATA are called *intra-LATA calls.* The MFJ further stated that calls between LATAs (*inter-LATA calls)* must be carried by an IEC, not by a LEC. The 1996 Telecommunications Reform Act replaced the 1984 MFJ and opened the door to competition in this area. It allows either type of call to be carried by either a LEC or an IEC.

Local Exchange Carrier (LEC)   The provider of local telephone services. Prior to the 1996 Telecommunications Reform Act, the LEC was your local telephone company.

With the passage of that act, the local telephone company is now called the *incumbent local exchange carrier (ILEC)* and its competitors are called *competitive local exchange carriers (CLECs).*

MCI Decision of 1976   A ruling by the federal court in MCI's favor against AT&T; it allowed MCI and other common carriers to handle long distance services for the general public.

MCI Ruling of 1969   The ruling in 1969, by the Federal Communications Commission, that forced AT&T to allow private-line customers of Microwave Communications Inc. (MCI) to use local telephone lines for access to MCI's private-line network.

Modified Final Judgment (MFJ)   The 1984 agreement—reached on August 24, 1982, between AT&T and the Department of Justice—to settle an antitrust suit brought against AT&T by the Justice Department, on November 20, 1974. This judgment modified the 1956 Final Judgment. The major provisions of the 1984 MFJ were deregulation of long distance services and divestiture of the Bell Operating Companies by AT&T.

Monopoly   A form of market where one company is the sole provider of goods and/or services. There is no competition. Market demand does not set the price of goods or services. In a nonregulated monopoly, price is set by the monopoly service provider. In a regulated monopoly, price is set by the government agency charged with regulating the monopoly service provider.

Network   An overused term. In telecommunications, the word *network* has many meanings, usually determined by the context. The term *public switched telephone network (PSTN)* refers to the

interconnection of switching systems in the PSTN. *Switching network* refers to the component inside a switching system that switches one circuit to another circuit. The *network in a telephone* refers to the hybrid network, which performs a two-wire to four-wire conversion process.

Point of Presence (POP)   The point at which the local exchange carrier and interexchange carrier facilities meet each other.

Preferred Interexchange Carrier (PIC)   Determined as follows: A customer tells the local exchange carrier who it wants to use as its long distance service provider. The LEC makes an entry in the database of the switching system serving this customer. That database entry will inform the switch which IEC should be used on long distance calls placed by that customer. With the 1996 Telecommunications Reform act, customers must be able to choose a carrier for toll calls within a LATA as well as the same or a separate carrier for long distance service.

Private-Line Services   Services provided by a common carrier to a private organization to help that organization establish its own private network. A private network cannot be accessed or utilized by the public. A private network is exactly that— it is private and can only be accessed and used by the private organization.

Public Utilities Commission (PUC)   The state government agency that regulates telecommunications within the state.

Regulated Monopoly   A sole provider of goods or services; regulated by a government agency. A regulated monopoly must seek approval from the government agency for anything it wishes to do. The government agency sets the price that a monopoly can charge for its goods and services.

Revenue Sharing   The sharing of toll revenue between the Independent telephone companies and AT&T prior to the 1984 MFJ.

Rural Electrification Administration (REA) Department of the federal government established during the Depression under President Franklin D. Roosevelt by the Rural Electrification Act of 1936. This "act" put people to work bringing electricity to rural America. The "act" was amended in 1949 to bring telephone service to rural America. This department was renamed the Rural Utility Services Department.

Specialized Common Carrier Decision of 1971   An extension of the 1969 MCI Decision by the FCC. This ruling allowed any common carrier to handle private-line networks.

Station Equipment   The largest segment of station equipment is the telephone, but the term *station equipment* has been broadened to include anything a customer attaches to a telephone line. The most common piece of station equipment is the telephone. A modem, CSU/DSU, personal computer, and keysystem are all referred to as station equipment.

Tariff   The document filed by a common carrier that defines in detail any service proposed by the carrier and the charge proposed for the service. For interstate service, the tariff is filed with the FCC. For intrastate services, the tariff is filed with the PUC.

Telecommunications Reform Act of 1996 Effectively replaces the 1984 MFJ. This act is designed to accelerate competition for providing services in the local and long distance markets. It eliminates the franchised local territory concept. The LEC is no longer the only service provider in town. Other common carriers can come into an incumbent local exchange carrier (ILEC) territory and provide local telephone service. These new local service providers are called *competitive local exchange carriers (CLECs)*. Prior to this act, the older incumbent LECs (ILECs) were restricted to providing local services only. Under this act, if the ILEC can prove that it has not hindered competition in its local exchange territories, it can now offer long distance services.

3

Telecommunications PSTN Technology (1876–2000)

KEY TERMS

Alternating Current (AC)
Central Office
Circuit Switching
Class 5 Exchange
Communication
DACs
Direct Current (DC)
Exchange Code
Extended Area Service (EAS)
Frequency Division Multiplexing (FDM)

Full-Duplex Transmission
Half-Duplex Transmission
Interoffice Calls
Intraoffice Calls
Line
Line Relay
Local Exchange
Medium
Multiplexing
Numbering Plan Area (NPA)

Receiver
Register (Dial Register)
Simplex Transmission
Sine Wave
Tandem Exchange
Telecommunications
Time Division Multiplexing (TDM)
Time Division Switching
Transmitter

OBJECTIVES

On completion of this chapter, you should be able to:

1. Explain the differences between simplex, half-duplex, and duplex transmission.
2. Have a basic understanding of the telephone set, how the transmitter converts sound waves into electrical waves, and how the receiver converts the electrical waves back into sound waves.
3. Understand the history and evolution of central exchange switching.
4. Understand why a city has many central exchanges and how exchange boundaries are established.
5. Understand the differences between intraoffice and interoffice calls.
6. Explain extended area service calling plans and pricing.
7. Identify the major standard-setting organizations.
8. Discuss why circuit switching rather than packet switching is used for voice calls.
9. Describe the North American Numbering Plan.
10. Explain dual-tone multifrequency signaling.
11. Explain how a dial register works and its purpose.

12 Explain the differences between in-band and out-of-band signaling.

13 Explain the importance of Signaling System 7 and its uses.

Telecommunications is the science or technology of communication by telephone or telegraph. *Communication* can take on any of several meanings. For our purposes, it is a means of sending messages, orders, and responses by electronic, electrical, or electromagnetic means. The communication process allows information to pass between a sender (transmitter) and a receiver over some medium. The three parts of any communication system are:

- Transmitter
- Receiver
- Medium

The *transmitter* is the device responsible for sending information or a message in a form that the receiver and medium can handle.

The *receiver* is the device responsible for decoding or converting received information into an intelligible message. In a conversation between two people, one person will be transmitting information while the other is receiving the information.

The *medium* is the device or substance used to transport information or a message between the transmitter and receiver. If a conversation is face to face, the medium used to carry the message is air. Television and radio broadcasting also use air as a transmission medium. The signal from the station is connected to a transmitter. The transmitter is a very high-powered electromagnetic wave generator located next to a high tower. The signal generated by the electromagnetic wave generator is sent to an antenna mounted on the high tower. The power of the signal is so high that the signal will leave the antenna and radiate out into the air. The signal is pulled out of the air using a receiver. The receiver includes a tuning device to allow tuning in, or selecting one station's signal out of the many different signals arriving at the antenna of the receiver. When air is used as a transmission medium, the message is sent in all directions and is called a *broadcast* (scattered or spread over a wide area) *signal.* Broadcast signals can be picked up by anyone with a receiver tuned to the broadcaster's transmitting signal frequency. Using air as the transmission medium is the most economical way for a radio or television station to reach everyone.

3.1 SIMPLEX, HALF-DUPLEX, AND FULL-DUPLEX TRANSMISSION

The radio and television stations transmit signals, and our television sets and radios receive them. We have no way of transmitting signals back to the radio or television station. Transmission in one direction only is called *simplex transmission.* When we have a conversation with someone, we transmit for a short time and the other person receives our message. We will then switch positions as transmitter and receiver several times during the conversation. This type of transmission, where transmitters on each end of a medium take turns sending over the same medium, is called *half-duplex transmission.* Citizens Band (CB) radio is a common application that uses half-duplex transmission. If a transmission system allows signals to be transmitted in both directions at the same time, the system is called a *full-duplex transmission* system. Often a full-duplex system is actually made up of two simplex systems. To take an example, on a highway we have traffic lanes going in opposite directions. Traffic can only go in one direction on one of the lanes (simplex). Used together, the two lanes allow traffic to flow in both directions at the same time (full-duplex).

3.2 TWISTED COPPER WIRE AS A MEDIUM

We can use copper wire instead of air as a medium. Cable television companies and local telephone companies use wire as a medium, which gives the transmitter more control and security over the messages being transmitted. Local telephone companies use two wires to connect each phone to their switching office. Each set of two wires is referred to as a *pair*. The telephone company refers to wires serving a local telephone as the *local loop*.

3.3 INVENTION OF THE TELEPHONE—1876

As noted earlier, the telephone was invented by Alexander Graham Bell in 1876. He used a pair of wires to connect two telephones to each other and used one device for both the transmitter and receiver (referred to as a *transceiver*). This device required no external power and could convert sound waves into low-current electrical waves. It could also convert the electrical energy received from the distant phone back into sound waves at the receiving phone.

3.4 INVENTION OF THE CARBON TRANSMITTER

One year later, in 1877, Thomas Edison invented the carbon microphone. This transmitter required a battery for the supply of electrical energy. Edison's transmitter did not produce electrical energy; it consisted of carbon granules enclosed in a capsule containing a diaphragm. As sound waves struck the diaphragm, it would compress and decompress the carbon granules, causing the resistance of the granules to decrease and increase. As this happened, it caused the electrical current flowing from the battery to increase and decrease in unison with the speaker's voice. This varying electrical current was sent over the phone line to the distant phone, and the receiver converted the varying electrical current into sound waves. Edison's transmitter and battery supply generated a much stronger signal than the Bell transceiver. This type of carbon microphone is still used today as the transmitter in many telephones. The original transmitter/receiver device invented by Bell now serves only as the receiver of the telephone. Transmitters and receivers fall into a category of equipment called transducers. A *transducer* is a device that converts one form of energy into a different form. The transmitter coverts sound-wave energy into electrical energy. The receiver converts electrical energy into sound waves (see Figures 3-1 and 3-2)

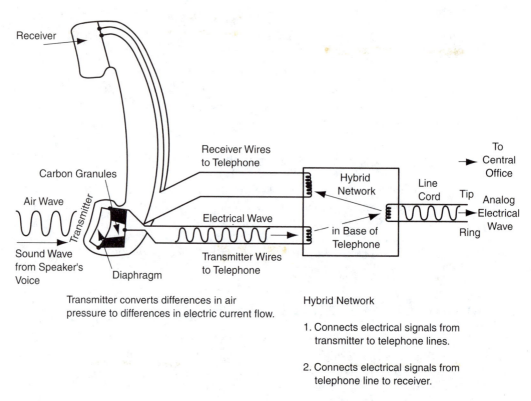

Transmitter converts differences in air pressure to differences in electric current flow.

Hybrid Network

1. Connects electrical signals from transmitter to telephone lines.

2. Connects electrical signals from telephone line to receiver.

3. Connects DC voltage from telephone line to transmitter.

FIGURE 3-1 Telephone transmitter and hybrid network.

3.5 LOCAL POWER SUPPLY AT EACH TELEPHONE

As telephones were introduced to the public, each phone had its own battery. The battery that powered the telephone was composed of two large $1^1/_2$-V DC cells connected in series to make up a 3-V DC battery. A *direct current (DC)* battery supply maintains a constant voltage that gradually gets lower as the battery discharges.

Many readers will not need to know more about DC voltage beyond the fact that a DC voltage is used to supply power for the telephone. Today the central office supplies the DC voltage. If the reader desires to delve deeper into DC battery construction and operation, this subject is discussed in Appendix A (Sections A.7 to A.13). The first telephones were equipped with very big "flashlight" batteries that supplied 3 volts to the transmitter of the telephone. Today, telephones no longer contain a local battery supply. The voltage for the transmitter (and DTMF circuit of the touch-tone dial) is supplied from the central office. The voltage at the central office battery is 52-Volt DC. If the phone is on-hook, no electric current will flow and we would also measure 52 volt at the telephone.

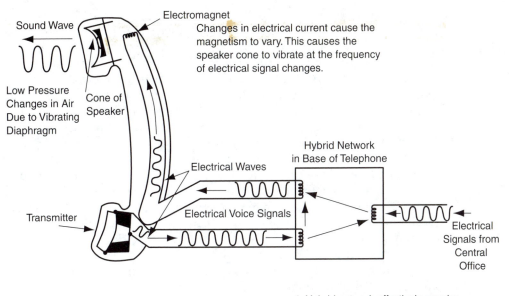

FIGURE 3-2 Telephone receiver and hybrid network.

1. Hybrid network effectively couples electrical signals from the central office to the receiver.

2. A small amount of the electrical signal from the transmitter is coupled to the receiver.

When the telephone is taken off-hook to make or answer a call, electric current will flow. As the electric energy travels over the wire pair between the telephone and the central office, some of the voltage is lost as heat in the wire pair. We refer to the use of electric energy as a voltage drop. The amount of voltage drop in the wires depends on the resistance of the wire and the amount of current flowing in the wire pair. The resistance of the wire depends on the size (gauge) and length of wire used. This in turn depends on how far the telephone is from the central office. Every wire pair used to connect telephones to a central office will connect through an electronic circuit (called a line circuit) to the central office battery. When a telephone is off-hook and electric current is flowing through the line circuit, wire pair, and telephone, each will use electric energy. Thus, each of these will have a voltage drop. The amount of voltage used up by the line circuit and wire pair will leave a voltage between 8 and 24 volts for the telephone. The variance in voltage at the phone is because longer wire pairs will have more voltage drop (use up more voltage) than shorter wire pairs. For the longest wire pair, the telephone will get about 8 volts. On a very short wire pair, the telephone will get about 24 volts. For additional information refer to Appendix B (Sections B.9 and B.10).

Each phone was equipped with a ringer and a hand crank connected to a magneto. A *magneto* is a device containing a coil of wire inside a magnetic field.

When the wire coil is turned so that the wire of the coil cuts through a magnetic field, electric energy is created in the wire. This is the same principle used by commercial electric companies to generate electricity for our homes. This type of electricity is referred to as **alternating current (AC)** electricity. Operating the hand crank on a phone would cause a 90-V AC signal to go out over the telephone line to the distant phone. The AC signal would cause the ringer to operate on the receiving telephone. An AC voltage supply generates a voltage that is constantly changing and reversing polarity. The voltage rises from zero volts to a positive value, then gradually becomes zero, then drops to a negative value, and finally returns to zero. This trip from zero to maximum positive, to zero, to maximum negative, and back to zero volts is called a *cycle.* The number of cycles an AC signal goes through in one second is called the *frequency* of the signal and is measured in cycles per second (cps). Cycles per second are often called *Hertz* as a tribute to Heinrich Hertz, who discovered electromagnetic radiation (the birth of radio waves) in 1886. The frequency of the signal generated by a magneto depends on how fast the crank could be turned. *Frequency* pertains to a process that repeats itself; the term denotes how often the process repeats. An event that occurs at noon every day can be said to have a frequency of once a day, or a frequency of 30 times a month (sometimes 28, 29, and 31 times a month), or a frequency of 365 times a year. The time frame used in electrical frequency measurements is the second. The more cycles completed in one second, the higher the frequency. Figure 3-3 shows two AC (sine) waves. The **sine wave** in (a) completes two full cycles in one second. The sine wave in (b) completes four full cycles in one second.

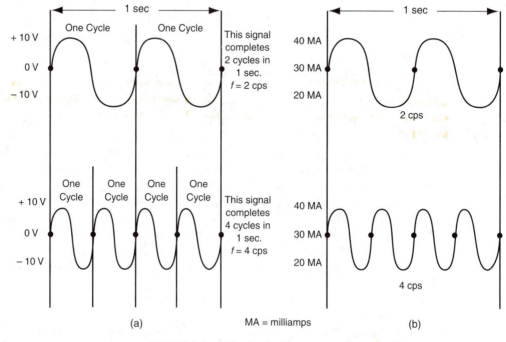

MA = milliamps

FIGURE 3-3a, 3-3b Analog electrical signals.

Varying DC Signal

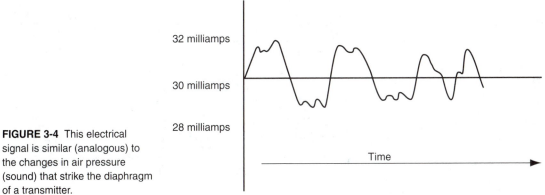

32 milliamps

30 milliamps

28 milliamps

Time

FIGURE 3-4 This electrical signal is similar (analogous) to the changes in air pressure (sound) that strike the diaphragm of a transmitter.

Voice signals are composed of several different sine-wave frequencies that are mostly between 300 and 4000 cps mixed together. If we hold our voice signal at a steady output (such as when we whistle), our voice approaches a pure sine wave. As we speak, the sounds we make are the result of continuously varying the frequency of the signal. Therefore, speech is not a pure sine wave but is a variable signal composed of several sine waves mixed together. A speech signal looks like the signal in Figure 3-4.

The original telephone unit was self-contained. Each phone had its own power supply furnished by batteries similar to the batteries that power a large camping lantern, and each phone could generate a high-voltage ringing signal to alert the called party. Telephones were connected to each other using iron wire strung on poles between the two phones. A separate pair of wires and phone was needed for each different location you wished to call. This cumbersome and expensive situation led to the development of the central wire center and manual switching system commonly referred to as *central*.

In Figure 3-5a one pair of wires connects one telephone directly to another telephone. In Figure 3-5b, all telephones are wired to the "central" wire center. At the central wire center, all lines are connected to an operator's switchboard. The operator can connect any line to any other line. This manual switching was the birth of switching systems.

One serious drawback to the early phone system was the need for the transmitting party to have a separate (individual) phone for everyone they wanted to call along with a different pair of wires for each phone. If you wanted to be able to call ten different people, you needed ten phones connected to ten different pairs of wires. The other end of these pairs of wires would go to phones at the ten different locations you wanted to call. A large business would have several phones in place and still have limited calling capabilities. The first step toward solving this problem was to wire all telephones through a central location. The next step was to install a manual switchboard and wire all lines coming into the central location to this switchboard. Manual switching systems were introduced to allow the connection of one phone to any other phone connected to the system.

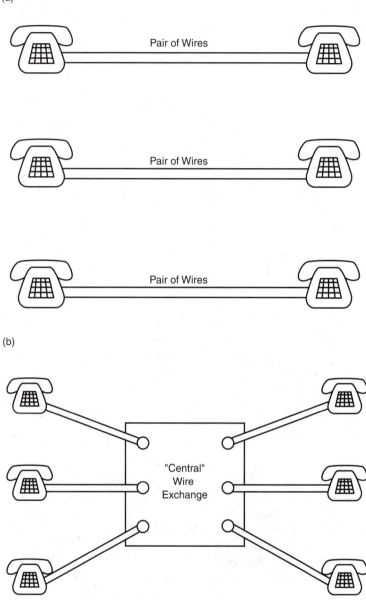

FIGURE 3-5 (a) One pair of wires connects one phone directly to another phone. (b) Each phone is connected by a pair of wires to the wire center. An operator can connect any phone line to any other phone line.

3.6 CENTRAL EXCHANGE—1878

The first telephone exchange was installed in 1878, two years after Bell invented the telephone. The Bell Company recognized the need for a method of connecting one telephone to any other telephone. It also recognized the cost savings to be achieved by wiring all phones through a central location. The central location is referred to as *central office,* *central, central wire center, central exchange,* or *exchange.* To provide everyone who had a phone with the capability of calling any phone connected to the central office, an operator switchboard (patch panel) was added to the system. All wire pairs coming into the central office were connected to individual jacks, signal relays, and lights on the patch panel. When someone wished to place a call, the person would use the hand crank at the phone to send a ringing signal to the operator. The ringing signal would operate the relay and lamp on the patch panel associated with this particular telephone line. The operator would answer the call through a plug-in and jack arrangement that would connect a pair of wires from her telephone to the incoming line.

The wires of the operator's phone were connected to a plug. The wires coming in from phone lines were connected to jacks. Each phone line could have telephones from as many as ten residences connected to the line. A private line had telephones at only one location (a residence or business) attached to the line. If a town had 200 telephone lines, the patch panel at the operator's switchboard had 200 jacks, relays, and lamps. The operator could answer any one of these by plugging her phone into the jack of the calling line. The operator answered calls using the word "central." On being told who the caller wished to call, the operator would plug her phone into the jack of the person being called and ring their line using the crank on her phone. When the called person answered the phone, the operator would remove her phone from the jack and connect the jack of the calling party to the jack of the called party using a wiring cord that had a plug on both ends of the cord. This cord was called a *patch cord.*

The phone lines were wired to jacks similar to the headset jack on a portable AM/FM cassette tape player. The plug on the operator's phone and the plugs on the patch cord are similar to the plug on the end of the wires connected to a cassette player's headset. If you look closely at the plug on the headset of a portable cassette player, you can see a copper metal tip on the plug, a band of plastic around the plug behind the tip, then a narrow copper ring followed by another band of plastic. In the telephone industry, the wire that connected to the tip of the plug was called the *Tip lead* and the wire that connected to the copper ring was called the *Ring lead.* The wires on the jacks were also called *Tip* and *Ring* according to which part of the plug they made contact with (Figure 3-6).

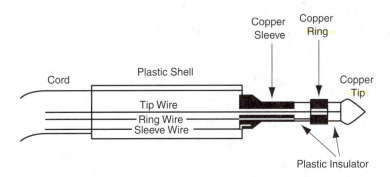

FIGURE 3-6 Switchboard cord at operator's switchboard.

3.7 OPERATOR SWITCHBOARD

Before the invention of automated switching systems, all calls were handled by an operator. An incoming call on the manual switchboard was identified by a lit light above a jack. When a caller wanted to get the operator, it was necessary to turn the hand crank on the telephone or lift the receiver off the hook. This action would operate a line relay in the central office and cause the light above the individual's jack to light. The operator would take a rear patch cord and plug it into the jack below the light (which would cause the light to go out) and find out who the caller wished to call. On being told who to call, the operator would take the front cord, directly in front of the cord used to answer the call, and plug it into the jack of the desired line. She would then operate the ring key to ring the called telephone. The lamps behind the cords on the desktop are called *supervisory lamps.* The rear lamp would light if the calling party hung up. The lamp just in front of the rear lamp would light if the called party hung up. The operator used the key in front of the patch cords to select which party she wanted to talk to when a private conversation was desired. This key was also used to select which cord the ringing signal would be sent over, when the ring key was operated.

3.8 LONG DISTANCE OPERATOR SWITCHBOARDS

In addition to the jacks at the manual switchboard that connected to wires going to local telephones, there were also jacks connecting to wires that ran to switchboards located in other towns and cities. The local operator could plug a patch cord into one of these jacks and ring the distant operator. AT&T and the Bell Companies installed operator switchboards in several large cities in each state for the purpose of handling only long distance calls. The long distance centers were wired to each other. The wires going between them terminated on jacks at the long distance operator's switchboard position. Each town in a state had several jacks on its switchboard wired to the nearest long distance switchboard. The local operator would place all long distance calls to the Bell long distance operator for completion over the AT&T network. On some long distance calls, several long distance operators would be used to connect from one long distance center to another before reaching the called party's local operator.

3.9 DIVISION OF LARGE CITIES INTO
SEVERAL LOCAL EXCHANGE AREAS

In large cities such as New York and Chicago, the local exchange would be split up into several exchanges scattered around the city. This reduced the congestion of wires coming into each exchange. Instead of one local exchange with thousands of wires coming to it, each of the local exchanges had a few hundred wires coming in. The operator at each exchange had access to jacks wired to the other local exchanges around the city for *interoffice calls.* During this era, few problems were encountered using this arrangement, because most local telephone calls were made within one office. Few lines were needed between offices because there were few interoffice calls. As the number of interoffice lines grew, it became necessary to distinguish

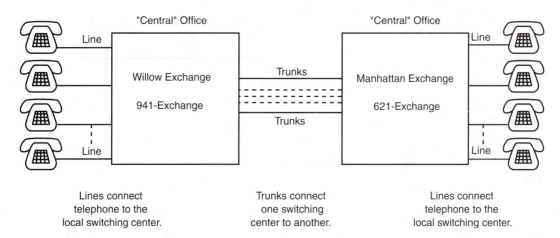

FIGURE 3-7 Interexchange call from customer at Willow to customer at Manhattan or vice versa.

local phone lines from the lines going to other offices. The lines going to other offices were called *interoffice trunks* or just *trunks.* This change in terminology helped avoid confusion when informing a repair person of phone problems. If a line was out of service, the repair person knew only telephones attached to that line were out of service. When told a trunk was in trouble, the repair person knew that this outage was impacting all interoffice calling going over that trunk to another exchange (see Figure 3-7).

To differentiate telephone numbers in one local exchange from the numbers in another, each local exchange was given the name of the local area. Some names were Manhattan, Downtown, Bronx, Willow, Riverside, Little Chicago, and so on. If a caller were placing a call to telephone number 4567 in the Manhattan exchange, the caller would tell the operator they wanted Manhattan 4567. The operator knew which local exchange to place the call to from the name of the exchange given. If the caller's phone was connected to the Willow exchange, the operator at the Willow exchange would place a patch cord into a trunk jack wired to the Manhattan exchange switchboard. The Willow operator would tell the Manhattan operator to connect the call to number 4567. The number 456 would tell the Manhattan operator to plug into line jack 456. The last digit would tell the operator which party on the line to ring. In our example, party 7 on the 456 line would be rung by the Manhattan operator.

Even when offices were distributed around the city, some of the local exchanges would be so large that hundreds of switchboards and operators were needed. Each switchboard would have specific lines cabled to its jacks; incoming calls could not be accessed from all switchboard positions. Therefore, it was necessary to add interposition jacks to the switchboards. Switchboard 1 had several jacks wired to jacks at switchboard 2, 3, 4, and so on. Every switchboard position had jacks wired to jacks at other positions. This allowed the operator to place calls to lines not appearing at her position. On interposition calls, the operator plugged into a jack associated with the position where the called line was terminated and asked that operator to ring the desired line.

The room in which all operator switchboards were installed was called a *switchroom.* The BOC tried to keep the size of the switchroom to 50 positions or less by using more than one exchange in large cities. In larger cities it was impossible to keep the downtown exchange small. Each city also had only one long distance operator center. In a large city the switchboards in these long distance centers were so long that management personnel sometimes used roller skates to move around the room and assist operators.

3.10 LINE AND TELEPHONE NUMBERING

Operators remembered which line each person was on in small towns. A caller would just ask the operator to ring someone by name and the operator knew which line to ring. As more people were added to the phone system, it became more difficult to remember who was on which line. To make calling easier in large cities, it was necessary to give the operator line numbers (telephone numbers) instead of a name. By the late 1950s, the growth of telecommunications led to the development of the *North American Telephone Numbering Plan.* This plan assigned three-digit area codes to each area of North America and assigned a unique three-digit exchange code to each central exchange within an area code. With this ten-digit numbering plan, every telephone in North America had its own unique identifier.

3.11 INTRODUCTION OF CENTRAL POWER AND LINE EQUIPMENT

One of the first improvements made in the early telecommunications system was to eliminate the batteries and hand crank at the telephone. The hand crank was used initially to generate a signal that would cause the ringer of the distant phone to operate. The batteries were used to provide electrical energy for the telephone transmitter. The hookswitch of the phone would connect the battery to the transmitter and telephone line when the receiver was lifted from the hookswitch. When the receiver was hung up on the hookswitch, the contacts of the switch would open the circuit to the battery and telephone line. Thus, the battery was only connected when the phone was in use to conserve the battery and lengthen its life.

With the introduction of central exchanges and operators, a phone was now rung by the switchboard operator. Engineers came up with a new way to signal the operator instead of using the hand crank. They equipped the hookswitch with additional contacts that would operate when the receiver was lifted and connect the transmitter and receiver directly to the telephone line. The engineers installed a 24-V DC battery in the local exchange office. A 24-V battery is the equivalent of two 12-V car batteries connected together in series. The negative lead of the 24-V battery was wired to one side of a device called a *line relay.* The other side of the line relay was wired to the ring wire of a local phone line, at the jack on the operator switchboard. There was one relay for every line connected to the switchboard. The positive side of the battery was connected through a second winding of wire in the line relay to the Tip wire of the phone line.

When the telephone receiver was lifted, the hookswitch closed the Ring wire to the transmitter and receiver. The other side of the transmitter and receiver was closed by a separate set of contacts on the hookswitch to the Tip wire. When the hookswitch was operated by a phone going off hook, an electrical circuit was completed. Electrical current then flowed from the negative terminal of the 24-V battery through the wiring of the line relay, out over the Ring wire to the telephone, through the hookswitch contacts (which were closed), through the transmitter, back to the other closed hookswitch contacts, back over the Tip wire, and returned through the second winding of wire in the line relay to the positive terminal of the 24-V battery. Electrical current flowing through the wiring of the line relay attached to the line will cause this relay to close a set of contacts together. The operation of the contacts on this relay caused a miniature light to turn on above the jack on the operator's switchboard and sound a buzzer. The operator heard the sound and saw which lamp was lit. The operator answered the incoming call by plugging into the jack below the lit lamp. When the operator plugged into the jack, contacts on the jack opened, causing the lamp to go out and the buzzer to stop sounding (see Figure 3-8).

The 24-V battery was still connected to the phone and would supply the energy necessary to power the transmitter. Consequently, the batteries at the phone were no longer needed. When the receiver was hung up, the hookswitch returned to normal. This would disconnect the transmitter and receiver from the ring lead and connect the ringer of the phone to the ring lead. The ringer of a phone would only respond to an AC signal and looked like an open circuit to a DC voltage. When the receiver was hung up and the hookswitch was normal, the path for DC current flow was opened. With no current flowing through the relay attached to

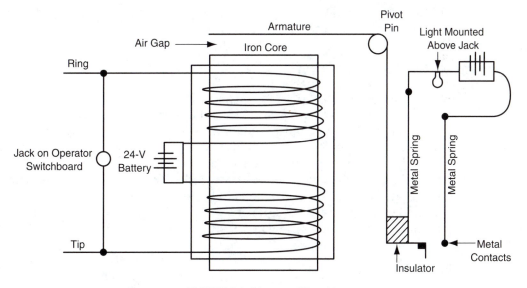

FIGURE 3-8 Diagram of line relay.

this line at the switchboard, the relay would release and turn off the switch contacts that lit a lamp and caused the buzzer to sound. With the ringer at the phone attached to the line, the operator could ring the phone when necessary. Because all telephones attached to a party line had the same ringers in them, every phone on the line would ring when the operator sent a ringing signal on the line. All parties on a party line were assigned a certain code for their ring. One party could be assigned a ring of two long rings followed by one short ring, another party's ring would be three short rings, and so on. The operator would use these codes and people would only answer when they heard their code.

Later advances in operator switchboard design added a third lead to the plug of the patch cord and to the jacks. This third lead was called the *sleeve lead*. From the jack, this third lead was wired to a contact on the line relay that was closed to ground when the line relay was released. The sleeve of the patch cord would make contact with the sleeve of the jack when the operator inserted the plug into the jack. The other end of the patch cord had the sleeve lead wired to a lamp next to the switch. The other side of the lamp was connected to the battery. Now as long as the customer had the phone off hook and the line relay was operated, the lamp would stay dark. When the customer hung up the phone, the relay released. This would place a ground on the sleeve lead and the lamp would light. Operators no longer needed to monitor the call but could watch the lamps (called *monitor lights* or *supervisory lamps*) to see when a call was over.

Manual operator switchboards were also used at business locations. When installed at one business location to handle all incoming and outgoing calls, the system was called a *private branch exchange* or *PBX*. The operation of a PBX was much the same as the operation of the switchboard in a central exchange. If you wished to call a large business, you would ask the local operator to connect you to the business. The PBX operator would answer the call and patch you through to the telephone of the person you wanted to talk with.

3.12 AUTOMATED SWITCHING

As the number of telephones in use grew, it became necessary to explore ways to automate the switching process. If we did not have automated switching today, it would require that all of our population work as switchboard operators or that we reduce by 90% the number of people allowed to have or use phones. In 1892, Almon B. Strowger invented the first automated switching system. This system was designed so the telephone customer rather than an operator could give directions to the switch. The instructions to the switch were provided by a dial. A dial was added to the telephone so that it could be connected to the automated switching system. This automated switch had stepping relays that followed the dial pulses. At the completion of a digit, the stepping relay connected the caller to another stepping relay. The completion of a call was done by using several stepping relays, and the system was called a *step-by-step switching system* or *Strowger switch*. Although the Strowger switch automated the switching process, these switches were not deployed in any great numbers until about 1925. Many rural areas did not replace manual switchboards with automated systems until after 1950. In the early 1960s, I installed many Strowger switches in Wisconsin to replace manual switchboards. The invention of automated switching systems was one of the greatest developments of the 20th century.

The Strowger switching systems have been replaced by computerized systems. All switching systems manufactured for use as public switching systems now use computers and software programming to control the switching of calls. These switching systems are called *stored program control (SPC)* switching systems. The SPC system makes it possible to offer the enhanced services customers want, such as three-way calling, call waiting, caller identification, last number redial, and so forth. These systems also make it possible for the LEC to provide Feature Group D (equal access). The SPC system requires almost no maintenance and takes much less floor space than its electromechanical predecessors.

When Strowger switching systems were used in a private company to replace manual PBXs, they were known as *private automated branch exchanges (PABXs)*. SPC switching systems have now replaced Strowger switches in private branch exchanges. SPC systems designed for use as a PABX are called *computerized branch exchanges (CBX)*. Today, the acronym PBX no longer refers to a manual private switchboard, because these switchboards no longer exist. All PBXs manufactured today are actually computerized PABXs (CBXs). It is understood that PBX now means the same thing as PABX and CBX. The only major differences between the SPC used for a private switching system (PBX) and the SPC used for a public switch is the speed of the central processor, the amount of memory, the size of secondary storage devices, and the number of peripheral devices attached to the system. The SPC switching system used for a PBX will be quite small and usually occupies less space than a desk or file cabinet. The SPC switching system utilized in a public central office usually takes about 400 square feet for a large office (10,000 lines) and about 100 square feet for a small office (1000 lines).

3.13 ADVANCES IN OUTSIDE PLANT DESIGN

Technology has also improved tremendously in the area of outside plant. Outside plant at the turn of the century was nothing more than wire on a fence post. As more people started using telephones, the outside plant employed the same wire and poles utilized for telegraph wire and telegraph poles. This type of wiring for telephones was called *open wire*. Iron wire was used and was attached to the telephone poles with glass insulators. This wire was exposed to weather; when wind speed was high, the wire would rust, deteriorate, and break in two. With the development of paper insulation, cables were created with many wires inside the cable. To insulate it from other wires, each wire was coated with shellac and wrapped in paper. The invention of plastic led to the use of a plastic covering on each wire as an insulator; these cables were called *plastic insulated cable (PIC)*. The cables could be buried, which eliminated damage from bad weather.

3.14 ADVANCES IN MULTIPLEXING TECHNOLOGY

A third area where technology has improved tremendously is multiplexing technology. *Multiplexing* is the placement of more than one telephone call over the same facility. Several multiplexing techniques were used, but the most favored technique until the 1970s was frequency division technology. As AT&T developed and continuously improved this technology, it was able to produce a *frequency division multiplexer (FDM)*, which could place 10,200 telephone calls over one coaxial cable pair. In the 1970s, another multiplexing technology took hold and is now favored over FDM: *time division multiplexing (TDM)*. It is really not a new technology; it was developed when vacuum tubes were state-of-the-art technology. Vacuum tubes could not handle TDM efficiently. The development of transistors and their evolution into integrated circuits has made efficient use of TDM possible. Basically, TDM was a technology that emerged ahead of its time, but today it is the technology underlying switching and transmission systems.

TDM technology is being advanced in the local telephone services arena. Integrated Services Digital Network (ISDN) is a TDM technology. The facility between a telephone and a central exchange is predominantly copper wire. Improvements are being made to TDM technology that uses copper wire as a medium. TDM technology will allow the local phone line to handle more than a telephone call. One TDM facility between the central exchange and your house will handle telephone calls, Internet access, alarm systems, and control systems. Copper wire can meet these needs easily. If coaxial cable or fiber optic cable is used as the facility to the central exchange, it is also possible to carry television signals on the TDM system. TDM technology will continue to improve, and it will not be long before copper wire can also be used to carry television signals to our homes as part of a TDM system.

3.15 AUTOMATED CENTRAL EXCHANGES

Innovations like automated switching systems, fiber optics, and multiplexing techniques for placing numerous conversations over one medium represent the greatest advances in telecommunications. We can thank AT&T Bell Labs for most of these developments in telecommunications technology. Bell Labs invented the transistor. The integrated circuit chips used in computers, and the signal logic processors used in a modem, would not exist if Bell Labs had not invented the transistor. Bell Labs not only made these advances but improved on the technology invented by others. Bell Labs developed and enhanced the fiber optic technology used by IECs and LECs. It also invented and improved the automated electromechanical crossbar and the computerized, stored program control, switching systems.

The automated switching systems allow us to reach anyone in the world simply by dialing the person's telephone number. This automated system can interpret dialed digits as a specific address location on the PSTN and connect one caller to anyone else. Automated switching systems were placed in the central exchanges to automate the functions previously handled by operators. Throughout this book, references are made to these switching systems as either central offices, exchanges, or central exchanges. All three terms mean the same thing and are interchangeable.

3.16 THE LOCAL EXCHANGE

All telephones connect to switches (central exchanges) usually located within a few miles of the telephone. We call the switch that a telephone connects to the *local exchange* or *local office*. A local switching exchange is also referred to as a *class 5 exchange*, or the *end office*, because it is the lowest and last switching system in the PSTN five-level hierarchy. The class 5 exchange is the node on the PSTN used by a telephone to access the PSTN. All telephones connect to the end offices using line circuits in the end offices. The end office connects to other class 5 and class 4 offices using trunk circuits. The switching system for an end office must be capable of connecting any of its line circuits to any other line circuit for a local *intraoffice call.* An *intra*office call is a call completed within one class 5 exchange. Calls that cannot be completed within the originating exchange but require connections via another central office are called *inter*office calls. *Intra* means "within" and *inter* means "between."

Figure 3-9 illustrates an intraoffice call between two telephones serviced by the same local exchange. The exchange prefix for both telephone numbers is 941. Telephone number 941-3333 has called 941-2222. The switch translates the dialed number into switching instructions and connects the line circuit serving 941-3333 to the line circuit serving 941-2222.

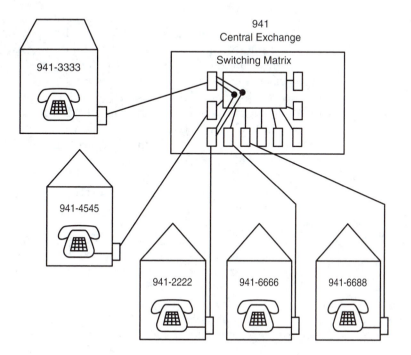

FIGURE 3-9 Intraoffice call.

3.17 LOCAL EXCHANGE LINE

The local exchange line that connects our telephone to the central office will undergo major changes in the next few years. The Telecommunications Reform Act of 1996 allows almost any common carrier to provide local telephone service. Many IECs and cable television companies have formed strategic alliances to offer phone service. Many IECs are also using PCS technology to provide local telephone service. AT&T bought McCaw Cellular Telephone Company and will use this technology as a vehicle to provide local telephone service. AT&T has announced that this system has 70 million potential customers. The LECs must do something to protect their business. Most competition for local phone service will be in the major metropolitan areas. The LECs will accelerate deployment of digital services in these areas to provide more enhanced services in the competitive environment. I have seen several news releases that state that LECs are deploying fiber and coaxial cable facilities to the customer's doorstep. The days of a twisted pair of wires providing analog service are numbered. The services to our homes and business will soon be furnished by a digital subscriber line facility with capacity to furnish much more than local phone service. Some business owners are already using ISDN as a vehicle to carry the signal from a surveillance video camera to a remote monitor. The remote monitor may be at the home of the business owner, or at any other location they desire.

3.18 TOLL CALLS OVER THE PUBLIC SWITCHED TELEPHONE NETWORK

All *inter*state toll calls must involve a toll office and an IEC to complete the call. With the passage of the Telecommunications Reform Act of 1996, LECs can also perform IEC functions and can complete interstate calls. When a customer places a long distance call, the local exchange will connect the caller to the class 4 toll office serving the originating class 5 exchange. If the toll call covers less than 30 miles, the class 4 toll exchange will also be serving the terminating class 5 office. Most of these short distance calls are within the same LATA. In this case the caller is connected from the class 5 office to the toll office, and the toll office will connect the call to the class 5 office serving the called number (see Figure 3-10).

The use of a five-level hierarchy for the toll network allows the PSTN to use far fewer trunk circuits in the network than if a four- or three-level network was used. There are almost 20,000 class 5 central exchanges in the United States. If some type of hierarchy did not exist, each class 5 exchange would need multiple trunk circuits to all other class 5 exchanges—a costly and antiquated network design. Just as the first central exchange was put in place to eliminate having to wire each phone to every other phone, the hierarchy design was implemented to eliminate the need for direct connections between exchanges.

The design of any network involves a trade-off between the number of transmission facilities (circuits) and the number of switching stages used. It would be physically impossible or at least extremely costly to have a direct connection between each and every local switching exchange in the United States. A much better arrangement is to connect several class 5 offices to a class 4 office. The five-level

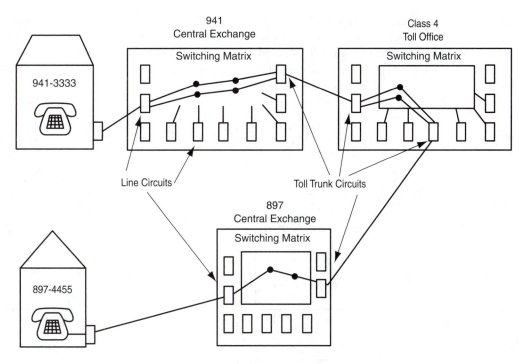

FIGURE 3-10 Interoffice call.

hierarchy design used by AT&T was put in place to handle toll calls. All calls between exchanges were toll calls. The class 4 office was called a *toll office*. When someone dialed a 1 or 0 for a long distance call, the person was connected from the class 5 office to the class 4 toll office via an interoffice toll trunk.

There are about 950 class 4 toll offices in the United States. Most class 4 toll offices are owned by one of the BOCs. If the call is an inter-LATA call (from one LATA to another), it is handed to the originating caller's preferred IEC. With divestiture in 1984, the separation of AT&T and the BOCs left the ownership of the class 3 primary centers with AT&T. Therefore, inter-LATA calls handled by AT&T are passed from the LEC's class 4 toll center to AT&T's class 3 primary center. There are about 150 AT&T primary centers in the United States. If AT&T's primary center cannot complete the call, it will pass the call on to a class 2 sectional center. AT&T has about 50 class 2 sectional centers. If the class 2 sectional center cannot complete the call, it passes the call to a class 1 regional center. AT&T has about seven class 1 regional centers in its network.

When Sprint built its toll network, it was able to take advantage of the latest in switching and transmission technologies to create a flatter network. Sprint has about 48 toll centers, which are combined class 3, class 2, and class 1 exchanges. These 48 toll centers are computer-controlled switching systems and are interconnected with fiber optic cables. When a toll call is handed from the LEC to Sprint, it is passed from the LEC's class 4 toll office to one of Sprint's 48 toll centers.

If the called telephone number is more than 1000 miles away from the originating exchange, the caller must use several layers of the PSTN network. Figure 3-11 is a diagram showing the completion of a toll call from Wausau, Wisconsin, to Raeford, North Carolina, using AT&T's five-level network.

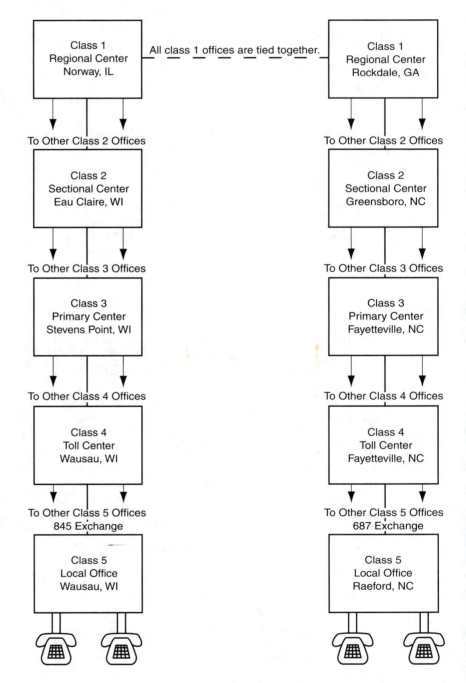

FIGURE 3-11 The five-level PSTN network.

3.19 EXTENDED AREA SERVICE IN THE PUBLIC SWITCHED TELEPHONE NETWORK

Over time, customers insisted they be able to place calls between some exchanges without having to pay a toll charge. Sometimes a town or city has several central exchanges located around the city. Customers demanded the ability to place calls between these exchanges without a toll charge. This free interoffice calling between local exchanges is called *extended area service (EAS)*. EAS is not really free. Its cost is included in the local service rate charge. As additional EAS points are added to a particular local exchange, the local service rate for the exchange is increased. The more telephones that can be called without a toll charge from a local exchange, the higher the local rate is. This is why the rates for telephone service in a city are higher than those in a small town.

In a small town there is only one central exchange, which serves the town and the adjacent rural area. If another town is close by, EAS calling may be allowed to it. When two towns are close to each other and have a high degree of community interest between them, they will petition the PUC for EAS. Most small towns do not have EAS. When they do have them, only a few exchanges will be involved. When the number of exchanges involved in an EAS plan is small, these class 5 central exchanges are connected directly to one another via EAS trunk circuits (see Figure 3-12).

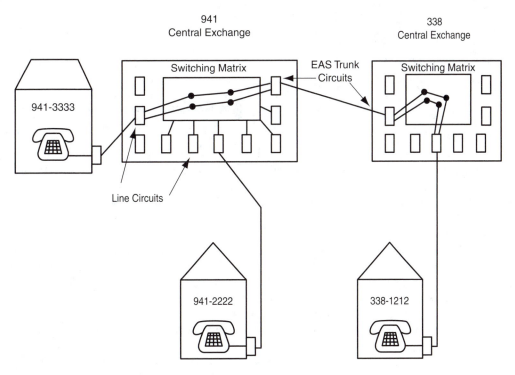

FIGURE 3-12 An extended area service call direct to another exchange.

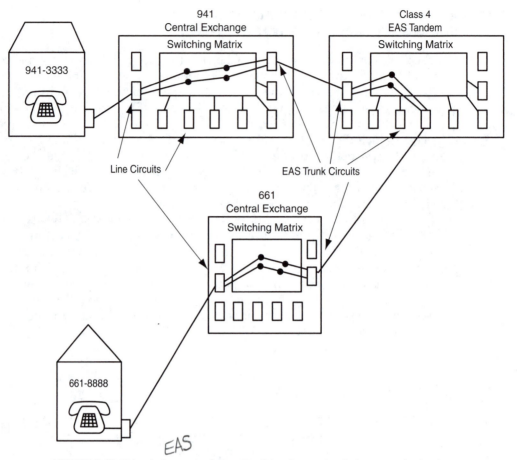

FIGURE 3-13 Extended area service connections via an extended area service tandem.

In a larger city, there are many local exchanges within the city that must be able to connect calls between them without a toll charge. A person originating a call from an exchange in the southern part of New York City must be able to call a telephone in a northern exchange (or any of the 50-plus exchanges serving New York City) without a toll charge. When many class 5 switching systems exist in an EAS network, it is cheaper to use a class 4 switching exchange and connect the EAS calls between class 5 exchanges via the class 4 exchange than to use direct circuits between each office. The use of a class 4 switch to connect calls between class 5 exchanges significantly reduces the number of trunk circuits that would be needed to connect many class 5 exchanges directly to each other. This reduces the cost of the EAS network.

The class 4 switch used for EAS calls is called the *class 4 EAS tandem exchange.* Each of the local exchanges is connected to the EAS tandem using EAS trunk circuits. Local calls between exchanges will be connected from the originating exchange to the EAS tandem, which in turn will connect the call to the appropriate class 5 exchange containing the called number (see Figure 3-13).

Some ILECs, such as Illinois Bell and Sprint/Centel of Des Plaines, Illinois, limit the EAS area in a city the size of Chicago to a few exchanges that are within a few miles of each other. They charge a toll for calls that cross several exchange territories. Thus, to call from an exchange located 25 miles from downtown Chicago to a telephone located downtown, a toll charge is applied. This is an area ripe for competition, and CLECs have begun to converge on Chicago to offer lower-cost alternatives.

3.20 CIRCUIT SWITCHING VIA TANDEM EXCHANGES IN THE PUBLIC SWITCHED TELEPHONE NETWORK

The switching systems in the PSTN connect (or switch) input circuits to output circuits. These switches have the ability to connect any circuit attached to the switch to any other circuit attached to it. This type of switching is called *circuit switching*. *The PSTN is a circuit-switched network.* As stated earlier, the central exchange was developed to allow any telephone to be connected to any other telephone linked to the exchange. When a local call is made within a switch, it is called an *intraoffice call.* The class 5 switching system connects one line circuit to another to complete an intraoffice local call. Because the call is completed within the switch, it is also called an *intraswitch call.* When a called number is not found within the originating switch, the call must be sent to another switch for completion of the call. If the call is a toll call, it will be sent to a class 4 toll switch for handling. The class 5 switch will connect the line circuit of the caller to a trunk circuit going to the class 4 toll office.

The class 4-to-class 1 toll exchanges do not have subscriber line circuits attached to them. These exchanges only have trunk circuits that connect to other central exchanges. All class 1, 2, 3, and 4 exchanges are called *tandem exchanges.* Here *tandem* means "in conjunction with" or "associated with." Each class 5 office connects to a certain class 4 office. Each class 4 office connects to a specific class 3 office. Several class 5 offices connect to each class 4 office. Several class 4 offices connect to each class 3 office. Likewise for class 3, 2, and 1 offices in a five-level hierarchy network. Each of these is associated with specific offices below and above its level. The PSTN has been flattened with the introduction of fiber optics and computerized switching. Figure 3-14 illustrates the connection of our theoretical long distance call from Wausau, Wisconsin, to Raeford, North Carolina, using the flatter Sprint Network.

3.21 ESTABLISHMENT OF THE NORTH AMERICAN NUMBERING PLAN

Establishment of seven-digit dialing was part of the North American Numbering Plan. North America was divided into many *numbering plan areas (NPAs).* Each NPA was assigned a three-digit number called the *area code.* Within each NPA, each central office exchange is assigned a three-digit number called the *exchange code.* This three-digit number represents the first three digits of the seven-digit telephone number. In numbering plan assignments, if a number could be any value from 1 to 0, it was designated by the variable X. If a number could be any value from 2 to 9, it was designated by the variable N. If a number could be only a 1 or 0, it was designated by the variable 1/0. Exchange codes were of the form NNX. Area codes were of the form $N1/0N$.

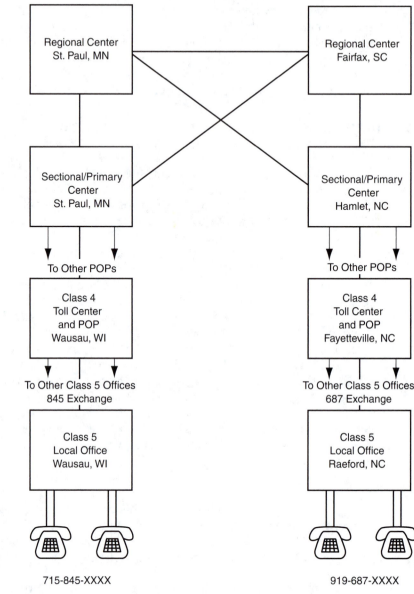

FIGURE 3-14 The flatter four-level network.

715-845-XXXX

919-687-XXXX

The area code had to be of the form $N1/0N$ to assist electromechanical switching systems (such as SXS and XBAR) in the switching of calls. The use of a 1 or 0 as the second digit of an area code allowed electromechanical switches to distinguish between an area code and an exchange code. Area codes could be identified because the first and third digits could be any number except a 1 or 0, while the second digit must be a 1 or 0. Central office exchange codes were designated by numbers of the form *NNX*. A central office exchange code could not have a 1 or 0 as either a first or second digit. Additionally, each NPA or area code could not have more than one central office exchange code with the same code. Every exchange within an area code had to have its own unique code. Central office numbers (exchange codes) could be reused in many different area codes, but within any particular area code or NPA, each central office exchange must have a unique *NNX* identifier.

When the North American Numbering Plan was implemented, existing exchange names were converted to exchange codes. The Willow exchange, for example, was converted to 941. Look at the telephone dial and you will see that the letter *W* is also the digit 9. The letter *I* is the digit 4. Exchange names were converted to exchange codes according to which number on the dial contained the letter. Only the first two letters of the exchange code were converted using this technique. Have you ever heard of Glenn Miller's song "Pennsylvania Six-Five Thousand" (Pennsylvania 65000)? This was the telephone number for the Pennsylvania Hotel (which later became the Statler Hotel) in New York City. Miller played in the hotel's Cafe Rouge, and people dialed the hotel's number for reservations. This famous number was 65000, located in the Pennsylvania exchange of New York City.

With the North American Numbering Plan, Pennsylvania 65000 became 736-5000. This is not nearly as glamorous sounding as using exchange names. Note that *P* = 7 and *E* = 3 on the telephone dial. This method of assigning exchange names allowed people to dial a number using the first two letters of the exchange name. Exchange names were also used in the directory (for example, PE 6-5000). After people had accepted seven-digit dialing, the abbreviated names were replaced with the corresponding digits. Local central offices are referred to by their exchange names and/or exchange code. The central office serving telephone numbers in a certain part of South Kansas City is called the Willow Exchange and is also called the 941 exchange.

Using area codes of the form *N1/0N* made it possible to have 162 ($9 \times 2 \times 9 = 162$) area codes or NPAs. North America ran out of NPA area codes in 1995, and the requirement that the second digit of an area code be a 1 or 0 was dropped. Fortunately, by this time all electromechanical toll offices had been replaced by SPC offices, and dialed numbers were no longer sent over the voice switching path. Switching connections between offices are now established over SS7 networks. The new area code numbering system was put in service in 1995. Alabama had run out of central office codes. The area code for Alabama was 205; a new area code of 334 was added on January 15, 1995, for Montgomery and the southern half of the state. This change allowed the northern half of Alabama (which includes Birmingham) to reuse the exchange codes that had been reassigned to the 334 area code. The new 334 area code will be able to grow by using the exchange codes that had been left behind in the old 205 area (see Figure 3-15).

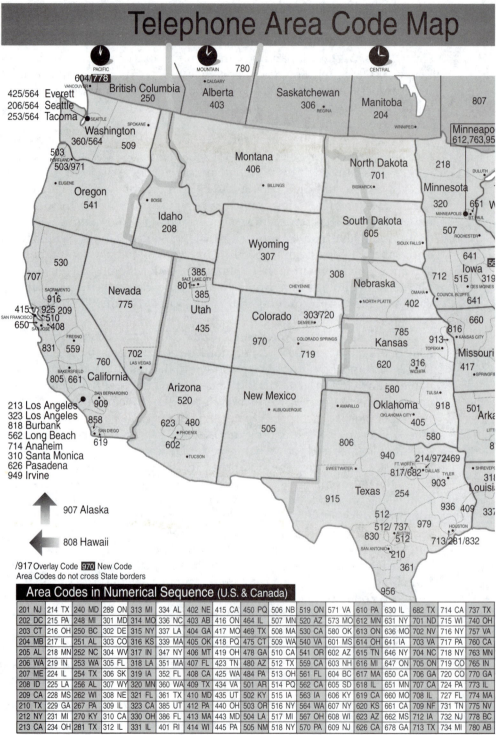

FIGURE 3-15 Area code map. (Courtesy of Sprint)

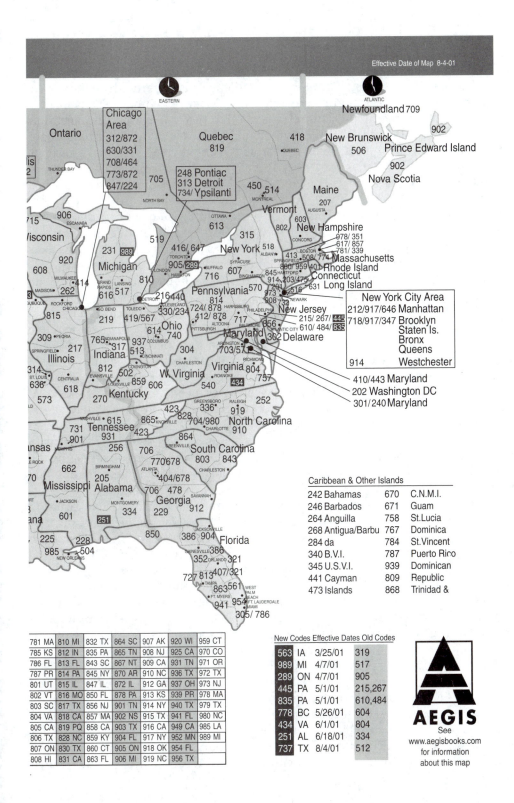

Effective Date of Map 8-4-01

EASTERN

ATLANTIC

Newfoundland 709

Ontario

Chicago Area
312/872
630/331
708/464
773/872
847/224

248 Pontiac
313 Detroit
734/ Ypsilanti

Quebec
819

418

New Brunswick
506

902

Prince Edward Island

902
Nova Scotia

lis
2

THUNDER BAY

705

NORTH BAY

OTTAWA

613

450 514
MONTREAL

Maine
207
AUGUSTA

715

906
ESCANABA

519

315

603
CONCORD

New Hampshire

Wisconsin

231 989

416/ 647

New York 518
ALBANY

978/ 351
617/ 857
781/ 339

608

920
MILWAUKEE

Michigan

905 289
TORONTO

LONDON

HAMILTON

BUFFALO

SYRACUSE

BINGHAMTON

607

Vermont

802

413
SPRINGFIELD

BOSTON
508/ 774
HARTFORD
860/ 959 401
845
914 203/475
201 516
908

Massachusetts

Rhode Island

Connecticut

631 Long Island

414

MADISON

262

GRAND
RAPIDS

810

LANSING

716

570

973

NEWARK

3

262

DUBUQUE

616 517

216 440
CLEVELAND

Pennsylvania

814

NY2

New Jersey

New York City Area
212/917/646 Manhattan
718/917/347 Brooklyn
Staten Is.
Bronx
Queens
914 Westchester

815

ROCKFORD

CHICAGO

SO. BEND

219

DETROIT

TOLEDO

330/234

724/ 878

HARRISBURG

412/ 878

PHILADELPHIA

215/ 267/ 445
610/ 484/ 835

309

PEORIA

765
INDIANAPOLIS

419/567

Ohio

614

717

ATLANTIC CITY

856

302 Delaware

217

SPRINGFIELD

Illinois

317

937
COLUMBUS

740

Maryland

703/571

ARLINGTON

410/443 Maryland
202 Washington DC
301/240 Maryland

314

ST. LOUIS

812

502

513
CINCINNATI

304

CHARLESTON

Indiana

Virginia

RICHMOND

804

636

618

859
LOUISVILLE

EVANSVILLE

606

COVINGTON

540

ROANOKE

757

573

270

Kentucky

434

GREENSBORO

RALEIGH

252

423

336

919

731

615
NASHVILLE

865
KNOXVILLE

828

704/980
CHARLOTTE

910

North Carolina

901

931

423

864

910

ansas

256

706
GREENVILLE

South Carolina

LITTLE ROCK

MEMPHIS

Tennessee

770 678
ATLANTA

803

843

CHARLESTON

662

205
BIRMINGHAM

404/678

70

Mississippi

Alabama

706

478

Georgia

SAVANNAH

912

601

JACKSON

334
MONTGOMERY

229

Caribbean & Other Islands

242 Bahamas 670 C.N.M.I.
246 Barbados 671 Guam
264 Anguilla 758 St.Lucia
268 Antigua/Barbu 767 Dominica
284 da 784 St.Vincent
340 B.V.I. 787 Puerto Rico
345 U.S.V.I. 939 Dominican
441 Cayman 809 Republic
473 Islands 868 Trinidad &

251

850

386 904
JACKSONVILLE

Florida

225

228

985

504

NEW ORLEANS

GAINESVILLE

352

386

ORLANDO

321

727 813

407/321

TAMPA

863 561

FT. MYERS

941 954

305/ 786

WEST PALM BEACH

FT. LAUDERDALE

MIAMI

na

781 MA	810 MI	832 TX	864 SC	907 AK	920 WI	959 CT
785 KS	812 IN	835 PA	865 TN	908 NJ	925 CA	970 CO
786 FL	813 FL	843 SC	867 NT	909 CA	931 TN	971 OR
787 PR	814 PA	845 NY	870 AR	910 NC	936 TX	972 TX
801 UT	815 IL	847 IL	872 IL	912 GA	937 OH	973 NJ
802 VT	816 MO	850 FL	878 PA	913 KS	939 PR	978 MA
803 SC	817 TX	856 NJ	901 TN	914 NY	940 TX	979 TX
804 VA	818 CA	857 MA	902 NS	915 TX	941 FL	980 NC
805 CA	819 PQ	858 CA	903 TX	916 CA	949 CA	985 LA
806 TX	828 NC	859 KY	904 FL	917 NY	952 MN	989 MI
807 ON	830 TX	860 CT	905 ON	918 OK	954 FL	
808 HI	831 CA	863 FL	906 MI	919 NC	956 TX	

New Codes Effective Dates Old Codes

563	IA	3/25/01	319
989	MI	4/7/01	517
289	ON	4/7/01	905
445	PA	5/1/01	215,267
835	PA	5/1/01	610,484
778	BC	5/26/01	604
434	VA	6/1/01	804
251	AL	6/18/01	334
737	TX	8/4/01	512

AEGIS
See
www.aegisbooks.com
for information
about this map

3.22 CIRCUIT SWITCHING

Each wire pair leaving a central office is referred to as a *line* if it connects to a telephone. The 52-Volt battery supply at the central office passes through an individual electronic circuit in the central office for each line. Most switching systems include a relay in this electronic circuit. This relay is referred to as a *line relay*. The electronic circuitry and line relay is called a line circuit and resides on a printed circuit board that is called a *line card*. As discussed in Chapter One, switching systems come in all sizes from the small switches used for a Private Branch Exchange (PBX) in a business to the larger switching systems found in a local central office. Line cards in a PBX contain from eight to 32 line circuits. Line cards in a central office used to contain from 16 to 32 line circuits. This can be a problem. If we need to replace a line card because one circuit is bad, we will take 16 to 32 telephone lines out of service when we pull the card out of the system. Most central office switching systems have changed their line card architecture and adopted a technology that uses very small line cards with only one circuit on a card. PBXs still use line cards with 32 to 64 line circuits on a card to reduce cost.

A line circuit can be described as the actual electronic circuit card that a telephone line connects to in the central office. Many times when people refer to a line circuit, they not only mean the electronic circuit on the line card but the twisted wire pair between the telephone and the central office as well. One should think of a circuit in voice and data telecommunications as one voice or data path. The wire pair between a telephone and central office carries one conversation or data signal and is a line circuit. One wire pair, with associated electronic circuitry, that connects one central office to another central office is called a *trunk circuit*. Line circuits connect telephones to a local central office. Trunk circuits connect switching systems to one another. If a call is made from one telephone to another telephone served by the same switch, the switch connects the originating line circuit to the terminating line circuit. Figure 3-16 shows two telephones connected to one another via the antiquated SXS switching system. This is similar to Figure 3-9 that was used earlier to depict an intraoffice call. The advantage of Figure 3-9 is it shows the switching matrix without detail; thus, Figure 3-9 is representative of any switching system, from the antiquated space division SXS of yesterday, to today's stored program controlled (SPC), time division, switching system.

If the originating telephone and the called telephone reside on different switches, then the originating switch will connect the originating line circuit to a trunk circuit as shown in Figures 3-10, 3-12, and 3-13. The trunk circuit provides a path to the receiving central office. The terminating central office connects the incoming trunk circuit to the line circuit of the called telephone. If no multiplexing is involved, the switching system used must be a space division switching where each conversation occupies its own wire path between the calling and called party such as shown in Figure 3-16 for an older Strowger SXS switching system.

The circuit between a calling and called telephone actually consists of many different circuits connected together much the way highways are connected together. If we wish to drive from Topeka to St. Louis, we would drive north on highway I-35, and at the intersection with I-70E in Kansas City, we would switch

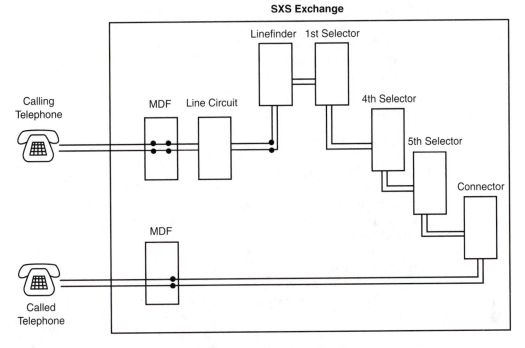

FIGURE 3-16 Step-by-step exchange.

over to highway I-70E because I-70E connects Kansas City to St. Louis. There are two circuits, or highways, involved (I-35N and I-70E). The job of the switching system (driver) is to switch the car from one highway to another highway to reach the desired destination.

In telecommunications, the job of the switching system (SPC switching system) is to switch voice or data circuits from one voice highway to another voice highway to get the voice to the desired destination. Voice paths (voice highways) and signaling paths (SS7) connect switching systems to each other. The signaling paths are used to inform switching systems how the voice paths should be switched. Each switching system then connects the appropriate voice paths (circuits) together so that an end-to-end circuit is achieved. Because switching systems in the PSTN are used to connect one line or trunk circuit to another line or trunk circuit, the PSTN is referred to as a circuit switched network.

3.23 WHAT IS A CIRCUIT?

What is a circuit? It is a path that information, voice, data, video, etc. can flow over. In the telecommunications network, a circuit is a path that allows communications to take place between a transmitter and a receiver. Earlier, we defined the path between a transmitter and a receiver as a medium. In its simplest form, a circuit is a medium. But we do not always use the simplest form. In many instances, we have many circuits using one media through multiplexing technology to place many circuits onto one media. Most telephones connect to a central office using individual wire pairs. In this case, the physical medium or media (wire pair) has one circuit on it. No multiplexing is involved here. Each wire pair is one circuit. If we use the one-wire-pair-per-circuit philosophy for trunk circuits to connect calls between central offices, we would need a lot of wire pairs between central offices. In order to reduce the number of wire pairs needed between central offices, we use multiplexing technology to place many trunk circuits on one physical medium or media.

3.24 TIME DIVISION MULTIPLEXING (TDM)

The most predominant multiplexing method used in telecommunications today is Time Division Multiplexing (TDM). Let's use the highway system to explain TDM. Suppose we have a two-lane highway between Chicago and New York with one lane in each direction, and we do not allow cars to share the highway. If cars cannot share the highway, the next user at Chicago cannot get on the highway until the previous user gets off the highway in New York. To support 32,256 cars traveling at the same time without sharing the highway, we would need 32,256 highways. In order to support travel in both directions (full-duplex transmission), these highways have one lane for each direction. We can readily see that dedicating a highway between Chicago and New York for only one car in each direction would be very inefficient. How do we make more efficient use of a highway? We use time division multiplexing. For auto travel, each car is at a specific point on a highway at a different time. If more than one car occupies the same space at the same time, we have a wreck. Therefore, we have a protocol that states each car must not occupy space at the same time it is occupied by another user.

If we wanted to connect 32,256 telephone calls between Chicago and New York without multiplexing, we would need 32,256 cable pairs. Each wire pair would carry voice signals in both directions. The example used previously for highways points out how inefficient and costly it is not to use TDM on our highways for cars. It is also very inefficient and costly to dedicate a media (voice highway) to one telephone call and not use TDM on our voice highways.

The most optimum use of the highway occurs when all cars travel at the same speed. This does not happen on highways used by cars, but it does happen on the highways (fiber media) used by telecommunications. So for our analogy, let's run all the cars on our highway at the same speed. We could put cars bumper to bumper on the highway between Chicago and New York. With TDM, instead of having only one car using the highway between cities and having to wait until that car reaches its destination before putting another car on the highway, we can place millions of cars on one highway. The instant one car leaves, we can put another car on the highway. Each car has its own time slot on the highway. With all cars traveling at the same speed, we have an orderly procession of cars down the highway. As long as the cars all travel at the same speed, we will never have a wreck regardless of whether they are going 10 mph or 10,000 mph. The faster they travel, the more cars the highway can handle per minute, hour, day, etc. The hardest function to accomplish is getting cars on and off the highway in an orderly fashion. The faster the cars travel down the highway, the harder it is to get them onto and off of the highway without causing a wreck.

For voice and data communications, if we use a multiplexer that can combine 32,256 voice circuits, we only need two glass fibers between Chicago and New York. We need two fibers because we do not normally engineer fiber media to carry full-duplex communications. They are engineered as simplex media, and it takes two simplex paths to make a full-duplex path. Again, this is like a highway. There is a simplex path with traffic flowing in one direction. There is also a simplex path with traffic flowing in the opposite direction. Together the two simplex paths form a full-duplex path. The major difference in analogy between a highway for cars and a fiber highway for telecommunications is that we can transport everything in our car (time slot on the highway) at the same time and only need one time slot but we cannot transport all voice or data communication in one time slot.

To use TDM for voice, every voice signal is first converted into a digital signal. The digitized voice signal consists of 64,000 bits (1or 0) per second. The process used to convert the analog voice signal to a digital voice signal results in 8000 little pieces of the voice signal per second, with each piece consisting of 8 bits. These little pieces of the voice signal are called samples. Each 8-bit sample of the voice signal is assigned an individual time slot on a voice highway. Since each voice signal is broken down into 8000 samples per second, we must provide 8000 time slots per second on the fiber highway for each 8-bit sample of the voice signal.

We put one sample of a voice signal on a fiber highway and then each 1/8000 of a second later we put the next sample for that voice signal on the fiber highway. Each sample of voice consists of 8 bits, and the space these 8 bits occupy is referred to as a time slot. If the speed of the samples on the highway is 8000 samples per second, the highway will only support one call at a time because each voice signal is made up of 8000 samples. If the speed for the time slots that samples can be put in is increased to 80,000 time slots (samples) per second, we could support 10 calls over the same fiber highway. If we increase the speed of the time slots to 258,048,000 time slots per second, the fiber highway will handle 32,256 voice circuits. The faster we can send samples (time slots) down a fiber highway, the more samples it can support. The more samples it can support, the more voice or data calls, it can support. The faster the samples travel, the more voice/data circuits the fiber highway can handle per minute, hour, day, etc. In telecommunications, the speed of the TDM system is governed by how fast the multiplexers can get voice samples onto and off of the media and the capacity of the media. Up to 96 voice circuits can be multiplexed over twisted pair wire media. Today, we use microwave radio or fiber optic cable if we wish to multiplex more than 96 circuits. In the past, coaxial cable was used as well for these applications, but coaxial cable has gradually been replaced by fiber. We still use coaxial cables inside select central offices for connections between some 672-circuit multiplexers, but even this application is moving toward the use of fiber as the connecting media.

3.25 WHAT IS A CIRCUIT IN TDM?

What is a circuit? If only one voice is placed on a media, the circuit is essentially the media. If more than one voice is placed on a media using TDM, a circuit is a time slot on the media. Many people refer to these time slots as virtual, or logical circuits as opposed to physical circuits. The physical circuit (fiber media) discussed previously supports 32,256 logical circuits. Because the logical circuits consist of time slots that no one else is allowed to use (remember each voice sample occupies its own specific and individual time slot), everyone appears as if they have their own physical circuit. No one realizes they share the physical media with others.

3.26 TIME DIVISION SWITCHING

Figure 3-16 illustrates space division switching where each voice path must have its own separate pair of wires from the calling party to the called party. The switching system connects one pair of wires (circuit) to another pair of wires (circuit). This is circuit switching where each circuit occupies its own space (pair of wires). Space division switching systems still exist in keysystems, but all central offices are now time division switching systems. These time division switching systems simply connect a time slot on one voice highway to a time slot on a different voice highway. Again let's use the highway analogy. With TDM, all time slots are going down the fiber highway whether they have voice samples in them or not. This greatly simplifies the TDM system as each multiplexer knows where each and every time slot is at all times. This is referred to as synchronous time division multiplexing.

Think of synchronous TDM as cars going down the highway whether they have people in them or not. Okay maybe this is not a good idea. Maybe a better analogy is to think of this in terms of as time slots reserved for each car we wish to place on a highway. If we do not put a car in a specific time slot, there will be space between the previous car placed on the highway and the next car placed on the highway. The space between these two cars will equal the number of time slots that we did not put cars into.

Suppose we have 32,256 time slots on our fiber highway that repeat every 1/8000 of a second. Think of the time slots as cars that we can put a voice sample in. The cars are numbered 1 to 32,256 and every 1/8000 of a second, we start with car number 1 again. The receiving switch is getting 32,256 time slots every 1/8000 of a second. Now suppose that we want to establish a voice conversation from Kansas City to Chicago but have no direct route and must go via St. Louis. The switching system in Kansas City will use the SS7 network to find out which time slots between Kansas City and St. Louis are available and which time slots between St. Louis and Chicago are available. Let's assume that time slot 300 between Kansas City and St. Louis is available and time slot 500 between St. Louis and Chicago is available. Instructions are sent to the switching system in St. Louis to use time slot (300) on the fiber media between it and Kansas City. Instructions are sent to the St. Louis and Chicago switches telling them to use time slot (500) on the fiber media between them.

Instructions are issued to the Kansas City switch telling it to connect the voice signals from the calling party onto the assigned time slot (300) to St. Louis for this call. Instructions are issued to the St. Louis switch telling it to take the voice signals coming in on the specified time slot (300) from Kansas City and put them on the time slot (500) specified for the fiber highway to Chicago. At Chicago, the time slot (500) from St. Louis is connected to the called telephone. In time division switching

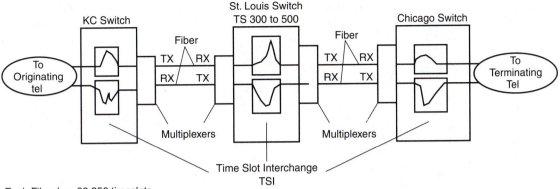

Each Fiber has 32,256 timeslots

FIGURE 3-17 TDM Switching. The TSI in St. Louis connects time slot 300 from Kansas City to time slot 500 to Chicago and in the return direction connects TS 500 from Chicago to TS 300 to Kansas City.

the circuits are time slots and we simply connect one time slot to another as shown in Figure 3-17.

In Figure 3-17, the switches contain a Time Slot Interchange (TSI). The TSI is nothing more than memory (RAM) with gate circuits to gate 8-bit voice/data samples into and out of the RAM. In our previous example, we multiplexed 32,256 circuits. Each circuit consisted of time slots, and we put one voice sample (8 bits) in each time slot. For a switching matrix designed to handle 32,256 conversations, we must have 32,256 RAM locations and each location must be able to store 8 bits. Therefore the TSI consists of a 32,256 Byte (or 258,048 bit) memory. The time slots on the fiber media are gated, or connected, to their respective memory location as samples are received. One memory location exists for each time slot. When time slot one arrives at the switch it is gated or connected to RAM location 1. When time slot two arrives, it is gated to RAM location 2, etc.

In our example, when time slot 300 arrives from Kansas City, the 8 bits in this time slot will be stored in RAM location 300 on the TSI. Only one fiber can gate samples into a TSI. If we need to support more than one fiber, we must have multiple TSIs (one TSI for each fiber). This is one side of the puzzle. All incoming time slots come in on a receive fiber and get mapped into a TSI memory location. How do we get voice samples onto the transmit fibers that leave a switch? This is very simple. The TSI contains gates to connect the TSI to all transmit fibers. Thus, only one receive fiber can gate into a TSI, but we can gate out of the TSI to any transmit fiber. For the TSI at St. Louis that is dedicated to the incoming Kansas City fiber, we get voice samples and store them in this TSI. We can then gate them out onto any fiber. We can gate them on a fiber to New York, Atlanta, Chicago, etc. For further discussions, let's assign a number to this TSI store for the incoming Kansas City fiber and call it TSI number 5. In our example, we have instructed the St. Louis switch to gate (make a connection) between the Kansas City fiber and the RAM memory of TSI number 5 at time slot 300. At this point in time, RAM location 300 is connected to the Kansas City fiber. The 8 bits in time slot 300 will be received and stored (written) in memory location 300 of TSI number 5.

The St. Louis switch has also been instructed to close the gate (make a connection) between TSI number 5 RAM location 300 and transmit fiber to Chicago when time slot 500 occurs on the Chicago fiber. The contents of store 300 will be read out onto the Chicago fiber at time slot 500. The switching system in St. Louis has connected an incoming voice sample in time slot 300 from Kansas City to time slot 500 going to Chicago. Since the switch receives 8000 samples per second, the gate into memory location 300 operates 8000 times a second and the contents of TSI number 5 RAM location 300 is changed every 1/8000 of a second. TDM switching requires very fast gating circuits to gate all time slots into the appropriate memory location at the right time and then be able to gate the contents of memory onto the appropriate transmit fiber at just the right time slot.

The example just used is very simplified for illustrative purposes. It does not represent the actual path, nor does it include all switches that would be used for a Kansas City to Chicago call. To place a Kansas City to Chicago call, about six switches would be involved. Typically these would be the local exchange of the calling telephone, the equal access toll office in Kansas City, the IEC switch at Kansas City, the IEC switch at Chicago, the equal access toll switch at Chicago, and finally the local exchange serving the called telephone.

The discussion on the use of TSIs for time division switching also applies to devices called *Digital Access Cross-connect systems (DACs)*. Fibers can be connected to a DAC, and we can program the DAC to connect any time slot of one fiber to any time slot of another fiber. The DAC contains TSIs, and we simply program the gating into and out of a specific memory location for each circuit we cross connect. Cross connect means to connect one circuit to another circuit such that the two circuits remain connected to one another all the time. This is referred to as a nailed-up connection. If we program the DACs to connect time slot 200 of one fiber to time slot 400 of another fiber, these time slots will always be connected to one another until the DACs is programmed to change the nailed-up connection.

With switching systems, connections between time slots is dynamic. Each connection is set up at the beginning of a call and remains that way until the call ends. At that time the connection between time slots is released and the switch can assign the time slots in a different manner for the next call it receives. In our example, the next call on time slot 300 from Kansas City to St. Louis may be destined for New York instead of Chicago. This requires different gating instructions to the TSI. With a DAC, connections between time slots are static. The DAC keeps the time slots connected forever (or at least until the DAC is reprogrammed). A DAC would be used to establish a private line circuit that is leased to a business customer and not accessible to the public.

A DAC is also used to connect high-speed fiber media into lower-speed connections to a switch. If the switch does not have a 32,256 TSI, we cannot connect a fiber directly to the switch. Suppose the switch has TSIs that have 672 RAM locations on each TSI. This will require 48 TSIs for each fiber. In this case the fiber is run to a DAC and the DAC is programmed to connect the first 672 time slots to an input to the switch for TSI number 1. The next 672 time slots are programmed onto the second input of the switch and go to TSI number 2, etc. In this scenario, we have one fiber into a DAC and 48 coaxial cables from the DAC to the switch. Each coaxial cable connects to a DS-3 interface card at the switch. DS-3 is a multiplexing system that places 672 circuits onto one media. DS-3 will be discussed in the chapter on multiplexing.

3.27 DIAL PULSES

Figure 3-18 illustrates dial pulses. These pulses can be depicted as either pulses of current or pulses of voltage. The top diagram shows that when a telephone is off hook, a certain amount of current flows in the circuit to the central office. The amount of current depends on the length of the local loop. In our example, 40 milliamps of current is flowing when the telephone is off hook. To dial the number 3 using a rotary dial, the caller inserts a finger in hole 3 of the dial and pulls the dial around until the finger hits a finger stop. The finger is removed from the dial and the dial spins automatically back to its normal at-rest position. As the dial returns to normal, the loop is opened three times. Each time it is opened current flow will cease (0 milliamp). After opening the loop three times, the dial will be back at normal and the loop remains closed with a steady current flow of 40 milliamps in the local loop and central office equipment. The equipment will recognize each absence of current flow as one pulse of a digit. The equipment counts the number of open-circuit conditions encountered and stores the count in a register as the dialed digit.

Figure 3-18 also shows the dial pulses as voltage conditions. Each time the dial opens the circuit, a voltmeter would read the 48 V of the central office battery placed on the loop. When the dial contacts close the loop, current will flow and voltage drops occur across each component in the circuit (that is, the central office equipment, the local loop, and the telephone). The voltage drop read across the telephone will be between 8 and 13 V (depending on the length of the local loop). In Figure 3-18, we are reading the voltage across the telephone, and in our example the voltage changes from 10 V (closed circuit) to 48 V (open circuit) for each open-circuit condition—that is, each pulse.

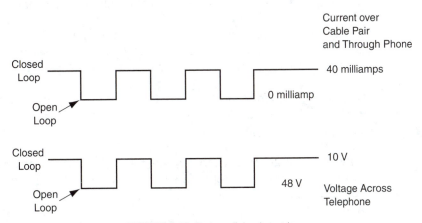

FIGURE 3-18 Rotary dial pulse string.

3.28 DUAL-TONE MULTIFREQUENCY DIAL OR TOUCHTONE DIALPAD

Telephones manufactured today are made with a dial that use tones for dialing signals. This dial is called a *dual-tone multifrequency dial or DTMF dial*. The dial is a keypad containing 12 or 16 buttons (see accompanying table). The buttons are arranged into four rows and three or four columns. A separate tone is assigned to each row and column. When any button is depressed, it sends out a unique set of tones to the central exchange.

	1209 Hz	1336 Hz	1477 Hz	1633 Hz
697 Hz	1	2	3	A
770 Hz	4	5	6	B
852 Hz	7	8	9	C
941 Hz	*	0	#	D

Depressing button 1 will cause a 697-cycle (Hertz) tone and a 1209-cycle tone to be sent to the central exchange. The receiver in the register at the central exchange decodes the different combinations of tones received into the equivalent dialed digit. The dial register stores the dialed digit in a storage device. The storage device for older electromechanical switches, such as SXS or XBAR, was a relay. The storage device for SPC switching systems is RAM. The DTMF pad used on a telephone has 12 buttons: *, #, and 1 through 0. It would have been possible to use a single tone for each button pushed rather than the dual-tone design. This would require the ability to generate and detect 12 tones (16 if you include *A, B, C, D* buttons). The dual-tone design requires fewer tone detectors in the dial register of the central office than a single-tone design would. The use of fewer tones also reduces the sensitivity requirement of the receiver. A dual-tone receiver is better than a single-tone design in its ability to recognize tones distorted by abnormal conditions on a local loop that cause noise on the cable pair.

All push-button telephones are not DTMF telephones. Some push-button telephones do not send DTMF tones; they convert the button pushed to rotary dial pulses. You can recognize this type of push-button telephone because you can hear the clicks of the rotary dial in the phone receiver. You can also tell when a push-button phone is not a DTMF phone since this type of telephone will not work with voice messaging systems. It will work with the central office switching system because the dial register in the central office is designed to work with either a rotary or a DTMF dial. Voice messaging systems must receive DTMF tones to inform the system of what action it should take.

Rotary dial signaling pulses cannot pass through a central exchange. DTMF tones will pass through the exchange to the messaging system attached to the called number. Because a dial register in a central office can recognize both rotary dial pulses and DTMF tones, both a rotary dial telephone (called a *500 set*) and a touch-tone phone (called a *2500 set*) will work with any central office exchange. DTMF tones must be present for at least 40 milliseconds (ms) in order for the register at the central office to recognize the tones. A pause of 60 ms is required between digits.

The total time for transmission of one digit is 100 ms; therefore, it is possible to transmit 10 digits in 1 sec. DTMF is faster than rotary dial, which takes an average of $1^1/_2$ sec per digit or 15 sec for 10 digits.

3.29 STORAGE OF DIALED DIGITS IN A REGISTER

In a common control switching system, such as XBAR and SPC, the dialed digits are received by a device called a *dial register* or simply the **register.** The register recognizes a rotary dialed digit by counting the number of open-circuit conditions (pulses) sent by the dial (Figure 3-19). The pulses are counted by a stepping relay, which steps once for each open circuit (pulse). The register also contains tuned circuits so it can detect the different tones sent by a touchtone keypad.

3.30 COMPUTER-CONTROLLED SWITCHING

The first computer-controlled switches, introduced in 1960, retained the XBAR switching matrices and simply replaced the electromechanical common control units of the XBAR switch with electronic circuitry. The marker was replaced by a

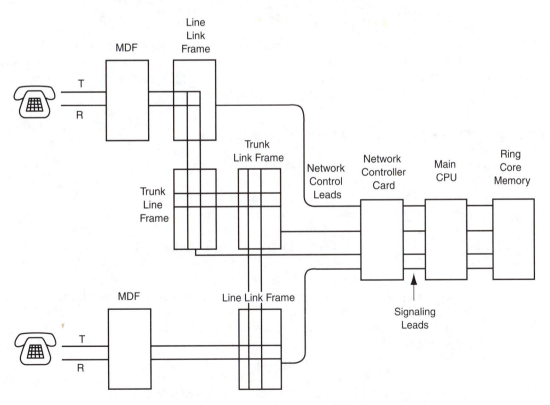

FIGURE 3-19 Early SPC with analog XBAR matrix.

computer. The computer controlled the XBAR switching components. The previous section on XBAR switching was included in this book because computer-controlled switching was initially based on the XBAR switching system. The early computer-controlled switching systems were space division switching systems that employed a XBAR switching matrix or a reed relay switching matrix. Computer-controlled switching is also referred to as *stored program control* because switching is controlled by a software program. The computers used to control switching in 1965 were solid-state design using transistors and printed-circuit technology. The computer was composed of numerous circuit cards placed in approximately six cabinets. Each cabinet was about 7 ft high and 2 ft wide. The memory for the computer was composed of ring core memory. The operating program resided in one ring core field, and a second ring core field contained the database necessary to translate dialed telephone numbers into an equipment location (see Figure 3-19).

3.31 SWITCHING CONTROL BY A MICROPROCESSOR

Since 1965, advances in electronic circuit design have resulted in the microprocessor on a chip used to control today's switching. The switching programs are relatively simple and can be controlled by an 8086 microprocessor. The ring core memory has been replaced by random access memory (RAM), tape drive storage, and disk storage. Additionally the XBAR switching matrix has been replaced by *time division switching* technology. The marriage of computer control and time division switching technology is used on all automated switching systems manufactured today. This chapter has provided only an overview of past switching systems because they are rapidly becoming extinct. All switching systems manufactured today are computer-controlled switching.

3.32 STORED PROGRAM CONTROL TRANSLATORS

The translators in SPC toll tandem offices look up numbers in a database and are much smarter than their electromechanical predecessors. The setup of switching paths is no longer done over the voice network but over a special signaling network called *Signaling System 7 (SS7)*. SS7 can establish a complete switching path in the voice network between several central office exchanges in milliseconds. The development of SS7 has meant that the originating toll office can wait until all digits have been dialed before calling in the translator. When a toll call is being placed, the caller dials a 1 as the first digit. The end office (class 5 local exchange) register stores the first digit as a 1, and the translator of the local exchange instructs the switching network to connect the calling line to a toll trunk going to the class 4 toll office. All ten digits of the called number are received by the register in the toll office switching system. The translator knows the first three digits are an area code. The middle digit no longer has to be a 1 or 0 for the translator to recognize it as an area code. That requirement existed when we were allowed to place long distance calls within an area code without having to dial the area code of our home NPA.

Now that the area code must be dialed on all long distance calls, even those within the same NPA, all long distance calls are ten-digit calls. The translator in a toll switch is programmed to recognize the first three digits stored in a register as an area code regardless of what the digits are. This has eliminated the requirement for an area code to be of the form N1/0X; an area code can be of the form NXX. This provides 800 possible area codes instead of the previous limit of 160. The PSTN was running out of area codes when the new codes were added in 1995. Once the translator in a toll switch has determined where the call should be sent, the SS7 network is notified to establish the switching path necessary to connect the call.

3.33 IN-BAND SIGNALING

The ability to establish call setups using the SS7 network has been a tremendous benefit to the LECs and IECs. Prior to the establishment of the SS7 network, all call setup was done over the voice network. Call setup and signaling used the same path that our conversation for the call would use. This type of signaling was called *in-band signaling*. Signaling was done within the voice band. The signaling used over the toll system was a dual-tone multifrequency system called *MF signaling*. MF signaling was developed prior to the introduction of DTMF for telephones. The two systems cannot use the same tones for signaling; DTMF tones for touchtone telephones are significantly different from MF tones. MF signaling used combinations of tones per the accompanying chart:

Digit Sent	Frequencies Sent
1	700 Hz + 900 Hz
2	700 Hz + 1100 Hz
3	900 Hz + 1100 Hz
4	700 Hz + 1300 Hz
5	900 Hz + 1300 Hz
6	1100 Hz + 1300 Hz
7	700 Hz + 1500 Hz
8	900 Hz + 1500 Hz
9	1100 Hz + 1500 Hz
0	1300 Hz + 1500 Hz
KP	1100 Hz + 1700 Hz
ST	1500 Hz + 1700 Hz
IDLE	2600 Hz single tone

*If a trunk circuit was not in use, the trunk would place a 2600 cycle tone on the facility. A failure of the facility would cause a loss of tone and all circuits would go busy to prevent their use on calls.

When dialing signals were transmitted, the dialed digits were preceded by *KP* and followed by *ST*. For example, KP3125551212ST would be sent over the toll network to reach telephone number 555-1212 in the 312 area code.

This in-band signaling system was hit with heavy fraudulent use in the 1970s. Some people learned the signaling scheme for MF signaling and made equipment that could duplicate these tones for each digit. These devices were called *blue boxes*. A person using them would make a direct distance dialed (DDD) call to an 800 number. The originating toll office would register that an 800 number was being called and would not bill the call when it was finished. After calling the 800 number, a speaker on the blue box device was held close to the transmitter of the phone and a key was depressed to transmit a 2600-cycle tone to the speaker. The 2600-cycle tone was heard by equipment in the toll center, and the tone forced the toll circuit to go idle and drop the connection to the 800 number. The circuit between the originating phone and toll office was not disconnected, nor was the trunk between the class 5 office and the class 4 toll center. These facilities were using loop current signaling, not MF signaling, and the circuit to the toll center would stay up until the phone was hung up. The trunk between the class 4 toll center and the class 3 toll center used MF signaling and was forced idle by the 2600 tone. All connections from the class 4 toll center on out were disconnected.

After forcing the toll trunk idle, the caller would still be connected to the toll office and a toll trunk going into the class 3 toll office. The caller would now press the *KP* key. The tones representing *KP* were sent to the speaker of the blue box and were picked up by the transmitter of the phone. The *KP* tones went out over the voice path to the class 3 toll office. When the class 3 toll office saw *KP*, it assumed this was the start of a new call and attached a dial register. The caller now keyed in the desired area code and telephone number, followed by *ST*. The dial register accepted the dialed digits and the call to that number was completed. At the end of the call nothing was billed because the billing register in the DDD machine of the originating office still had the 800 number stored as the called number.

3.34 OUT-OF-BAND SIGNALING (CCIS)

This fraud involving in-band signaling is what led AT&T to develop out-of-band signaling called *common channel inter-office signaling (CCIS)*. CCIS separates signaling from the associated voice call and carries the signaling for a call over a different facility than used by the voice call. Signaling signals for many different voice calls are combined and sent over a common channel. CCIS has evolved into the signaling system 7 (SS7) network.

3.35 STANDARDS ORGANIZATIONS

Standards for the telecommunications industry are generally established by the American National Standards Institute (ANSI), the Electronic Industry Association (EIA), and the Consultative Committee on International Telephone and Telegraph (CCITT). CCITT is part of the International Telecommunications Union (ITU). The ITU is chartered by the United Nations. CCITT standards are now being changed to the name of the parent organization and are called *ITU standards.* ANSI established most formal standards used by the telecommunications industry in the United States. ANSI does not actually create and write standards; it coordinates and sanctions the activities and standards written by other organizations. The Exchange Carriers' Standards Association, which is recognized by ANSI, produces some of the U.S. standards for telecommunications. After the 1984 divestiture, the RBOCs formed the Bell Communications Research Company (Bellcore) to do joint research and development activities. AT&T Labs and Bellcore participate in ANSI and CCITT standards activity. There are many informal or de facto standards that were set by AT&T's Bell Labs.

There are no set standards for telephones, central office switches, PBX switches, and key telephone systems. But, telephones must be able to work on any switching system. This can only be accomplished if the telephone set manufacturer designs the telephone to meet established interface standards. Bell Labs published a document known as the *Local Switching Systems Generic Requirement (LSSGR)* outlining the basic interface design criteria for local (class 5) central offices. Knowing the interface requirements allows independent manufacturers to design to that interface. There is no standard established for the telephone, but it must be designed to work with the central office.

When party lines existed, many different types of ringers were available in phones, which caused problems when trying to buy a phone that would ring properly. The FCC established a ringer equivalency code for different types of ringers; this code is also called the *ringer equivalency number (REN)*. When buying a telephone, people first had to check with their LEC to find out what the ringer equivalency code was for their telephone number. When they purchased the telephone, they would look at the REN number on the bottom of the telephone for a match. The elimination of party lines has led to the elimination of the requirement for the central office to be able to send many different ringing signals. Telephones manufactured today do not use a mechanical bell for the ringer. They use an electronic circuit that detects the presence of any ringing signal. Quite appropriately, this circuit is called a *ring detect circuit*. When this circuit detects a ringing signal, it will send an audible warble tone out through a speaker. This ability to ring on any frequency of ringing signal has led to the elimination of the need for ringer equivalency codes.

The formulation of standards is necessary for equipment made by different manufacturers to work together. A standard such as EIA 232 allows a manufacturer to design equipment intended to work with a PC's serial port simply by knowing the EIA 232 standard. The PC manufacturers design their computers to send and receive data over the serial port in conformity with the EIA 232 standard. Open standards allow many manufacturers to make equipment. Some large companies try to establish their own proprietary standards, but this stifles growth in the industry. When ISDN was first introduced, several proprietary standards were in use. Northern Telecom employed a different standard than AT&T. An ISDN phone purchased for use on an AT&T switching system could not be used at a later date on a Northern Telecom switch. With the adoption of an international standard, the phones are now designed to one standard and will work on AT&T as well as on other manufacturers' central offices. Prior to deregulation, the LEC provided all station equipment, and it was the LEC's responsibility to ensure compatibility between devices. The LEC is no longer responsible for end-to-end service, and the interconnection of equipment from many manufacturers and carriers is only possible by having published standards.

IBM used proprietary standards on its first PCs and the systems did not sell well. Only after making proprietary standards public and adopting existing standards did the PC business soar. Now if you buy an IBM PC, you do not have to buy all your hardware and software from IBM. If you want to add a disk drive or new software program, you have many competing manufacturers to choose from. With published interface standards in the telecommunications industry, many manufacturers can supply telecommunications equipment. The equipment from different manufacturers can be used together because the interface between them meets the same standard. While there is no standard for a telephone, and it can be designed as a manufacturer sees fit, it must be designed to meet the interface standards for a local central office exchange or it will not work. When ISDN was first introduced, it was introduced with each manufacturer using proprietary standards. Therefore, equipment from different manufacturers would not work together. You could not use an ISDN station device manufactured for use on an AT&T switch on a Northern Telecom switch. CCITT published ISDN standards in 1984 and an update was issued in 1988. American companies were slow to adopt these standards, which delayed acceptance of ISDN.

Most of the standards developed for use in the PSTN voice network are V. (V dot) standards. Modem standards are V. such as V.34 for 28,800 bps, V.34bis for 33,600 bps, and v.90 for 56,000 bps. Most of the standards developed for use in the packet switched network are X. standards such as the X.25 standard used to interface a device to the PSN and the X.75 standard to interface switches in the network. Standards will be discussed in detail as we encounter devices having standards in later chapters.

3.36 SUMMARY

The implementation of central exchanges with switching capabilities was the single biggest factor in the rapid growth of telecommunications at the turn of the 20th century. Most manual switchboards were staffed by young boys when they were introduced. The service provided by young boys was less than professional and they were quickly replaced by female operators. "Central" was a word heard by many people for the first half of the 20th century. Manual switchboards were not replaced in many rural areas until after World War II. The introduction of a central power supply and line equipment was the first step toward automating the switching process. This automation of switching would lead to the loss of jobs for local exchange operators but would be a necessary step in the growth of telecommunications. If manual switching were still used in the local exchange, every person in the United States would have to work as an operator in order to handle the volume of local calls being placed today. The next major development in telecommunications was the ability to put many conversations over one facility. This multiplexing technology was first deployed in transmission facilities connecting central offices together. It was later deployed as part of the central office and PABX switching system in computer-controlled switching systems. Accompanying the development of multiplexing technologies was the development of media such as coaxial cable and cables containing glass fibers to handle high levels of multiplexing. This multiplexing technology has grown to a point that it allows 516,096 conversations to be placed over one pair of glass fibers. Multiplexing technology is now being extended from the central office to the residence.

The greatest advances in telecommunications over the past 25 years have occurred in the areas of switching systems, fiber optics, and multiplexing technologies. These technologies have existed for more than a half century, but significant changes and improvements have unfolded in the last 25 years due mainly to research work at AT&T Bell Labs. The development of transistors and very large-scale integrated circuits has made it possible to implement technologies for switching and multiplexing that were not feasible with vacuum tubes and mechanical switching. The SXS switch was a critical and necessary step toward the automation of switching but could not possibly serve the demands made on today's switching systems. Thanks to continuing research and development by the telecommunications industry, we have switching systems and transmission facilities that allow us to communicate with one another instantly. These developments led the way in the trend toward an information explosion. In fact, many things we take for granted today—such as PCs, real-time stock quotes, up-to-date news, the Internet, and so on—would be impossible without the technological innovations created by AT&T Bell Labs and the rest of the telecommunications industry.

CHAPTER REVIEW QUESTIONS

1. What are the three parts of a communications system?
2. What functions must a receiver perform?
3. What is the difference between half-duplex and full-duplex?
4. Why does the LEC use wire as a transmission medium?
5. Who invented the carbon transmitter?
6. What type of electricity is used to provide power for the transmitter of the telephone?
7. What type of electricity is used to provide power for ringing the telephone?
8. What is the central wiring center called?
9. When was the first central exchange installed?
10. How many wires connect the telephone to the central exchange?
11. What are the names given to the wires connected to the phone?
12. What is a line?
13. What is a trunk?
14. Why were exchange names needed?
15. In large cities with thousands of lines on a switchboard, how were calls placed to lines not appearing on the position of the originating line?
16. What were the advantages of a central exchange?

17. What standards organization provides most standards for telecommunications in North America?
18. What standards organization provides international standards?
19. What company is responsible for most de facto standards in telecommunications?
20. What factor is the major impediment to the deployment of new technologies?
21. When was the first automated switching system invented?
22. What are computer-controlled switching systems called?
23. What major factor delayed implementation of TDM technology?
24. What is plastic insulated cable called?
25. What have been the greatest developments in telecommunications over the past 25 years?
26. Who invented the step-by-step switching system? When was it invented?
27. What advantage does a common control system offer over SXS?
28. What is the PSTN?
29. Why was it necessary to modify the North American Numbering Plan to include new area codes?
30. When was the crossbar switching system introduced?
31. Can a SXS receive DTMF digits?
32. What is the name given to the device that receives dialed digits from a rotary phone or tones from a DTMF keypad?
33. What is the name given to switching systems using microprocessors for the control of switching functions?
34. When was the North American Numbering Plan changed to include area codes of the form *NXX*?
35. What are the primary common control components in a XBAR switch?
36. Which component controls all functions in the XBAR?
37. What is the primary purpose of a translator?
38. How many tones are used by a DTMF pad to transmit one digit?

PROJECTS AND CASE STUDIES:

Project #1

The deployment of SPC central offices has allowed LECs to drastically reduce operating expenses. Engineering and maintenance costs have gone down tremendously. Check the web site for your state public utility organization to see if rate reductions have ever been ordered in your state because the cost to supply services has gone down. Discuss some of the advantages that a computer controlled switching system has over the older electromechanical switching systems.

GLOSSARY

Alternating Current (AC)   A signal that is continually alternating the direction of current flow due to changes in the polarity of the voltage of the signal. In other words, it is a signal that starts at zero voltage, rises to a maximum positive potential, declines to zero, and continues the decline until it reaches a maximum negative potential, then returns to zero to complete one cycle of a signal. The number of cycles the signal completes in one second is the frequency of the signal. This is the type of electricity supplied by the local power company to our homes and businesses.

Central Office   The central wire center or central exchange. All telephones in a small geographic area are wired to a central exchange, which serves all telephones in that area.

Circuit Switching   The process of connecting one circuit to another.

Class 5 Exchange   Also called the *end office*. The lowest-level switch in the PSTN hierarchy; the local exchange.

Communication   A process that allows information to pass between a sender (transmitter) and a receiver over some medium.

Concentration   The use of fewer outlets than inlets.

Digital Access Cross-connect (DAC)   A device used to connect fiber circuits or cable circuits to each other. The cables or fibers connect to ports on the DAC. The ports contain a time slot interchange (TSI). The TSI at each port connects via TDM links to all other ports. The TSIs are programmed to connect any time slot (channel) of one port to any time slot of any port.

Direct Current (DC)   The electric current flow is at a constant rate and flows continuously in one direction. The direction of electric current flow (outside the battery) is from the negative terminal of the battery, through a device attached to the battery, to the positive terminal of the battery. The amount of electric current flowing is measured in amperes.

Exchange Code   The first three digits of a seven-digit telephone number.

Extended Area Service (EAS)   A class 4 switching system used in a metropolitan area to connect class 5 switching systems together on EAS calls. Extends the rate base of a local exchange so that calls to neighboring exchanges are not toll calls but local calls.

Frequency Division Multiplexing (FDM)   The process of converting each speech path to different frequency signals and then combining the different frequencies so they may be sent over one transmission medium.

Full-Duplex Transmission   If a transmission system allows signals to be transmitted in both directions at the same time, the system is called a *full-duplex transmission system.*

Half-Duplex Transmission   A type of transmission where transmitters on each end of a medium take turns sending over the same medium.

Interoffice Calls   Calls completed between telephones served by two different central exchanges.

Intraoffice Calls   Calls completed between telephones served by the same central exchange.

Line   The circuit or cable pair which connects a telephone to the local central office switching system.

Line Relay   An electromechanical device attached to a telephone line at the central exchange. When a subscriber took the handset of the telephone off hook, electric current flowing in the relay caused it to operate and signal the operator.

Local Exchange   The switching system that telephones are connected to.

Medium   The device used to transport information or a message between the transmitter and receiver.

Multiplexing   In telecommunications, a process combining many individual signals (voice or data) so they can be sent over one transmission medium.

Numbering Plan Area (NPA)   An area represented by an area code.

Receiver   The device responsible for decoding or converting received information into an intelligible message.

Register (Dial Register)   The device in a common control switching system that receives dialed digits.

Simplex Transmission   Transmission of signals in one direction only is called *simplex* communication.

Sine Wave   A graphic representation of an AC signal along a time line.

Tandem Exchange   Class 1, 2, 3, and 4 exchanges that are part of the PSTN.

Telecommunications   The communication of voice or data over long distances using the public switched telephone network or privately owned networks.

Time Division Multiplexing (TDM)   The process of converting each speech path into samples, then combining the different samples by transmitting each sample at a different time, so they may be sent over one transmission medium.

Time Division Switching   A form of switching (stored program control or SPC) where each conversation occupies its own distinct and separate time slot on a common wire path through the switching system.

Transmitter   The device responsible for sending information or a message in a form that the receiver and medium can handle. The device in a telephone that converts the air pressure of a voice signal into an electrical signal that represents the voice.

Customer-Provided Equipment: Telephones, Keysystems, PBXs, and Personal Computers

KEY TERMS

Address Bus
Alerters
Analog Signal
Applications Program
Bridged Ringer
Centrex
Computerized Branch Exchange (CBX)
Computer Telephony Integrated (CTI) Systems
Data Bus
Decimonic Ringing
Digital Signal
Digital Signal Processor (DSP)
Direct Inward Dial (DID) Trunk
Dual Simutaneous Voice and Data

Dual-Tone Multifrequency Dial (DTMF)
Feature Phone
Foreign Exchange Line
Grounded Ringer
Harmonic Ringing
Hookswitch
Hybrid Keysystem
Hybrid Network
Keysystem
Magneto
Network Interface Card (NIC)
Operating System
Private Automatic Branch Exchange (PABX)
Private Branch Exchange (PBX)
Proprietary Telephone
Receiver

Ring Detector
Ringer
Sidetone
Software Defined Network (SDN)
Station Equipment
Straight-Line Ringer
Superimposed Ringer
Telecommunications Application Program Interface (TAPI)
Transmitter
Tuned Ringer
2500 Telephone Set
Universal Asynchronous Receiver Transmitter (UART)
Varistor
Wide Area Toll Service (WATS)

OBJECTIVES

On completion of this chapter, you should be able to:

1 Describe how the telephone works.
2 Have a basic understanding of sidetone and why it is needed.
3 Describe the differences between 500, 2500, feature phone, and proprietary telephone sets.
4 Understand how an ISDN telephone works.
5 Understand how keysystems work.
6 Understand how features on a PBX are accessed.

7 Understand the various features available with a PBX.

8 Be able to recommend when to use a keysystem or PBX.

9 Describe the Telecommunications Application Program Interface (TAPI) used to allow a PC to serve as a telephone and answering machine.

10 Understand the differences between a private-line network and a software-defined network.

11 Discuss how the Dialed Number Identification System can be used.

12 Know the difference between vertical and horizontal cable.

13 Discuss the purposes of a hybrid network.

14 Explain the differences between a ringer and ring detector.

15 Know the differences between an IDF and an MDF.

In 1981, the Computer Inquiry II decision by the FCC mandated that all equipment on the customer's premises had to be owned by the customer or some interconnect company. The telephone companies had to establish a separate interconnect company if they wanted to sell station equipment to their customers. This equipment is called *customer-provided equipment* or *customer-premise equipment (CPE).* CPE includes the telephone, keysystem, PABX, computer, computer terminal, modem, and the customer service unit (CSU) used for connection to a data line. The 1996 Telecommunications Reform Act allows telephone companies to sell CPE directly to end users; they no longer have to use a subsidiary for this purpose.

The telephone set is the primary device employed in placing a telephone call over the PSTN. Keysystems are multiline telephones used to provide voice communication capabilities to small businesses, and PABXs are used by larger businesses, for their voice communication needs. The PC was originally utilized for word processing and spreadsheet applications, but its capabilities have continued to expand. Applications programs have been developed that allow the PC to replace the telephone, keysystem, and PABXs. It can be used as a data terminal and/or voice terminal. PCs are far more expensive than telephone sets and most voice calls will still be made using telephones, while almost all data communication will involve a computer. Most of the time the computer used for data communication will be a PC.

A dispute over the invention of the telephone set marks the beginning of telecommunications. As we noted earlier, several inventors were working on devices that could convert audio waves into electrical signals. Two of these inventors were Elisha Gray and Alexander Graham Bell. Both individuals filed for a patent on February 14, 1876. The patent application and disclosure forms they filed described a variable resistance transmitting device as part of the telephone. Gray's description was included in the body of the formal disclosure document. Bell's formal document did not originally contain the disclosure of a variable *transmitter;* remarks about a variable transmitter were handwritten in the margin of the document. This was a critical component of the invention. Did Bell learn of Gray's device and hurriedly add these comments? Evidently some of the Supreme Court justices ruling on the patent believed he did. The Supreme Court was split in its decision upholding Bell as the inventor of the telephone, but the majority gave him priority. When someone invents something, the person is always faced with other people or companies trying to steal the technology. It may well be that people were trying to steal the invention from Bell. On March 10, 1876, Bell and his assistant, Thomas Watson, successfully demonstrated a working model of the variable resistance transmitter.

Bell's original concept for the telephone did not use this type of transmitter but used a small electromagnetic receiver as both the transmitting and receiving device (a combined transmitter/receiver is referred to as a *transceiver*). You can test this device yourself. Instead of talking into the transmitter on your next telephone call, talk into the *receiver.* Cover the transmitter of the telephone with your hand and speak into the receiver. You can make yourself heard, but the other person will tell you the volume is very low. Using the receiver device as a transmitting device produced small levels of current and could not produce a loud signal in the far-end receiver. By using a variable resistance transmitter as proposed by Gray (and Bell's marginal notes), a much higher signal level was attained. The variable resistance transmitter consisted of an acid and water solution between two wires attached to the transmitter. When sound hit the diaphragm, it moved the wire attached to it. This movable wire would move up and down in the acid solution. The more immersed it became, the less resistance there was between the two wires.

Unlike the transceiver, which can generate its own signal and does not need external power, the variable resistance transmitter cannot generate a signal. It must have power supplied by a battery. The voice signal hitting the diaphragm will cause the resistance of the transmitter to vary. The resistance varies in unison with the speaker's voice, in turn causing current flow from the battery to vary. The varying electrical signal is similar to the varying sound signal. For this reason, it is called an *analog signal.* The electrical signal is analogous (similar) to the sound signal. All electrical signals with varying amplitudes are called analog signals in telecommunications. This varying signal is coupled by wires from the transmitter to a receiver. Varying current in the receiver will cause a varying sound output in unison with variances in the electrical current.

4.1 CARBON GRANULE TRANSMITTER

The liquid acid variable resistance transmitter was not practical for use by the general public. Bell continued to investigate different devices for use as a transmitter. In 1877 Thomas Edison invented a variable resistance transmitter that utilized carbon powder. In 1878, Henry Hummings, a British inventor, developed a transmitter using carbon granules. In 1886, Edison improved Hummings' design. When air waves from your voice hit the transmitter, they cause a thin metal diaphragm to move back and forth at the frequency of the voice signal. This movement puts more or less pressure on the carbon granules. Higher pressure packs the carbon granules tighter and closer together and lowers the resistance of the transmitter. Less air pressure leads to looser granules and a higher resistance. The carbon granule transmitter is the transmitter used in most telephones. The transmitter and receiver for early telephones were wired in series between the two wires of the local loop. With this arrangement, the signal of the transmitter was reduced by loss in the receiver before the signal reached the local loop. The telephone design was changed to improve the amount of transmitter signal reaching the local loop. A transformer with three windings was added to the phone. One winding connected to the two wires of the local loop, one winding connected to the transmitter, and one winding connected to the receiver. The design of the transformer couples voice signals between windings in such a manner that maximum signal transfer occurs between the transmitter winding and the winding attached to the local loop. The transmitter winding was also physically attached to the loop winding in order to pass the DC voltage from the loop to the transmitter.

4.2 SIDETONE

The transformer in a telephone couples a small portion of the transmitter signal to the receiver. This signal coupled to the receiver is called *sidetone.* Sidetone is necessary so that people talking into the transmitter hear what they are saying. Try talking and covering one ear with your hand. You will hear yourself only in the uncovered ear. Since you cannot hear with the covered ear you sound strange, as if you are in a well. Coupling sidetone to the receiver helps eliminate the problem. The receiver is coupled to the local loop winding so that maximum signal transfer occurs on signals coming from the central exchange to the phone. A device called a *balancing network* is also included between transformer windings. This device matches the impedance of the phone to the impedance of the cable pair to improve signal energy transfer. The combination of the transformer and balancing device is known as a *hybrid network.*

The word *hybrid* means a combination of things or technology. The hybrid network in the telephone combines a four-wire transmission system and a two-wire transmission system. A four-wire transmission system has one pair of wires for transmission of signals and a separate pair of wires for receiving signals. Each path is a simplex path. The transmitter of the telephone has one pair of wires connecting it to the hybrid network. The receiver of the telephone has a pair of wires connecting it to the hybrid network. Four-wire transmission systems provide better control of signals and signal quality is higher. The transmission path between the telephone and the central office is a two-wire circuit. This two-wire circuit connects to the hybrid network. The job of the hybrid network is to connect transmit signals from the transmitter to the cable pair (line circuit) and connect signals being received over the cable pair from the line circuit to the receiver.

We do not use four wires from the telephone to the central office because it would cost twice as much to connect telephones to the switching system. We use a hybrid circuit on each end of the cable pair. At the telephone we connect the four wires of the transmitter and receiver to the cable pair. At the central office, the line circuit also contains a hybrid network to interface signals from the cable pair to the codec (coder/decoder). Signals that were received from the transmitter of the telephone will be connected via the hybrid to a Coder. Signals from the Decoder will go through the hybrid, out on the cable pair, through the hybrid in the telephone, to the receiver.

Four-wire transmission systems are superior to two-wire transmission systems but are more costly. We do two-wire to four-wire conversion to improve transmission quality and we do four-wire to two-wire conversions to save money. The device that handles the conversion is a hybrid network.

4.3 INVENTION OF THE RINGER

While work was proceeding on improvements to the telephone transmitter, Bell was also looking for a way of alerting people that someone wanted to talk to them over the phone. His assistant, Thomas Watson, invented the device called a *ringer.* Each phone was equipped with a hand-cranked electric *magneto* and a ringer. When a call was placed, the originating party would generate a high-voltage AC signal by using the hand-crank magneto. This signal caused the ringer of all phones connected to the line to ring. Watson patented his ringer invention in 1878. Improvements have been made in the original ringer, but the ringer used in many of today's telephones is basically the same device Watson invented. If the ringer in your telephone is the old-style electromechanical ringer, it is Watson's ringer.

4.4 RING DETECTORS

The current trend in telephone design is not to use electromechanical ringers. The latest trend uses a solid-state transistorized device, called a *ring detector,* to detect ringing signals. When a ringing signal is detected, the ring detector sends a signal to activate a sound from a small speaker. The sound from the speaker is usually a loud shrilling tweeter. Some devices being marketed have microprocessors and ROM chips to include different messages that can be sent to the speaker. When the ring detector senses an incoming ringing voltage from the central exchange, it sends a signal to the microprocessor chip. The chip will select one of the messages stored in ROM and send it to the speaker. These messages can be almost anything you desire, from jingles to poems (for example, "pick up the phone, pick up the phone, you know you are home so pick up the phone"). The electromechanical ringer is being replaced by speaker-type audible alerters.

When switching systems had to support party-line service with as many as ten customers on one line, the switching system had five different frequencies of ringing signals. On these party lines one side of the ringer was wired to earth ground. Five customers had the other side of their ringer wired to the Tip side of the cable pair. The other five customers had the other side of their ringer wired to the Ring lead. Thus, by using five different ringing signals and placing the ring signal on either the Tip or Ring of the cable pair, we could ring ten different telephones and only the ringer desired would ring.

Because we used five different ring signals in the past and have a lot of telephones in service with ringers that are designed to ring on only one specific signal, many rural central offices still have multiple ring frequencies. Today we do not have ten party lines. Two party services are still available but very few people subscribe to it. With one or two party service, the ringer is not wired to ground. One side of the ringer connects via the hookswitch to the Tip and the other side of the ringer connects to the Ring lead of the cable pair. New telephones are designed for single party service and will ring on any ringing frequency sent by the local switching office. The mechanical ringers and ring detect circuits are designed to work on frequency of ring signal as long as it is greater than 60 volts AC. Most telephone companies will gradually phase out the different ringing signals and migrate to one ringing frequency of 20 cycles per second at 90 volts AC.

4.5 RINGING SYSTEMS

Different types of ringing signals are used in central exchanges. Some central exchanges use ringing frequencies of 20, 30, 40, 50, and 60 cps (referred to as *decimonic ringing*). Some exchanges use $16\,^2/_3$, 25, $33\,^2/_3$, 50, and $66\,^2/_3$ (usually referred to as *harmonic ringing*). These multiple ringing frequency schemes were utilized primarily by independent telephone companies serving rural areas. Using these ringing schemes in rural areas, it was possible to assign different ringing frequencies to each party on a party line. When a call arrived for a particular party on the line, the appropriate ringing frequency was sent over the line and only one phone would ring. Bell used mostly 20-cycle ringing exclusively because most lines in a Bell office were either one- or two-party lines. For party lines, Bell used a technique called *superimposed ringing.* The ringing signal would be imposed on top of either a negative 130-V or positive 130-V DC battery supply. The telephone had a diode wired in series with the ringer. If the proper 130-V polarity was present for that phone, only that phone would ring. This type of ringer was called a *superimposed ringer.*

The ringer in the telephone must be compatible with the ringing signals used by the central exchange. Each telephone ringer was assigned a code by the FCC. Before buying a telephone, it was necessary to check with the local telephone company to find out what code ringing signal was assigned to your telephone number. Different ringing frequencies were originally designed for party-line service. Each party on the line had a distinct *tuned ringer* that would only ring on one frequency of ringing signal. If each phone on a party line had a different ringer, it was possible to send out different ringing signals for each phone on the line. Only the called phone would ring on a party line using multiple ringing signals.

The elimination of party-line telephone service eliminated the need for multiple ringing frequencies. This also eliminated the need for different types of ringers in the phone. Telephones manufactured today are made with a ringer that will ring on any frequency of signal. This ringer is called a *straight-line ringer.* This eliminated the need to check ringer codes when buying a telephone. Most new telephones have done away with the mechanical ringer and use a ring detect circuit instead. A ring detect circuit is an integrated solid-state electronic device designed to detect the presence of a 90-V AC signal at any frequency. This circuit is often called an *alerter.* When the circuit detects a ringing signal it causes a beeper, buzzer, or tone to sound. The central exchange must retain multiple ringing signals in rural areas because of the need to remain compatible with all the old telephones still in service. As time passes most exchanges will phase out the old ringing signals. The exchange will use mostly 20-cycle ringing signals in the future.

4.6 STATION EQUIPMENT

The telephone set belongs to a broader classification known as station equipment. *Station equipment* is basically anything that connects to a local loop (telephone, answering machine, teletype, fax machine, modem, and PC). Telephones come in a variety of designs, colors, and styles. Many telephones are designed to work with the local telephone company's central office (some phones are designed to only work on private switching systems). Most telephones work on one cable pair or line from the central exchange and are referred to as single-line phones. The telephone receives its power from the central office 52-V DC battery. If the single-line phone has the older rotary dial, it is referred to as a *500 telephone set.* If the phone uses a touchtone keypad (dual-tone multifrequency or DTMF) for dialing, it is called a *2500 telephone set.* The current telephone set design has been arrived at after years of improvements. The primary functions of the telephone are to: (1) signal the local central exchange that a call is being originated or has been answered (switch-hook), (2) signal the customer of an incoming call (using a ringer or *alerter*), (3) provide a signal to the telephone exchange that will indicate who the caller wishes to contact (rotary or DTMF dial), (4) convert voice signals into electrical signals (transmitter), (5) convert electrical signals into sound signals (receiver), (6) provide a means of matching the impedance of the telephone to the local loop (hybrid network), and (7) provide just the right level of sidetone to the receiver regardless of loop length (hybrid network).

4.7 TELEPHONE STANDARDS: LSSGR

There are no official standards for the telephone. However, the phone must be able to work with the equipment in the central exchange. The central exchange and line interface requirements can be found in the *Local Switching Systems Generic Requirements (LSSGR)* by AT&T and Bell Communications Research. These recommendations serve as a de facto standard for the telecommunications industry. The line interface requirements specify how long the local loop can be, what type of signaling can be used, and what type of ringer is required. The length of the local loop determines how much current will exist in the local loop, when the phone is off hook. Telephones containing a carbon granule transmitter require a minimum of 23 milliamps of current for the transmitter to work properly. The phones containing new transmitters require 20 milliamps. The DTMF pad is also powered over the local loop. It requires a minimum of 20 milliamps to work properly. The ringer must be capable of operating with a 60V to 150V AC signal between 16 cps and 66 cps.

4.8 FEDERAL COMMUNICATIONS COMMISSION REGISTRATION

When a telephone is designed, it must be registered with the Federal Communications Commission (FCC). This requirement was brought about as part of telecommunications deregulation in the 1980s. Prior to deregulation, the local telephone company included one telephone as part of a subscriber's basic service and would rent additional phones for $1.75 a month. The telephone company owned all phones. Maintenance of phones was included as part of the basic subscription fee.

With deregulation, a person no longer has to rent phones from the local telephone company. You can buy your phone from anyone and are responsible for all maintenance. When telephones were provided by the local phone company, it made sure the quality of phones was high, and manufacturers were held to high standards of quality. To ensure manufacturers would continue to provide high-quality telephones, regulation was turned over to the FCC.

4.9 THE DIAL AND DUAL-TONE MULTIFREQUENCY PAD

When automated switching systems were invented, the dial was added to the telephone. The dial would interrupt the current flowing in the line, and the switching systems counted the interruptions of current to determine dialed digits. Telephones being manufactured today are made with a dial that use tones for dialing signals. This dial is called a *dual-tone multifrequency dial DTMF* or *dial* and was introduced by AT&T in 1964. The dialpad is a keypad containing 12 or 16 buttons arranged in four rows and three or four columns. A separate tone is assigned to each row and column. When any button is depressed it sends out a unique set of tones to the central exchange.

	1209 Hz	1336 Hz	1477 Hz	1633 Hz
697 Hz	1	2	3	A
770 Hz	4	5	6	B
852 Hz	7	8	9	C
941 Hz	*	0	#	D

The central exchange decodes the different combinations of tones received into the equivalent dialed digit. Because of the need to be compatible with all the old telephones with rotary dials, central exchanges will still work with these dials. There are some pushbutton telephones that do not send DTMF tones. They convert the button pushed to rotary dial pulses. Some phones allow you to select whether tones or rotary dial pulses will be sent by the keypad. Rotary-type dial pulses will not work with voice messaging systems that ask you to press a button on your phone. These messaging systems use DTMF tones to inform the system of what action it should take. DTMF tones can travel over the voice circuit to the called number. If you call your answering machine from another phone, it will respond to DTMF tones. The machine has a DTMF receiver that will decode the tones sent and follow the commands you give it using DTMF tones.

The major components of a telephone are: (1) the transmitter, (2) the receiver, (3) the hybrid network, (4) the dial or DTMF pad, (5) the ringer, and (6) the *hook-switch* (or *switch-hook*). A block diagram of the phone is shown in Figure 4-1. A circuit diagram for a rotary dial telephone (500 set) is included in Figures 4-2 and 4-3. A circuit diagram for a touchtone (2500) set is displayed in Figure 4-4.

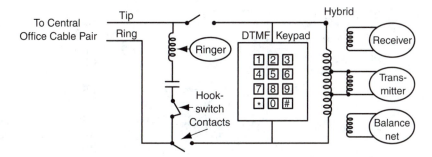

Hookswitch contacts shown when phone is on hook (hung up).

FIGURE 4-1 Diagram of a telephone set.

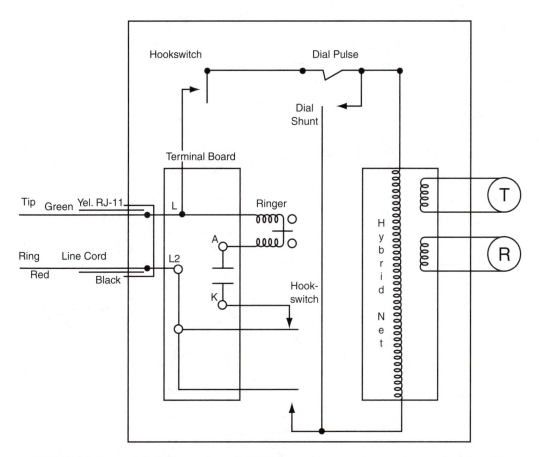

FIGURE 4-2 Rotary dial telephone schematic. Dial pulse springs open as many times as the value of the digits dialed, then remains closed. Dial shunt springs closed when dial is pulled to a finger stop and remains closed until the dial returns to normal. This prevents pulses from being heard in the receiver.

Customer-Provided Equipment

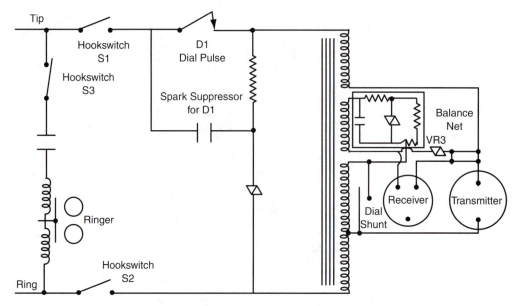

FIGURE 4-3 500 set (rotary dial) schematic.

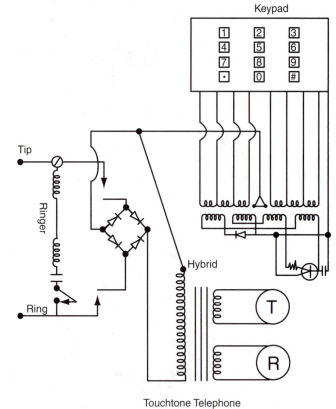

FIGURE 4-4 Touchtone set
(2500) schematic.

Touchtone Telephone

4.10 TELEPHONE CIRCUIT DESCRIPTION

The following information involves discussions about voltage and current and should be considered as advanced material. An overview of basic electricity principles is provided in Appendixes A, B, and C for readers who wish to understand the relationship of voltage and current in electrical circuits. If your particular course is not designed to delve into electrical concepts, you may choose to skip Section 4.10.

The hookswitch contacts S1 and S2 are in series with the Tip and Ring of the connecting line from the central exchange. Contacts on S1 and S2 are wired so that when the phone is hung up, the ringer is the only device attached to the line. The ringer can be bridged across the line by connecting the yellow wire to the green wire. When a ringer is connected between the Tip and Ring wires, it is referred to as a *bridged ringer.* When a telephone was used on an eight-party line, one side of the ringer was connected to ground. On an eight-party line, the yellow lead from the ringer was connected to a ground wire. When a ringer was connected between one of the wires of the local loop and ground, it was referred to as a *grounded ringer.* Later designs of telephones include a hookswitch contact in series with the ringer. When the handset of the telephone is picked up (the phone goes off hook), these hookswitch contacts open the ringer from the line and attach the transmitter, receiver, dial, and hybrid network to the line. With this arrangement the ringer will not be on the line during transmission and reception of voice signals, and will not reduce those signal levels.

A *varistor* (V22) is placed across the Tip and Ring leads to limit the amount of voltage applied to the transmitter. A shorter local loop will have less voltage drop in the loop and supply a higher voltage to the telephone than a longer loop with its higher loop loss. The varistor decreases the voltage supplied on a short local loop so that it will not be significantly higher than the voltage a phone on a long loop receives. A telephone attached to the longest local loop will receive about 9 V DC after loop loss. If a varistor were not used in the telephone, on a short loop it would receive about 20 V. This high voltage would cause signal currents of more than 60 milliamps in the local loop and would produce crosstalk in the cable. The varistor will limit the voltage across the phone to approximately 14 V and will limit current through the transmitter to 60 milliamps of current on the shortest loops. Thus, current levels for all loops will range from the minimum requirement of 20 milliamps to a maximum of 60 milliamps. These values are based on newer telephones with an internal resistance of 250Ω. The current flow will be below the maximum recommended value of 60 milliamps. A second varistor (VR3) is placed across the receiver to limit the voltage of the signal across the receiver so that the sound generated by the receiver is not loud enough to damage an eardrum.

The rotary dial contains two separate set of contacts. One set of contacts (D1) are normally closed and will open when a number is dialed. These contacts open and close once for each digit dialed. If a 5 is dialed, they open and close five times and remain closed at the end of the dialed number. Another set of contacts (D2) are called *shunt contacts* and are normally open. When the rotary dial is pulled to the finger stop, they operate to place a short (or shunt) across the transmitter and receiver. The shunt contacts keep a short across transmitter and receiver until the dial returns to normal. The dial pulses generated by D1 will not pass through the receiver but will bypass it. The resistor and capacitor (R & C) across the dial contacts (D1) act as a spark suppressor to prevent damage to these contacts. To dial a number, we put our finger in the hole of the number desired and pull the dial around until the finger hits the finger stop. When we pull our finger out of the dial, a spring returns it to normal. The spring is adjusted to return the dial to normal in one second when a 0 is dialed. The dial contacts (D1) will be opened from one to ten times (depending on the number dialed) as the dial returns to normal. The dial will open the contacts, allow them to close briefly, open them again, allow them to close briefly, open them again, and so on. Each open and closure is called a *pulse.*

Dialing a zero results in ten pulses in one second, and we say the speed of the dial is ten pulses per second (pps). At 10 pps, each pulse is one-tenth of a second (100 milliseconds or ms). Each time the contacts open they will be opened for 60 ms and closed for 40 ms. The open occurs when the contacts break open and is called a *60% break.* The open time is 60% of the pulse time. The dial contacts are adjusted to provide a 60-ms open interval for each pulse. With a 100-ms pulse, the contacts will close briefly (40 ms) between each open. Thus, the dial is adjusted to provide a 60/40 pulse. There is a small distance between the finger stop and the one hole of a dial. This design ensures that a small amount time is provided between digits. During this time the contacts remained closed. This time frame is designed to be more than 500 ms. When the equipment in the central exchange is receiving dial pulses, it recognizes each open interval (60 ms) as a pulse belonging to one digit and will count these pulses to determine the digit dialed. The switching equipment uses the length of the closed interval to determine when the end of a digit occurs. With a closure of 40 ms, the switch continues counting each open condition. When the closure is 500 ms, the switch recognizes this as a signal that the previously counted pulses should be stored as a digit and that the next open will be the start of a new digit. This 500-ms time frame between digits is referred to as the *interdigital time frame* (see Figure 4-5).

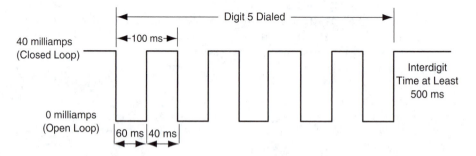

FIGURE 4-5 Dial pulse string. The loop current is 20 milliamps or greater when a phone is off hook. When a number is dialed, the dial opens the loop. The diagram shows how the current changes when the number 5 is dialed.

4.11 ELECTRONIC FEATURE PHONES AND ANSWERING MACHINES

When additional features have been added to the basic single-line telephone such as last number redial and speed dialing lists, the phone is called a *feature phone.* The feature phone has a microprocessor and memory contained in the telephone but is still powered by the standard 52-V DC voltage from the central exchange. This electronic phone can be designed to include special features such as an answering machine. The answering machine uses an electronic solid-state ring detector, which signals the microprocessor on each ring received. The microprocessor will count the number of rings received and can be programmed to answer on two or four rings, or not answer at all. The answering machine also includes a DTMF receiver chip. The owner can call into the answering machine from a remote location and use the DTMF dial, at the remote location, to provide directions to the answering machine. The directions are usually given by sending two digits as a code. The DTMF receiver in the answering machine receives the tones sent by the DTMF dial and is programmed to provide a specific function depending on the code received.

When the answering machine answers a call, it is programmed to look at the first code as a password. If it receives the correct password, it will honor any additional functions requested. These functions can include any of the following: turn on the machine, turn off the machine, play back all messages, play back only new messages, record a new announcement, fast forward, fast reverse, rewind, change the password, and so on. The use of DTMF tones to signal the answering machine allows a person to perform any administrative function on the machine from any remote location. The machine has programming that will automatically answer an incoming call, wait for and recognize a password code, return a recorded announcement, respond to DTMF signals and interpret the received code, and perform the function assigned to that code.

Software is available for your PC to allow it to function as an answering machine when it is equipped with a special modem and a multimedia sound card. The modem can recognize the start of an incoming call from the initial signal received when it answers a call. The initial signal will indicate whether the incoming call is a voice call, an e-mail, or a fax, and the modem will answer the call accordingly. When it detects that the incoming call is a voice call, it sends an answering message to the caller. This message is recorded using the PC's multimedia card and its attached transmitter (microphone). The message is converted by the card into digital format and stored in a file on the hard drive. The system answers a voice call by calling up this file and playing it back through the multimedia card and modem to the phone line. The message usually states that if you wish to leave a message for person A, press 1 on your phone, press 2 to leave a message for B, and so forth. The modem and software detect which digit was received and will store the message received in a file under that person's mailbox (directory). The multimedia card converts the received message into digital file for storage. Many businesses are beginning to use a PC with multimedia, a special modem, and appropriate software to answer calls after hours or to serve as a voice mail system.

4.12 PBX PROPRIETARY TELEPHONES

PBX switching systems can be equipped with line circuits to interface to a regular 500/2500 telephone set, and they can be equipped with special line circuits to interface to a proprietary telephone set. The *proprietary telephone* set is a specially designed electronic feature phone. These phones are proprietary telephones because they contain proprietary software developed by the switching manufacturer. The CPU in the telephone set is programmed to talk to the CPU of the switching system. Because they contain special programming to talk with a particular switching system, the proprietary telephone for one manufacturer will not work on another manufacturer's switch. These telephones work with a special line interface circuit and offer more features than a 2500 set. Since they are designed to provide access to several lines and features, they are often called *multiline telephones.* These telephones require four station wires between the line circuit and the telephone. Two wires are used for the Tip and Ring leads to carry voice as well as DTMF signals. The other two wires serve as signal leads carrying *digital signals.* One signal lead carries signals from the PBX line circuit to the telephone, and the other wire carries signals from the telephone set to the PBX line circuit. These signaling leads provide a duplex signaling path, allowing the CPU of the PBX and the CPU in the telephone to talk to each other. Using this out-of-band signaling path, digital messages can be sent between CPUs to provide instructions from one CPU to the other. If a button on the telephone set is pushed, the CPU in the telephone sends a message to the CPU in the PBX to inform it of the button pushed. The CPU in the PBX looks in the line translation's database for the line circuit associated with this phone to determine what line or feature has been assigned to that button. It will then activate that line, function, or feature.

A PBX has many features available. An electronic, multibutton, proprietary telephone allows those features to be accessed by merely pushing the button assigned to a feature. The electronic telephone enhances the operation of a PBX because it has features that complement those in the PBX. The programs running the CPU in the switch and the CPU in the telephone are proprietary. This is the major reason why the electronic telephone of one vendor will not work on another vendor's switch. When the user of an electronic telephone wishes to access a station line, the person merely pushes a button associated with a line. The CPU of the PBX receives a message from the telephone over the signaling lead. The CPU is informed by the line table that this button is assigned as a particular station number and issues instructions to the switching matrix of the PBX. Acting on these instructions, the switching matrix connects dial tone to the line circuit's voice path and attaches a DTMF register to receive dialed digits from the phone. The tones of the DTMF pad on the phone are transmitted over the voice path to the register in the switch. The switch processes the call based on the digits received.

4.13 HANDS-FREE PHONES VS. SPEAKERPHONES

Most electronic telephones have a feature called *hands-free.* These telephones have a small speaker that serves as a receiver if the phone is on hook. After pushing a button to access a line, dial tone is heard in the speaker. If the phone is taken off hook, the receiver of the handset will replace the speaker. Thus you can listen to the progress of a call being established through either the handset or the hands-free speaker. The speaker will allow you to hear the called phone ringing. You will also hear the voice of the called party when he or she answers, but that person cannot hear you through the speaker. You must speak into the transmitter of the handset. A hands-free telephone allows you to place a call without picking up the handset, but once the call is connected, you must talk into the transmitter of the handset.

Some people erroneously think that the speaker on a hands-free telephone will also transmit your voice, but it will not. It is hands-free in a receive mode only. The telephone that will transmit and receive in a hands-free mode is called a *speakerphone*. The speakerphone has a speaker that acts as a receiver and also has a small transmitting device usually placed on the front edge of the telephone. You can recognize the location of this transmitter. It will be covered by the case of the telephone. The case has two to three small slits in the plastic housing to allow voice signals to reach the transmitter. The hands-free telephone allows you to listen without lifting the handset, whereas the speakerphone allows you to both talk and listen without lifting it. Some speakerphones have electronic circuitry to detect when the user is talking. If the person is not talking, the transmitter output is very low. When the transmitter detects a voice, it reduces the output of the speaker. The operation of this electronic circuit prevents feedback to the speaker and eliminates squealing from it.

4.14 INTEGRATED TELEPHONE AND DATA TERMINAL

Some electronic feature telephones include a *visual display unit (VDU)* and a keyboard. The phone is usually equipped with two telephone lines and numbers. The VDU provides a visual display of speed calling lists and system options. The user can use arrow keys on the keyboard to select an option and then press the Enter key to activate that option. The keyboard and VDU enable the phone to serve as a dumb terminal. The phone has an EIA-232 connection for connection to a modem. The user can use one of the two telephone numbers for placing voice calls. You would use the modem (which has a connection to its own telephone number) for placing calls to a computer. The dumb terminal and modem allow using this device for e-mail.

4.15 INTEGRATED SERVICES DIGITAL NETWORK TELEPHONES

The telephones discussed above are designed to convert voice signals into analog electrical signals. ISDN telephones convert voice signals into 64,000-bps digital signals. An ISDN phone contains a codec to perform this conversion; it connects to the central exchange using a *digital subscriber line (DSL)*. The DSL operates at 160,000 bps and provides the user access to 144,000 bps. An ISDN telephone separates the 144,000 bps into three logical channels. Two of these channels are called *Bearer (B)* channels and one is called the *Delta (D)* channel. The ISDN telephone uses one of the B-channels for the telephone's voice circuitry. The other B-channel can be used to service another device or can be used to serve a second telephone number. The D-channel carries signaling between the telephone and the central exchange. It can also carry packet data. ISDN telephones also have an EIA-232 port, which allows connecting the serial port of a PC to the ISDN phone. The PC can use the second B-channel to send data over the switched network. This allows for the transmission of data without the use of a modem if the receiving end is also using ISDN lines. (ISDN telephones are discussed in more detail in Chapter 9.)

4.16 KEYSYSTEMS

When a business needs more than one telephone number, either a multiline phone or keysystem can be used. Multiline telephones can handle three lines; these telephones have four buttons (one button for each line and a hold button). If a business needs three or more lines, a keysystem is preferred over a multiline phone. *Keysystems* use special electronic telephones that can access several telephone lines and are also referred to as *multiline phones.* Many keysystems have names indicating the number of central exchange lines and telephones the system can handle. A 308 keysystem can have three local telephone numbers and eight multiline phones attached to the system. A 612 keysystem can handle six exchange lines and twelve multiline telephones. Because of the extensive use of electronic components and microprocessors, today's keysystems are called *electronic key telephone systems (EKTS).* The part of the keysystem that controls the system is called the *keysystem unit* or *KSU.* Telephone lines from the central office are connected to the KSU. All keysystem multiline telephone sets are also connected to the KSU. The KSU contains a space division switching matrix under the control of a microprocessor. The size of the switching matrix determines how many exchange lines and telephones the system can handle. The switching matrix in the KSU connects telephone lines from the central office to the multiline telephone that answers the call.

Telephones for a keysystem are special purpose phones, and each keysystem manufacturer makes phones that will only work with its keysystems. A single-line telephone (2500 set) will not work on a keysystem because the keysystem uses proprietary signaling protocols. This is the same reason multiline phones from different manufacturers will not work on keysystems made by another company. Each keysystem manufacturer uses a different signaling protocol. Keysystems have just begun appearing from a few manufacturers that do allow the use of a regular 2500 set instead of a proprietary telephone.

A 2500-type telephone has only one pair of wires connecting it to the central office. Everything is done over this pair of wires. This one pair of wires carries the off-hook signal, dial signals, ringing signals, on-hook signals, and the voices of a conversation. A keysystem uses four wires to connect a telephone to the KSU. Two wires are utilized to carry voice and DTMF dial signals. Two additional wires are used to carry signals between the telephone and a microprocessor located in the KSU. The phone will have one button for each exchange line coming to the keysystem. For a 308 system, the multiline phone will have three buttons for outside lines. It will also have a hold button and an intercom button. When the multiline phone is taken off hook, nothing will happen unless a line button on the phone is depressed. When the phone is off hook and a line button depressed, a microprocessor in the phone will send a signal over the signal pair of wires to the microprocessor in the KSU. This signal will cause a switching matrix in the KSU to connect the telephone to the desired exchange line. On outgoing calls, a dial tone will be received from the exchange, and the station

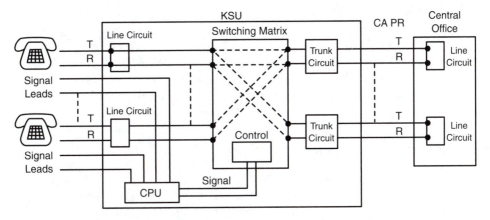

FIGURE 4-6 Keysystem signaling. The transmitter and receiver of any telephone can be connected to the transmitter and receiver of any trunk circuit by the switching matrix. The user presses a button on the keysystem telephone to select the trunk desired.

can send DTMF tones over the voice pair of wires using the DTMF pad on the multi-line phone. (see Figure 4-6).

4.17 KEYSYSTEM AUTOMATED ATTENDANT

In any keysystem, one of the stations will be designated as the attendant position or station. The system is configured to ring this station on all incoming calls. The attendant can answer incoming calls, put them on hold, and then use the intercom to signal a station to pick up the appropriate incoming line. Many keysystems are capable of automatically answering incoming calls and requesting callers to identify which department they want. An auto parts store using this system will prompt callers to press 1 for service, 2 for parts, and 3 for tires, and instruct callers to stay on the line if they are not calling from a touchtone phone. The microprocessor will detect the DTMF tones received and ring the appropriate station. If no tones are received within a few seconds, the call will ring in on the attendant's phone. Each station is assigned an intercom number. To call another station, the user will press the intercom button and then press the digit on the DTMF pad for the station desired. A DTMF receiver in the intercom circuitry will detect the station number dialed, and ask the microprocessor of the KSU to connect the two stations. The CPU of the KSU operates the switching matrix in the KSU and rings the called station.

4.18 KEYSYSTEM CABLING

Keysystems come in a variety of sizes and with a wide assortment of features. Early electromechanical keysystems such as Western Electric's 1A2 keysystems required a 25-pair cable between each telephone station and the KSU. The *electronic key telephone systems (EKTS)* manufactured today only need two pairs of wires between the station and KSU. The reduction in cabling has been a boon to keysystem use. Keysystems are found in almost all small businesses that require more than two exchange lines. Auto parts stores, small department stores, home improvement stores, as well as the offices of doctors, lawyers, and dentists are prime candidates for a keysystem. Many businesses only need a small EKTS, but it is possible to purchase an EKTS that will handle as many as 24 central office trunks with 100 stations. However, it is usually advisable to use the more efficient (though much more costly) PBX when a business needs this many central office lines and/or stations.

4.19 LARGE KEYSYSTEMS (HYBRIDS)

Today's EKTS offers many PBX-type features. Large keysystems are usually called *hybrid keysystems.* They are considered a hybrid of a keysystem and a PBX. With a large keysystem, it is impossible to put keys for all central office lines on the phones, so central office lines are accessed by dialing 9 (the type of access used by a PBX). A regular small keysystem offers access to central office lines by pushing a button on the keysystem phone associated with the line desired. The CPU of the EKTS is controlled by software burned into a read-only memory (ROM) chip. A small portion of the CPU is controlled by an electrically erasable programmable read-only (EEPROM) chip. This chip contains features that can be programmed by the customer from the attendant's telephone. The customer can program which central office lines can be accessed by each phone and which phones will ring on incoming calls. Each line on each phone can be programmed individually to ring or not ring. All programming is changed in the EEPROM chip controlling the CPU of the KSU. Each station also has a microprocessor (CPU) for controlling signaling to the KSU. This microprocessor is also controlled by ROM and EEPROM. The EEPROM is used to store speed-calling entries that can be programmed by the station user. A station user also has access to system speed calling lists stored in the KSU's EEPROM by the attendant station.

Early keysystems required all calls to be answered by an attendant. After answering a call and determining who the call was for, the attendant would put the caller on hold and ring the intercom station of the person needed. Once the station user answered the intercom, the attendant would advise them which incoming line they should answer. Today's keysystems have automated this attendant function and allow callers to direct their own calls. The KSU answers the call with a recorded announcement instructing the caller to depress a DTMF key on the phone for the department required (for example, "Thank you for calling Cole Auto, please press 1 for tires, 2 for parts, 3 for service, and so on").

4.20 FEATURES OF KEYSYSTEMS

The features available in keysystems depends on how old the technology of the particular system is. Some features are provided by the KSU; others are provided by the station instrument itself. Current technology offers:

1. The ability to pick up or access any central office line.
2. The ability to put any line on hold.

3. Automatic recall of a line that has been on hold for a while. The time a line can be on hold before automatic recall is activated is programmable by the attendant.
4. Last number redial (this is a feature provided by the telephone).
5. Speed calling (some by the telephone and some by the KSU).
6. Hands-free answer (by the telephone).
7. Intercom (some systems provide several conference lines).
8. Music on hold.
9. Power fail transfer (the KSU transfers line 1 to a 2500 phone when a power failure occurs).
10. Privacy. Unlike the old 1A2 keysystem, today's keysystems will not allow multiple phones to pick up on the same line unless the first station to use the line authorizes it via the establishment of a conference call.
11. Programmable features and restrictions such as the ability to restrict which stations can use outside lines and which stations can place toll calls.
12. In some cases, station instruments that have displays for time, date, number dialed on current call, amount of time on current call, station number of caller, and speed call list. Many of these features are provided by the phone itself, but some—such as calling station—require input from the KSU.
13. Distinctive ringing, which provides a different ringing signal for incoming lines and intercom calls.
14. Supervisory signals to lights on each phone for each line and intercom button. The lights are usually off when the line is not in use but are lit steadily when in use. If a line is on hold, the light will blink. The lights are usually inside the button used to access a line or intercom.
15. Paging (offered by most keysystems). A person can initiate a page from any station to any individual station or have the page broadcast to all stations.

All current keysystem telephones have a hands-free speaker. Today's telephones no longer have a ringer, and the ringing signal is converted to an audible twitting sound emanating from a small speaker. The EKTS takes this speaker technology one step further. The speaker allows a caller to place a hands-free call and hear the call progress in the speaker. The speaker will not pick up the station user's voice. Once the called party answers a call from the keysystem user, it is necessary for the handset to be used by the keysystem caller. It is possible to buy a true hands-free speakerphone that will pick up the station user's voice and allow conversation both ways over the speaker. These "speakerphones" cost more than a "hands-free phone" and are usually only provided to senior management stations.

4.21 WHEN SHOULD YOU USE A KEYSYSTEM?

A keysystem provides a business with access to multiple central office trunk lines and many features typical of a *private branch exchange (PBX)* at a small portion of what a PBX will cost. A keysystem is the best solution to the telecommunications needs of a small- to medium-sized business. Keysystems are not designed to handle a large volume of calls at the same time; PBX or hybrid keysystem is required for large calling volumes. When a business has more than ten incoming trunks from the central office, a hybrid is preferred over a regular keysystem because a regular keysystem requires one button on the phone for each trunk. When a business has more than 24 trunks or more than 100 stations, a PBX is preferred. Regular keysystems are the perfect station equipment for small retail stores, restaurants, and doctors' and dentists' offices.

4.22 CBXs

Large companies have used private switchboards since the turn of the 20th century. Originally these private switchboards were manual switchboards, and an operator established connections between the calling and called parties. When automated switching systems replaced the manual central office exchanges, smaller versions were developed to supplant private switchboards. These systems were called *private automatic branch exchanges (PABXs).* When stored program control (SPC) switches were created to replace central office switching, smaller versions were developed to replace the PABX. These smaller systems are called *computerized branch exchanges (CBXs).*

The CBX can provide many features because the features can be implemented by software. The CBX basically consist of line circuits, trunk circuits, a switching matrix, and common equipment. The line circuits are used to connect telephones to the CBX switch, the trunk circuits connect trunks from other switching systems (such as the local central office switching system) to the CBX switch, and the switching matrix connects these circuits to one another under directions from the common equipment. The common equipment consists of a CPU, memory, switching matrix, dial registers, and secondary storage devices such as a tape drive or disk drive. The predominant trunk circuits in a CBX are the trunk circuits that connect the CBX station user to the local central office on dialing 9. These trunks are called *central office* or *C.O. trunks* (see Figure 4-7). If the CBX trunk circuit is not a *direct inward dial (DID) trunk* circuit, it will connect to a line circuit at the class 5 central exchange.

A DID trunk circuit allows anyone in the PSTN to call a CBX station directly by dialing the seven-digit PSTN number assigned to the CBX station. The first three digits will connect the caller to the class 5 central exchange that hosts the CBX. The fourth digit dialed will cause the Class 5 exchange switch to connect the caller to a DID trunk circuit that connects to the CBX. On seizure of the DID trunk, the CBX connects a dial register to it and will collect the last three digits dialed by the caller. These digits will be used by the CBX to determine which station the DID trunk should be connected to. If a business has more than 1000 stations, it will be assigned its own exchange code. Callers will then be connected to the CBX after dialing three digits, and the CBX will collect the last four dialed digits to determine which station the caller wants to be connected to (see Figure 4-8).

The CBX performs call switching like a central office switch. Every telephone must connect to a line circuit in a switching system. Regular telephones connect to

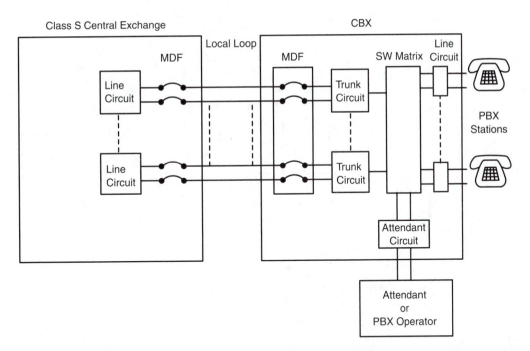

FIGURE 4-7 CBX connections to stations and central exchange.

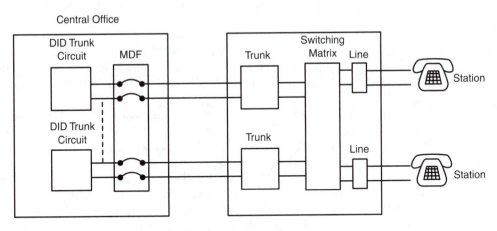

FIGURE 4-8 Direct inward dial trunks.

line circuits in the local class 5 central exchanges. CBX stations connect to line circuits in the CBX. When a line goes off hook to make a call, the line circuit senses loop current being supplied to the telephone and sends an off-hook signal to the CPU in the form of an interrupt request. The CPU sees the request and will jump to the program address that starts serving a line request for service. The execution of

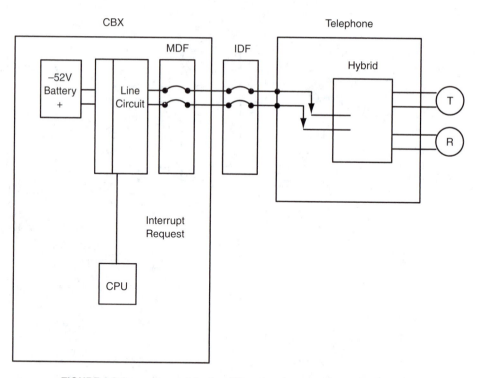

FIGURE 4-9 Loop current detection. When the phone is off hook, electric current flows and interrupt is issued by the current detector of the line circuit to the CPU.

the program results in an idle dial register being found and assigned to collect dialed digits from the telephone. The CPU will issue instructions to the switching matrix to connect the line requesting service to the dial register assigned to handle the call (see Figure 4-9).

After a dial register collects a digit, it issues an interrupt request to the CPU. The CPU will notice that the request for service is from a dial register and jumps to the program address that starts serving a register request for service. The register will be asked for the value of the digit dialed. If 0 was dialed, the switching matrix will be instructed to connect the caller to the switchboard attendant. If 9 was dialed, the caller will be connected to a central office trunk circuit and can dial a local number. If the caller has dialed 2, 3, 4, 5, 6, 7, or 8 as the first digit, the program will be written to handle the call as desired (see Figure 4-10).

Sometimes an 8 as the first digit is used to connect a caller to special wide area toll service (WATS) trunks for long distance calls. Some companies will also have dedicated trunks called *tie lines* or ***foreign exchange lines*** connected to the PBX. Dialing one or two digits may access these special trunks. Normally, a 2, 3, 4, or 5 as the first digit is used to access telephone stations connected to the CBX. With four-digit station numbering, this provides for 4000 stations. If three-digit station numbering is used in the scenario above, the number of stations that could be connected to the CBX would be 400.

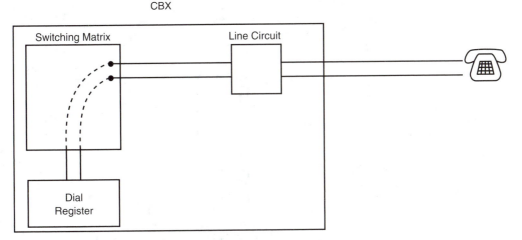

CBX

FIGURE 4-10 Connection of station to dial register.

CBXs are designed to permit the use of regular 2500 single-line telephones. Manufacturers of CBXs also allow for the connection of the manufacturer's proprietary telephone sets. The proprietary sets are usually multiline sets, as opposed to the single-line 2500 set. Single-line telephones access features by code dialing. After getting a dial tone, if the user flashes the hookswitch, this indicates to the dial register and CPU that the next digits dialed will be a code to access a particular feature. Each feature is assigned a two- or three-digit code. Proprietary phones are usually multibutton phones that allow you to have more than one PBX telephone number assigned to your phone. Some of the buttons will be assigned a particular feature; you can access the feature by pressing the button.

The beauty of a CBX is the flexibility it offers. You can establish any numbering plan that suits your needs and can create any access schema you wish in order to access special services and trunks. The CBX allows you to establish a private network to connect to other company locations and can provide numerous features, outlined below. A CBX will also save the company money. Instead of installing several hundred business lines from the local central exchange to each of the telephones in a large business, a CBX will require between 5 and 10 central office trunks per 100 stations to connect the local central office to the CBX. Not only will this significant reduction in central office trunks/lines save a lot of money, it will also allow stations to call one another using a three- or four-digit station number.

CBXs are evaluated for a particular application depending on the maximum number of lines and trunks they can provide. This is usually stated in terms of the number of different devices that a CBX can support—known as *port capacity*. CBXs are also evaluated in terms of their ability to grow (*expansion capacity*), the maximum number of telephone calls they can handle in one hour (*calling capacity*), and the maximum number of telephone calls they can handle simultaneously.

The functions offered by a standalone PABX or CBX can be provided by a service sold by the LEC that is called *Centrex*. The LEC can equip its local central exchange

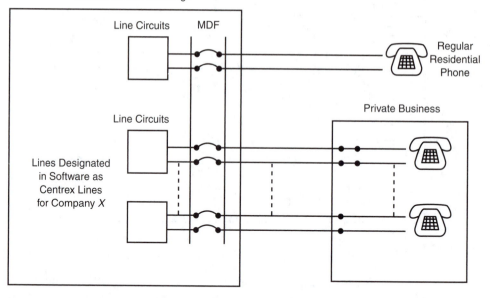

FIGURE 4-11 Centrex.

switching system with software and hardware that allows the local switch to provide PBX services. Certain line circuits in the class 5 switching system are defined in software as line circuits that are part of Centrex service to a particular Centrex customer. The 1981 Computer Inquiry II ruling dealt a near-fatal blow to Centrex. There was no way for the LEC to separate the Centrex costs from the local switching costs, and this severely limited the offering of Centrex. The 1996 Telecommunications Reform Act removes the last barrier to Centrex (see Figure 4-11).

Many businesses opt to use Centrex because they can get all the services offered by a CBX without having to buy one. If they do not own a CBX, they do not need maintenance or administrative personnel for it. This can help reduce operating expenses. The downside to using Centrex is a higher monthly cost for service. Every telephone at the business location must be connected to cable pairs to the class 5 central exchange. This results in a higher cost than using a few trunk lines to connect a class 5 exchange to your private CBX. Since the LEC is providing the switching system for Centrex services, you also pay for use of the switching system.

You can buy Centrex services from the LEC and let it do all maintenance and administration for stations connected to Centrex or you can buy administrative service capability, which allows the people in the telecommunications department of your company to administer moves, adds, and changes to station sets. The price of CBX technology has decreased to the point that a company may be better off buying a CBX rather than leasing Centrex service. A telecommunications manager must evaluate each situation and application when it occurs. Technology continues to change, and this demands that we continually review every option on an annual basis.

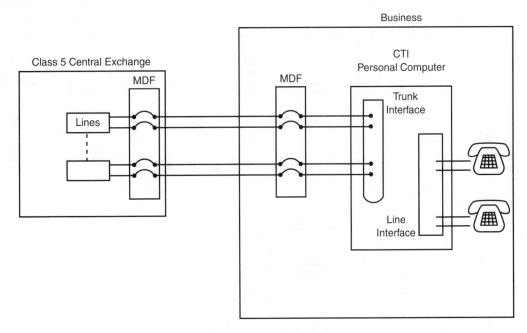

FIGURE 4-12 CTI-based PBX.

Computer-based PBX systems continue to get better and offer more options. These *computer telephony integrated (CTI) systems* come in many sizes and price ranges. The digital time division switching methodology used in central office switches, CBXs, and CTI-based PBXs is available on semiconductor chips that keep getting smaller and cheaper. This has allowed for the development of printed circuit cards containing tremendous numbers of voice/data channels on TDM media. Some CTI-based systems use cards designed to plug into expansion card slots inside the PC. Other CTI-based systems use the PC to control cards placed inside an expansion unit separate from the PC and connected to it via a serial or parallel port. We are also starting to see the PBX services provided via CTI-based systems in a communication server that sits on a local area network (LAN). Voice communications are then carried over the LAN from the communication server to the workstation (see Figure 4-12).

The wiring plan in a CBX environment consists of wiring all telephone jacks (usually an RJ-11 jack), in a certain area of the building (such as the 3rd floor north wing of a building), to a connecting block (a type 66 block or a BIX block), mounted in a janitor's closet in the same area. The 66 or BIX block is then referred to as an *intermediate distributing frame (IDF)*. The station cable that is used to connect the RJ-11 jack to the IDF is regular four-conductor station wire. This wiring between the RJ-11 jack and the IDF is often called the *horizontal cable* because it runs horizontally in the space above ceiling tiles to the IDF (see Figure 4-13).

A 25-pair cable is run from the IDF down to the room in which the CBX resides. At the CBX room the cable is connected to another BIX or 66 block mounted

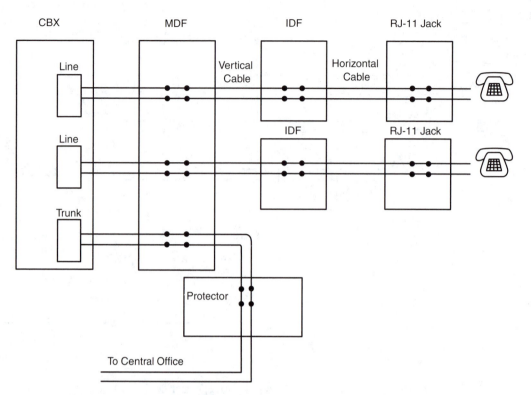

FIGURE 4-13 MDF and IDF.

on a large piece of plywood. The blocks mounted in this area are collectively re-ferred to as the *main distributing frame (MDF)*. The cables between the IDF and the MDF are called *vertical cables*. Line circuits from the CBX are also cabled to blocks on the MDF. The installer can run a short piece of a wire pair between connecting blocks to connect any telephone to any line circuit. Any wire pair in a vertical cable can be connected to any wire pair in the cable going to the CBX by using a short piece of twisted-pair wire, commonly called a *jumper wire*.

4.23 PRIVATE NETWORKS AND WATS

Some PBXs use trunk circuits to establish private networks. The trunk circuit uti-lized is the same trunk circuit used for central office trunks. Instead of connecting to the local central exchange line and trunk circuits, they connect to special circuits such as *wide area toll service (WATS)* or connect to a *foreign exchange (FX)*. WATS was developed by AT&T as a method of providing discounts on toll to high-volume business users. Prior to WATS, all calls on the long distance facilities were billed the same for everyone. WATS lines bypass the billing machine. These lines have meters that keep track of the number of calls placed and the total conversation time for each month. Businesses could purchase flat-rate WATS or measured-service WATS. They could also purchase incoming and outgoing WATS lines. Interexchange carriers

(IECs) provide WATS for out-of-state calls. Local exchange carriers (LECs) provide WATS for in-state calls. This is due to the 1984 MFJ, which states that inter-LATA calls must be handled by the IEC and intra-LATA calls must be handled by the LEC.

The 1996 Telecommunications Reform Act makes it possible for IECs and LECs to offer both interstate and intrastate calls. IECs provide six different WATS lines, called *bands*. A band 1 WATS can call into adjacent states. A band 2 WATS can call everywhere a band 1 line can, plus a few more states adjoining the adjacent states. Basically, out-of-state WATS bands radiate out in a circular fashion from the originating state. The circle that encompasses adjacent states is band 1. The circle that has a little longer radius is band 2. This continues with each circle having a little longer radius until a radius is long enough to cover all states. This is band 6 WATS.

For in-state WATS, the LECs divided the states up into three bands. One WATS line could not be used to dial both an in-state and an out-of-state telephone number. A WATS line was needed from the LEC for in-state calling, and one was needed from the IEC for out-of-state calls. This area is ripe for competition. Why have two separate lines? I expect the IECs will aggressively pursue an opportunity to provide both services with one WATS line. It will be more difficult for LECs to provide out-of-state WATS, but as they get into the regular long distance market, I am sure they will also offer both services over one line. Changes are presently occurring in this area. Check with your IEC and LEC to see what they offer. WATS lines are often accessed by dialing 8. When PBX has more than one band, additional access codes must be used. When a company has multiple WATS lines, it often equips the PBX with the *least-cost routing (LCR)* feature. LCR determines from the called number how the call should be placed.

4.24 SOFTWARE-DEFINED NETWORKS

WATS has not gone away, but the IECs do offer attractive alternatives. They provide *software-defined networks (SDN)* and T1 systems that can be used to build private networks. When a business is large enough to have many PBXs scattered across the United States, it often ties them together using tie trunks. These trunks are accessed by dialing a special prefix that varies at each location to achieve a uniform dialing plan. One PBX may use numbers 2100 to 2900, a second location can use 3100 to 3900, and a third location may use 4100 to 4900. The first location would use access digit 3 to access trunks to the second location, and access digit 4 to access trunks to the third location. These locations would access the trunks on an outgoing basis using access digit 2. Thus, any telephone number in the company can be accessed by any phone in the company simply by dialing the four-digit telephone number. The telephone numbers at each PBX start with a different thousands digit. The thousands digit is used as a routing digit and identifies a unique PBX location on the private network.

4.25 FOREIGN EXCHANGE TRUNKS

Some PBXs may use a trunk circuit to access the line circuit in another PBX or distant central exchange. These trunks are FX trunks. When the access digit is dialed at the PBX, the outgoing trunk circuit is seized. It will use a rented T1 channel from an IEC to seize the line circuit in the distant exchange or PBX. The distant exchange or PBX will return dial tone and local numbers can be dialed. If a PBX in Kansas City has an FX line to a Chicago central office, the station user in Kansas City can dial the access code for the FX trunk and then dial any number in Chicago. Additionally, the Chicago LEC would assign a local telephone number to the line circuit. This would allow customers in Chicago to dial a local Chicago telephone number and not have to pay a toll charge to reach the Kansas City PBX. Incoming calls from the Chicago telephone number are answered by the Kansas City PBX attendant. This process can be reversed to provide dial tone from the PBX to a distant location. When the PBX provides dial tone to a distant telephone rather than a distant exchange, the circuit is called an *off-premise extension (OPX)*. An OPX is used to provide extensions off of a PBX to a branch office and sometimes to a senior manager's home. The OPX can be provided with access to any feature of the PBX. An OPX connects to a line circuit in the PBX.

4.26 MANAGING THE PRIVATE NETWORK

As the size of a business grows, so does the task of managing the company's telecommunications and data communication needs. Large companies employ one manager for their voice network and another for their data network. As businesses jump onboard the "right-sizing" or "downsizing" craze, they are forcing more work onto fewer people to increase productivity. This has to result in increasing productivity. No management skill is needed to improve productivity. No pep rallies and no motivational techniques are needed, just reduce the workforce. The workload that was done by a laid-off employee will be picked up by the remaining workforce, or will be treated as unnecessary and discontinued as a cost-cutting measure. The remaining workers are forced to work 50- or 60-hour workweeks at top management's directives (explicit or implicit).

Thankfully not all companies have embraced downsizing as a substitute for managing their business. As new technologies and training lead to improved productivity and result in reduced personnel requirements, some companies look to see how they can make use of their workforce assets in other ventures. Executive-level management in these companies continually looks for future growth opportunities. Instead of laying employees off, it transfers them to other, newly created job opportunities within the company. If top management of your company is more interested in downsizing the company than in growing it, you may be faced with a tremendous challenge in managing both voice and data networks. This is one of the effects of downsizing as applied to a telecommunications manager.

The use of central office trunks, WATS trunks, TIE trunks, OPX lines, and leased data lines to build a private network may be too overwhelming for the manager. You must continually monitor the traffic at all PBX locations, do a traffic analysis and a cost analysis of the various services being used, evaluate the different service offerings of many IECs, LECs, and independent carriers, and determine how the networks should be engineered. Some PBX locations may not have enough call volume to justify TIE lines or WATS lines. Continually analyzing many different network components to achieve maximization of company resources can be a tremendous undertaking. The manager must downsize this workload by placing more of it on the vendors. This can be done by opting for a *virtual private network (VPN)*, also called *software-defined network (SDN)*.

VPN and SDN are names used to identify a private network for business that uses portions of the PSTN facilities to carry traffic for the private network. These virtual networks are offered by the IECs. The manager does not need to design and manage TIE lines, WATS lines, and 1+ long distance lines. These are all eliminated. The IEC will design your interoffice and long distance network into its database and provide you with this VPN/SDN. Calls between PBXs are connected by trunk circuits from the PBXs to the IEC. The IEC decides how the call should be handled and processes it accordingly. Access between the trunk circuits of the PBX and the IEC can be switched or dedicated. The LEC can provide access to the IEC or private VANs can provide dedicated access to the POP, for connection to an IEC.

4.27 SDN AND SS7

VPN and SDN are services that have been made possible with the deployment of SS7. The IEC receives the calling number and called number from the originating PBX. The calling number of a VPN/SDN customer will be in the IEC's database and identifies the originating customer as a virtual network customer. The IEC forwards this information over SS7 to a database (SCP). The database at the SCP contains entries on which telephone numbers are part of the VPN and which calls are off-net calls. The SCP translates the dialed number for a VPN customer into an area code + 7 digit number and provides the instructions for completing the call over the IEC's regular network (PSTN). Billing is applied for the call according to whether it is an on-net to on-net, or on-net to off-net, call. The VPN/SDN can also be set up to allow receiving calls from an off-network location. Company employees who travel can use special access numbers and authorization codes to access the VPN. Customers can also be included as part of your VPN.

The use of VPN/SDN has become popular. TIE lines and WATS lines are usually more expensive than paying for access lines to the IEC and paying the VPN charges. Managing the network is much easier since the IEC handles all calls. The manager gets all billing and traffic data from the IEC. This simplifies the data gathering and analysis process. The manager does not have to worry about a TIE line going down because they do not exist in a VPN. If problems occur in the IECs' network, they have alternate routing capabilities to keep the VPN in service. The IECs even offer telecommunications managers limited access to the SS7 network and the SCP so they can manage the VPN. This allows the manager to reroute calls from one PBX location to another in the event a weather problem, equipment problem, or work stoppage closes a PBX location.

4.28 PBX: PURPOSE AND FUNCTIONS

A PBX is a private switching system that provides switching capabilities for stations connected to the PBX. It allows the construction of a private network and provides users with a wide array of features. Most features are developed in software. The primary purpose of any PBX switching system is to connect a calling station to another station on the PBX or to connect the calling station to the desired trunk circuit. This switching function is part of the basic call processing software and is part of generic 1 software. Each time new features are developed for a PBX, they are added to the existing generic program and the number of the generic program is changed.

Generic program 31 will contain features and improvements not found in generic 30. As new features are needed, the generic is modified or added to and a new generic is released that includes the needed feature. Switching manufacturers love to see new features developed because a good portion of their revenues comes from selling existing customers the latest software release. Microsoft continually adds new bells and whistles to its Windows software for PCs. Each new release carries a different version number. Switching manufacturers do the same thing with their software.

4.29 SYSTEM FEATURES

Features developed for a PBX system are called *system features*. These features are developed to improve the efficiency of the PBX itself (such as least-cost routing and DID) or to provide enhanced administrative features (such as call detail recording). Least-cost routing is sometimes called *automatic route selection.* Each station can be assigned to a particular class of service and trunk group restriction. This will determine which trunk routes they are allowed to use. Least-cost routing will look at the trunk routes allowed by a station and place the call over the circuit that results in the lowest cost per minute. If you use a software-defined network, automatic route selection may not be required. With SDN, you usually have only two trunk routes. One group of trunks connects to the local exchange (C.O. trunks) for local calling. The second group of trunks connects to the IEC SDN and handles all long distance calls. The SDN will do automatic route selection instead of the PBX.

4.29.1 Direct Inward System Access

Direct inward system access (DISA) is a system feature that allows company employees to access the company's private network by dialing a special number and entering a password. The special number provides trunk access to the company's PBX. Once they are connected to the PBX, they must enter an authorization code and password. After access is approved by the system, employees can dial anywhere that a regular PBX station can. This allows them to access WATS lines and the SDN to place long distance calls.

Sometimes the DISA feature is accessed by dialing an 800 number. This number will connect via the PSTN to a trunk circuit in the PBX. After dialing the 800 number, employees can enter the DISA code to access all the capabilities of the PBX. They can place long distance calls using the SDN rather than using a more expensive calling card. They can also access the e-mail and voice mail system to leave messages. Security of DISA is a problem. Unauthorized people have gained access to a company's private network and have used it to place long distance calls. Telecommunications managers should change authorization codes and passwords often. They should also keep a close watch on billing invoices and on the call detail recording system to detect abuse.

4.29.2 Dialed Number Identification System

This system feature allows the PBX or automatic call distributor (ACD) to receive information from the central exchange about which particular station or group of stations the call should be connected to. This feature is used most of the time on an ACD system. The ACD will send the call to a particular group of ACD agents based on the *dialed number identification system (DNIS)* information. This could also be accomplished by assigning a different telephone number to each group of agents. A telemarketing group will often handle calls for many companies. A different in-WATS (800) number could be assigned to each company. When someone dials the 800 number, the call is routed to the appropriate agent group. DNIS performs a similar function and uses one group of telephone numbers for receipt of all calls. The PSTN will interpret a called number and change it into the number of the telemarketing group. The call is placed over the PSTN to the ACD, along with information identifying who the caller was calling.

4.30 STATION FEATURES

All features developed for users of the PBX are called *station features* (such as call transfer and call forward). Before station features can exist, they must be included in the system's generic load. Most PBXs have a system feature that allows more than one customer to use a PBX. This concept became popular in the 1970s. Real estate companies leased office space and telecommunications facilities to tenants. Each customer (tenant), as well as the lines belonging to that customer, was identified in the database of the PBX. Features accessible by a particular customer were also identified in the database. A station could only have access to the features identified for that specific customer. Thus, all station features had to be identifiable at the system level and the customer level before they could be added to the line translations database for that station and become a station feature.

A clear line of demarcation cannot always be drawn between system and station features. A feature can only exist for a station if it is a feature the system offers. Stations do not have access to some system features. System features should be thought of as features available to all station users or used by the system itself. For example, "dial-9 access for a central office trunk" is a standard feature on all PBXs and exists for all stations. This feature must be restricted on a station-by-station basis when stations are denied this system feature. A station feature can be thought of as a feature that must be specified in the line translations table for each station.

4.30.1 Call Forward

Call forward is both a system feature and a station feature. It is a system feature because the system will automatically forward unanswered calls to a voice mailbox or an attendant. It is a station feature because stations can be given this feature and allowed to program a specific location for forwarding unanswered calls. The system or the station can also provide for forwarding calls when the station is busy. "Call forward busy" can be the same forwarded location or a different location than used by "call forward no answer."

4.30.2 Call Transfer

Call transfer is a station feature. When users wish to leave their desk, they can program a location for the transfer of all calls. Call transfer will override the call forward no answer feature. Call forward no answer will forward calls after a certain number of rings (this number is flexible and programmable). Call transfer will send calls immediately to the transferred location, and that station receives ringing immediately on receipt of a call in the system for the transferred station. As mentioned earlier, station features for 2500 sets are accessed by code dialing. A proprietary set accesses features when a button is pushed. To activate call transfer, station users must press the call transfer key on their proprietary sets, or flash the hookswitch and dial the access code for call transfer. The number of digits for call transfer is also programmable. This allows the administrator to restrict call transfer to four digits. Station users cannot transfer calls to their home telephone or any seven-digit number unless the administrator programs the line for seven-digit transfer. Four-digit transfer will allow call transfer to any station on the PBX.

4.30.3 Ring Again

Ring again, or camp-on queuing, allows a station user to camp on a busy station. This is a PBX feature and only works when the called number is a station on the PBX. If the called PBX station is busy, the caller activates the ring-again feature and hangs up. When the called station is finished with the present conversation and hangs up, the line is idle. The CPU will send a special ringing signal to the calling station's phone. The party knows by the special ring that it is the call camped on busy. The calling party picks up the telephone. When this happens, the PBX will ring the station that was camped on. This feature prevents callers from having to continuously place calls to a busy phone and allows them to make more productive use of their time. The ring-again feature will continuously test the line. When it becomes idle, the caller is notified.

4.30.4 Conference Call

PBXs can be ordered with conference call capabilities. This feature requires the addition of hardware. A PCB must be supplied that provides a conference bridge. This card is designed to connect many different time division slots. The number of slots it can connect determines how many stations can be connected in a conference. This PCB ensures that an adequate level of transmission is maintained for all connected stations. Most PBXs allow up to eight stations on one conference. The initial purchase of a conference call feature must take future needs into account. The number of conference cards to provide as well as the number of stations each PCB can connect must be considered. This feature cannot be changed without purchasing new hardware as well as software. An upgrade to handle more stations on one conference or to increase the number of conference circuits can be costly.

Establishing a conference call is easy. Flashing the hookswitch and dialing the access code or pushing the conference button on a proprietary phone provides dial tone. The user dials the station number desired. When the called station answers, the calling party announces that a conference is being established. Although most PBXs allow a private consultation with the called party at this time, I would still be careful about what is said. The original party you were talking to should not be able to hear your conversation with the station that was dialed for a conference. This feature allows you to consult privately with a third party. Once you have consulted with the third party and decide a conference is in order, you can bridge all parties together by pushing the conference button again. With 2500 sets, you would have to flash the hookswitch to link all three stations. If additional stations must be added to a conference, the same procedure is followed again: press the conference button (this puts the original party on hold and provides your station with dial tone), dial the desired station, consult or announce the conference when the called party answers, press the conference button again, and the called party is added to the existing conference.

4.30.5 Speed Dialing

PBXs have two versions of speed dialing; the system provides a system speed dial as well as station speed dial. System speed dial numbers are programmed by the attendant and can be accessed by all stations provided with the system speed dial feature in their line table. If the line table includes the station speed dial feature, the station can program frequently called numbers that are accessible only by that station. System speed dial numbers usually have two digits. The tens digit identifies the number as a system speed dial number. For example, the attendant could program speed dial for 1-312-555-6622 and assign this number to speed dial code 92. When any station activates the speed dial feature and dials 92, it will be connected to 1-312-555-6622. If the station activates speed dial and dials a single digit such as a 3, it will connect to the telephone number it programmed at its station as speed dial code 3.

System speed dial is used to store numbers many station users call frequently. These may be telephone numbers for a branch office, the main office, the credit union, and so on. Station users program their individual speed dial locations for their home, a customer or vendor they deal with on a regular basis, and so forth. Most PBXs provide the capability of having ten speed dial locations for each station and ten for the system speed dial. Advanced PBXs offer the capacity for more than ten speed dial numbers. Of course it is also possible to have station sets with speed dial capabilities. This provides the user with more flexibility but is somewhat redundant and unnecessary.

4.30.6 Call Pickup Groups

Most telephones are arranged in groups. All engineering phones are assigned to one group number. Accounting is in a separate group, operations has its own group, marketing and human relations have their own group number, and so on. The group number a phone is assigned to is an entry in the line translations table (line table). When a phone is added or someone moves to a different department, the group number is one entry made in the line table. When any telephone in a group is ringing, it can be answered using the call pickup feature. This feature allows anyone in your group to answer a call coming in to your department. Calls cannot be picked up for other groups. If your neighbor is working for accounting and you are working for engineering, you cannot pick up calls for your neighbor unless the system administrator assigns both phones to the same group. Sometimes groups are assigned on a geographic basis rather than a departmental basis to allow for this type of pickup. This feature is helpful when people leave their desk without activating call transfer. If the system has call forward no answer activated, the incoming call will be forwarded if someone does not use call pickup to answer the call. Remember that call forward no answer forwards calls to a predetermined station after a certain number of rings.

4.30.7 Call Park

If you answer a call at your desk and need to transfer the call to another location, you could use call transfer. Sometimes this is not practical. Suppose you need to go to the basement to retrieve information the caller is asking for. There is a phone in the basement area but no one is there to answer the call. You can use call park in this situation. Call park is like a huge parking lot in which a call can be parked. If you do not specify a parking slot when you activate the call park feature, the call will be parked in the slot corresponding to your station number. If you wish, you can specify any parking slot desired. The parked call can be picked up at any telephone by dialing the call park retrieve access code. You could park your caller and then go to the basement telephone and retrieve the call from the parking lot. This feature is also used to transfer calls when you do not know the location of the desired party. On receiving a call for Joe Smith, you can park the call and use the paging system to notify Joe Smith. You would tell Joe he has a call and what parking slot the call is in—for example, "Joe Smith, please call code 64."

4.30.8 Intercom Groups

When a group of people call each other a great deal, the stations can be put in an intercom group. Many different intercom groups can be defined in most PBXs providing this feature. This allows dialing other stations in the group with a one- or two-digit number. Special features and restrictions can also be set up for each group.

4.30.9 Multiple Appearance Directory Numbers

When a telephone number appears on more than one telephone, it is referred to as a *multiple appearance directory number (MADN)*. This feature can allow a single-line telephone to have more than one telephone number. It also allows the use of single-line telephones to answer an incoming call. The call will cause all assigned station sets to ring unless they are on another call.

4.30.10 Executive Override

Executive-level managers are usually provided with access to all trunks and stations. Many PBXs also offer the executive override feature. If your station has this feature, you can override a busy signal. When you get a station busy signal, you would activate the executive override feature. Activation of this feature should cause a brief warning tone to be placed on top of the called station's conversation. You would then be connected into the existing call. Of course, this is not good phone etiquette, and since executives do not want someone doing this to them, they are provided with a privacy feature. This feature prevents anyone from using executive override to barge in on an executive's conversation. This feature also prevents the attendant from monitoring an executive's conversation. Although the privacy feature is assigned primarily to executive phones, it is also used on data lines. This prevents the interruption of a data transfer due to a call waiting tone. When the executive override feature is activated, the station receives a warning tone. This tone will cause an error on data handled by a modem. To prevent this from happening, the modem-equipped line is assigned the privacy feature.

4.31 PERSONAL COMPUTERS

The PC has become a major station equipment component used for voice and data communication. The PC has become an integral part of our lives. I know that we were able to run our businesses and personal lives in the past without the PC, but today I wonder how we did. It is so easy for me to log onto the Internet and get up-to-date news, weather, and financial information. I now have the world's libraries accessible via my Internet connection. If I type a document on my PC, the spell checker keeps everything accurate. If I need to make changes, they are so much easier to do on the PC than on a typewriter. Project management software allows us to use a PC to track the status of various projects. The actual expenditures and time used can be checked against our budgeted figures and provides a control mechanism for a project manager. Spreadsheets can be used to perform various financial scenarios quickly and help us determine the best financial actions to pursue. I have only touched the tip of the iceberg. The PC provides us with a tremendous tool to increase our productivity and profitability.

The PC is a relatively simple device built on rather complex technology. Thanks to its design, users do not need to understand what makes it work. They only need to know how to use a particular applications program written to run on the PC. When the PC was first developed, users had to write their own programs. To be a user, you had to be a programmer. A few programmers developed some programs that would control the input and output operations of a PC. These programs were called *operating systems.* This allowed programmers to develop *applications programs* that could run on top of the operating system. As more and more of these applications programs became available, they made PCs easier to use and more people bought them.

Many people think that IBM invented the PC because we are always using the phrase "IBM or IBM compatible." *IBM did not invent the PC.* What IBM brought to the PC world was its wealth and marketing expertise. I have read several articles that go on and on about the IBM 13 locked up in Atlanta and later in Boca Raton as being supervisionaries that gave us the PC. This is simply not true. The people we refer to as *hackers* gave us the PC. Intel and Motorola were in the forefront of supplying hardware to these visionaries. What IBM did do is make Bill Gates' *disk operating system (DOS)* the standard operating system for IBM PCs.

The two predominant operating systems were CP/M (computer program for microcomputer) and DOS. With the marketing dominance of IBM, its selection of DOS as an operating system dealt a death blow to CP/M. Many people think CP/M was a superior system. In 1998, the Justice Department pursued allegations of antitrust actions by Microsoft. This was way too late. IBM and Microsoft's alliance killed the major competitor to DOS a long time ago. A few applications programs were written to run on CP/M-based systems, but when it became clear that the whole IBM and IBM-compatible world was going to be DOS, software manufacturers stopped supplying applications for CP/M. If you cannot get the applications, why buy the operating system? PCs came with DOS included as part of the package deal. Why buy a CP/M operating system when you could get DOS for free?

A PC is just an expensive boat anchor without software. The operating systems software brings the PC alive, and the applications programs turns it into a useful tool. The number of applications programs available continues to grow. The CTI-based PBX is accomplished by loading an applications program on a PC that will turn the PC into a CBX. Several applications will allow you to use your PC as a telephone. One of these is the *Telecommunications Application Program Interface (TAPI).* When you use dial-up networking or Hyper Terminal in Windows, you are using TAPI. Microsoft Phone also uses TAPI to provide the interface needed to turn your PC into a telephone answering machine.

Dual simultaneous voice and data (DSVD) modems can connect to your PC via an internal card slot or via an RS-232 serial port. These modems include jacks that allow you to connect a microphone and speaker to the modem. You can then place voice calls over your modem. With the Microsoft Phone software controlling your modem, you can have it automatically answer calls for you, digitize the caller's voice mail, and place the digitized message on your hard drive. A section of the hard drive will be set up as mailboxes by the Microsoft Phone software. Several packages provide this capability. Fax Talk messenger by Thought Communications, Inc. provides an applications package that allows you to set up separate mailboxes for voice and for faxes. You can select the auto-detect option, and the software will

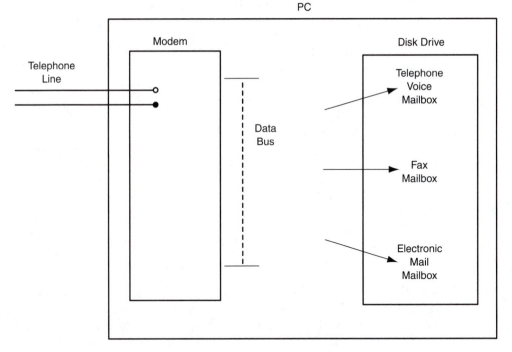

FIGURE 4-14 PC answering machine.

work with the modem to determine if it is receiving a voice call or a fax call. It will then answer the call as appropriate and place the message in the appropriate mailbox (see Figure 4-14).

The PC can connect to other computers directly via a special cable called a *null-modem cable* or they can connect to each other over a LAN using a **network interface card (NIC),** but most connections from one PC to another occur via modems. This allows us to connect the PC via the modem to our telephone line. We can then use the PSTN to connect to other computers on the PSTN or on the Internet.

As stated earlier, the PC is a relatively simple device. It consists of input and output devices and a system unit. The output devices are a visual display unit (VDU), a printer, or a secondary storage device (disk drive, tape drive, zip drive, and so on). The input devices are a keyboard or mouse. Sometimes we will get input from files stored on a disk drive or CD ROM. Storage devices store output until we need it, then we can retrieve it as input. Serial ports are also used for input and/or output. The keyboard and mouse connect to the system unit via their own individual serial port. Additional serial ports can be used for other devices such as a modem.

External modems connect to the system unit via serial ports on the back of the PC system unit. Internal modems contain the hardware for the serial port on the modem card but require the assignment of port addresses as well as an *interrupt*

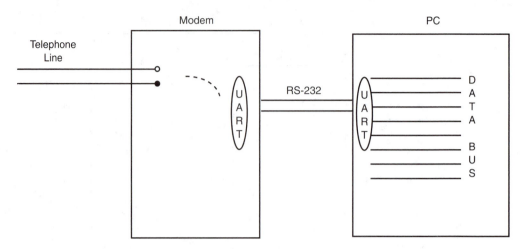

FIGURE 4-15 UART connections.

request (IRQ) to the internal modem card. The hardware device associated with a serial port is called a ***universal asynchronous receiver transmitter (UART)***. The functions of the UART can also be implemented as part of a ***digital signal processor (DSP)***. The UART functions are to basically connect a multilead parallel transmission medium to a single-lead serial transmission medium (see Figure 4-15).

Inside the system unit of a PC is the CPU. An operating system program controls the CPU. Applications programs are written to run on top of the operating system and use the operating system's program to deliver commands to the CPU. Programs and data are moved into and out of the CPU over data leads that are in parallel. These leads form what is called the ***data bus.*** The more leads a data bus has, the faster information can be moved to and from the CPU. The CPU also has an ***address bus*** that connects it to all peripheral devices and to memory storage. It uses the address bus to send data on the data bus to the right location in memory or to the right device.

Early PCs had 8 leads on the data bus; this was quickly expanded to 16. The 486 CPU has 32 leads and the Pentium has 64 leads on the data bus. Even though we can move data faster into and out of the CPU, we cannot take full advantage of this with all peripheral devices. Anything that has to connect to the PC via an *industry standard architecture (ISA)* bus will be limited to a 16-lead data bus. The newer PCI bus has 32 data leads. If we connect our computer to another computer via a modem and our phone line, the UART function must be implemented to interface the multiple-lead data bus of the PC to the single transmit lead going to the distant modem.

Although we may have 64 data leads that we can transfer data over within the PC, there is only one lead that we can transfer data over to an external modem, and a UART is required. We may not have a physical transmit lead to connect an internal modem to the PC, but we must still have the UART function performed.

An internal modem card will usually plug into an ISA slot (16-data lead bus). This gets data to the internal modem card very fast but the modem connects to a phone line, which is one circuit. We will only have one circuit path through our dial-up path within the PSTN. We cannot have 16 paths through the PSTN. Therefore, the data must be sent one signal at a time over the PSTN circuit. This requires that we perform the UART function prior to delivering the data to the modulator of the modem.

The PC continues to get better and faster. The developments in Internet technology and telecommunications software for the PC continue to provide us with more ways to use the PC for voice, data, and video communication. The PC is gaining popularity as the station terminal of choice because it allows us to do more than a voice call. The telephone will continue to serve most users of the telephone network because of its low cost and ease of use, but the day is fast approaching when new homes will be built with speakerphones in all rooms connected to a central PC. When a call rings in, the user will tell the PC to "answer phone," and the person will be connected to the caller. If they wish to place a call to Mom, they will simply say: "call Mom." The PC will activate Microsoft Phone (or some similar application). It will look up Mom's number in the phone list the user created and call her.

4.32 SUMMARY

This chapter has focused on the various analog telephone sets. These devices are used to convert voice signals into analog electrical signals. Most of these telephones are connected over a local loop to a digital central office. The central office interface to a local loop cable pair is called a line circuit. In a digital central office, the line circuit includes a codec to convert the analog signal into a digital signal. Telephones are being manufactured that include the codec in the telephone. Most of these are manufactured for use on PBX switching systems. The telephone manufactured with a codec for use on the local loop is called an Integrated Services Digital Network (ISDN) telephone. The ISDN telephone converts a voice signal into a digital signal for transmission over the local loop, but the local loop must be a DSL. This limits the range of an ISDN line to 3 miles unless repeaters are added to the DSL. ISDN technology is discussed in greater detail in Chapter 9.

This chapter also provided an introduction to CBXs and PCs. CBXs are used by large businesses to connect their phones within one location together. All telephones in the building are wired to line circuits in the CBX. Users can dial each other by dialing the four-digit extension number assigned to the telephone they wish to call. If they wish to call someone in the PSTN, they dial 9 to get an outside line (C.O. trunk circuit). These lines are called outside lines because many trunk circuits in the CBX will actually connect to line circuits in the central office. The CBX offers many system and station features that make use of the telephone more productive. A PC using a CBX applications program can now be used to implement a CBX system. These systems are known as computer telephony integrated (CTI) systems and are often referred to as UnPBX systems. The tremendous growth of the Internet has fueled the growth in PCs. PCs can be used as voice, video, or data terminals.

CHAPTER REVIEW QUESTIONS

1. Name an applications program that turns a PC into a telecommunications terminal.
2. What was the key element in Elisha Gray's design as well as in Alexander Graham Bell's design?
3. What was Thomas Edison's contribution to the telephone?
4. How much current does a carbon granule transmitter need?
5. What type of voltage is required for power to the standard single-line telephone set (2500 set)?
6. Which standards agency sets the standards for telephones?
7. What larger classification of telephone equipment does the telephone belong to?
8. Why must station equipment be registered with the FCC?
9. What is a multiline phone?
10. What advantages does a keysystem offer over a simple multiline phone?
11. What is an easy way to tell if a keysystem is a hybrid system?
12. What is the difference between a hands-free phone and a speakerphone?
13. What is the difference in cabling for the old 1A2 and today's EKTS?
14. What type of switching matrix does an EKTS use?
15. How are system features programmed?
16. How are station features programmed?
17. What is the purpose of a hybrid circuit in the telephone?
18. What is sidetone and what purpose does it serve?
19. How many tones are sent to the central office when one DTMF button is depressed?
20. Why is a varistor used in the telephone?
21. What is DID?
22. How are C.O. lines accessed by a keysystem? By a hybrid keysystem? By a CBX?
23. What is an IDF? An MDF?
24. Station wiring to the IDF is also called _____ wiring.
25. What are the two most predominate quick-connect blocks used in the telephone industry?
26. What is CTI?
27. What are the three major output devices on a PC?
28. What is the major advantage of a parallel bus compared to a serial bus?
29. What hardware device exists between the serial port on the back of a PC and the system's data bus?
30. How does a PC control where it reads or writes information from/to?

PROJECTS AND CASE STUDIES:

Project #1

Jackson manufacturing has decided to locate a facility in your town. They will install a PBX with 250 extensions and 20 central office trunk circuits. They will need five Tie Lines connecting to their headquarters in Chicago and will need five WATs lines. They will need a Local Area Network for connecting 200 PCs together and will need a data connection to Chicago. You can investigate costs for each item pertaining to voice and data if you wish, but for this exercise let's find out how much it would cost to have a T1 line from your town to Chicago. We could place the data and tie lines on this T1. You might also want to find out from a PBX service provider what a 250 station PBX would cost. Tell them you want all multiline feature phones, 20 Central Office Trunks, and the standard features such as call forwarding, call transfer, etc.

Case Study #1

Residential Real Estate Company has 30 agents working out of a midtown office. What would be the best system to handle their voice calls?

Case Study #2

Medical Pharmaceutical has a headquarters in Dallas, Texas. They employ 500 people. Which voice system is most appropriate for this firm? What are some station and system features they would need?

GLOSSARY

Address Bus   The physical circuit path used by a PC (or other computer) to notify other components where data should go. It controls the sending or receiving of data to and from the appropriate device or memory location.

Alerter   A ring detector circuit that results in a chirping sound from a speaker.

Analog Signal   An electrical signal that is analogous (similar) to a voice signal. The signal continuously varies in amplitude and frequency. An analog signal has an infinite number of values for voltage, current, and frequency.

Applications Program   A program developed for a special application such as word processing, project management, data management, and so on.

Bridged Ringer   A term indicating that the ringer is wired between the Tip and Ring leads. Bridged ringers were used on private lines as well as on two- and four-party lines.

Centrex   A service offered by local exchange carriers as a direct replacement for a PBX. The Centrex service uses a special software program and the line circuits and switching equipment of the local class 5 switching office to provide Centrex.

Computerized Branch Exchange (CBX)
Standalone PBX switching systems implemented via stored program control switching systems. Almost all PBXs in service today are CBXs or CTI-based systems.

Computer Telephony Integrated (CTI) Systems
Switching systems implemented using a PC with special applications software and special circuit cards that can interface to telephones and central office lines. CTI-based systems are available to serve as a PBX or an automatic call distributor (ACD).

Data Bus The physical circuit path inside a computer that data travels over to get from one location to another.

Decimonic Ringing A ringing system used by Independent telephone companies. The system could send out five different ringing signals (10-, 20-, 30-, 40-, and 50-cycle signals).

Digital Signal An electrical signal that has two states. The two states may be represented by voltage or current. For example, the presence of voltage could represent a digital logic of 1, while the absence of voltage could represent a digital logic of 0.

Digital Signal Processor (DSP) A very large-scale integrated circuit chip that controls the sending and receiving of electronic signals. DSPs are used to perform data transfer functions in regular modems, ADSL modems, ISDN devices, PCs, remote access servers, and so on.

Direct Inward Dial (DID) Trunks Trunk circuits at the class 5 central exchange that allow someone on the PSTN to dial directly into a particular station on a CBX.

Dual Simultaneous Voice and Data Allows a device to transmit voice and data signals at the same time over the same media.

Dual-Tone Multifrequency (DTMF) Dial The dial on a 2500 set that sends out a combination of two tones when a digit on the keypad is depressed.

Feature Phone A telephone containing electronics that allow it to provide features. When additional features have been added to the basic single-line telephone such as last number redial and speed dialing lists, the phone is called a feature phone.

Foreign Exchange (FX) A term used to refer to all class 5 exchanges that cannot connect to your particular telephone on a local call basis. If I have a telephone in Kansas City, then St. Louis, Chicago, New York, and so on are foreign exchanges to me.

Grounded Ringer A term indicating that the ringer is wired between the Tip lead and ground or between the Ring lead and ground. Grounded Ringing was used on eight- and ten-party lines.

Harmonic Ringing A ringing system used by Independent telephone companies. The system could send out five different ringing signals (162/3-, 25-, 332/3-, 50-, and 662/3-cycle signals).

Hookswitch Often called a *switch-hook.* This is the device in the telephone that closes an electrical path between the central office and the telephone, by closing contacts together, when the receiver is lifted out of its cradle.

Hybrid Keysystem A keysystem designed to handle more than 24 central office lines and 40 telephone sets. Central office lines are accessed by dialing 9 for access.

Hybrid Network A network that consists of a transformer and an impedance matching circuit. The transformer performs a two- to four-wire conversion and vice versa. In a telephone, the hybrid network connects the two-wire local loop to the four-wire transmitter/receiver.

Keysystem   A telephone system used by small businesses to allow several central office lines to terminate at each telephone attached to the system.

Magneto   A device that generates an AC voltage by turning a coil of wire inside a magnetic field. The old hand-crank telephone had a magneto attached to the hand crank.

Network Interface Card (NIC)   Interfaces a PC to the media used by a local area network. The NIC contains DSPs that buffer data between the PC and LAN. Each NIC has a unique hardware address and will monitor the LAN media for data addressed to it.

Operating System   Software program that controls the basic input/output operations of a computer.

Private Automatic Branch Exchange (PABX)   A switching system that resides in a business location and is owned by the business (or third-party supplier). It connects telephones within the business to each other when the originator dials an extension number or connects the originator to the local central office switch when a "q" is dialed.

Private Branch Exchange (PBX)   A switching system designed to serve a business and is located in the building that houses the business. All PBXs sold today are computerized branch exchanges (CBXs) or CTI-based systems.

Proprietary Telephone   A telephone designed by its manufacturer to only work with certain keysystems or PABXs made by that same manufacturer.

Receiver   The device responsible for decoding or converting received information into an intelligible message.

Ring Detector   An electronic solid-state transistorized device designed to detect the presence of a ringing signal (a 90-V AC signal).

Ringer   An electromechanical device (relay) that vibrated when a 90+V AC signal was received by the telephone. The vibrating device would strike metal gongs to create a ringing sound.

Sidetone   In a telephone set, some of the transmitted signal is purposely coupled by the hybrid network to the receiver so you can hear yourself in the ear covered by the telephone receiver. This signal is called sidetone.

Software-Designed Network (SDN)   A service sold by IECs to serve as a replacement for a business's private-line network. SDN allows you to replace WATS, FX trunks, and tie lines with connections via the IEC's SDN.

Station Equipment   The largest segment of station equipment is the telephone but the term station equipment has been broadened to include anything a customer attaches to a telephone line. The most common piece of station equipment is the telephone. A modem, CSU/DSU, personal computer, and keysystem are all referred to as station equipment.

Straight-Line Ringer   A ringer that will operate on any ringing signal. This type of ringer is used on single-party private lines.

Superimposed Ringer   The type of ringer used by Bell on four-party lines. The ringer included a diode. Ringing signals were superimposed on top of a DC voltage. The diode only allowed the signal with the correct polarity of DC for that station to pass to the ringer.

Telecommunications Application Program Interface (TAPI)   A special application program that allows the PC to act as a telecommunications device.

Transmitter   The device responsible for sending information or a message in a form that the receiver and medium can handle. The device in a telephone that converts the air pressure of a voice signal into an electrical signal that represents the voice.

Tuned Ringer   A ringer that will operate on only one ringing signal. These ringers were used on party lines. They were tuned to only ring on the signal assigned to that party.

2500 Telephone Set   The standard single-line telephone, which contains a DTMF dialpad.

Universal Asynchronous Receiver Transmitter (UART)   A specially designed DSP chip that interfaces a parallel data bus to a serial data lead and vice versa.

Varistor   A specially designed resistor. As the voltage across the resistor increases, the resistance increases.

Wide Area Toll Service (WATS)   A service sold by LECs and IECs. The LECs traditionally have sold intrastate WATS and the IECs have sold interstate WATS. With the 1996 Telecommunications Reform Act, LECs and IECs can sell both services. A customer pays so much a month or so much per minute of billing (or both) to be able to place long distance calls within a particular geographic region. The larger the region a business desires access to, the higher the cost will be.

5

The Medium

KEY TERMS

Branch Feeder Cable
Carrier Serving Area (CSA)
Crosstalk
Dense Wave Division
   Multiplexing (DWDM)
Distribution Cable
Drop Wire (Drop Cable)
Exchange Boundary
Facility
Feeder Cable

Jumper Wire
Load Coil
Loading
Local Loop
Loop Extender
Loop Treatment
Main Distributing Frame (MDF)
Main Feeder Cable
Modified Long Route Design
   (MLRD)

Multiplexer
Outside Plant
Outside Plant Engineer
Plastic Insulated Cable (PIC)
Resistance Design
Revised Resistance Design
   (RRD)
Subscriber Line Carrier-96
   (SLC-96)
Voice Frequency Repeater (VFR)

OBJECTIVES

On completion of this chapter, you should be able to:

1 Explain why the two wires connecting the central exchange to a telephone are known as the local loop.

2 Explain why twisted-pair copper wire is used instead of fiber for the local loop.

3 Know the color code for the wires inside a cable.

4 Understand the principles behind the design of the local loop.

5 Explain revised resistance design.

6 Explain modified long route design.

7 Explain when to use revised resistance design or modified long route design principles in designing a local loop.

8 Explain what a load coil is and why we use them.

9 Explain why loops longer than 18,000 feet need to be loaded.

10 Explain what a jumper wire is and what it is used for.

11 Explain why the serving area for a central exchange located in the city is kept under 3 miles.

12 Explain what a carrier serving area (CSA) is.

13 Explain why higher-frequency signals are attenuated more than low-frequency signals when they are transmitted over a nonloaded pair.

14 Explain the differences between feeder, distribution, and drop cables.

15 Explain what a SLC-96 TDM system is.

16 Explain why we might use SLC-96 instead of twisted-pair copper wire to serve as the medium to connect telephones to the central exchange.

17 Explain when we would use a loop extender and/or VFR.

18 Explain why the medium for SONET must be fiber optic cable

Every communication system requires a medium connecting the transmitter to the receiver. Wire was the only available medium when the telephone was invented; it continues to be the preferred medium for connecting the telephone to the local switching exchange. The medium of choice for long distance transmission was coaxial cable and microwave radio until the 1980s. Microwave radio uses air as its medium and carries signals composed of electromagnetic radiation. Fiber optic cable has now become the medium of choice for the long distance networks. This type of cable is made of glass strands and is used to carry signals composed of light energy.

Neither air nor glass will conduct electricity. Wire is an excellent conductor of electrical energy, which is why it is used to carry electricity to our homes and is also why we use it to connect our telephones to the central exchange. The telephone receives its electrical energy power from the central exchange, and the signals traveling from the telephone to the exchange are electrical signals. At the central exchange, we can convert these electrical signals into light signals or electromagnetic waves and then use other media such as fiber optic or microwave (see Figure 5-1). The wiring used to connect telephones in a home or business to the central exchange is called the *local loop.* This local loop will remain mostly twisted-pair copper wire for the next 10 to 20 years, but experiments are being conducted using fiber optic to serve as a local loop. Use of fiber to serve a telephone requires an expensive device to interface the telephone to the fiber, and the telephone must have a local power supply since it cannot receive power from the central exchange over the glass fiber. Cable television companies have begun to provide telephone service using their coaxial cables. Some LECs and IECs have established joint ventures with cable television companies to use their coaxial cable in certain local service area locations.

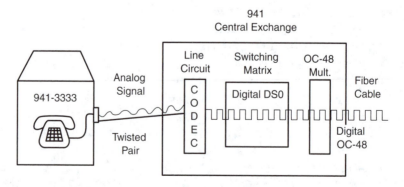

FIGURE 5-1 Analog phone line connected to a digital central exchange.

5.1 LOCAL LOOP

The twisted-pair local loop is the weakest link and most costly portion of the telecommunications network. But describing the twisted-pair local loop as the weakest link does not mean it is easy to break. The term *weakest link* refers to the inability of the local loop to handle high-frequency signals. The twisted-pair local loop has a narrow bandwidth. The other portions of the PSTN are composed of wide-bandwidth *facilities*. Voice and data travel across the PSTN on ribbons of fiber, which can handle many high-bandwidth signals, only to arrive at the local loop, which restricts communication to low-frequency analog signals. The local loop is usually from a few hundred feet to several miles in length, but it is often referred to as the "last mile." The heart of the PSTN contains fiber optic transmission media carrying high bit rate digital signals, but our connection to the PSTN is the low-bandwidth local loop. Think of the fiber optic transmission medium as a very large pipe and the local loop as a garden hose attached to the pipe. It is the size of the garden hose that determines how fast water can flow from a faucet.

The twisted-pair local loop is like a small pipe or garden hose. It limits our communication to low-frequency audio signals. It is technically cost prohibitive to replace the existing wire cables with fiber optic cable and compatible phones in the local loop. To use fiber optic in the local loop, an interface device will be required at every phone to convert electrical energy into light waves. The device could be incorporated in the phone, but this would require customers to buy new phones. Twisted-pair copper wire has provided us with a low-cost medium to connect telephones to a central exchange, and the twisted-pair local loop has served as an excellent transmission medium for analog voice signals.

Increasing demand for the ability to place high-speed data signals over the local loop will eventually lead to the deployment of fiber optic facilities to replace all, or a portion, of the local loop. The first phase of fiber introduction into the local loop has been to use fiber as a main feeder cable to serve a segment of the exchange territory. *Multiple subscriber line carrier systems (SLC-96s)* are placed on a **multiplexer** in the central office. Each SLC-96 connects to 96 line circuits. Demultiplexers are placed at several locations along the fiber route, and several SLC-96 field units are placed on the demultiplexer. The SLC-96 field units are powered by a local AC power source at the field terminal.

The SLC-96 field units convert the commercial AC power into DC voltage to power the telephone. Distribution wire cable pairs are connected from the SLC-96 units to telephones in the area. The field unit provides power over the wire pair to the telephone (see Figure 5-2). Using fiber as a feeder cable only, allows customers to keep the telephones they currently have and does not require them to purchase an interface device. The telephone company may choose to take fiber directly to the house, but then the customer will be forced to buy new telephones or an interface device.

Originally both the local loop and toll loop were constructed using wire as a facility. At the turn of the 20th century, plastic had not been invented. Since plastic insulated wire was not available, it was necessary to string individual strands of wire on telephone poles and attach them to the poles with glass bulb insulators. This type of wiring is called *open wire.* Two wires were needed for each telephone

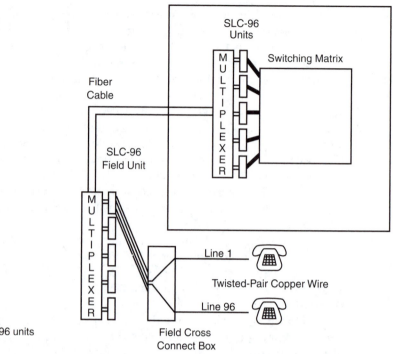

FIGURE 5-2 SLC-96 units over a fiber.

line circuit. Iron wire was the wire of choice for the local loop due to its strength and low cost. Because iron is not as good a conductor of electricity as copper, a thicker wire was needed when using iron wire. The wire was a little thicker than the wire in a heavy-duty clotheshanger. Copper wire was the wire of choice for trunk circuits between central exchanges. In the telephone industry, the engineering and construction of these facilities are the responsibility of **outside plant engineers,** located in the *outside plant engineering department.*

When more than one circuit was needed, a crossarm was added to the pole, making it look like a cross. A crossarm could support ten wires (five circuits). For every five circuits, additional crossarms were added. As the number of telephones grew, the local plant in large cities became a mess (see Figure 5-3). In small rural communities, open wiring systems were small; there were usually up to ten telephones on each line. Most residences in a small town were clustered together close to the central office. Serving the farms in the country required only a few lines leaving the central exchange in each direction.

Because only a few lines were needed to equip a small rural exchange with ten-party lines, open wire served many rural areas until the late 1950s. During the 1950s, most independent telephone companies aggressively pursued the elimination of party-line service. Party-line elimination was done in phases. First all these lines were reduced from ten- to four-party lines; a few years later, they were further reduced to two-party lines and private lines. To convert a ten-party line to ten private

FIGURE 5-3 New York City 1890—Open wire used for outside plant.

lines required nine more pairs of wires from the central office to the area converted. Party-line reduction and elimination required conversion of the outside plant from open-wire to aerial and buried cable.

5.2 INTRODUCTION OF PLASTIC INSULATED CABLE

The use of insulated wires makes it possible to put many wires in one cable, and they will be electrically isolated from each other by the insulation surrounding each wire. When insulated wires were introduced, copper wire was chosen over iron. Copper conducts electricity better than iron, and the choice of copper allows the use of a smaller-diameter wire. Smaller wire allows more wires to be placed inside one cable.

The first cables were made with copper wires coated with shellac. Each shellac-coated wire was then wrapped with paper to insulate the wires in a cable from each other. The wires were grouped together and surrounded with a lead covering to protect them from damage. Lead-covered cables served the local loop until the introduction of *plastic insulated cable (PIC).* Each wire is now covered with a colored plastic material. The colors of the plastic can take on 50 different combinations of hues. This makes it possible to have a 25-pair cable with each wire in the cable identified by a unique color scheme (see Figure 5-4).

5.3 PLASTIC INSULATED CABLE COLOR CODE

The colors used in PICs are blue, orange, green, brown, slate, white, red, black, yellow, and violet. The white, red, black, yellow, and violet colors are referred to as *mate* (or *primary*) colors. The blue, orange, green, brown, and slate colors are called *secondary* colors. The color of plastic around each wire will be composed of two colors: a mate color and the secondary color. The Tip wire of a pair of wires will have a wide band of mate color and a narrow band of secondary color. The Ring wire will have a wide band of secondary color and a narrow band of mate color.

The cable pair designated as pair 1 in a cable is composed of one wire that is predominately white with a narrow band of blue (this is the Tip lead of pair 1), and one wire that is predominately blue with a narrow band of white (this is the Ring lead of pair 1). The colors of pair 1 are referred to as white-blue and blue-white. White-orange and orange-white are the colors for pair 2. The mate color of white is used with the secondary colors of green, brown, and slate for pairs 3–5. Pairs 6–10 use red as the mate for blue, orange, green, brown, and slate. Black is the mate color for blue, orange, green, brown, and slate on pairs 11–15. Yellow is the mate for pairs 16–20, and violet is the mate for pairs 21–25 (see Table 5-1).

If a cable contains more than 25 pairs, the first 25 are grouped together and wrapped with two plastic strings. One string is white and the other is blue. The second 25 pairs use the same color codes for the wires, but these 25 pairs are wrapped with a white-and-orange string. These plastic strings are called *binders.* Employing the same scheme used in coloring the wires, it is possible to have 25 different binders from white-blue to violet-slate. Twenty-five binders, with 25 pairs each, makes it possible to have 25×25 or 625 pairs in a given cable. Larger cables are constructed by wrapping from 100 to 625 pairs in a white-blue binder, the next 100 to 625 pairs in a white-orange binder, and repeating the same process over and over.

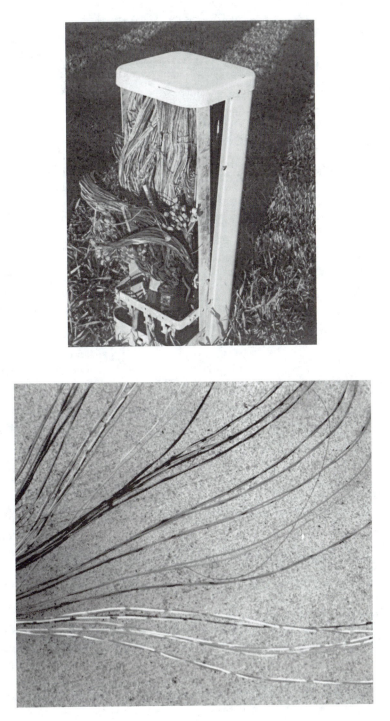

FIGURE 5-4 Plastic insulated cable.

Table 5-1 Plastic Insulated Cable (PIC) COLOR CODE

Mate	Color	Pair Number in Cable	Tip Wire	Ring Wire
White	Blue	1	White-blue	Blue-white
Red	Orange	2	White-orange	Orange-white
Black	Green	3	White-green	Green-white
Yellow	Brown	4	White-brown	Brown-white
Violet	Slate	5	White-slate	Slate-white
		6	Red-blue	Blue-red
		7	Red-orange	Orange-red
		8	Red-green	Green-red
		9	Red-brown	Brown-red
		10	Red-slate	Slate-red
		11	Black-blue	Blue-black
		12	Black-orange	Orange-black
		13	Black-green	Green-black
		14	Black-brown	Brown-black
		15	Black-slate	Slate-black
		16	Yellow-blue	Blue-yellow
		17	Yellow-orange	Orange-yellow
		18	Yellow-green	Green-yellow
		19	Yellow-brown	Brown-yellow
		20	Yellow-slate	Slate-yellow
		21	Violet-blue	Blue-violet
		22	Violet-orange	Orange-violet
		23	Violet-green	Green-violet
		24	Violet-brown	Brown-violet
		25	Violet-slate	Slate-violet

Insulated cables have replaced iron wire and glass insulators on poles. This not only significantly improved the looks of the *outside plant* but also reduced maintenance. The next major improvement in outside plant was to remove the cable from elements of weather by burying the cable. Buried cable requires little maintenance. Most damage to buried cables comes from excavation activity when a road is widened, a new water system established, cable for television placed, or new telephone cable buried. Not surprisingly, the telephone company itself does a lot of damage when it buries a new cable. Cable cuts can be prevented by calling the telephone company a few days ahead and asking it to find its cables and mark their location. In many localities, all utilities have one number listed for cable-locate call-ins. A contractor need only call this number, and all utilities will locate and mark their buried facilities. Contractors and individuals can be assessed fines up to several thousand dollars if they damage cables while digging if they did not call for cable locate services.

5.4 OUTSIDE PLANT RESISTANCE DESIGN

When designing outside plant, the engineer must provide a wire thick enough to furnish proper power to the telephone. The older telephone transmitter needs 23 milliamps of current to work properly. With a central office battery supply of 48 V, SXS central office equipment was originally designed to work with a maximum loop resistance of 1000Ω. Later, the SXS design was improved to handle loops up to 1200Ω. The outside plant engineer would provide the thickness of wire needed to ensure that the resistance of the longest local loop did not exceed 1000/1200Ω. Designing the local loop so that it does not exceed a stated resistance value for the longest loop is referred to as *resistance design.* 26- and 24-gauge wire is the predominant wire size used for most local loops. 19- and 22-gauge wire may be used when the loop is extremely long (7 to 20 miles) but this is rare.

5.5 SXS DESIGN CRITERIA: 27 MILLIAMPS OF LOOP CURRENT

In the *Strowger automatic step-by-step* switching system, the relay that recognized when the called party answered (called a *Ring trip relay*) required 27 milliamps of current to operate. Each telephone had approximately 400Ω of resistance, and the Ring trip relay had 350Ω of resistance. The electrical circuit of the longest local loop consisted of 48 V DC applied from the negative terminal of the central office battery, through the Ring trip relay (350Ω), over the Ring lead of the local loop (500Ω), through the telephone (400Ω), back over the Tip lead of the local loop (500Ω), to the positive battery terminal. Total resistance in this series electrical circuit is 1750Ω (350 + 500 + 400 + 500).

The flow of electricity in a circuit depends on the amount of voltage applied to the circuit and on how much opposition or total resistance the circuit has to the flow of electricity. This is expressed as a formula called *Ohm's law* (refer to Appendix A, Section A.16). Current is equal to total voltage applied to a circuit divided by the resistance of the circuit (current in amps = volts/ohms). In the local loop above, the current in the loop will be equal to: 48 V/1750Ω = 0.02748 amps. This is 27.48 milliamps and satisfies the requirements of the Ring trip relay in the central exchange. Of the 1750Ω total resistance, 1000 was in the local loop.

5.6 TRANSMITTER DESIGN CRITERIA: 23 MILLIAMPS AND 20 MILLIAMPS OF LOOP CURRENT

Over time, improvements in the design of the central office and telephone set have allowed the use of longer loops. Central exchange Ring trip relay design in electromechanical switching systems was improved so the relay could work on 20 milliamps of loop current. This would allow the relay to work on loops as long as 1600Ω, but the transmitter of the telephone required 23 milliamps of loop current and limited the loop to 1200Ω. Engineers worked on the carbon granule transmitter and developed an improved transmitter that will work with 20 milliamps of current.

This development took place about the same time that computerized *stored program control (SPC)* switching systems were introduced. A stored program control switching system is supplied power from a 52-V battery. The combination of higher voltage at the central office and the lower 20-milliamp transmitter current requirement allows the longest loop of a SPC exchange to be 1800Ω under resistance design.

In today's environment the amount of current needed by central office equipment is no longer the requirement governing outside plant design. Electronic, SPC, central offices are designed to work with electrical currents of less than 20 milliamps. SPC switching equipment manufacturers tout designs that will work on loops as long as 1700 and 1800Ω when the telephone is equipped with a 20-milliamp transmitter.

The requirements of the telephone transmitter have become the governing factor in outside plant design. The Bell LECs engineer loops based on the newer 20-milliamp requirement. Most rural telephone companies engineer to the old transmitter requirement of 23 milliamps. When calculating resistance for a local loop in a SPC environment, a 52-V battery supply is used for calculations. Ohm's law can be used to find the maximum resistance possible at 20 milliamps, using a 52-V battery supply, as follows: Resistance = applied volts/amps of current needed. Therefore, 52 V/0.020 amp = 2600Ω. Allowing 400Ω for the line relay at the central exchange and 400Ω for the telephone will leave 1800Ω (2600 – 800 = 1800) as the longest local loop we can use for a SPC exchange. Using a telephone with the newer 20-milliamp transmitter and a central office battery feed of 52 V allows the design of a longer loop (1800Ω). This is the criterion used by manufacturers when they state that their switching system will support a local loop of 1800Ω.

5.7 REVISED RESISTANCE DESIGN

The design concepts for outside plant have been revised, and outside plant engineers now use a concept called *revised resistance design (RRD)* to engineer the local loop. Under revised resistance design, the engineer uses the length of a local loop to determine how the local loop will be designed. Any loop up to 18,000 ft is engineered with nonloaded cable for a maximum loop resistance of 1300Ω. Loops between 18,000 and 24,000 ft are engineered with loaded cable for a maximum loop resistance of 1500Ω. Loaded cable is cable that has additional inductance added to the cable to improve its ability to handle voice signals on long loops. The technique used to load a cable is covered in Section 5.14.

5.8 MODIFIED LONG ROUTE DESIGN

Loops longer than 24,000 ft are engineered using *modified long route design (MLRD)*. MLRD involves placing *range extenders* (also called *loop extenders*) and voice frequency repeaters (VFR) on loaded cable pairs at the central office. The use of these devices is referred to as *loop treatment.* The loop treatment is wired between the line circuit of the switch and the cable pair of the local loop (see

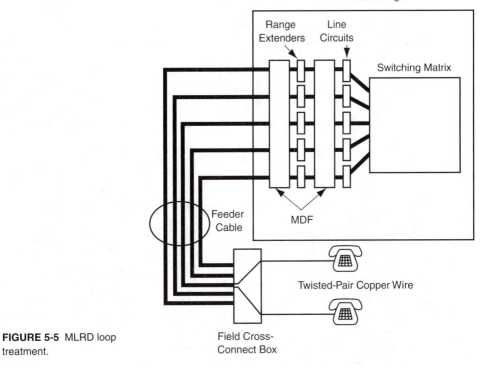

Central Exchange

Range Extenders

Line Circuits

Switching Matrix

Feeder Cable

MDF

Twisted-Pair Copper Wire

FIGURE 5-5 MLRD loop treatment.

Field Cross-Connect Box

Figure 5-5). The range extender boosts the voltage supplied to the line from 52 V to either 78 or 104 V. The amount of voltage boost to use depends on the length of the cable pair served by that particular loop extender. This voltage boost will ensure that the transmitter and DTMF pad receive adequate current. It also ensures enough current for the signaling requirements of the central office equipment. The VF gain circuit is used to amplify voice frequency (VF) signals to ensure that the overall power loss of the circuit is between 4 and 8 dB. In rural areas, some telephones are so far from the office that both loop treatment and 19-gauge wire are used to serve the telephone.

MLRD can be utilized instead of RRD for loops between 18,000 and 24,000 ft. When these longer loops are designed according to RRD, 19- or 22-gauge cable will be required to meet the resistance requirement of 1500Ω. MLRD can be employed on these loops to allow the use of smaller wire (such as 24- or 26-gauge) to serve the same area. Using the smaller wire with a range extender on the line is cheaper than using a thicker wire to serve the same telephone. The range extender is an electronic device that enables a switch to handle local loops exceeding the design requirements of the central office switching system. A range extender can effectively double the range of a SPC central office from 1800 to 3600Ω.

5.9 SERVING AREA OF A CENTRAL OFFICE

The serving area of a central office depends on the geographic area it is situated in. The serving area of a central office in an urban area is smaller than the area served by a rural office. Exchanges (central offices) are often called *wire centers*. One design objective in metropolitan areas has been to limit the physical size of the wire center by limiting the number of wires (cable pairs) served by a center. All cable pairs terminate in a central office on the *main distributing frame (MDF)*. The more wire pairs coming into the exchange, the longer the MDF must be to accommodate all pairs.

The MDF is designed so that cables can be wired to terminal blocks mounted on the front and back sides of the MDF. Terminal blocks are mounted vertically on the back side of the MDF and horizontally on the front side. All outside plant cable terminates on the vertical side of the MDF. All central office equipment (line circuits and trunk circuits) is wired to the horizontal side of the MDF. A short length of wire (called a *jumper wire*) is used to connect a cable pair to a line circuit. The use of a jumper wire allows the connection of any cable pair to any line circuit.

When an MDF serves more than 20,000 lines, many jumpers become very long. With many jumpers lying alongside each other, the possibility of **crosstalk** between adjacent jumper wires becomes a problem. Therefore, most central office exchanges are equipped to serve 10,000 lines, since this is the quantity of telephone numbers available to the *NNX* code assigned to an exchange. Large metropolitan areas, such as Wall Street, will have more than one *NNX* code assigned in the same switching center due to the sheer volume of telephones concentrated in a small geographic area. In this case, the switching center may serve two to five exchange codes and 20,000 to 50,000 lines.

Metropolitan areas have a high population density per square mile, so it is easy to have 10,000 people in an area of 9 to 25 sq mi. Central office exchanges in large cities are generally placed about 4 to 6 mi from each other. Thus, each exchange has a serving radius of 2 to 3 mi in a large city. This means all cables are nonloaded and designed for 1300Ω loops. Exchanges in large cities are designed to serve a small geographic area to keep the quantity of lines served by each exchange to a manageable level and to keep the length of the longest line served under 3 mi. Placing central offices within 6 mi of each other meets this requirement. All loops are shorter than 18,000 ft, and 26-gauge nonloaded cable can be used for all loops. Twenty-six-gauge wire is much cheaper than the coarser gauges (22-, 24-, and 19-gauge).

It is cheaper to use more exchanges and smaller wire in the loop than to use fewer exchanges and coarser cable for the long loops. Theoretically, it is possible to place one central exchange in the heart of a city and have it serve all telephones within a 30-mi radius using MLRD and 19-gauge cable for the longest loops. But the cost of cable would be tremendous. A 19-gauge wire contains almost five times as much copper and costs almost five times as much as a 26-gauge wire. The sheer number of wires coming into the exchange would be unmanageable. A much better design is to use many switching systems scattered around town and to establish small serving areas for each exchange (see Figure 5-6).

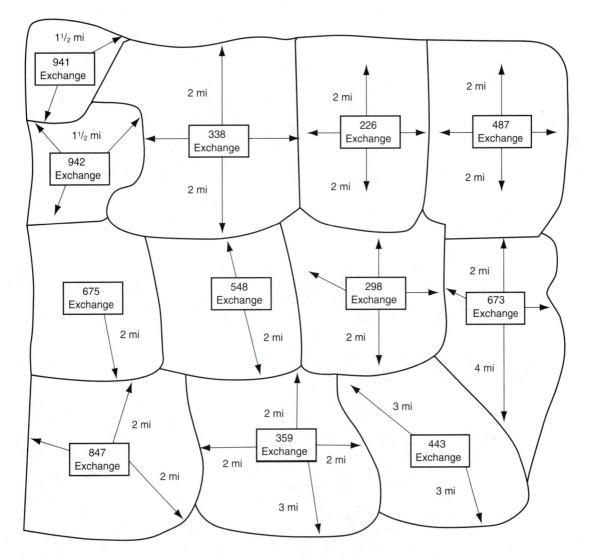

FIGURE 5-6 Typical exchange boundaries in a city.

Cities have a high population density per square mile. When the city is subdivided into smaller exchanges, each exchange still serves thousands of telephones. The investment in the switching equipment of the central exchange can be spread over a large base of customers. On the other hand, the size of a serving area for a rural central office must be large enough to serve several hundred people in order to reduce the switching system cost per customer. This requirement means a rural exchange must serve a large area. We try to get as many lines as possible into each exchange. This reduces the capital expenditures needed for buildings and central office equipment (see Figure 5-7). Designing the outside plant for a maximum loop

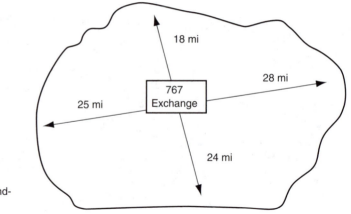

FIGURE 5-7 Exchange boundaries in a rural area.

resistance of 1500Ω, using RRD, will allow for an exchange to serve telephones that are 17 mi away if 19-gauge wire is used. Nineteen-gauge wire has a resistance of about 42.5Ω per mile (8.04Ω per 1000 ft) at 68°F. Since two wires are needed for the loop, *the total resistance for a two-wire loop using 19-gauge wire is about 85Ω per mile.*

5.10 RESISTANCE DESIGN: DETERMINING THE GAUGE OF WIRE TO USE

The size of wire is referred to as the *gauge* of the wire. The larger the number, the smaller the diameter of the wire. A 22-gauge wire has a diameter of 0.025 in. The resistance of a wire varies inversely with the cross-sectional area of the wire. The cross-sectional area of a wire is cut in half when the gauge of the wire goes up by 3. A 22-gauge wire has half the cross-sectional area of a 19-gauge wire; since a 19-gauge wire is twice as large as a 22-gauge wire, the 19-gauge wire has half the resistance of the 22-gauge wire. Twenty-two-gauge copper wire has a resistance of 16Ω per 1000 ft, and 19-gauge copper wire has a resistance of 8Ω per 1000 ft. The resistance will vary slightly as the temperature increases or decreases. Twenty-two, twenty-four, and twenty-six gauge wire are the gauges used predominantly in the local loop. Refer to Appendix B, Table B-2 and Section B.9.

Since two wires are needed in a local loop when using 22-gauge wire, the loop will have a resistance of 2 × 16 = 32Ω per 1000 ft or approximately 170Ω per mile. Earlier, it was determined that the local loop of a SPC exchange cannot have a resistance greater than 1500Ω using RRD. At 170Ω per mile, using 22-gauge cable, we can go approximately 8.8 mi in all directions from the central office. The use of 22-gauge cable allows one central office to serve all telephones within 8.8 mi. Lines longer than 8.8 mi will receive loop treatment.

The total resistance for a two-wire loop using 19-gauge wire is about 85Ω per mile. A loop using 22-gauge wire has a resistance of about 170Ω per mile. Going up on the gauge of a wire by 3 (22 − 19 = 3) doubles the resistance of the wire. The larger the gauge, the smaller the wire. A local loop using 24-gauge wire has a resistance of about 275Ω per mile (26Ω per wire per 1000 ft), and *a loop using 26-gauge wire has a resistance of about 440Ω per mile at 68°F.* A local loop constructed of 26-gauge wire is limited to serving a 3-mi radius. This is another reason that the serving area for a metropolitan office has a 3-mi radius. It allows the use of 26-gauge cable in the local loop. The smaller the wire (the higher the gauge number), the cheaper the cable. LECs can save a tremendous amount of money by using the smallest-diameter wire possible in the local loop.

A rural exchange can also use nonloaded 26-gauge cable pairs to serve customers within 3 mi of the exchange, but will require expensive 19-gauge loaded cable to serve a 17-mi radius. These long loops could be served with range extenders and VF gain devices (MLRD) to allow the use of 22-gauge cable. RRD could be used on loops that are 1500 to 2400Ω, and MLRD could be used on the loops between 2400 and 2800Ω. The engineer selects the design that will result in the lowest cost to install, which will depend on the number of lines involved. Nineteen-gauge loaded cable can be used with range extenders and VF gain to serve a 34-mi loop. This effectively doubles the range of an 1800-Ω SPC exchange to 3600Ω. Outside plant engineers usually limit the maximum loop resistance of the longest loop to 2800Ω. By using 19-gauge cable and loop extenders, a 2800Ω loop can extend 32 mi from the central exchange. The use of 22-gauge cable (with loop treatment) will cut the serving distance in half (to 16 mi). The maximum distance that can be covered, using range extenders and VF gain, with loaded 24- or 26-gauge cable is 10.2 and 6.4 mi respectively. The outside plant engineer can select RRD or MLRD to serve each area of an exchange. The use of 22-gauge loaded cable will allow serving all lines within 8.8 mi of the exchange without the use of loop treatment (that is, range extenders and VF gain). Going beyond 8.8 mi requires the use of 19-gauge cable or design of the loop according to MLRD.

Another option available to the outside plant engineer is to use a subscriber carrier system for serving a particular area of the exchange. Subscriber carrier will require the use of two nonloaded cable pairs with repeaters placed every mile along the route. Subscriber carrier is available to serve from one line to hundreds of lines. Most subscriber carrier systems are designed to accommodate 96 lines and are called *subscriber line carrier-96 (SLC-96).* A SLC-96 can be placed as far from an exchange as desired.

Subscriber carrier is often used to provide service to a rapidly growing part of the exchange serving area. If a new housing development springs up in an area that has only a few vacant cable pairs, subscriber carrier can be used to provide service until new cable can be added to the area or until the area can be served permanently with subscriber carrier. As mentioned early in this chapter, many SLC-96 systems can also be multiplexed over a fiber facility to serve a remote area of the exchange (see Figure 5-7).

5.11 CARRIER SERVING AREA

So far we have discussed the serving area of a central office. The serving area for each central office is established based on the longest loop that the engineer wishes to have the exchange handle. The outermost customer establishes the *exchange boundary* for a central office. Another term used in outside plant engineering is *carrier serving area (CSA).* This is a distant area of the exchange that can support access to DS0 digital service and ISDN without special loop treatment. A CSA is served by a subscriber carrier system. A CSA cannot have any loops longer than 12,000 ft from the subscriber carrier field terminal, using 24-gauge or coarser wire. If 26-gauge cable is used, the loop cannot be longer than 9000 ft. All loops must use nonloaded cable pairs.

Establishing an area distant from the central exchange as a CSA, is done through the use of subscriber carrier. The design rules for CSA apply to the cable pairs extending out from the subscriber carrier to telephones (see Figure 5-8). The CSA requirements are also met by all lines within 2 mi of the central office wire

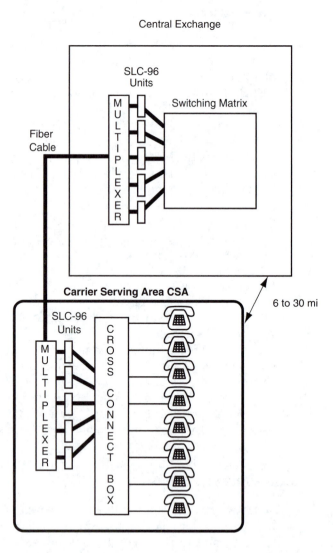

FIGURE 5-8 Carrier serving area.

center, but these lines are not part of a CSA because they are not served by subscriber carrier. These lines will support digital services and are administered the same way as lines in a CSA.

5.12 VOICE SIGNALS (AC) IN THE LOCAL LOOP

The resistance of wires limits how far a phone can be placed from a central exchange. We noted earlier that the new carbon transmitter needs 20 milliamps of current to work properly. Using Ohm's law to find voltage, we can calculate the amount of voltage at the phone: volts = current × resistance. Thus, volts = 0.020 amp × 400Ω (resistance of the phone) = 8 V. We apply 52 V DC to a line at the central office. As current flows through the line relay and local loop, we lose power and less voltage is available at the phone than is applied at the central exchange.

Resistance design of the outside plant ensures that the DC power loss over the local loop is within limits. But DC power loss is not the only factor to consider when using wire as a transmission medium. Voice signals are converted by the transmitter of the phone into variances in electric current. These variances look like AC signals. If a local loop has 30 milliamps of current flowing through it when no one is talking, the current will vary when someone talks through the transmitter. The current changes may be 29.95, 30.05, 29.96, 30.04, or almost any value between 27 and 33 milliamps. The varying DC signal looks like an AC signal that has a center point of 30 milliamps.

5.13 CAPACITIVE REACTANCE CONSIDERATIONS

Wire has properties other than resistance that must be considered when an AC signal is placed over the wire. A wire pair acts like a capacitor. A capacitor is made by separating two conductors with an insulating material. The two wires of our local loop (Tip and Ring) are very close together in a cable and are separated by the plastic insulation around each wire. The wires of the local loop have a capacitance of approximately 0.083 microfarad (µf) per mile. The longer the loop the greater the capacitance. When an AC signal is placed on a wire pair, the capacitance between the two wires will affect the signal. The capacitance between the two wires allows some of the voice signal to leak from one wire to the other. The amount of signal leakage that occurs is governed by the amount of opposition the capacitance between the two wires offers to the signal. Capacitive reactance measures the opposition that a capacitor has to an AC signal.

Capacitive reactance is inversely related to the frequency of an AC signal, and inversely related to the amount of capacitance between the wire pair. The higher the frequency of an AC signal, the lower the opposition of a capacitor to the signal. The higher the capacitance a device has, the less opposition it offers to an AC signal. The lower the frequency of a signal the higher the capacitive reactance. This is why a capacitor blocks a DC signal. The frequency of a DC signal is 0 and opposition to the signal is maximum. The longer the cable pair, the higher the capacitance will be. The longer a local loop is, the more signal loss we have due to leakage through the capacitance between the wires.

On local loops longer than 3 mi, capacitance of the cable pair becomes great enough to degrade a voice signal. Voice signals consist of frequencies between 300 and 3300 Hz. On loops longer than 3 mi, capacitive reactance causes a significant power loss to signals above 1000 Hz. The higher the frequency, the greater the loss. This effect causes the level of signal heard at the called party's receiver to go up and down as the frequency of the signal changes. This can be annoying. To prevent this from happening, more inductance is purposely put into the local loop to offset the effects of capacitive reactance. Inductive reactance and capacitive reactance tend to offset each other. The extra inductance is designed into a device called a *load coil*.

5.14 LOADING CABLE PAIRS TO IMPROVE VOICE TRANSMISSION

Cable pairs shorter than 3 mi do not require *loading.* For cable pairs longer than 18,000 ft, *load coils* are used. What type of load coil to use and how far apart to place them depends on the capacitance between the wires and the gauge of the wire. The type and thickness of insulation used will determine the capacitance between a pair of wires. This capacitance between the two wires (Tip and Ring) of a line circuit is called *mutual capacitance.* Cables manufactured for the local loop have 0.083 μf of capacitance per mile regardless of the gauge of wire in the cable. Some cables are designed for special high-frequency applications, such as for use with T1 carrier systems. These cables have a capacitance of 0.066 μf per mile.

Most loaded local loops will use 22- and 19-gauge wire. If a loop needs to be loaded, it is longer than 3 mi. Loops longer than 3 mi require the use of 22-gauge wire. Loops longer than 8.8 mi require the use of 19-gauge wire, or the use of loop extenders. Regardless of the gauge of wire used, the mutual capacitance is 0.083 μf per mile. To offset this mutual capacitance, load coils having 88 millihenries (mh) of inductance are placed at 6000-ft intervals on the cable. The use of 88-mh load coils results in a cutoff frequency of approximately 3800 Hz. The cutoff frequency of a circuit is the frequency at which 70.7% of the input voltage reaches the output of the circuit. A 29.3% reduction in voltage level is a change of 3 dB. Notice in Figure 5-10 that the signal has experienced a drop of 3 dB (from –3 to –6 dB) at approximately 3800 Hz.

The cutoff frequency of the local loop can be raised to 7400 Hz by using 44-mh coils and placing them at 3000-ft intervals—in other words, using half the inductance of an 88-mh coil and placing twice as many coils on the loop. Therefore, 44-mh loading costs twice as much as 88-mh loading. Almost all loops are loading with 88-mh load coils. The use of 44-mh load coils is reserved for use on loops that must be specially conditioned to handle frequencies between 3800 and 7400 Hz. Since most local loop uses 88-mh load coils, loaded loops cannot effectively pass signals above 3800 Hz.

Spacing load coils 3000 ft apart is called *B spacing*, spacing 4500 ft apart is *D spacing*, and 6000-ft spacing is *H spacing*. A load coil designated as 19H88 means it is an 88-mh coil designed for use on 19-gauge cable and 6000-ft spacing. When the outside plant engineer determines that cables serving a particular area are going to be more than 3 mi from the central exchange, the cable is loaded. The first load coil will be placed 3000 ft from the central office. The second load coil is placed 9000 ft from the central office, and a third load coil, if needed, is placed 15,000 ft from the central office. The first load coil is placed at half the recommended distance for the coil being used.

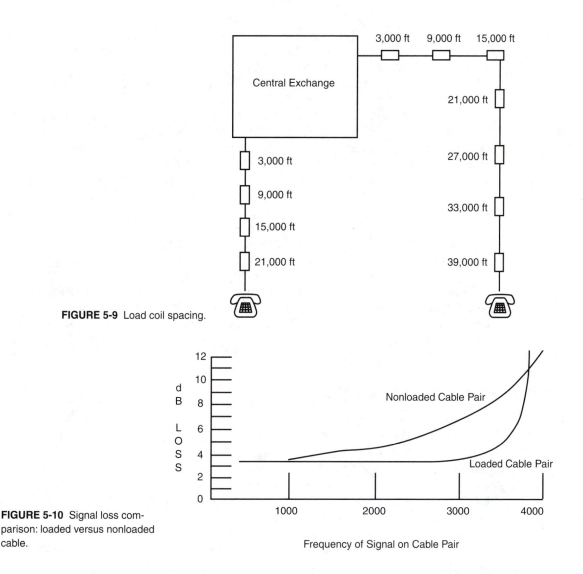

FIGURE 5-9 Load coil spacing.

FIGURE 5-10 Signal loss comparison: loaded versus nonloaded cable.

Load coils are enclosed in one case and connected to two-cable stubs that will protrude outside the case. When a cable needs to be loaded, it is cut. The load coil is spliced in between the end of the cable going to the exchange and the end of the cable that goes to telephones. To load a 200-pair cable, a load case containing 200 coils will be attached to a pole for aerial cable, or placed in a pedestal for buried cable, and then spliced into the cable. Note that the first load coil is placed 3000 ft from the central exchange. When a telephone call is established between two lines of the exchange, this results in a 6000-ft spacing. The first load coil of each line is 3000 ft from the exchange. When the exchange connects the two lines, these two load coils are 6000 ft apart. Thus, even though the first load coil is placed 3000 ft from the exchange, an H88 load coil is used for the first load coil, because the spacing between the two load coils of a connected call will be 6000 ft (see Figure 5-9).

Loading a cable flattens out the power loss for all signals below 3000 Hz (see Figure 5-10). Signals between 3000 and 4000 Hz will have increasing power losses, and signals above 4000 Hz will have so much power loss that the signal will be hard to hear. To achieve maximum transmission, the telephone should not be closer than 3000 ft, or more than 9000 ft from the last load coil serving the telephone.

The single biggest disadvantage of using wire as a medium is the narrow bandwidth of wire due to mutual capacitance and loading. It can only carry signals having frequencies between 0 and 3800 cps efficiently. Data signals are far higher frequencies and cannot be carried over cable unless they are regenerated every mile. Putting regenerators on every local loop would be very costly. It is possible to convert voice into a digital signal at the phone; an *Integrated Services Digital Network (ISDN)* phone does exactly that. However, ISDN phones do not work more than 3 mi from the exchange, unless the ISDN circuit is being provided over subscriber carrier or a specially conditioned cable pair.

Load coils improve the low-frequency response of a cable but add more loss to high-frequency signals. When the outside plant engineer knows a particular cable pair is going to carry high-frequency signals, an order is issued to take the load coil off of that pair, and the pair is spliced together without going through the load coil case. If the cable pair is going to be used for a digital signal, a signal regenerator is placed between the two pairs of wire at the point where the load coil is removed.

5.15 DATA ON THE LOCAL LOOP

The local telecommunications network was designed to handle voice frequencies. Data can be sent over specially designed local loops and central office equipment. When a business needs to transmit high volumes of data, it will lease these special facilities and the local loop will be conditioned with special equipment such as equalizers and amplifiers. Low volumes of data can be sent over regular phone lines and facilities by using a modem. The modem is placed between the digital output of a computer and the analog phone line. The modem converts digital signals received from the computer into analog signals between 300 and 3300 Hz. These signals are handled by the telecommunications network with no problem (see Figure 5-11).

When modems are used, they must be used in pairs, with one modem at the transmitting end and another at the receiving end. The modem at the receiving end converts the received analog signals into the appropriate digital signal for the computer on the receiving end. Digital data signals can use wire for transmission, but only for a short distance. If regular telephone wire (unshielded twisted pair) is used for data, the signals must be regenerated every 1 to 3 mi depending on the speed of data transmitted. ISDN lines use digital subscriber line circuits, which will work up to 3 mi. The customer supplies an ISDN device and attaches it to the digital line. If digital data lines are leased from the LEC, the customer must provide a digital interface device instead of a modem. The digital interface device for data is called a *customer service unit (CSU)* or *data service unit (DSU)*.

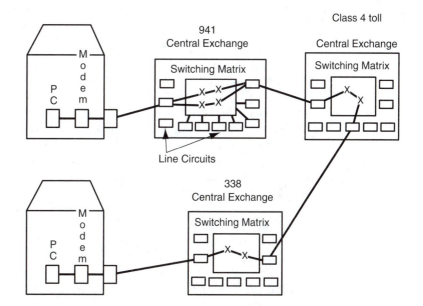

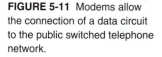

FIGURE 5-11 Modems allow the connection of a data circuit to the public switched telephone network.

5.16 POWER-LOSS DESIGN

The amount of power loss that any signal has over a medium is measured in decibels (dB). Decibels are used to compare two power levels that vary in a logarithmic fashion. A change in power level barely noticeable to the human ear is 1 dB. A decibel measures the relationship between two power levels; therefore, a decibel is not an absolute measurement. If one signal has twice the power of a second signal, the difference between the two signals is 3 dB. The difference between a 1-mW signal and a 2-mW signal is 3 dB.

Likewise, the difference between a 1-MW signal and a 2-MW signal is 3 dB. In the first case, a change of one-millionth of a watt is all that is necessary to double the power, but in the second case, a million watts is needed to double the power. In both cases the change in power is a change of 3 dB. The formula for calculating dB is: dB = 10 log P_2/P_1, where power = P. Anytime P_2 is twice the value of P_1, the value of P_2/P_1=2. This represents a doubling of power. The log of 2 is 0.3010 and 10 times 0.3010 equals 3.01. Thus, according to the formula, 3 dB equals a doubling of power. Every 3-dB change is a double of power. A change from 1 to 2 mW equals 3 dB. A change from 2 to 4 mW equals 3 dB.

To provide an absolute type of power measurement in telecommunications, all signals are measured with respect to a signal level of 1 mW when a test set is used. The test set used to measure a signal compares the measured signal to a 1-mW reference level. Decibels measured with reference to 1 mW are called *dBm*. A signal of 0 dBm = 1 mW. A signal of 3 dBm equals 2 mW. A signal of 6 dBm equals 4 mW, and so on. When a signal is measured it will result in a measurement expressed as dBm. If the signal ratio is a calculation rather than a measurement, it is stated simply as dB.

For example, the change of a signal from 4 to 8 mW would be a change of 3 dB. If the signals were measured, the measurement at the 4-mW point would be 6 dBm, and the measurement at the 8-mW point would be 9 dBm, since both are measured with respect to a power level of 1 mW. A gain in power is stated as positive dB and a loss in power as negative dB. Note that going from 4 to 8 mW is +3 dB, and going from 8 to 4 mW is –3 dB. The formula also produces the same result: dB = 10 log P_2/P_1 = 10 log 4/8 = 10 log 0.5. The log of 0.5 = –0.3, and 10 times –0.3 = –3. If the power of one signal is half the power of another signal, the relationship between the two is –3 dB. Thus, doubling the power is 3 dB and half the power is –3 dB.

For the local loop, this power loss over the wire between the central exchange and the telephone should not exceed 8.5 dB with a signal that is 1000 Hz. The loss at 500 and 2700 Hz should be within 2.5 dB of the loss measured at 1000 Hz. In each central office a test number is assigned to a piece of equipment that will generate a test tone of 1004 Hz for a few seconds, then transmit 500 Hz for a few seconds, and then transmit 2700 Hz for a few seconds. Each test signal is generated at 0 dBm (1 mW). After the 2700-Hz signal has been sent, the circuit becomes quiet until the device is released.

A repairperson can call into the test number with a special test set. The repairperson will connect the test set to the Tip and Ring of a local loop at the customer's house. The test set can measure loop current from the central office to determine if it is at least 23 milliamps. The repairperson can dial the test number to measure loss at the various frequencies. After the test number has sent all three tones and is quiet, the test set can be used to measure noise on the line. Noise should not be more than –60 dBm (30 dBmC0). As noted earlier, dBm is decibels as referenced to a power level of 1 mW (0 dBm = 1 mW). Also, dBmC0 is decibels as referenced to noise with C message weighting. Zero (0) dBmC0 is equal to –90 dBm (90 dB less than 1 mW). This level (0 dBmC0) is set as the starting (or reference) point for noise measurements.

Appendix B covers dB formulas in detail. We can rearrange the dB formula (dB = 10 log P^{-2}/P^{-1}) to find the power level at -90 dBm: dB = 10 log P_2/P_1. Rearranging the formula results in: antilog (dB/10) × P_1 = P_2. Substituting known numbers results in: antilog (-90/10) × 1 mW = P_2; antilog –9 × 0.001 = P_2; the antilog of –9 is 0.000000001 and the antilog of –9 × 0.001 W = 0.000000000001 W or 1 pW. Thus 1 pW equals -90 dBm or 0 dBmC0. This is the starting point for noise measurements. Noise should not exceed 30 dBmC0 on the local loop. This limit of 30 dBmC0 is 1 nW of noise. Noise has to be below this extremely low level because the power of the voice signals is extremely low (less than 1 mW).

Noise on the local loop can usually be traced to an unbalanced line, crossed lines, induction, or a ground fault. These problems are often caused by water entering a cable and damaging the insulation around the wires. Unbalanced lines can also be the result of a poor splice where two cables are joined. When party lines existed, they used grounded ringing, which caused unbalanced lines.

Induction noise results when the aluminum sheath surrounding all wires in a cable has been destroyed or improperly grounded. This can allow noise induction

FIGURE 5-12 Twisted-wire pair.

from power lines when a telephone cable is mounted on a power-line pole. Induction between wires within a cable would be a problem if the wires of a pair were not twisted around each other. When one wire is parallel and close to a wire carrying an electrical signal, it acts like the secondary winding of a transformer and will pick up the signal of the current-carrying wire. Twisting the wire pair prevents this source of crosstalk. The amount of twist per inch to use on cable pairs has been scientifically determined.

When current leaves the central exchange, it flows out over the Ring wire and returns over the Tip wire. Twisting will first cause the Ring lead of one pair to lie next to an adjacent wire, then the Tip will lie next to the adjacent pair. Any induced signal in the adjacent wire from the Ring lead will be offset by the signal induced in an opposite direction by the Tip lead. Wires must lie close to each other in order to have many wires in one cable. Twisting the wire pairs and then twisting the pairs around each other eliminates induction and enables wire pairs to be placed close together. Twisting also eliminates the inductive effect between the two wires of the same pair (see Figure 5-12).

5.17 FEEDER CABLES, DISTRIBUTION CABLES, AND DROP WIRES

The cables that leave a wire center (central exchange) are called *main feeder cables.* These are large cables containing 1800 to 3600 pairs of wires. The *feeder cable* has a direct route to the area of an exchange it will serve. The feeder cable leaves the central exchange through an underground conduit, which is a metal or plastic tube that comes in many diameters. The conduits used for feeder cables are usually plastic tubes with a diameter of 4 to 6 in. The conduit extends from the central exchange to a large manhole near the exchange. Conduits are placed under the streets in all directions from the central exchange; feeder cables are pulled through these conduits to serve different areas. Manholes are placed at intervals along the route to allow for splicing cables together. Sometimes a long length of cable will be pulled through several manholes. The cable will be passed through the manhole from one conduit to another without splicing.

There is a limit on how long the cable pulled through a conduit can be. If you try to pull a very long cable through several manholes, the pulling strain on the cable can be great enough to damage the cable. The length of cable that can be pulled without damage will vary according to its size and construction. Few cables can be run as one continuous cable from the central exchange to their final destination. The cable run is broken up into smaller segments, and these segments are spliced together in the manholes. It will also be necessary to take some cable pairs out of a cable and run them in a different direction. A feeder cable may arrive in the manhole at the middle of an intersection of two streets as a 3600-pair cable. The outside plant engineer may specify splicing: (1) 900 of the 3600 pairs to a 900-pair cable in a westbound conduit; (2) 900 pairs to a 900-pair cable in an eastbound conduit;

and (3) 1800 pairs to an 1800-pair cable in a conduit that continues in the same direction as the original feeder cable.

Manholes provide access to underground cable and serve as a junction point for many cables going in different directions. One or more of the cables coming into the manhole will be a feeder cable. The feeder cables start out as a large cable at the central exchange and are spliced into smaller cables called **branch feeders.** Each branch feeder serves a specific area of the exchange. Branch feeders will be spliced into smaller cables called **distribution cables** that are 25- to 400-pair cables. These distribution cables fan out in all directions from the branch feeder to serve homes and businesses (see Figure 5-13).

In older downtown locations, the distribution cables are aerial cables usually placed on poles in the alley. Because an alley is between two streets, the distribution cable will serve customers on the west side of one street and customers on the east side of the other street. Access points are provided on aerial cable for attaching the wires from a home or business. The access points are called *ready access terminals,* and the wires going to a home or business are called a **drop wire** or **drop cable.** An aerial drop consists of two wires encased and separated from each other in black rubber or plastic material.

Housing developments in a suburban location are usually served with buried distribution cable; it is buried in the backyard along the property dividing line between houses. Access to the cable is provided at devices called *pedestals.* A pedestal is a steel can about 18 in. high and 6 in. square. Pedestals are usually placed at the intersecting property line for four houses. The drop cables from the four houses are buried directly in the ground on a line-of-sight path between the house and the pedestal. Buried drops usually have eight wires in the cable. The cable is impregnated with a petroleum jelly to keep out moisture. Larger pedestals are used to gain access for splicing buried branch feeders to buried distribution cables. These pedestals are steel cans 3 ft high and 9 in. square (see Figure 5-14).

5.18 SUBSCRIBER CARRIER

Subscriber carrier is used in outside plant as a pair-gain device. In an urban area where a new subdivision or resort complex has been built, the engineer can choose between running new cable to the subdivision or using subscriber carrier. It is usually more economical to use subscriber carrier until the area builds up with more than 200 homes. Subscriber carrier is an electronic device operating like the carrier systems used to carry telephone calls between central offices. Carrier systems are devices that place many conversations over one cable pair; placing many conversations over one wire is called *multiplexing.* (The different ways that multiplexing is achieved are covered in Chapter 6. As many as 96 telephones can be placed on a subscriber carrier system. This system is often called SLC-96 (pronounced "slick 96").

SLC stands for subscriber line carrier (or subscriber loop carrier). The number following SLC indicates how many lines (referred to as *channels*) it will handle. The subscriber line carrier uses two cable pairs. The engineer will use two existing cable pairs for the SLC-96 and thus saves (or gains) 94 pairs. Most SLCs use time division technology to put 96 telephones on two cable pairs. (Time division technology is covered in greater detail in Chapter 6.)

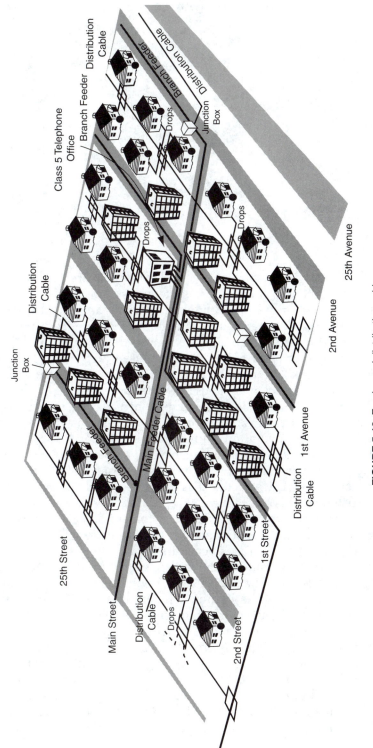

FIGURE 5-13 Feeder and distribution cables.

FIGURE 5-14 Picture of pedestal.

Smaller versions of SLC are available when needed. Single-channel and eight-channel systems, which will work over one cable pair, are used when it is necessary to serve fewer than 96 lines. These systems use amplitude modulation and frequency division multiplexing technologies, which are also covered in Chapter 6. These systems work like radio transmitters/receivers and require the use of a non-loaded cable pair. Single-channel systems must be used within 18,000 ft of the exchange, but eight-channel systems can be used anywhere. The radio transmitters/receivers work like an AM radio, and the systems are called *AML-1* and *AML-8* to designate how many lines they will serve.

AML-1 can be used to add a customer when all cable pairs serving the area are already in use. An area close to the central exchange may have experienced some growth that used up all vacant (unused) cable pairs and no pairs are available to serve the new customer. It is also possible that all pairs may be in use when one of the existing pairs goes bad and cannot be fixed. Instead of running a new cable, the engineer can use AML-1, selecting a cable pair currently in use by a neighbor of the affected customer. This cable pair will be used to serve both customers.

The drop wire serving the affected customer is attached to the same cable pair serving the neighbor. A radio transmitter/receiver is attached to each end of the circuit. A transmitter/receiver is attached to the drop wire serving the new customer. The mate for that transmitter/receiver is installed in the central exchange and attached to the cable pair. The new customer will be served by the radio, which uses the neighbor's cable pair as its transmission medium. The old customer will be served by the physical cable pair. A filter circuit is placed on the drop wire of the old customer to prevent the radio signal from causing noise (see Figure 5-15).

AML-8 will serve eight customers by using one cable pair for the system. The central office equipment for AML-8 has eight transmitters and receivers, each of which acts like a radio. Each radio operates on a different frequency. The AML-8 unit to be placed in the field is situated at the location where branch feeder cable meets distribution cable. One side of the AML-8 central office equipment is connected to eight line circuits, and the other side is connected to one feeder cable pair.

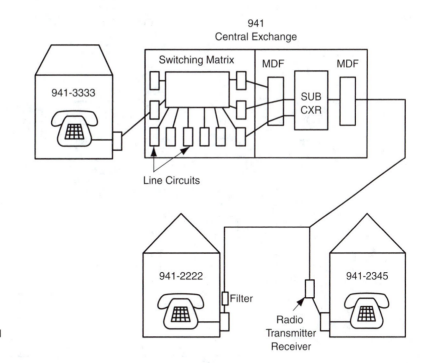

FIGURE 5-15 Single-channel subscriber carrier.

The field unit is also connected to the same feeder pair; the other side of it is connected to eight different distribution pairs.

Each of the eight customers attached to AML-8 will have their own unique radio within the AML-8—a technology similar to commercial broadcast radio technology. Many stations are transmitting at the same time over the same facility, but you can only hear the station your receiver is tuned to. To provide two-way communication, each end of the AML-8 system has eight transmitters and eight receivers. This requires the use of sixteen different radio frequencies. The voice signals placed over AML-8 will not interfere with each other; the person assigned to channel 1 is tuned to the channel 1 frequency and can only pick up signals for channel 1 (see Figure 5-16).

5.19 MEDIUM FOR LONG DISTANCE NETWORKS

When long distance networks were first established, the only medium available was wire. Toll trunks used a larger gauge of wire than the local loop. Most wires between cities were 16- or 19-gauge copper wire. These media were also equipped with *voice frequency repeaters (VFRs)* on each end. A VFR could amplify voice signals as much as 12 dB. The VFR was adjusted to compensate for all but 4 dB of a toll circuit's loss; thus, all toll circuits would have a total loss of 4 dB regardless of the distance between cities. When radio carrier systems were introduced to multiplex 12 or 24 trunk circuits together, copper wire was also used as the medium for these multiplexing systems.

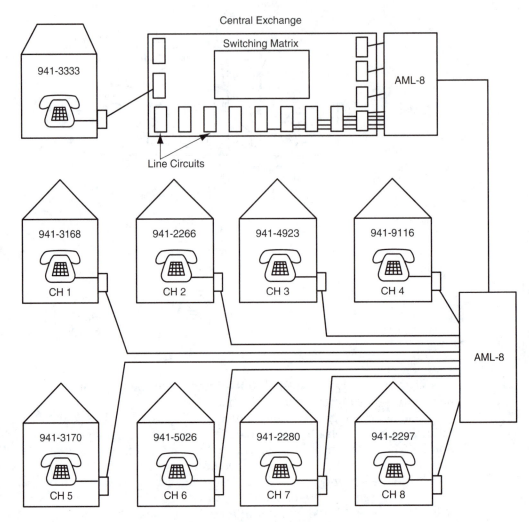

FIGURE 5-16 Eight-channel subscriber carrier.

Later, AT&T developed a frequency division multiplexing (FDM) carrier system that could multiplex 600 conversations or toll trunks onto one facility. This carrier system could not use regular copper wire but had to use coaxial cable as its medium. The cable used to connect a television set to the cable television company line is coaxial cable. AT&T continued to develop the capabilities of these systems and in 1978 had an L5E system that could place 13,200 trunk circuits on one coaxial cable. The coaxial cable used had 20 pairs plus 2 spare pairs. This cable could handle 10 L5E systems (132,000 two-way trunk circuits) between switching centers with repeaters or regenerators placed at intervals of 1 mi on the cable.

AT&T also developed carrier systems that used time division multiplexing (TDM) technology. The basic system was a 24-channel system called *T1*. A 48-channel (T1-C) and a 96-channel (T2) system were also developed by AT&T. These systems could use twisted-pair copper wire as their medium. The wire pair had to be un-loaded and regenerators placed at 1-mi intervals along the cable route. In addition, AT&T developed a 672-channel TDM carrier system. This system operates at 44.736 Mbps and consists of 28 T1 systems multiplexed together. The system is called a *T3 carrier system,* and the signal level (44.736 Mbps) is called a *DS3 signal.* The high speed of this signal limits the medium choice to coaxial cable, microwave, and fiber.

Microwave radio systems were also developed by AT&T. These systems could multiplex up to 28,244 trunk circuits. Microwave signals are referred to as *line-of-sight transmission.* The signals are very high frequency signals (4 to 11 gigahertz) that are focused by the transmitting antenna into a narrow beam and transmitted in a straight line. The curvature of the earth and the height of an antenna limit how far apart transmitters and receivers can be. The microwave towers used in the PSTN are placed 20 to 30 mi apart depending on the terrain. The signal received at one of the relay towers is demodulated and fed into a transmitter, which transmits to the next tower down the line.

Microwave signals can be impacted by rain, but different frequencies are affected by different rain patterns. To prevent signal fade due to rain or other atmospheric conditions, two different transmitters (thus two different frequencies) are connected to each dish on the microwave tower. Equipment at the receiver monitors and compares the two received signals. The signal of highest quality is demodulated and used as the input to the transmitters, which feed the signals to the next tower in line. Microwave and coaxial cable have given way to fiber optic cable as the medium of choice in the toll network. AT&T still has L5E coaxial systems and microwave systems in use but has converted much of its long distance network to fiber. Sprint's long distance network uses fiber exclusively as the transmission medium.

5.20 FIBER OPTIC CABLE

A fiber optic strand is made by surrounding a thin fiber of glass with another layer of glass called *cladding.* The cladding acts like a mirror and reflects light down the fiber. The glass core and cladding are surrounded by a jacket to prevent stray light from entering the fiber. A fiber optic communication system consists of a light pulse generator, the glass fiber, and a receiver that can detect changes in the light pulses.

Fiber optic systems are noise free. Noise is caused by electromagnetic induction on top of electrical signals. Electrical signals are susceptible to noise, but fiber does not use these signals; it utilizes light signals, which cannot be impacted by electromagnetic induction. The fiber has been improved over time, and today most high-capacity systems use a fiber known as *single-mode fiber.* This fiber does not provide multiple paths for reflected light, and the light travels along the axis of the fiber. Single-mode fiber uses lasers as the light source.

AT&T introduced fiber optic technology in 1979 using graded-index fiber. This system multiplexed 672 trunk circuits onto one fiber using a signal rate of 44.736 Mbps (digital signal level 3 or DS3). In 1984 AT&T introduced a single-mode fiber system that multiplexed 2688 trunk circuits on one fiber. In 1986, a single-mode fiber was handling 6048 trunk circuits with repeaters spaced at 20-mi intervals. Using a standard OC-48 time division multiplexer, the standard single-mode fiber today is carrying

32,256 trunk circuits using a signal rate of 2.488320 gigabits per second. Repeaters are spaced at 27-mi intervals. Repeaters on fiber optic cables placed in the ocean are spaced about 75 mi apart. Repeaters can be spaced farther apart at lower speeds.

Sprint and AT&T have upgraded their fiber routes to handle as many as 516,096 trunk circuits (channels). This has been accomplished by employing *dense wave division multiplexing (DWDM) technology.* Instead of using strictly TDM technology and trying to extend this technology to multiplex many more signals in a TDM mode, DWDM technology combines FDM technology and TDM technology. The light spectrum of a fiber optic facility is broken down into multiple frequencies. Light waves are composed of many frequencies. Blue light has a different frequency from red light, orange light, green light, and so on. Transmitters and receivers for fiber optic facilities have been developed that transmit and receive at a specific frequency.

At present, DWDM uses 16 or 32 different frequencies. An individual OC-48 TDM connects to each of the 16 or 32 different transmitters in a dense wave division multiplexer. By splitting the light spectrum into 16 different frequencies and attaching an OC-48 to each transmitter, the fiber optic facility can carry 516,096 channels (32,256 × 16). The next advance will be to use an OC-96 or OC-192 as the TDM device on a DWDM. Each OC-192 can carry 129,024 channels. By placing OC-192s on a 16-channel DWDM, the fiber can carry 2,064,384 channels (16 × 129,024 = 2,064,384) (see Figure 5-17).

A fiber optic strand carries signals in one direction only. For transmission in both directions at the same time, two fibers are needed for each fiber optic circuit. The basic fiber cable for long distance will have 72 fiber pairs and one steel wire. The steel wire is used to pull the cable between manholes. Six of the 72 pairs are reserved as spares. With 66 available pairs, this cable can handle 2,128,896 trunk circuits (66 × 32,256) when OC-48 multiplexers are used. The same cable can carry 34,062,336 channels (16 × 66 × 32,256) when the facility uses OC-48 TDM and DWDM technology. By using OC-192 TDM and DWDM, the same 72-pair cable can carry 136,249,340 calls.

It is estimated that a single-mode fiber can handle 3 million trunk circuits when the transmitting lasers and receiving devices are developed that can generate and detect a signal of 250 gigabits per second (250 billion bps). This is ten times the signal rate of an OC-48 multiplexer. This signal rate has the capacity to handle more than 3 million voice circuits or transmit 25,000 books in one second. The OC-48 capacity of 32,256 voice circuits over one fiber is equivalent to transmitting about 250 books per second. By combining a 250-gigabit TDM multiplexer and DWDM technology, a fiber could carry 50 million circuits.

As improvements are made in the laser transmitters and receivers to achieve greater capacity, improvements in their design have also led to wider spacing for repeaters. A major improvement made in fiber optic transmission systems was the simple adoption of standards. Prior to standardization, different vendors' systems could not be connected directly to each other. The adoption of synchronous optical network (SONET) standards was a major boost to the deployment of fiber transmission facilities and networks. The basic SONET signal rate in the United States is OC-1 (672 trunk circuits).

DWDM

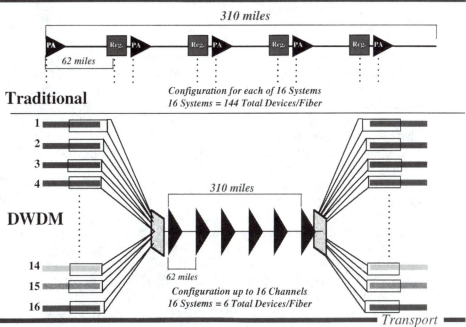

FIGURE 5-17 Dense wave division multiplexing. (Courtesy of Sprint.)

A T3 carrier system uses a DS3 signal rate of 44.7 Mbps to multiplex 672 trunk circuits. An OC-1 SONET system uses a STS-1 (synchronous transport signal-1) rate of 51.8 Mbps to multiplex 672 trunk circuits. The difference in signal rates is due to the additional overhead in the STS-1 signal. The SONET level of OC-48 (48 OC-1 systems or 32,256 trunk circuits) (48×672) was the standard for telecommunications but is rapidly being replaced by OC-192s on a wave division multiplexer. SONET standards permit the interconnection of different vendors' equipment and allow stripping out a DS0, DS1, or DS3 signal without having to demultiplex the whole bit stream. When DWDM is used, it is necessary to use a DWDM receiver that will demodulate one complete frequency in order to have access to the OC-48 or OC-192 TDM. The required channels can then be pulled out of the OC-48 or OC-192 TDM stream.

5.21 FIBER TECHNOLOGY

When optical communications was first introduced, light waves were sent through the air from a transmitter to a receiver. When fiber optic communications was introduced in the lab, light waves were sent through glass strands, and transmission was limited to a few yards due to severe attenuation of the signal by the fiber media. Glass fibers caused signal loss of more than 1000 dB per meter of glass fiber. It was evident that the key to commercial application of fiber optic transmission was to significantly reduce this attenuation. The major source of attenuation was impurities in the glass media. The key to reducing attenuation was to reduce the amount of impurities in the glass fiber.

Researchers and scientists at Corning Glass Corporation found ways to manufacture glass fiber with low levels of impurities and soon produced fibers with less than a dB of loss per meter. Today they manufacture glass fiber that has less than 10 dB of loss per 50 kilometers (approximately 30 miles). LECs and IECs use these glass fibers to construct transmission facilities in the PSTN. Fiber optic media can also be constructed from plastic instead of glass, but since it causes more attenuation than a glass fiber, the use of plastic fiber is limited to short distances such as Local Area Networks (LANs).

The capacity of a signal to carry information is directly related to the frequency of the signal. Higher-frequency signals have higher capacity than lower frequencies. Copper can be used for low-frequency signals with low attenuation, but as the frequency of a signal increases so does the attenuation. A twisted-pair copper wire can carry signals up to 4000 Hz for approximately six miles with around 8 dB of loss. Twisted-pair copper wire can be used on LANs at 100 Mbps, and because the distance is short (100 meters or less), the attenuation is low. T1 carrier systems can use twisted-pair copper wire but the signal must be regenerated after 6000 feet because of attenuation of the signal. ADSL is used on copper wire but distance is limited to 18,000 feet. The signals used in ADSL are high-frequency signals and experience a tremendous loss in power, but the ADSL modem is smart enough to recover signals that have experienced high attenuation.

For higher-frequency signals, twisted-pair copper wire is not the media of choice due to high attenuation of the signals by the twisted-pair media. Coaxial cable was used for high-frequency signals in the megahertz range, while microwave radio was used for signals in the gigahertz range. Many coaxial and microwave systems have been replaced by fiber optic systems that can handle even higher-frequency signals in the terahertz range. Electromagnetic signals in the terahertz range include ultra-violet, visible, and infrared light waves. Plastic fiber can be used to carry light waves and ultraviolet waves with less loss than infrared light waves, but glass fibers are best for infrared light waves. Infrared light waves are invisible, and the invisible light waves just below the visible light spectrum are referred to as near-infrared light waves.

Since light waves are high-frequency signals, they have very high capacity. Since the best media to use for light waves is glass fiber, fiber optic transmission has a capacity advantage over other systems. Since the glass fiber cannot conduct electricity, it is unaffected by Electromagnetic Induction (EMI). This makes fiber a noise-free media because most noise on telecommunication circuits is caused by EMI. Fiber also has the advantages of being low weight and small in size. Fiber, impervious to wire-tapping, provides high security for data being transmitted.

Light waves quite naturally travel at the speed of light, but the speed will vary depending on the media through which the light is passing. The speed of light in free space (vacuum) is 3×10^8 meters per second. Light travels slightly slower in the atmosphere (air) but is slowed a great deal when it passes through glass or water. A measure that is used to compare how well a material allows light to pass through it is the refractive index. The refractive index, represented by the symbol n, is found by dividing the speed of light in a vacuum (represented by the symbol C) by the speed of light of the subject material (represented by the symbol V).

$$n = C/V$$

The refractive index for a vacuum will naturally be 1.0. The refractive index for air is 1.003, water is 1.33, and glass is between 1.4 and 1.9 depending on the amount of impurities in the glass. By rearranging the formula to $C = n \times V$, we can readily see that light travels from 1.4 to 1.9 times faster in a vacuum and air than it does in glass. From earlier formulas, we concluded the wavelength of a signal equals the speed of light divided by frequency of the signal. Since a light wave travels slower in glass than in air, the wavelength will be longer in glass than in air. When a light wave passes from one media to another with a different refractive index, the wavelength will change, causing the light wave to be refracted. Some of the light may also be reflected away from the second media. The wavelength in any material can be found by dividing the wavelength in a vacuum by the refractive index of the material: $W(x) = W(vac) / n(x)$.

The amount of light that is either refracted or reflected when light passes from one media to another depends on the refractive indexes of the two media as well as the angle at which the light wave passes through the boundary separating the two media. The angle at which a light wave strikes a material is referred to as the angle of incidence and, in our example, will be represented by angle I. If an imaginary line is drawn perpendicular to the boundary between the two media, the angle between this perpendicular line and the light ray is the angle of incidence (angle I). Figure 5-18a

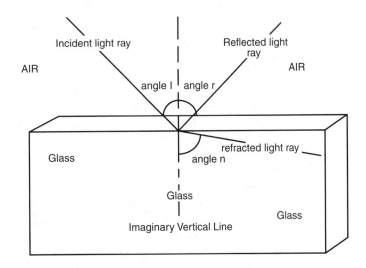

FIGURE 5-18a Angles of incidence, reflection, and refraction.

(continued on next page)

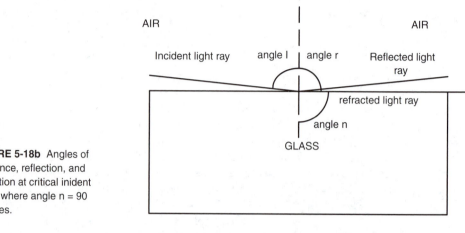

FIGURE 5-18b Angles of incidence, reflection, and refraction at critical inident angle where angle n = 90 degrees.

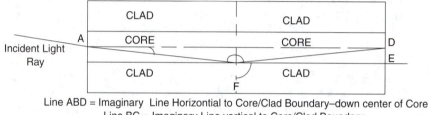

Line ABD = Imaginary Line Horizontial to Core/Clad Boundary–down center of Core
Line BC = Imaginary Line vertical to Core/Clad Boundary

Angle of Propagation = BAC   Angel of Incidence = BCA
Angle of Reflection = BCD   Angle of Refraction = FCE

FIGURE 5-18c Angles of incidence, reflection, and refraction at critical inident angle where angle n (FCE) = 90 degrees.

illustrates the angle of incidence (angle I), the angle of refraction (angle n), and the angle of reflection (angle r). Notice that the angle of the reflected light wave equals the angle of the incident light wave. Angle r = angle I. The angle of the refracted light-wave depends on the angle of the incident light wave and the difference of the refractive index of the two materials that light is passing through. SIN of Angle I times refractive index of media 1= SIN of Angle n times refractive index of media 2.

As the angle of incidence becomes greater, an angle is reached where the refracted wave travels along the border between the two materials. This angle is known as the critical angle of incidence. If we increase the angle just slightly beyond this angle of critical incidence, the entire incident light ray is reflected back to the first media. This is how fiber optic media is designed. In Figure 5-18a, one media is air and the second is glass. With a fiber optic media, both media are glass. One is the glass core and the second is a glass clad that has a slightly lower refractive index.

The clad surrounds the core. If light enters the core at just slightly more than the critical angle of incidence, when it tries to pass from the core to the clad, all light is reflected back to the core. This total internal reflection is what keeps light inside an optical fiber.

In order to achieve total internal reflection of light waves in a fiber, the refractive index of the core material must be greater than the refractive index of the cladding material, and the light wave must enter the core at an angle greater than the critical angle of incidence. We mentioned earlier that the angle of incidence is the angle between an imaginary line drawn perpendicular to the boundary between the two media and the light ray. If we were to draw a line horizontal to the boundary of the two media, the angle of the light ray to the horizontal line is referred to as the propagation angle. When the light ray is at the critical angle of incidence, the propagation angle is referred to as the critical propagation angle. It is necessary to direct light rays into a fiber at the critical propagation angle or less in order to achieve total internal reflection (see Figure 5-18b). Notice in Figure 5-18c, the horizontal line drawn down the center of the core and the vertical line drawn perpendicular to the core/cladding boundary. The light ray forms a right triangle ABC with the horizontal and vertical lines. A triangle contains 180 degrees. In a right triangle such as triangle ABC, angle A and angle C must equal 90 degrees; therefore, angle A must equal 90 degrees – angle C. From this we conclude that the critical angle of propagation equals 90 degrees minus the critical angle of incidence.

If the refractive index of the core is 1.5 and the refractive index of the clad is 1.48, what will be the critical angle of propagation? To achieve total internal reflection, we use the formulas provided to find the critical angle of incidence. Note that at the angle of critical incidence, angle n (the angle of refraction) = 90 degrees and the SIN of 90 degrees is 1.

SIN angle I × n of core = SIN angle n × n of clad
SIN angle I × n of core = 1 × n of clad
SIN angle I = 1 × n of clad / n of core
SIN angle I = 1.48/1.50 = .98666É.
Angle I = SIN^{-1} (.9866)
Angle I = 80.633 degrees

The angle of propagation = 90 degrees – angle of incidence; therefore, the critical angle of propagation = 90 – 80.633 = 9.367 degrees.

5.22 SUMMARY

The medium used to convey voice signals from the telephone to the central exchange is mostly twisted-pair copper wire. This facility is called the local loop. The local loop was designed to effectively carry signals in the 0 to 4000 Hz range. Most of the intelligence contained in an analog voice signal falls in this range of frequencies. When the loop is longer than 18,000 ft, the cable pairs must be loaded to compensate for the signal loss caused by the mutual capacitive reactance of the cable pairs. Loops shorter than 18,000 ft do not have to be loaded. Load coils adversely affect digital signals. When these signals must be carried by the local loop, the length of the loop is limited to 18,000 ft, unless the cable pairs are unloaded and digital signal regenerators are placed on the cable pair.

Most loops in a city are less than 18,000 ft. These loops do not have load coils and can carry low-speed digital signals without line treatment. High-speed digital signals, such as the 1.544 Mbps of a T1 carrier, will require signal regenerators every mile along the route. Wire cable has been replaced by fiber optic cable for most T1 systems and other high-speed data signals. The medium of choice in the long distance area is fiber optic cable. Most LECs and IECs have replaced T1 multiplexing systems with SONET OC-48 multiplexers and fiber optic cable.

Because of the tremendous investment LECs have in the twisted-pair local loop, they will seek to employ technologies that extend its lifespan. Exchange boundaries for exchanges in a city have a radius of less than 3 mi from the central exchange. Thus, these lines can be served by twisted-pair copper wire and still carry digital signals. Most LECs will phase fiber into the local loop by first deploying fiber as feeder and distribution cables. By multiplexing many SLC-96 systems over one fiber, the LECs can establish a CSA and provide DS0 capabilities to customers located many miles from the central exchange. These capabilities are critical to the deployment of ISDN and other digital services in the remote areas of a rural exchange.

The rural exchange must serve a much larger area than an exchange in a city and requires the application of RRD and MLRD design concepts to determine the size of wire cable to use in serving the area. Rural areas cannot have ISDN and other digital services without unloading these long loops and placing signal regenerators along the cable route. Establishing a CSA through the use of subscriber carrier is a more cost-effective way to provide digital capabilities to remote areas. But because a large investment would be needed to convert all long loops from wire cable to a CSA, it will be some time before all rural areas are converted to fiber feeder cables and a CSA. In the past, many rural customers did not desire digital services, though today more of them are demanding these services. As the demand increases, the LECs will selectively deploy fiber and a CSA concept to the areas with a high demand for digital services.

CHAPTER REVIEW QUESTIONS

1. What is the medium of choice for the local loop? Why?
2. What is the medium of choice for the toll network? Why?
3. What device governs design of the local loop? Why?
4. Outside plant engineers use what principle in designing the local loop?
5. What are the two wires connected to our telephone called?
6. What does *PIC* stand for?
7. Where does the power for the telephone come from?
8. How much voltage does the central exchange apply to the local loop?

9. How far can a telephone be located from the central exchange before loop treatment is needed?
10. What does *SLC-96* stand for?
11. When would subscriber carrier be used?
12. What is mutual capacitance?
13. How does the outside plant engineer offset the effects of capacitive reactance in the local loop?
14. What is D spacing?
15. At what points is a local loop cable loaded?
16. How many simultaneous conversations can OC-48 support?
17. What does *SONET* stand for?
18. Why are wire pairs in a cable twisted?
19. What is the primary source of noise and crosstalk in a local loop cable?
20. Why is fiber optic cable not susceptible to noise from electrical interference?

PROJECTS AND CASE STUDIES:

Case Study #1

Your firm is looking at establishing a research and development facility at Blue Sky, Montana. You wish to offer a relaxed work atmosphere to attract employees to your firm. So you have decided to build an office complex for 500 employees on a 300-acre tract about 30 miles from a large city. The central office for this area is 10 miles away. Discuss some of the challenges you will face in order to provide telecommunications services for this office complex from the LEC, ILEC, and ISP.

Case Study #2

Joe Smith lives six miles from the central exchange. Can he get ADSL? If he is served by SLC-96, can he get ADSL? What would be the options available in the rural area outside the city you live in?

GLOSSARY

Branch Feeder Cables   Cables that connect distribution cables to main feeder cables.

Carrier Serving Area (CSA)   The creation of DS0 capabilities in a remote area of the exchange by using subscriber carrier to establish a remote wire center.

Crosstalk   An undesirable condition where a circuit is picking up signals being carried by another circuit. For example, a customer hears other conversations or noise on a private line. This is usually caused by a breakdown in the insulation between wires of two different circuits.

Dense Wave Division Multiplexing (DWDM)
Technology that splits the light spectrum into many different frequencies. A time division multiplexer is then assigned to each of the different light wave frequencies. DWDM is used to multiplex multiple OC-48 (or OC-192) systems over one fiber pair.

Distribution Cable   A cable that is fed by a branch feeder cable from the central exchange and that connects to telephones via drop wires. The distribution cable is usually a small cable of 25 to 400 pairs.

Drop Wire (Drop Cable)   A pair of wires connecting the telephone to the distribution cable. One end of the drop wire connects to the protector on the side of a house or business, and the other connects to a terminal device on the distribution cable.

Exchange Boundary   Pertains to the limits of the serving area. The outermost customers being served by a particular exchange establish the boundary of the exchange.

Facility   A term typically used to refer to the transmission medium.

Feeder Cable   Includes various types. Cables leaving the central exchange are called *main feeder cables.* The main feeder cables connect to branch feeder cables. The branch feeder cables are used to connect the main feeder cables to distribution cables.

Jumper Wire   A short length of twisted-pair wire used to connect wire pairs of two different cables. A jumper wire is a very small gauge of wire (usually 24 or 26 gauge). Except for its small size and much longer length, a jumper wire is similar to the jumper cable used to start a car with a dead battery.

Load Coils   The devices that introduce additional inductance into a circuit. Load coils are added to cable pairs over 18,000 ft to improve the ability of the pair to carry voice signals.

Loading   Purposely adding inductance to a cable pair over 18,000 ft long to offset the mutual capacitance of the cable pair.

Local Loop   A term used to describe the facilities that connect a telephone to the central exchange. These facilities usually consist of a twisted pair of wires.

Loop Extender   A hardware device. Can be connected between the line circuit of the switch, and the cable pair serving a remote area of the exchange, to extend the range of a central exchange. A loop extender adds a voltage boost to the circuit. By doubling the voltage applied to a line, the distance served can be doubled.

Loop Treatment   A term indicating that a local-loop cable pair has been specially conditioned. Typically this term describes the addition of extra equipment to a local-loop cable pair. To extend the range of a central exchange, a device called a *loop extender* can be connected between the line circuit of the switch and the cable pair serving a remote area of the exchange. A loop extender adds a voltage boost to the circuit. By doubling the voltage applied to a line, the distance served can be doubled. Another device, called a *voice frequency repeater (VFR)*, is also added to long local loops. The VFR will amplify voice signals in both directions to compensate for the extra decibel loss of long loops.

Main Distributing Frame (MDF)   In a central exchange, the place where all cables that connect to the switch are terminated. Outside plant cables that connect to telephones and to other central offices are also terminated at the MDF. Short lengths of a twisted-wire pair (called a *jumper wire*) are used to connect wires from the switch to wires in the outside plant cable. In a PBX environment, the MDF is where all cables from the switch, IDFs, and cables from the central exchange are terminated and jumpered together.

Main Feeder Cable Cable that leaves the central exchange and connects to branch feeder cables.

Modified Long Route Design (MLRD) Outside plant design that involves adding loop treatment to cable pairs serving remote areas of an exchange. Local loops longer than 24,000 ft are designed using MLRD.

Multiplexer A device that can combine many different signals or calls (data or voice channels) so they can be placed over one facility. A multiplexer also contains a demultiplexer so that it can demultiplex a received multichannel signal into separate voice channels.

Outside Plant Cables, telephone poles, pedestals, and anything that is part of the telecommunications infrastructure and not inside a building.

Outside Plant Engineer The person charged with properly designing the facilities that connect telephones to central exchanges and the facilities that connect central exchanges together (outside plant designer).

Plastic Insulated Cable (PIC) Cable that contains wires electrically isolated from each other by plastic insulation around each wire.

Resistance Design Designing a local loop, which uses twisted-pair copper wire, so that the resistance of the wire serving a telephone does not exceed the resistance design limitations of the central exchange line circuit. Each switching system manufacturer will state how much resistance its switch is designed to support.

Revised Resistance Design (RRD) A design approach in which the outside plant engineer uses the length of the local loop, as well as the design limitations of the central exchange, to determine the local-loop design criteria. Loops up to 18,000 ft are designed to use nonloaded cable pairs with a maximum loop resistance of 1300Ω. Loops between 18,000 and 24,000 ft are designed to use loaded cable pairs with a maximum loop resistance of 1500Ω. Loops longer than 24,000 ft are designed according to modified long route design (MLRD).

Subscriber Line Carrier-96 (SLC-96) SLC-96 uses TDM to multiplex 96 lines over two cable pairs. Multiple SLC-96 systems are often multiplexed onto a fiber facility.

Voice Frequency Repeater (VFR) A device that will amplify voice signals in both directions to compensate for the extra decibel loss of long loops.

6

Multiplexing

KEY TERMS

Amplitude Modulation (AM)
Bandwidth
Channel Unit
Frequency Division Multiplexing (FDM)
Mixer (Modulator)
Multiplexing
Pulse Amplitude Modulation (PAM)
Pulse Code Modulation (PCM)

Sideband
Space Division Multiplexing
Statistical Time Division Multiplexer
Statistical Time Division Multiplexing (STDM)
Synchronous Optical Network (SONET)
Synchronous Time Division Multiplexer

Synchronous Time Division Multiplexing
Synchronous Transport Level 1
Time Division Multiplexing (TDM)
Wave Division Multiplexer
Wave Division Multiplexing (WDM)

OBJECTIVES

On completion of this chapter, you should be able to:

1 Discuss the differences between frequency division multiplexing (FDM) and time division multiplexing (TDM).
2 Discuss the various steps required to prepare a signal for TDM.
3 Understand amplitude modulation (AM).
4 Understand pulse code modulation (PCM).
5 Describe how a DS0 signal is formed.
6 Describe how a DS1 signal is formed.
7 Know the medium required for various levels of multiplexed signals.
8 Explain why all TDM systems have 8000 frames per second.
9 Explain the purpose of quantizing.
10 Explain wave division multiplexing (WDM).
11 Describe the various levels of SONET and know how many DS0 signals each contains.

The term *multiplexing* is derived from the word *multiple,* which means consisting of, or having, many individual parts. In telecommunications, multiplexing means to combine many individual signals (voice or data) so they can be sent over one transmission medium. The medium used is usually a cable pair, a fiber optic strand, or a microwave link. A multiplexer actually contains the equipment to do multiplexing and demultiplexing. A multiplexer/demultiplexer is attached to each end of a transmission medium. The equipment on each end is simply called a *multiplexer.* It is understood that a multiplexer also contains a *demultiplexer.* The multiplexer equipment on each end of the medium performs both functions.

It is important to note that many people view space division multiplexing as a form of multiplexing. In space division multiplexing, multiple communications are placed over many different wire pairs inside one cable. If the cable is considered the medium, multiplexing has occurred when many conversations are carried by a single cable. If a cable has 100 cable pairs in it, the cable can handle 100 calls and each call occupies its own pair (or space) within the cable. Cables connect the switching stages of a space division switching system, and each call will occupy its own set of wires between and within switching stages.

Space division multiplexing requires that each communication channel or voice path occupy its own set of wires for transmission of signals. If we consider the wire to be the medium instead of the cable, we cannot use this form of multiplexing, which by definition requires that each signal have its own transmission medium within the cable. To multiplex many conversations over one wire, we have to use a different multiplexing technique. We can use a process called *frequency division multiplexing (FDM)* or another process called *time division multiplexing (TDM)* to multiplex these conversations.

FDM is a process whereby each communication channel is made to occupy a certain band of frequencies and each channel is forced into different frequency bands. The human voice band is basically in the frequency band 0 to 4000 Hz. Through the use of a technique called *modulation,* we can force each communication channel to occupy a different band of frequencies. We cannot combine voice signals that are all 0 to 4000 Hz. We must change these signals to different frequencies for each communication channel before they can be multiplexed. For example, we can force one communication channel to be in the band from 60,000 to 64,000; a second channel is forced to be in the 64,000-to-68,000 band; a third channel is forced to occupy the 68,000-to-72,000 band; and so on. If the transmission medium we choose to use for the multiplexed signal is capable of carrying signals between 60,000 and 108,000 Hz, we could multiplex 12 channels of conversation over the medium using FDM. The band of frequencies from 60,000 to 108,000 Hz provides a 48,000-Hz bandwidth. This 48,000-Hz bandwidth contains 12 channels, and each channel has the 4000-Hz bandwidth needed by each voice channel (see Figure 6-1).

Bandwidth refers to the width of a signal, which is determined by subtracting the highest frequency of the signal from its lowest frequency. A voice signal is usually thought of as a signal between 0 and 4000 Hz. The bandwidth of this signal is 4000 Hz, and we say the voice signal has a nominal bandwidth of 4000 Hz. In the United States, AT&T designed its FDM systems to handle the band of signals between 200 and 3400 Hz. Thus, the FDM system used in the United States provides a bandwidth of 3200 Hz for voice (or data) signals. Voice signals below 200 Hz and above 3400 Hz will not be carried by the FDM system. We lose these signals, but this loss does not affect voice since most of the intelligence in a voice signal falls within the frequencies passed by FDM. The frequencies passed by a device or system is called the *passband.* The passband for FDM is 200 to 3400 Hz.

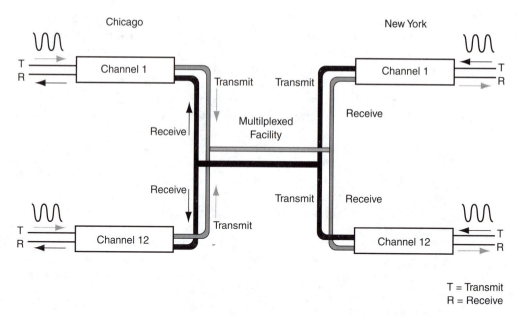

FIGURE 6-1 Twelve-channel frequency division multiplexing.

6.1 THE CHANNEL UNIT OF A FREQUENCY DIVISION MULTIPLEXING SYSTEM

Each signal fed to a FDM system interfaces to the multiplexer through a device called a *channel unit*. If a multiplexer can combine 12 different signals, it will have 12 channel units. The channel unit performs the function of interfacing the multiplexer to a specific type of input signal. The channel unit makes changes to the input signal so it can be combined (multiplexed) with other signals for transmission over the transmission medium. All voice signals arriving at a channel unit are between 0 and 4000 Hz. The channel unit contains a sharp cutoff bandpass filter to pass only the frequencies between 200 and 3400 Hz to a device called the *mixer* (or *modulator*). The modulator takes the 200-to-3400 Hz input signal and modulates a higher-frequency signal. Signals of the same frequency cannot be multiplexed together. The output of each modulator is a different frequency. This allows the signals to be multiplexed.

After the channel unit changes the frequency of the input signal to a higher frequency, the signals are combined. The 12 different output signals of the channel units are multiplexed and the composite signal is sent over the transmission medium to the multiplexer on the other end of the medium for demultiplexing. As the signal is received at the distant end, it is demultiplexed into 12 different signals and assigned to the same channel number used at the originating end. If an input signal comes in over channel 1 on the transmitting end, it will be received by channel 1 on the receiving end (see Figure 6-1).

6.2 TRANSMIT AND RECEIVE PATHS: THE FOUR-WIRE SYSTEM

If multiplexers are placed on a medium between two cities such as New York and Chicago, the signal sent from New York will be placed over a transmit medium to Chicago. Since Chicago is receiving the signal, the transmit medium from New York is attached to the receive terminal of the multiplexer at Chicago for demultiplexing (see Figure 6-1). This allows the Chicago receiver to hear the New York transmitter. For New York to hear Chicago, it is necessary to attach the transmit terminal of the Chicago multiplexer to a transmit medium to New York. This transmit medium from Chicago is attached to the receive terminal of the New York multiplexer for demultiplexing. Thus, multiplexing involves attaching multiplexers to two transmission media. When cable is used as the transmission medium, we have one pair of wires for each transmit/receive direction. Each of these two paths is usually placed in separate binder groups in the cable to reduce the possibility of crosstalk between the two pairs.

The first multiplexers utilized wire for the transmission media. One pair of wires was used for the signal going in one direction (that is, east to west) and a separate pair of wires for the signal going in the opposite direction (west to east). Thus, four wires were needed to connect one multiplexer to another. When multiplexers are used on fiber optic cable strands, two strands are needed (one strand for each direction). When microwave is employed as the transmission medium, multiplexers are assigned to a different transmitting frequency in each direction.

6.3 WHY USE MULTIPLEXING?

Multiplexing was first used to reduce the number of transmission media needed between cities or towns. If there were 24 trunks between two cities, 24 cable pairs were needed. By using two 12-channel multiplex systems, only four cable pairs were needed (two pairs for each system). Multiplex systems greatly reduced the number of media used between cities. This resulted in significantly reduced costs for trunk circuits. If multiplexers did not exist, we could not use fiber optic cable to carry many circuits. Fiber optic cable allows the multiplexer to combine as many as 6 million signals in one direction on one fiber strand. The multiplexer on fiber optic cable combines many signals and transmits the signals as pulses of light. The receiving multiplexer uses a photodiode to detect the individual pulses.

6.4 AMPLITUDE MODULATION

The different sounds of our language such as *a, e, i, o, u, by, my, car, far, tar,* and so on are due to the different frequencies of the sounds. Spoken words are nothing more than a string of signals that continually vary in frequency. When we speak into a transmitter or microphone, these devices convert the varying sound frequencies into varying electrical currents that have the same frequency as the sound. In FDM, these varying electrical signals are input to a device in the channel unit called a *mixer* (or *modulator*). Another signal, called the carrier signal, is also input to the

AM Mixer

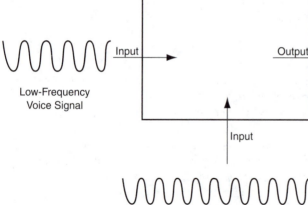

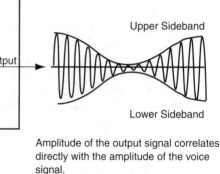

Low-Frequency
Voice Signal

Input

Output

Upper Sideband

Lower Sideband

Input

High-Frequency
Carrier Signal

Amplitude of the output signal correlates
directly with the amplitude of the voice
signal.
Frequency of the output signal
equals the frequency of the carrier
signal.

FIGURE 6-2 Amplitude modulation.

mixer. A *carrier signal* is a signal having a much higher frequency than the voice
input signal. In **amplitude modulation (AM),** the *amplitude of the carrier frequency*
coming out of the mixer will vary according to the changing *frequency of the input
voice signal* (see Figure 6-2).

Amplitude modulation is used by AM broadcast stations. These stations
transmit on carrier frequencies between 540 kilohertz (540,000 Hz) and 1.65 megahertz (1,650,000 Hz). How far these signals can be transmitted depends on the
power of the transmitter. Most AM transmitters used by commercial broadcast stations have sufficient power to send a signal 40 miles. This allows the FCC to reuse a
frequency in many parts of the United States, being careful not to assign the same
frequency to radio stations within 100 miles of each other. Many stations will exist
in each specific geographic area of the country, but each station will be transmitting
using different carrier frequencies. All these carrier frequencies arrive at the antenna of your radio. The radio has a tuning device that allows it to tune in and pick
out the carrier signal desired. The radio will then demodulate the signal we have
tuned to. The demodulation circuitry in the radio will recognize the changes in amplitude of the carrier signal and convert the changing amplitudes into an electric
current that changes in frequency. This signal is coupled to a speaker where the frequency of electrical energy is converted to vibrating sound waves of the same frequency. The radio listener hears the sound from the speaker, and it is the same as
the sounds spoken into the microphone at the transmitter.

The signal emanating from a radio tower is possible due to an effect called
electromagnetic radiation. When a high-frequency signal is attached to an antenna,
the signal will radiate off the antenna. The higher the frequency and power of the
signal applied to the antenna, the more easily it will radiate into the air. Broadcast

stations must use high-frequency signals, and high power, to force the signal off the transmitting antenna and out into the air. By using a modulation technique, we can take the intelligence from a low-frequency signal and insert the intelligence into a higher-frequency signal easily radiated into space.

In telecommunications, we use wire as a transmission medium instead of air. We do not want our signals broadcast to many receivers; we only want one receiver to receive our transmitted signal. The use of wire to connect the transmitter and receiver allows us to establish a point-to-point communication link. Since we do not want the signal to radiate off the wire, we use low-powered transmitters. We utilize amplitude modulation in telecommunications to gain many different voice channels. Every voice signal is 0 to 4000 Hz. If we are going to combine many voice channels over one transmission medium, each signal must be different. We can convert the voice signals into different signals by using amplitude modulation. This conversion process takes place in the channel unit of a multiplexer system. With 12 input voice signals, we need 12 channel units, where each channel unit has a different frequency of carrier signal input to its mixer. The output of each channel unit will be the higher carrier frequency, modulated by the voice input signal. We will have 12 different output signals, and each signal will contain the intelligence of the original voice signal fed to the input of the channel units.

6.5 FREQUENCY DIVISION MULTIPLEXING

Using FDM, many telephone calls can be multiplexed over two pairs of wires. If we wish to multiplex 12 calls over two pairs of wires between cities A and B, we hook up a multiplexer containing 12 transmitters to the wire pair in city A. Each of these transmitters uses a different carrier frequency. At the other end of the cable pair, we attach 12 receivers in city B each tuned to receive only one of the 12 transmitted carrier signals, since this is a simplex circuit from A to B. We would also use 12 transmitters in city B, hook them up to another pair of wires between A and B, and place receivers in city A. This gives us a simplex circuit from B to A. The two simplex circuits together make up a full-duplex circuit between A and B. As we have seen, the use of multiple carrier frequencies to provide multiple conversations over two pairs of wires is called FDM (Figure 6-1). A prerequisite for FDM is that all individual signals must be at different frequencies. In telecommunications, we use the AM process to change signals so they are different frequencies.

6.6 HYBRID NETWORK

The inputs to the channel units of a multiplexer usually come from a local telephone circuit. This circuit is a pair of wires that carries voice signals in both directions. The circuit from a telephone arrives at the input to the channel unit of a multiplexer as a two-wire circuit. The channel unit contains a device (*hybrid network*) that interfaces the two-wire input to a four-wire transmit/receive path. The transmit path connects to the transmitter for this channel, and the receive path connects to the receiver for this channel. The hybrid network connects the one-directional, transmit, and receive

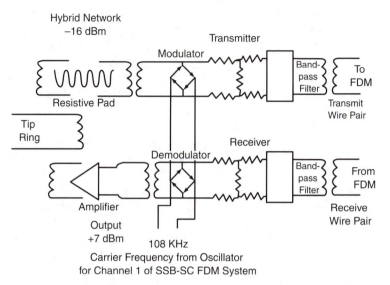

Channel Unit

FIGURE 6-3 Channel unit.

circuits of the multiplexer to the two-directional telephone circuit. The channel units contain the hybrid network, modulator, transmitter, receiver, and demodulator (see Figure 6-3).

6.7 AMPLITUDE MODULATION TECHNOLOGY

The first multiplexer to combine as many as 12 different signals into one signal for transmission over one medium in each direction used amplitude modulation technology. *Amplitude modulation* is a technique in which the amplitude of a high-frequency signal changes when the high-frequency signal is mixed with a low-frequency signal in a device called a mixer or modulator (mentioned earlier). The high-frequency signal is termed the *carrier frequency*; the low-frequency signal is called the *modulating signal.* When the carrier frequency and the modulating signal are mixed together, the output of the mixer will be four signals. One signal will be at the frequency of the carrier frequency and one will be at the frequency of the modulating frequency. The third signal output by the mixer has a frequency representing the sum of the carrier and modulating frequencies. The fourth signal will have a frequency that is the difference between the carrier frequency and the modulating frequency. For example, if a 64,000-Hz signal is modulated by a 4000-Hz signal, the outputs of the modulator will be 64,000 Hz; 4000 Hz; 68,000 Hz (64,000 + 4000); and 60,000 Hz (64,000 − 4000). The two additional signals generated by the mixing process are called **sidebands.** The higher signal is called the *upper sideband.* In our example the frequency of the upper sideband is between 64 and 68 KHz. The lower signal is called the *lower sideband;* in our example it is between 60 and 64 KHz. Both these sidebands are replicas of the original input signal but at a higher frequency (see Figure 6-2).

6.8 BANDWIDTH AND SINGLE-SIDEBAND SUPPRESSED CARRIER

The *bandwidth* of an AM signal is the difference between the highest frequency and the lowest frequency. In the example used, the bandwidth is 8000 Hz (68,000 – 60,000). The bandwidth of an AM signal is twice the modulating frequency. When a low-frequency signal is present as an input to the mixer, the amplitude of both sideband signals changes as the frequency of the input signal changes. We say the sidebands have been modulated (changed) by the input signal. Since both sidebands have been modulated, they both contain the frequency changes of the modulating signal. We only need to demodulate one of these sidebands to recapture the intelligence of the modulating signal. Since only one sideband is needed to demodulate the modulating signal, some systems will transmit only one sideband with the carrier frequency. These systems are called *single-sideband (SSB)* systems. Systems that do not transmit the main carrier frequency, but transmit the sideband only, are called *single-sideband suppressed carrier (SSB-SC)* systems.

6.9 CCITT STANDARDS FOR FREQUENCY DIVISION MULTIPLEXING

Telecommunications standards are set by several organizations. The predominant organizations are the American National Standards Institute (ANSI), which establishes standards for North America, and the Consultative Committee on International Telegraphy and Telecommunications (CCITT) organization within the International Telecommunications Union (ITU), which creates worldwide standards.

In the previous edition of this book, I discussed SSB-SC as a way to explain Frequency Division Multiplexing. The SSB-SC systems used by AT&T for their long distance network were analog systems. These systems have been replaced by fiber optic and time division multiplexing. I have discontinued the lengthy discussion of SSB-SC in this chapter, but have kept the material as part of a new Appendix G where I have tried to consolidate historical perspectives of technology.

6.10 APPLICATIONS FOR FREQUENCY DIVISION MULTIPLEXING

By now you should have an understanding of what FDM is. In order to do FDM, every signal must be at a different frequency. The process most often used in analog systems to make similar signals of the same frequency different is to take the signals through their own channel unit where Amplitude Modulation (AM) is used to develop signals at different frequencies. This process was widely used in the old analog long distance network, but has been replaced by *Time Division Multiplexing (TDM)* networks. A TDM system is used to transmit many signals over one media. Each signal is placed on the media for a brief instant of time at recurring intervals of time. So why study FDM? We need to understand FDM because we now use FDM on fiber optic facilities to place multiple TDM systems on the fiber.

When fiber optics was first introduced, we used one light signal on the fiber. The light signal was from a TDM system. The TDM system would turn the light brighter for a 1 and dimmer for a 0. Thus by rapidly varying the brightness of the light, the TDM system could transmit 1s and 0s very rapidly. The receiver would look at the brightness of the light and decipher the brightness of the received light into 1s and 0s. With this arrangement the fiber handled one TDM system.

Today, instead of connecting the TDM system directly to a fiber, we can connect the TDM system to channel units of a *Wave Division Multiplexer* and the wave division multiplexer is connected to the fiber. The wave division multiplexer can generate many different lightwaves. Each lightwave is a different frequency. Each channel will generate a different frequency of light that will be placed on the fiber. The use of many different frequencies is FDM, but when FDM is used on fiber optics, we refer to it as *Wave Division Multiplexing (WDM)*. One channel unit may transmit at a frequency of 192.100 THz, a second channel at 192.200 THz, a third at 192.300 THz, etc. up to 196.000 THz with a 100 GHz separation between frequencies. This would provide 40 different frequencies. This would allow us to connect 40 different TDM systems to one fiber. WDM systems have been developed that have 180 different frequencies to allow one fiber to support signals from 180 TDM systems.

6.11 WAVE DIVISION MULTIPLEXING

The light waves used in fiber optic technology are very narrow light beams. When we speak of the frequencies of light waves, we are referring to one of the electromagnetic properties of a wave. Light is also discussed in terms of photons. Just as we have small particles of electric energy called electrons, we have small units of light referred to as photons. Photons can be generated by the activities of electrons, such as when electrons flow through the filament of a light bulb. Electron flow within a Light Emitting Diode (LED) or within a Laser circuit generates a flow of photons.

LEDs are used on Local Area Networks that use fiber optic media. Lasers generate the light that is used on fibers in the long distance network. The laser of a transmitter used in fiber optic transmission sends photons out onto the glass fiber media. The quantity of photons transmitted is controlled by the digital signals at the input of the multiplexer. If a digit 1 is present at the input, the laser emits a high number of photons. If a 0 is present at the input, the laser emits a low number of photons. The receiver in a fiber optic system detects the intensity of light by detecting the quantity of photons it receives. These photons are extremely small and travel at the speed of light. Many photons can be transmitted in a very narrow beam because they are so small.

Most people are familiar with laser light pins or pointers that are used to highlight a slide show presentation on a screen. Although the laser pointer pen uses a small power supply, it generates a very visible light that travels quite a distance. This is because the light beam is very narrow. Light waves used in fiber optic transmission are much narrower beams. These narrow beams travel over glass fibers very easily. It should be noted that there is a difference in the laser pointer and the laser used in telecommunications. The laser pointer generates a visible red light while the laser used in telecommunications generates an invisible light ray.

The first fiber optic systems used in the long distance network did not use wave division multiplexing (WDM). Fiber optic systems used a transmitter that transmitted only one frequency of light. The transmitters used amplitude modulation (AM) to change the intensity of light being transmitted. A low intensity of light equals a 0, and a high intensity of light equals a 1. As mentioned earlier, the amount of information that can be transmitted using TDM depends on the speed of the transmitter, receiver, and capacity of the medium between them. A glass fiber can handle up to 50 Tera (Trillion) bits per second. Trillions of 1s and 0s can be transmitted over a fiber, with the number of digits transmitted depending on how fast the intensity of light being transmitted can be changed and how quickly the receiver can detect a change in the light intensity.

Today's electronics can actually change the intensity of light transmitted by a time division multiplexer's transmitter faster than other electronics in the multiplexer can look at all the inputs to the multiplexer. Until recently, the speed at which TDM gate circuits could switch to connect the inputs of the TDM to the transmitter for output to a fiber was around 1.5 billion times per second. This switching speed limited TDM to multiplexing 32,256 DS0 channels (each channel at 64 Kbps) together. The multiplexer that does this is an OC-48 multiplexer. The resulting line speed of OC-48 is 2.48832 Gbps.

Advancements made in TDM electronic circuitry have improved, and today it is possible to multiplex 129,024 DS0 channels together for TDM over a fiber using an OC-192 multiplexer operating at 9.95328 Gbps. Notice that this line rate is still well below the fiber's theoretical capacity of 50 Tbps. In fact it is only about 1/5000 of the fiber's capacity. Thus, the current state of time division multiplexer technology cannot take full advantage of the capacity of the fiber media. Multiplexers are used in pairs with a multiplexer/demultiplexer at each end fiber. This requires the use of two fibers. One fiber connects the transmitter of multiplexer A to the receiver of the distant demultiplexer B. The second fiber connects the distant transmitter of multiplexer B to the receiver of demultiplexer A.

The first commercial use of WDM was to use different frequencies of light for transmitters on each end of one fiber. By using different frequencies of light for each time division multiplexer's transmitter, it is possible to connect more than one multiplexer to a fiber. Multiplexer A can transmit at frequency A, while multiplexer B— on the other end of the fiber—can transmit at frequency B. The frequencies used in this type of WDM were 230 THz and 193.5 THz. This type of WDM is referred to as Broadband WDM.

Before proceeding further with WDM discussions, we need to discuss the relationship of frequency to wavelength because most discussions about fiber optic technology refer to wavelength rather than frequency. The higher the frequency of a signal, the more cycles per second the signal has. To achieve more cycles per second, each cycle must take less time. To take less time the length of each wave must be shorter. It was stated earlier that period (time for one cycle) and frequencies are reciprocals. $P = 1/F$ and conversely $F = 1/P$. The period of 230 THz lightwave is found by using the formula: $P = 1/F$; $P = 1/230,000,000,000,000$; $P = .0000000000000004347826086957$ of a second. Light travels at a speed of 300,000,000 meters per second. $.0000000000000004347826086957 \times 300,000,000 = .000001304347826087$ meters is the length of one cycle of a 230 THz signal. This is 1304.3478 nanometer (nm). From the above calculations, we can conclude that $P \times$ the speed of light = the length of one wave. Since $P = 1/F$, then $1/F \times 300,000,000$ meters = the length of one wave in meters. This becomes $300,000,000/F = W$. Since $300,000,000 = 3 \times 10^8$, the formula for wavelength can also be stated as: $W = 3 \times 10^8/F$. Please note that "W" is not the symbol used by most to represent wavelength. The symbol for wavelength looks like an upside-down "y," but I prefer to use "W" to represent wavelength.

The wavelength of one cycle of a 230 THz signal is slightly more than 1304 nm. The wavelength of a 193.5 THz signal is approximately equal to 1550 nm. Thus, the WDM technique referenced in our example used light signals of 1300 and 1550 nm. The transmitter at one end of the fiber used a 1300 nm lightwave, and the transmitter at the other end of the fiber used a 1550 nm lightwave.

6.12 DENSE WAVE DIVISION MULTIPLEXING (DWDM)

Dense Wave Division Multiplexing (DWDM) is referred to as Narrowband WDM. Dense Wave Division Multiplexing uses multiple different transmitter frequencies at one end of a fiber and multiple receivers at the other end of the fiber. This technique requires two fibers for each multiplexed circuit with a multiplexer/demultiplexer attached at the ends of both fibers. One fiber serves to connect the transmitters of multiplexer A to the receivers of demultiplexer B. The second fiber connects the transmitters of multiplexer B to the receivers of demultiplexer A. This arrangement makes use of two simplex circuits to provide full-duplex functionality.

It is too difficult to use DWDM in a purely TDM arrangement. The lightwaves used in fiber can only travel about 30 miles before they are attenuated to a point that they must be regenerated. In a TDM world, we must have a regenerator for every lightwave frequency used. This is costly and complicated. The development of erbium-doped fiber amplifiers (EDFA) made DWDM possible. These amplifiers use a different technique than regenerating the signal; they receive the attenuated signal and then amplify it. As we learned earlier, digital signaling is noise-free because we regenerate signals. The old analog signals had to go through an amplifier after so many miles to overcome attenuation, but amplifying the signal also amplified the noise. So why can we now say amplification of the analog lightwave is the right thing to do? Won't we amplify the noise? Yes, we will. But there is a major difference in a fiber media. The fiber media is not affected by electromagnetic induction (the major source of noise on a copper media); therefore, there is little noise present on the lightwaves. The extremely small amount of noise present is eliminated by circuitry within the amplifier before amplification of the signal. The gain of EDFA is between 20 to 40 dB. 20 dB of gain is amplification of power by a factor of 100, 30 dB is 1,000 times as much power, and 40 dB is amplification by a factor of 10,000.

Another advancement that made DWDM possible was the development of integrated lasers on one chip. Multiple lasers on one chip mean that we only have to connect the output of one chip to the fiber. It stands to reason that this is much simpler than trying to connect the output of many individual laser transmitters to one fiber. This multiple laser chip receives inputs from multiple time division multiplexers. If we have a 16 laser chip, it can have a time division multiplexer attached to each laser. Thus, we can attach 16 TDMs to the 16 lasers on the transmitter chip. Each laser is transmitting at a different lightwave frequency out onto the fiber optic strand.

The visible spectrum of light is approximately 428.5 THz (700 nm) to 750 THz (400 nm). Infrared light is defined as light between approximately .3 THz (1,000,000 nm) and 428.5 THz (700 nm). Infrared light is the invisible portion of the lightwave spectrum. Lightwaves used in DWDM are between 1530 nm (196 THz) and 1610 nm (186 THz). These wavelengths are in the infrared portion of the lightwave spectrum. Due to characteristics of EDFAs, they cannot achieve a linear gain for all frequency of signals between 196 THz and 186 THz. Most EDFAs work in the 1530 to 1560 nm range. This is referred to as the "C Band" and is also referred to as the "Red Band." This description has nothing to do with color. Remember, these are invisible light frequencies—they have no color. "C" stands for conventional wavelength. EDFAs have also been developed with high gains between 1560 nm and 1610 nm. The frequencies in this band are referred to as the "L Band" or "Blue Band." "L" stands for long wavelength.

Since the bandwidth of C Band EDFAs is 1530 nm to 1560 nm, the total bandwidth is 30 nm. Remember from earlier discussions that wavelength is the speed of light divided by the frequency of the signal: $W = 3 \times 10^8/F$. If we use a symbol "C" to stand for the speed of light, the formula is: $W = C/F$. This can be changed to $WF = C$. The relationship between change in frequency and change in wavelength is stated in the formula: $(\Delta F = -(C\Delta W)/W^2$. The relationship between change in wavelength and change in frequency is stated in the formula: $(\Delta W = -(C\Delta F)/F^2$. Suppose we want to place 40 different frequencies of light over a fiber using signals between 1530 nm and 1560 nm. To get equal spacing between 40 signals, the spacing should be 4 THz /40 = .1 THz (or 100 GHz). The distance from 1530 nm to 1560 nm is 31 nm and 31 nm/40 = .775 nm or approximately .8 nm channel spacing. (See Table 6-1.) The formula provides the following: $\Delta F = -(C\Delta W)/W^2$ and using 1550 as the wavelength and .8 nm as ΔW, we get $\Delta F = -(300,000,000 \times .0000000008)/.000001550^2 = .24/.000001550^2 = 99,895,941,727$ Hz or approximately 100 GHz.

The bandwidth of a C Band system is 4 TrillionHz. We can fit a lot of signals in a 4 THz spectrum. DWDM is available with transmitters that use 4, 8, 16, 32, and 180 laser transmitters. Each laser operates at a different frequency. In order for equipment from different manufactures to be compatible, the industry strives for standards. The standard frequencies for use in a 16-channel DWDM are chosen from a list developed by ITU that is referred to as the ITU Grid. Note that this grid uses wavelengths between 1560.61 nm and 1528.77 nm with 100 GHz separation.

6.13 TIME DIVISION MULTIPLEXING USING PULSE AMPLITUDE MODULATION (PAM) SIGNALS

A third type of multiplexing is *time division multiplexing (TDM).* Early TDM signals multiplexed samples of the human voice from different channel units over one common facility. These were analog voice samples, and the technique used to achieve these voice samples was called *pulse amplitude modulation (PAM).* An engineer named Nyquist made some studies of voice sampling techniques and developed the Nyquist theorem. According to this theorem, if the sampling rate of a signal is twice the highest frequency of the signal, the signal can be reconstructed using the samples. Transmitting samples of a signal instead of the complete signal makes it possible to use a medium for transmitting many samples. A certain amount of time is set aside for the transmission of each sample. Each sample is

TABLE 6-1 ITU Grid

Frequency (THz)	Wavelength (nm)
196.100	1528.77
196.000	1529.55
195.900	1530.33
195.800	1531.12
195.700	1531.90
195.600	1532.68
195.500	1533.47
195.400	1534.25
195.300	1535.04
195.200	1535.82
195.100	1536.61
195.000	1537.40
194.900	1538.19
194.800	1538.98
194.700	1539.77
194.600	1540.56
194.500	1541.35
194.400	1542.14
194.300	1542.94
194.200	1543.73
194.100	1544.53
194.000	1545.32
193.900	1546.12
193.800	1546.92
193.700	1547.72
193.600	1548.51
193.500	1549.32
193.400	1550.12
193.300	1550.92
193.200	1551.72
193.100	1552.52
193.000	1553.33
192.900	1554.13
192.800	1554.94
192.700	1555.75
192.600	1556.55
192.500	1557.36
192.400	1558.17
192.300	1558.98
192.200	1559.79
192.100	1560.61

transmitted over a common medium during its allotted time slot. The common medium is used to transmit all voice samples, but only one sample at a time. Each channel is multiplexed over the common facility by dividing the time available on the medium between all channels. This is TDM.

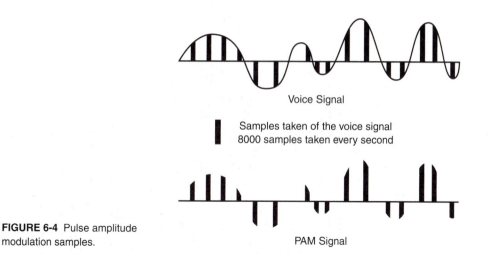

Voice Signal

Samples taken of the voice signal
8000 samples taken every second

FIGURE 6-4 Pulse amplitude
modulation samples.

PAM Signal

When a sample is taken of a voice signal, the voltage level of the signal at that point in time is stored in the channel unit. A 24-channel-unit TDM system would assign each channel to the TDM highway for 1/24 of a second. The transmitter on one end of the cable pair and the receiver on the other end were synchronized. Channel 1 of both ends was attached to the medium at the same time for 1/24 of a second. Then channel 2 on each end was attached to the cable pair, and so on up to channel 24. Then the system would start again on channel 1. When channel 1 was attached to the cable pair, the transmitter on one end would read out the voltage of the stored sample. This voltage traveled over the transmit cable pair to the distant channel 1 receiver. The channel 1 receiver would take the received voltage samples and perform a dot-to-dot connection between the received voltage levels as it fed them through the hybrid circuit to the output of the channel unit. This reconstructed the original signal. Figure 6-4 illustrates samples taken from a signal. Note how the original signal is reproduced when a line is drawn connecting the PAM samples in the bottom drawing. Figure 6-5 provides an example of a three-channel PAM/TDM system.

6.14 TIME DIVISION MULTIPLEXING USING PULSE CODE MODULATION (PCM) SIGNALS

Early TDM systems used PAM, and the systems were analog TDM systems. Since these systems used pulse amplitude modulation (PAM), they were susceptible to noise interference. A method was developed to convert the PAM signals into codes. Codes were established for 256 different amplitude levels. These codes, from 1 to 256, were stored in binary form (00000000 to 11111111). An 8-bit binary code allows for coding 256 distinct numbers (0 to 255) because the number of different combinations that 2^8 provides is 256. The conversion of an analog to a digital signal is done by first converting the signal to PAM and then converting the PAM sample into a digital code. The industry standard method for converting an analog into a

Time slots

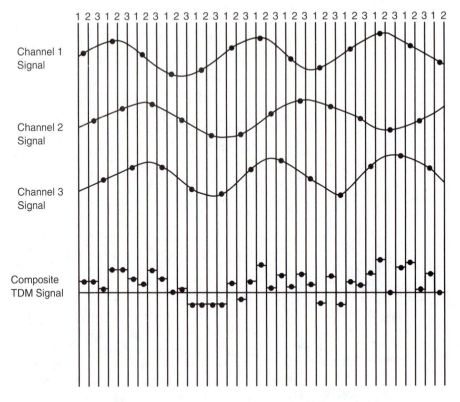

FIGURE 6-5 The creation of a time division multiplexing signal.

digital signal is called *pulse code modulation (PCM).* A TDM system using PCM signals transmits the digital code for the sample instead of the actual amplitude of the sample. The electronic device that uses PCM to convert an analog signal into a digital signal is called a *coder/decoder (codec).*

All TDM systems now use PCM, not PAM. PCM/TDM is a type of multiplexing that requires the conversion of incoming analog voice signals into digital signals. These digitized voice signals are then multiplexed using TDM technology. PCM/TDM is digital transmission, and digital transmission has many advantages over analog FDM and PAM/TDM. These advantages include:

1. Digital signals are less susceptible to noise.
2. Digital circuitry lends itself readily to integrated circuit design, which makes digital circuits cheaper than analog circuits.
3. Digital-to-digital interface is easily achieved.

6.15 PULSE CODE MODULATION

Most of the calls handled by the PSTN are voice calls, which are analog in nature. It is necessary to convert the analog voice signal into a digital signal to achieve the benefits of TDM. The industry standard method for converting one analog voice signal into a digital signal is PCM. There are five basic steps to PCM: sampling, quantization, companding, encoding, and framing.

6.15.1 Sampling

Sampling refers to how often measurements are taken of the input analog signal. The sampling rate in telecommunications is represented by the *Nyquist theorem*, which states that an analog signal should be sampled at a rate at least twice its highest frequency. In telecommunications, the network was designed to handle signals between 0 and 4000 Hz. Therefore, according to Nyquist, sampling at twice the highest frequency of 4000 Hz means a sampling rate of 8000 times per second. Every 1/8000 of a second (125 μs) a voltage measurement is taken on the input signal.

6.15.2 Quantization

In PCM, each voltage measurement is converted to an 8-bit code, and the 8-bit code is sent instead of sending the actual voltage. An 8-bit, two-state (1 or 0) code limits the conversion process to the recognition of 256 codes ($2^8 = 256$). The input signal can be any level. The input signal is not a discrete-level signal. There are billions of discrete levels between –1V and +1V, such as 0.0008362, 0.0794236, –0.998789, and so on. Since we only have 256 discrete codes available, the input signal must be adjusted so that it will be one of the 256 input levels that can be coded. This adjustment process is called *quantization*.

6.15.3 Companding

As noted, quantizing a signal will result in some distortion because we do not code the exact voltage of the input signal. This distortion is called *quantizing noise;* it is greater for low-amplitude signals than for high-amplitude signals. To overcome quantizing noise, the signal is compressed to divide low-amplitude signals into more steps. When the signal is decoded at the receiving system, it is expanded by reversing the compression process. The combination of compression and expansion is called *companding*.

6.15.4 Encoding

After quantization and compression of the input signal, it will be one of the 256 discrete signal levels that can be assigned an 8-bit code. The process of assigning an 8-bit code to represent the signal level is known as *encoding*.

6.15.5 Framing

The encoded 8-bit signal is time division multiplexed with 24 other 8-bit signals to generate 192 bits for the 24 signals. A single framing bit is added to these 192 bits to make a 193-bit frame. The framing bits follow an established pattern of 1s and 0s for 12 frames and then repeat in each of the succeeding 12 frames. This pattern is used by the receiving terminal to stay in sync with the received frames.

The integrated circuit chip that converts an analog input signal into a digital PCM signal is a coder/decoder or codec. The codec is a four-wire device. It accepts incoming analog voice signals and codes them into digital signals placed on the transmit leads to the TDM system. It also accepts digital signals over the receive leads from the TDM system and decodes them into analog outputs. A codec contains an analog-to-digital (A-to-D) converter, a digital-to-analog (D-to-A) converter, and a *universal asynchronous receiver transmitter (UART)*. The UART is necessary to convert the parallel 8-bit output of the A/D converter to a serial 8-bit stream that can be placed on the TDM highway. The analog side of a codec is typically coming from a telephone line that is using only two wires for both the transmit and receive signals. A hybrid network is part of the codec circuitry and is a device that converts a two-wire into a four-wire circuit. Two wires are used as an input from the analog line. Two of the output wires are used as a transmit pair of the coder circuitry. The other two wires serve as the receive pair of the decoder circuitry. The codec circuit card is part of the channel unit in a carrier system using TDM. If the TDM system is not interfacing to a two-wire analog circuit, but is interfacing to a four-wire digital circuit, a different channel unit will be used. This special channel unit has the circuitry necessary to interface to the digital signal.

6.16 NORTH AMERICAN DS1 SYSTEM

The basic building block for digital transmission standards begins with the DS0 signal level. DS0 is the 64-Kbps signal generated by a codec converting an analog voice signal to a digital signal. DS0 is a single voice channel. To convert an analog signal into a digital signal, the analog signal is sampled 8000 times each second and each sample is converted into an 8-bit code. Therefore, when one analog voice signal is converted to a digital signal using PCM, the digital signal is 64,000 bits per second or bps (8000×8). The basic digital multiplexing standard established in the United States was set by Bell Communications Research and is called the *Bell System Level 1 PCM Standard* or *Bell T1 Standard* (it is recognized by CCITT as CCITT Recommendation G.733). This standard is the standard for multiplexing 24 digital voice circuits over a pair of physical facilities. The CCITT Level 1 PCM standard for worldwide use is different from the Bell Standard because it involves the multiplexing of 32 DS0 circuits (30 are voice channels and 2 are signaling/synchronization channels).

The Bell T1 Standard is a 24-channel PCM/TDM system using 8-level coding (256 codes). The T1 system can be equipped with channel units that interface to analog signals, or it can be equipped with channel units that will interface to a digital signal. When the T1 system uses channels that interface to analog signals, the channel unit contains a codec to convert the analog signal into a DS0 signal. The sampling rate

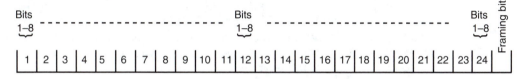

8 Bits in Each Channel

24 Channels × 8 Bits per Channel = 192 Bits + 1 Framing Bit = 193 Bits per Frame

1 Frame = 125 μsec

8000 Frames per Second

FIGURE 6-6 One frame of a DS1 signal.

for each channel is 8000 times per second, and each of the samples is converted into an 8-bit code. The output of the codec is 64,000 bps (8 bits per sample × 8000 samples per second = 64,000 bps). Each analog voice channel is converted to 64 kilobits per second (Kbps). This is referred to as *digital signal level zero (DS0).* When multiplexing 24 channels, one sample (8 bits) is taken from each channel one after the other and sent over the transmission medium. After sending the 8 bits for the 24 channel, one framing bit is sent to start a new frame and maintain synchronization between the transmitting and receiving multiplexers (see Figure 6-6). A frame consists of 193 bits (8 bits per channel × 24 channels + 1 framing bit = 193 bits). Since we are taking 8000 samples per second on each voice channel, it is necessary to transmit 8000 frames per second. The total bits transmitted when multiplexing 24 channels is 1,544,000 bps (1.544 Mbps). That is, 8000 frames per second × 193 bits per frame = 1.544 Mbps. This rate is also arrived at by the following method: 8000 samples per voice channel × 8 bits per sample × 24 channels + 8000 framing bits = 1.544 Mbps. Digital signal level 1 (DS1) = 1.544 Mbps. Finally, DS1 = 24 DS0 channels + 8000 framing bits.

It may help you understand multiplexing if you think of the time required to perform one operation. It would be possible to put two signals over a medium using TDM if we let each signal occupy the transmission medium for half a second. It is possible to allow ten signals to use a time-shared facility if each uses the facility for one-tenth of a second. The number of signals that can be multiplexed in a TDM system depends on the quality and design of the transmitter, receiver, and medium. If the receiver needs to see a signal for one-tenth of a second in order to properly recognize the signal, then the transmitter must place the signal on the medium for at least one-tenth of a second and only ten signals could be multiplexed. If a receiver can detect a signal change in 0.000000001 sec, then 1,000,000 signals can share the common facility, one at a time.

The number of events that can occur in a given time frame is the inverse of the time required for one event. The reverse is also true: the time required for one event is the inverse of the number of events in a given time frame. For example, if the time required for one event is one-tenth of a second, then we can do ten events in one second. Frequency (of events) = 1/time (of one event). Thus, $F = 1/{}^1/10 = 10$. Time required for one event: $T = 1/F$, so that $T = 1/10$. In telecommunications, we usually state time in seconds and the number of events as the number that can occur in one second. Metric system prefixes derived from Greek, Latin, and French are used to designate quantities. Refer to Chapter 1, Section 1.11.

Earlier, it was stated that the PCM sampling rate for a voice signal is 8000 times a second. The time for one sample = $1/f = 1/8000 = 0.000125 = 125$ microseconds (μs). In PCM this sample consists of 8 bits. These 8 bits will be stored in a temporary store; 125 μs later a new sample code will be stored by writing over the old stored code. The stored 8-bit code will be available for readout to the TDM for 125 μs before it is lost. Because the TDM highway is running at 1.544 Mbps, it can read 193 bits in $193 \times 647.6684 \times 10^{-9} = 125$ μs. Recall that 193 bits are transmitted in each frame by sending 8 bits from channel 1, 8 bits from channel 2, and so on all the way up to 8 bits from channel 24 and then 1 framing bit for $(24 \times 8) + 1 = 193$ bits. Also, one frame equals a sample from every channel and these samples are being generated at a rate of 8000 per second. We have to send 8000 frames per second to send every sample generated at the PCM level. Since each frame is 125 μs long (1/8000 of a second), we can send 8000 frames per second.

Higher levels of multiplexing can build frames containing more bits because the receiver, transmitter, and medium used at higher levels of multiplexing can recognize smaller and smaller bits of time as a 1 or 0 digital state. All TDM systems must send 8000 frames per second, but the frame size contains more bits at higher and higher levels of TDM (it can contain more because each bit is smaller or takes up less time). The DS1 signal of a T1 carrier system is generated by reading 8 bits at a time from each channel unit. Higher levels of multiplexing read more than 8 bits at a time from each input. Each higher level of multiplexing must have 8000 frames per second in order to stay in sync with the 8000-per-second requirement at the DS0 level. Each frame has to contain one frame from each lower-level system, just as DS1 contains one DS0 sample from each channel. For example, a DS3 signal is composed of 28 DS1 signals, and each frame of DS3 must contain one frame from each of the DS1 signals. Since each DS1 frame is 193 bits, each frame of DS3 contains 5404 bits plus additional overhead for control. The DS3 can receive 193 bits from DS1 system 1, then 193 bits from DS1 system 2, and so on. DS1 and DS3 both place one bit at a time on the TDM medium because it is a serial bit stream process. However, in DS1 the TDM bit stream will contain data from each channel as 8 adjacent bits on the TDM medium. A DS3 signal is composed of 28 T1 signals multiplexed together. The DS3 signal rate is 44.736 Mbps (28×1.544 Mbps + 1.504 Mbps additional overhead).

In the American T1 carrier system, the signals multiplexed are digital signals (PCM). The T1 receiver is capable of recognizing a bit as being a 1 or 0 in 0.000000647884th of a second (647.884 ns). To do higher levels of multiplexing, the time required for generating and recognizing a pulse must be smaller. Remember that frequency = 1/time. If we want a higher-frequency bit stream, each bit must take less time. In the European TDM system, the procedure is to multiplex 32 channels that are 64,000 bits each. This gives a total of 2,048,000 bits each second. The time for each pulse must be $1/(2.048 \times 10^{+6}) = 1/2.048 \times 10^{-6} = 0.4882813 \times 10^{-6} = 488.2813 \times 10^{-9} = 488.2813$ ns.

6.17 T1 CARRIER SYSTEMS

The multiplexing system of choice to connect two central offices is a North American DS1 system called *T1 carrier*. This 24-channel multiplex system uses the Bell Standard DS1 signal rate of 1.544 Mbps. These systems are generally connected to wire pairs in a cable that runs from one central office to the other. Because a DS1 signal level is a high-frequency signal, it cannot travel more than a mile and a half before the signal is lost. It is necessary to remove load coils from the pairs being used for T1 carrier to improve the high-frequency response of the cable pairs. It is also necessary to add regenerators to the T1 cable pairs every 6000 ft. Because cable pairs are loaded at the first 3000 ft from the central office and then every 6000 ft thereafter, as the load coils are removed from a cable pair to be used for a T1 carrier system, the pulse regenerators are put on the pairs at these points. The pulse regenerators are called *repeaters*. Repeaters are also added to the T1 line (or cable pair) at each central office. A T1 system is connected to a passive repeater at the central office. The central office repeater does not regenerate outgoing transmit pulse streams, but merely passes the pulses from the T1 terminal to the transmit cable pair. The next repeater will be 3000 ft from the central office, then additional repeaters are placed at 6000-ft intervals. The repeater looks at the incoming signal (1s and 0s), and for each 1 received, it sends out a fresh high-level output. Since the maximum distance between repeaters is 6000 ft, the T1 repeaters keep the signal at a high level.

Many T1 carrier systems are still in use, although the trend in larger cities has been to replace direct T1 carrier routes with an OC-48 fiber ring technology. An OC-48 fiber ring can carry 32,256 trunk circuits. The fiber optic cable links all central offices in a big city. An OC-48 multiplexer/demultiplexer is located in each exchange and is programmed to strip out the specific channels assigned to that exchange. T1 is used in places where fiber ring networks have not been installed. When a T1 carrier system is used, the on-hook and off-hook signals for each channel are sent over the system by a technique called *bit-robbed signaling*. Since two states are needed for the signal (either on-hook or off-hook), they can be represented by a binary bit (1 or 0). Because the signaling state does not change very often or fast, we do not need a signaling bit in every frame. The first signaling technique used for T1 substituted a signaling bit for each of the 24 channels in every sixth frame. This signaling bit for each channel replaced the least-significant voice bit in every sixth frame on each voice channel. This causes some distortion to the voice signal, but the distortion is not noticeable. Of course, this bit-robbing technique would destroy data if the signal on the channel is a data signal. For that reason, data on a PSTN T1 system is limited to 7 bits per channel in every frame, not just every sixth frame. Further, 7 bits per frame × 8000 frames = 56,000 bits per second. Thus a T1 system used in the PSTN can only accept a data rate of *56,000 bps on each channel.* This is referred to as *56-Kbps clear channel capability.* There are some proprietary T1 systems that use a means other than bit robbing to do signaling. These systems can use all 64,000 bps of the channel for data. A T1 system used as the vehicle for ISDN does not use bit-robbed signaling; thus, the signaling for all channels is carried by the 24th channel. The 24th channel does not serve as a voice channel. It serves as a common channel

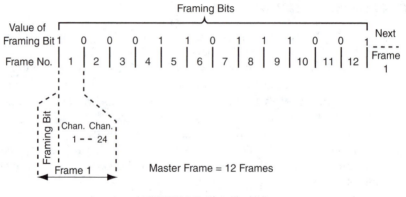

FIGURE 6-7 Superframe.

for signals and is called the *D-channel*. Therefore, a T1 system used to carry ISDN provides a clear channel capability of 64,000 bps.

Figure 6-7 shows the layout for a superframe. A superframe is composed of 12 frames. Each of the frames is 193 bits (twenty-four 8-bit voice channels plus 1 framing bit). Notice that the framing bit pattern over 12 frames is 100011011100. These framing bits allow the receiver to synchronize off the incoming bit stream. The least-significant bit of every channel in the 6th frame is used for signaling and is called the *A bit*. The least-significant bit of each channel in the 12th frame is also used for signaling and is called the *B bit*. Another type of signaling developed for T1 signaling is called *extended superframe (ESF)* signaling. The ESF consists of 24 frames; thus, there are 24 framing bits. Only 6 of the 24 framing bits are used to provide synchronization for the receiver. Six bits are used to provide a CRC error-checking protocol. The remaining 12 bits are used to provide a management channel called the *facilities data link (FDL)*. The ESF provides signaling bits in every 6th frame. The signaling bit in the 6th frame is called the *A bit*, the signaling bit in the 12th frame is termed the *B bit*, the signaling bit in the 18th frame is the *C bit*, and the signaling bit in the 24th frame is the *D bit* (see Figure 6-8).

When the transmitter of a T1 carrier system was first introduced, the maximum distance for sending the digital signal was 3000 ft. The sending of 1.544 Mbps looked like a 1.544-MHz signal to the cable pair. The design of the transmitter was changed to make a T1 signal look like a 0.772-MHz signal by alternating the polarity of every 1 (mark) transmitted. This technique is called *alternate mark inversion (AMI)*. AMI is a bipolar signal. The signal changes back and forth between +3 and −3V. A 0 is represented by no voltage at all. A 1 is represented by a +3-V signal. The next 1 transmitted is represented by a −3-V signal, and so on (see Figure 6-9). AMI provides for an error-detection scheme called *bipolar violation (BPV)*. Since the T1 line signal level will always alternate between either (1) a +3-V signal, then 0, and then −3V; or (2) +3V, then −3V, there should never be two consecutive +3-V signals or two consecutive −3-V signals. Figure 6-10 shows a bipolar violation occurrence. Compare the original signal of Figure 6-9 to the signal of Figure 6-10, which contains errors to the original signal.

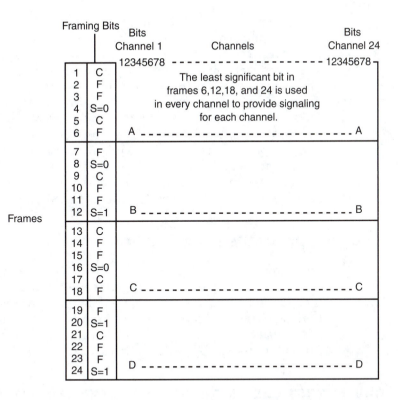

FIGURE 6-8 Extended superframe. ESF uses 6 bits (S) for synchronization (001011). C bits are used for CRC error checking and F bits comprise the facilities data link (FDL) channel.

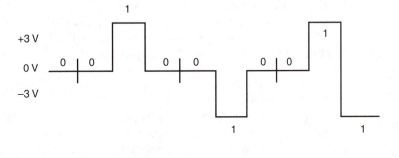

FIGURE 6-9 Alternate mark inversion.

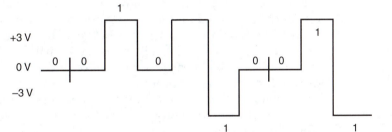

FIGURE 6-10 Bipolar violation.

The receiver is synchronized by each incoming 1. If a string of 0s are received, the transmitter could lose synchronization. To prevent a loss of synchronization due to many successive 0s, the *Binary 8 zero substitution (B8ZS)* technique was developed. The encoding of a voice signal will never result in more than seven successive 0s. If data is transmitted at 56,000 bps using 7 bits per channel time slot, the 8 bit in each channel time slot can be made a 1. This also prevents more than seven successive 0s. B8ZS was developed to keep a transmitter from sending more than eight consecutive 0s. When a string of eight consecutive 0s is encountered, the transmitter will induce a bipolar violation at the fourth and seventh 0s and AMI-compatible marks at the fifth and eighth 0s. The receiver detects the B8ZS code and reinserts eight 0s for the code. The receiver will receive enough 1s to stay in synchronization, either from actual 1s in the bit stream or from the 1s forced into the bit stream by B8ZS.

6.18 EUROPEAN TDM 30 + 2 SYSTEM

The European TDM system multiplexes 32 DS0 channels together. This system uses channel 0 for synchronizing (framing) and signaling. Channels 1–15 and 17–31 are used for voice. Channel 16 is reserved for future use as a signaling channel. Because framing signals are part of channel 0, additional framing bits are not required. The DS0 signals are 64,000 bps. The total signal rate for the European 30 + 2 system is 2.048 Mbps (64,000 bps × 32 channels = 2,048,000).

6.19 STATISTICAL TIME DIVISION MULTIPLEXING (STDM)

The T1 and E1 systems discussed earlier are referred to as *synchronous time division multiplexing* systems. In these TDM systems a channel time slot on the TDM highway is reserved for each input. Channel 1 input at one end of the TDM system will be carried by time slot 1 of the time division multiplexed facility and will be connected to the channel 1 unit at the other end of the facility. Time slot assignments are static and always reserved for a specific channel unit. It does not matter whether that channel has data to send. This can lead to a very inefficient use of the TDM facility.

The *statistical time division multiplexer* was developed to improve the efficiency of a TDM system. This type of multiplexer has many more inputs than the TDM facility can handle at the same time. A statistical time division multiplexer may have 48 channel units but use a 24-channel TDM highway running at 1.544 Mbps. Only 24 of the channel units can be in use at the same time. The channel units do not have reserved time slots on the 24-channel TDM highway. When a channel is in use, it will be connected to one of the 24 time slots of the TDM highway not in use. Time slots are dynamically assigned and any input channel can be mapped to any time slot. This means the frames on the TDM facility also need to carry addressing information in order to deliver the information sent by one channel to its corresponding channel on the other end of the TDM facility.

The scenario just presented was used so that you could compare *statistical time division multiplexing (STDM)* to a T1 TDM system. Statistical time division multiplexers are not used in the PSTN voice network, but in the data world; they are called *stat muxs, intelligent multiplexers,* and *asynchronous multiplexers.* They are asynchronous because if a channel has no data to send, no time slot is assigned or used on the TDM highway. Stat muxs use proprietary protocols, and you must ensure compatibility between the stat muxs utilized on each end of the TDM facility. The TDM facility will usually be a digital line leased from an ILEC, CLEC, IEC, or competitive access provider (CAP).

A typical arrangement using a stat mux might be several low-speed terminals running at 4800 bps. You might use a 16-port statistical time division multiplexer to connect these terminals to a leased 9600-bps facility. You can readily see that at 4800 bps, only two terminals can be connected at the same time to the leased line. The key to aggregating these terminals is the fact that they are interactive terminals. No one can type at 4800 bps; there is a great deal of dead time between inputs. Terminals are attached to an input port that contains a memory buffer. When the stat mux's CPU sees that an input port's buffer has information to send, it allocates a time slot and sends the data. The stat mux pays no attention to dead time at each input and effectively converts time on the TDM highway into all live data.

Statistical time division multiplexers are very popular for data-only applications, but their use is limited on multimedia applications, voice, and video, because these applications require time slots on demand. *Synchronous time division multiplexers* such as T1 or T3 are preferred when the applications transmit high volumes of data with little dead space between inputs.

6.20 HIGHER LEVELS OF TDM

The 24-channel T1 system developed in the early 1960s is the basic multiplex building block in North America. The next order of multiplexing is to combine two T1 systems into a 48-channel DS1C (T1C) system. Four T1 systems were combined into a 96-channel DS2 system in 1972. Twenty-eight T1 systems can be multiplexed into a 672-channel DS3 system, which is a 44.736-Mbps system. Six DS3 systems can be multiplexed into a 4032-channel DS4 system (274.176 Mbps). Wire can be used as the medium to carry DS1, DS1C, and DS2 systems. DS3 and DS4 must use coaxial cable, fiber optic cable, or digital microwave radio as a medium. Regenerators (repeaters) used on a fiber optic cable are placed 27 mi apart.

6.21 SONET STANDARDS

Fiber optics use *synchronous optical network (SONET)* standards. The initial SONET standard is OC-1. This level is known as *synchronous transport level 1 (STS-1).* It has a synchronous frame structure at a speed of 51.840 Mbps. The synchronous frame structure makes it possible to extract individual DS1 signals without disassembling the entire frame. OC-1 is an envelope containing a DS3 signal (28 DS1 signals or 672 channels). With SONET standards any of these 28 T1 systems can be stripped out of the OC-1 signal. The SONET levels OC-1, OC-48, and OC-192 are shown at the bottom of Table 6-2.

TABLE 6-2 TDM Comparisons

Designation	Number of DS-0 Circuits	Bit Rate
DS-0	1	64 Kbps
DS-1	24	1.544 Mbps
DS-1C	48	3.152 Mbps
DS-2	96	6.312 Mbps
DS-3	672	44.736 Mbps
DS-4	4032	274.176 Mbps
OC-1	672	51.84 Mbps
OC-48	32,256	2.48832 Gbps
OC-192	129,024	9.95328 Gbps

[Handwritten annotations: "T1" next to DS-1; "T2" next to DS-2; "T3" next to DS-3; "over copper wire { STS-1, STS-48, STS-192 }" next to OC-1, OC-48, OC-192; "Optical Carrier" bracketing OC-1 through OC-192]

OC-48 is 2488.32 million bps or 2.48332 billion bps. As noted, the prefix for billion is giga (G). Therefore, 2.48332 gigabits per second is the signal rate of OC-48. OC-48 is 48 OC-1 systems and contains 48 × 672 channels = 32,256 channels. One fiber optic strand will carry all 32,256 channels. An additional benefit to using fiber optic technology is that it will not pick up noise from any outside source influence. The glass strands of fiber are carrying light pulses produced by a LED or LASER. It is impossible for glass fiber to carry electrical energy. Noise in wire cables is caused by electrical currents being induced into the wire from outside electromagnetic radiation. This cannot happen with fiber optic cables because fibers cannot carry electrical signals. In many of the BOC territories, OC-48 has replaced T1 carrier systems. Instead of having multiple T1 routes from class 5 exchanges into a class 4 exchange, OC-48 technology and fiber optic cables link offices together in a ring network. Fiber will connect all offices in the metropolitan area. Each office is equipped with an OC-48 multiplexer/demultiplexer; each will only strip out the channels assigned to that office. This is the beauty of SONET standards. The whole multiplexed stream does not have to be demultiplexed to pull out one T1 system or even one DS0 channel. The OC-48 can be instructed to pull out one DS1 bit stream, and this stream can be connected to a T1 system.

Today's digital central offices have fiber optic multiplex loops inside the switch running at OC-1 levels or higher. With an OC-1 level switching loop, it is possible to have an OC-48 system at each central exchange on the fiber ring strip out an OC-1 signal of 672 trunk circuits and connect them directly into the switch. This is because SONET technology also allows stripping out one DS0 signal (one trunk circuit). The signals on the OC-48 fiber ring can be stripped out as needed one at a time and then assigned an appropriate slot on the OC-1 loop of the switch. Software in the switch and SONET multiplexers makes it easy to assign DS0 channels for use between any exchanges connected to the fiber ring network.

Sprint has upgraded its fiber routes to handle as many as 516,096 trunk circuits (channels). This has been accomplished by using wave division multiplexing. Instead of using strictly TDM technology and trying to extend this technology to multiplex many more signals in a TDM mode, wave division multiplexing technology combines

WDM

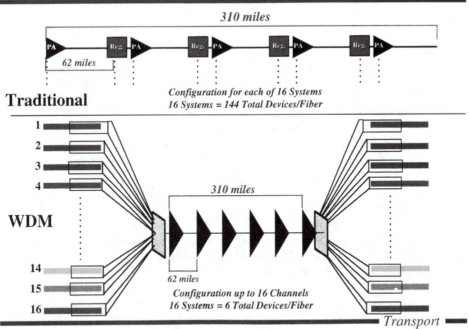

FIGURE 6-11 Wave division multiplexing. (Courtesy of Sprint.)

FDM and TDM technology. The light spectrum of a fiber optic facility is broken down into multiple frequencies. Light waves are composed of many frequencies. A blue light has a different frequency from a red light, an orange light, a green light, and so on. Transmitters and receivers for fiber optic facilities have been developed that transmit and receive at a specific frequency. Sixteen different frequencies are used. An individual OC-48 TDM connects to each of the 16 different transmitters in a wave division multiplexer. By splitting the light spectrum up into 16 different frequencies and attaching an OC-48 to each transmitter, the fiber optic facility can carry 516,096 channels (32,256 × 16). The next step in dense wave division multiplexing is to connect 32 OC-192 TDMs into a 32-channel wave division multiplexer. Each OC-192 can carry 129,024 DS0 channels. Within a year or so, we will see 160-channel wave division multiplexers. With 160 OC-192 time division multiplexers connected to this wave division multiplexer, one fiber can carry 20,643,840 DS0 channels. The total bit rate will be 1.6 terabits per second. This provides one fiber with the ability to transmit the contents of about 200,000 books per second (see Figure 6-11).

Note that in the example above, I used the colors of light for demonstration purposes only. Most people can readily understand the various colors of light and how they can be split into different frequency spectra. Fiber optic fibers carry light that has no color. The infrared light is below the visible light spectrum and is thus below frequencies that are red, blue, green, and so on. Because the light is from a laser, it is extremely dangerous to look into the end of a fiber. The light will perform laser surgery and damage to your retina.

6.22 SUMMARY

Multiplexing of signals onto one media can be accomplished using FDM and TDM. In order to do FDM, every signal must be at a different frequency. The process most often used to prepare signals for FDM is Amplitude Modulation (AM). This process changes the frequency of a signal. To do FDM every signal must have a different frequency. If two or more signals of the same frequency are to be multiplexed, they are taken through individual channel units. The output of the channel units will be signals at different frequencies, and they can be multiplexed onto one media. TDM simply takes a signal and places it onto a media for a brief instant of time. No two signals can be on the media at the same time. Each signal can be on the media only during its assigned time slot.

When fiber optics was first introduced, we used one light signal on the fiber. The light signal was from one TDM system. The TDM system would turn the light brighter for a 1 and dimmer for a 0. Thus, by rapidly varying the frightness of the light, the TDM system could transmit 1s and 0s very rapidly. The receiver would look at the brightness of the light and decipher the brightness of the received light into 1s and 0s. With this arrangement the fiber handled one TDM system.

Today instead of connecting the TDM system directly to a fiber, we can connect the TDM system to channel units of a Dense Wave Division Multiplexer (DWDM) and the DWDM is connected to the fiber. The DWDM can generate as many as 180 different lightwaves. Each lightwave is a different frequency. Each channel will generate a different frequency of light that will be placed on the fiber. The use of many different frequencies is FDM, but when FDM is used on fiber optics we refer to it as DWDM.

CHAPTER REVIEW QUESTIONS

1. What is multiplexing?
2. What is space division multiplexing?
3. What is frequency division multiplexing?
4. What is time division multiplexing?
5. What is the purpose of a channel unit?
6. What is a four-wire system?
7. What is amplitude modulation?
8. What is DWDM?
9. What is the difference between FM and FDM?

10. With a 32-channel DWDM and OC-192 TDM attached to each channel, how many DS0 circuits are being carried by the fiber?

11. How many voice channels are carried by a T1 system? A T3 system?

12. What type of modulation technique is used by a DWDM multiplexer?

13. What is the difference between PAM and PCM?

14. What is the industry standard method for converting an analog voice signal into a digital signal DS0?

15. What is the sampling rate of PCM?

16. How was the sampling rate of PCM determined?

17. What is quantization?

18. What device is used to do PCM?

19. What is the DS0 signal rate? What is the DS1 signal rate?

20. How long is each bit on the TDM path of a T1 carrier system?

21. How many bits are there in a DS1 frame?

22. How many frames per second does a DS1 system have?

23. When T1 is used on a cable pair, how many repeaters are needed?

24. What is the basic building block for SONET?

25. How many voice channels can be put on an OC-48 system?

26. What transmission medium is used for OC-48?

27. What are the major advantages of SONET standards?

PROJECTS AND CASE STUDIES:

Case Study #1

Our Bell Telephone Company has decided to deploy the latest in WDM technology. Quest has deployed 32-channel WDM with OC-192 TDM systems riding each channel. Does Nortel, Alcatel, or anyone else offer more than a 32-channel system? What is available?

GLOSSARY

Amplitude Modulation (AM)   Used by AM broadcast stations. In amplitude modulation, the amplitude of the carrier frequency coming out of the mixer will vary according to the changing frequency of the input voice signal.

Bandwidth   The width of a signal, which is determined by subtracting the highest frequency of the signal from its lowest frequency.

Channel Unit   Performs the function of interfacing the multiplexer to a specific type of input signal. The channel unit makes changes to the input signal so it can be combined (multiplexed) with other signals for transmission over the transmission medium.

Frequency Division Multiplexing (FDM) The process of converting each speech path to different frequency signals and then combining the different frequencies so they may be sent over one transmission medium.

Mixer (Modulator) An electronic device that has two inputs. One input is a low-frequency signal that contains intelligence. The other signal is a pure sine wave at a high frequency. This high-frequency signal is called the *carrier signal* because it carries our intelligent signal after modulation. The mixer combines the two signals so that we end up with a high-frequency signal that has the intelligence of the low-frequency signal imposed on it.

Multiplexing In telecommunications, a process combining many individual signals (voice or data) so they can be sent over one transmission medium.

Pulse Amplitude Modulation (PAM) The process used to take samples of analog voice signals so they can be multiplexed using TDM. The samples appear as pulses in the TDM signal.

Pulse Code Modulation (PCM) The industry standard method used to convert an analog signal into a digital 64,000-bps signal. PCM takes a PAM signal and converts each sample (pulse) into an 8-bit code.

Sidebands The two additional signals generated by mixing two signals. The upper sideband represents the sum of the carrier and modulating frequencies. The lower sideband will have a frequency that is the difference between the carrier frequency and the modulating frequency.

Both sidebands contain the intelligence of the modulating signal.

Space Division Multiplexing A process in which multiple communications are placed over many different wire pairs inside one cable. Each communication channel (voice or data) occupies its own space (occupies its own set of wires).

Statistical Time Division Multiplexer A hardware device that multiplexes many input channels of data or voice onto a common TDM medium between two STDM multiplexers but only places data into a TDM slot when the input has data to send. This makes the STDM multiplexer more efficient than a synchronous time division multiplexer because time slots are not reserved for each and every input channel. Since the STDM multiplexer must be smart enough to inform the distant STDM multiplexer which output channel should receive the data, these multiplexers are also referred to as *intelligent multiplexers.*

Statistical Time Division Multiplexing (STDM) A process of time division multiplexing where the time slots on the TDM medium between STDM multiplexers are dynamically allocated. Any input to the STDM multiplexer can be assigned to any time slot on the TDM medium. Time slots on the TDM medium are assigned to input channels that have data ready to send over the TDM medium to a distant STDM multiplexer. To accomplish dynamic allocation of any input to any time slot, the TDM multiplexer places a header on the data that each input channel has ready to transmit. This header informs the receiving STDM multiplexer which output channel is to receive the data.

Synchronous Optical Network (SONET) A network of fiber media connected together for the transport of digital signals.

Synchronous Time Division Multiplexer A hardware device that places many signals over one medium using time division multiplexing. A T1 system is a time division multiplexer.

Synchronous Time Division Multiplexing A multiplexer that has static assignments of time slots on the TDM highway between two multiplexers. Each time slot is permanently mapped to an input and is reserved only for that input regardless of whether the input channel has information to send.

Synchronous Transport Level 1 The basic building block signal level for SONET.

Time Division Multiplexing (TDM) The process of converting each speech path into samples, then combining the different samples by transmitting each sample at a different time, so they may be sent over one transmission medium.

Wave Division Multiplexer The hardware device that uses FDM to multiplex many TDM signals together over one medium.

Wave Division Multiplexing (WDM) Technology that splits the light spectrum into many different frequencies. A time division multiplexer is then assigned to each of the different light wave frequencies. Many TDM signals are multiplexed together so they can be placed on one fiber medium.

7

Analog and Digital Signals

KEY TERMS

Analog Signal

Asymmetrical Digital Subscriber
   Line (ADSL)

Crosstalk

Differential Manchester

Digital Signal

Integrated Services Digital
   Network (ISDN)

Manchester Coding

Nonreturn to Zero (NRZ)

Nonreturn to Zero Invert (NRI)

Protocol

OBJECTIVES

On completion of this chapter, you should be able to:

1 Describe the characteristics of an analog signal.

2 Describe the characteristics of a digital signal.

3 Explain the benefits of converting an analog voice signal into a digital signal.

4 Explain how analog signals are connected from a transmitter to a receiver.

5 Explain how digital signals are coupled from a coder to a decoder.

6 Explain what alternate mark inversion (AMI) is.

7 Explain what Manchester coding is.

8 Explain what differential Manchester coding is.

9 Explain nonreturn-to-zero (NRZ) and nonreturn-to-zero-invert signaling.

10 Explain the correlation between bandwidth and power loss over the local loop.

Electricity can be used to convey signals from one location to another using wire as the transmission medium. In earlier chapters, we saw that *telecommunications* means to communicate over a distance. The most common way of communicating over a short distance is to use electricity to represent voice or data signals. The predominant method of transmitting over a long distance is to use radio waves or light waves to carry the voice or data signal.

When using telecommunications to communicate over a short distance, various levels of electricity are utilized to represent voice or data signals. The level of electricity to use depends on the rules agreed to by the sender and receiver. The rule of communication used by devices to communicate with one another is called a **protocol.** Many different telecommunications protocols have been developed to transfer information from sender to receiver. The sender and receiver simply have to agree to use the same protocol.

7.1 COMMUNICATION SIGNALS AND PROTOCOLS

A communication protocol in telecommunications will specify what type of signal is to be used for communication, how the signal is to be manipulated, and how the signal is to be placed on the transmission facility. The communication protocol will specify whether to use simplex, half-duplex, or full-duplex transmission. Another part of the protocol will specify what electrical signals are to be used to represent the information to be communicated. Electrical signals use various levels of electricity to represent voice, data, or video. To gain a better understanding of electricity, you may wish to review the material in Appendixes A to C. In telecommunications, we employ two kinds of electrical signals: analog and digital.

An *analog signal* is an electrical signal with continuously varying amplitude (see Figure 7-1). Such a signal may consist of a single sine wave such as a 500-Hz

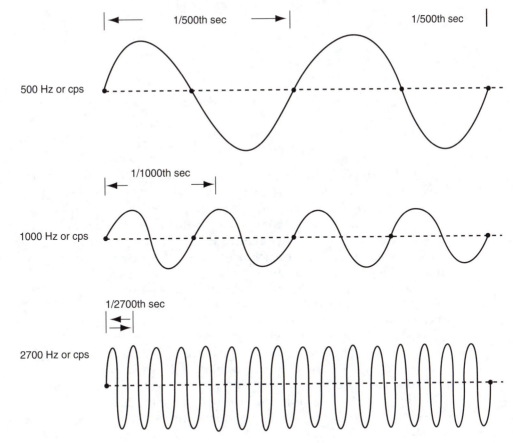

FIGURE 7-1 Sine waves.

FIGURE 7-2 Voice signal composed of many sine waves.

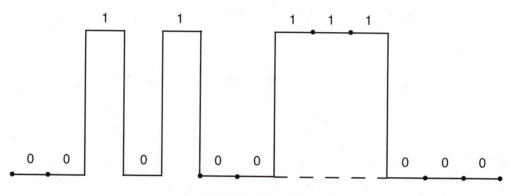

FIGURE 7-3 Digital signal.

signal, a 1000-Hz signal, a 2700-Hz signal, and so on. As stated in Chapter 2, the voice sound wave signal is composed of many different frequencies of signals mixed together (see Figure 7-2). When a sound wave strikes a telephone transmitter, it will cause the electrical analog signal from the transmitter to also be composed of many different frequencies of electrical signals mixed together. Analog signals have three basic characteristics that can be used to convey a message: amplitude, frequency, and phase.

A *digital signal* is a signal that can assume one of several discrete states. Most digital signals take on one of two discrete states (see Figure 7-3). The prefix *bi* means "two," and this two-state signal is also called a *binary signal*. Each state is represented by a different electrical signal. Two different electrical states can be used to represent the two binary digits (1 and 0). Because this two-state signal represents the digits 1 and 0, it is also called a *digital signal*. Information to be transmitted is converted into a digital code consisting of a specific 1s-and-0s pattern. This digital code is then transmitted over a facility to a receiver. The receiver follows the same protocol used by the transmitter and converts the digital signal back into the original information.

7.2 ANALOG SIGNAL

All electrical signals with varying amplitudes are called *analog signals* (*analog* is shorthand for "analogous"). When the electrical signal is similar to the signal it represents, it is termed an *analog signal*. When the telephone was invented, the device used to convert sound waves into electrical waves (and vice versa) was called a *transceiver*. The transceiver performed the functions of both transmitter and receiver. When you spoke into this device, it converted your sound waves into electrical signals, which had the same shape and frequency of the sound wave (see Figure 7-4). Since the electrical signal was similar to the sound wave, it was called an *analog signal*. This analog signal was conveyed by a pair of wires to a second transceiver. When it passed through the second transceiver, it converted the electrical signal back into sound waves.

The transceiver was a device that contained a coil of wire suspended inside a magnet. Thus the coil of wire was suspended inside a magnetic field. The suspension system was designed to allow the wire coil to move up and down when someone spoke into the transceiver. Moving wire inside a magnetic field will cause an electrical signal to be developed in the wire. The coil of wire would vibrate up and down at the frequency of the voice signal. The length of travel depended on the loudness of the voice signal. The transceiver developed a low-voltage signal, and a user had to shout into the transceiver to develop a signal with high enough voltage to travel very far.

The limitations of the transceiver were overcome by developing a different device to serve as the transmitter. The transmitter used in most telephones is a carbon granule transmitter developed by Thomas Edison. Unlike the transceiver, this transmitter cannot generate electricity; an external power supply is required to furnish power. The first commercially produced telephones used Edison's transmitter, and

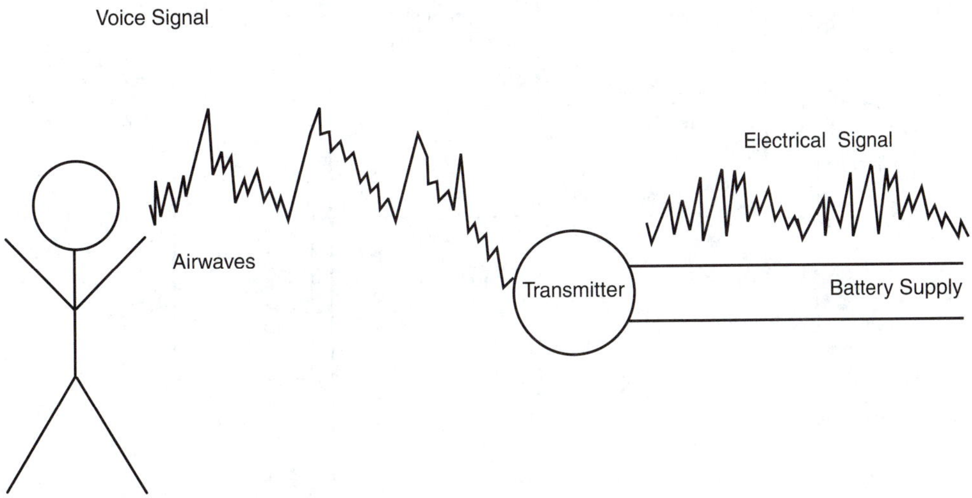

FIGURE 7-4 Conversion of airwaves into electrical waves by the transmitter. Because the transmitter is part of a series electrical circuit, the electrical wave occurs in both wires.

the power for it was supplied by a 3-V battery supply at each telephone. The 3-V battery consisted of two large 1.5-V cells wired in series. Each of these cells was about 2 in. in diameter and about 8 in. high.

The receiver of a telephone was stored in a hooklike device on the side of the telephone. This device was movable; when it moved, it caused a switch to operate. This combination was referred to as a *hookswitch* (some people call it a *switch-hook*). The operation of this switch connected the negative side of the battery power supply to the transmitter and connected the other side of the transmitter to a line wire. The switch also connected the positive side of the battery to a second line wire. When the telephone receiver was taken off the hook on which it was stored, the battery was connected between the transmitter and the line wire. The line wire was the wire used to connect to the other telephone on the call (see Figure 7-5).

The telephone transmitter contained many carbon granules packed inside a small enclosure. When the user spoke into the transmitter, the carbon granules would compress and decompress at the frequency of the voice signal. When compressed, the resistance between the two terminals connected to the transmitter decreased. When decompressed, the resistance between the two terminals increased. This compression and decompression caused the resistance of the transmitter to vary in unison with the frequency of the voice signal. Since the transmitter was in series with the 3-V battery supply and the line connected to another telephone, the current in this circuit would vary as the resistance of the transmitter varied. The varying current in the receiver caused it to reproduce the sound wave.

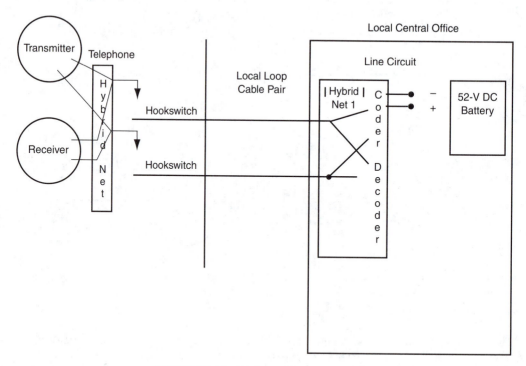

FIGURE 7-5 Electrical power to the transmitter.

The receiver in most telephones is a miniature version of the original transceiver, and significant improvements have been made in the design of the receiver. Today's receivers consist of a coil of wire around a magnet. As the current in the wire increases and decreases, it adds to and subtracts from the magnetic field. As the magnetic field varies, it causes a metal diaphragm suspended across the face of the magnet to vibrate back and forth at the frequency of the electrical changes. The vibrating diaphragm causes the air pressure in front of it to change at the frequency of the diaphragm vibration. This recreates the original transmitted voice signal (see Figure 7-6).

As mentioned earlier, devices that convert a signal from one form of energy into another are called *transducers*. The transmitter is a transducer that converts changes in air pressure (sound waves) into electrical signals. The receiver is a transducer that converts electrical signals into changes in air pressure (sound waves). The electric current used by these devices is very small; it will be somewhere between 20 milliamps and 60 milliamps. Louder voice signals cause more current to flow in the transmitter because the carbon granules are compressed tighter by the signal. Higher amplitudes for the electrical signal flowing through a receiver cause the sound waves generated by the moving diaphragm to be louder.

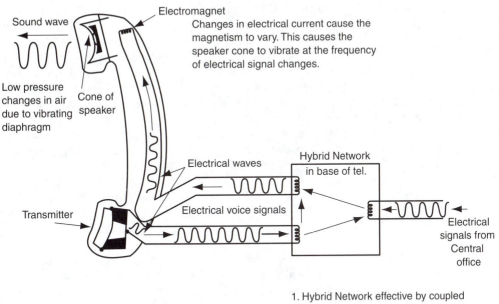

1. Hybrid Network effective by coupled electrical signals from central office to the receiver.

2. A small amount of the electrical signal from transmitter is coupled to the receiver.

FIGURE 7-6 Telephone receiver and hybrid network.

The louder we talk, the higher the amplitude of the electrical signal flowing through the transmitter; the more quietly we talk, the less electric current there is. Thus, the amount of electric current flowing captures the volume of the speaker's voice. Sounds we make are used to communicate with others. If others understand the protocol of our sounds, they can interpret the meaning of the sounds. If our protocol is English and the receiver can interpret the English protocol, communication is established. If we speak French and the receiver cannot interpret French protocol, communication cannot take place.

The telephone captures the sounds we make as analog electrical signals. These signals are conveyed by the medium to the distant receiver. At the receiver the sounds are reproduced. The telephone is a simple device used to convert sounds into electrical signals and vice versa. What the sounds mean is left to the user of the telephone to determine. The telephone has an established method in terms of how to convert sound waves into electrical signals and how to convert these electrical signals back into sound waves. Any higher-level protocol is left to the telephone user.

We can make different sounds by changing the pitch (frequency) of the sound. The loudness of our voice (or sound wave) determines how much electric current will flow through the transmitter, but the frequency of the sound wave conveys the intelligence or intended meaning of the sound. We are most interested in the frequency of the electric current changes, because frequency changes contain the intelligence of the signal. As we speak into the transmitter of a telephone, the carbon granules will compress and decompress at the frequency of changes in our voice signal. This action causes the electrical signal flowing through a transmitter to vary at the same frequency as the voice signal. When these variances reach a receiver, it will vibrate at the frequency of the received electrical signal. As the diaphragm vibrates in unison with the frequency changes in the electrical signal, it compresses and decompresses the air between the receiver and the ear of the listener at this frequency. The listener's ear detects the frequency of changes in air pressure as the sounds are generated by the speaker at the transmitter.

The transmitter does an excellent job of converting sound waves into electrical signals. The receiver does an excellent job of converting electrical signals into sound waves. The telephone simply converts sound waves at the transmitter into electrical signals and converts received electrical signals into sound waves at the receiver. The telephone users do the protocol conversion. So it does not matter if we speak English, Spanish, Russian, or Chinese; the telephone converts any of these sounds into its equivalent electrical signal and relies on the users of the telephone to employ a protocol that both are familiar with.

Every communication system contains a transmitter, communication medium, and receiver. The transmitter is responsible for converting messages into a signal that can be transported by the medium to a receiver. The receiver is responsible for converting received messages into signals that can be understood by the recipient. Our mouth, vocal cords, and brain serve as a transmitter when we speak to another person. The brain uses the vocal cords to convert thought signals from our brain into varying air pressure signals (airwaves). These changes in air pressure are easily carried by the selected medium (air) to the recipient. The eardrum and brain serve as the receiver. They convert messages received as airwaves into thought signals in the brain. Our brain decodes the signal into intelligible messages.

The telecommunications system also carries messages from a transmitter (or sender) to a receiver. For voice communication, the system uses wire as a medium to connect two telephones together. Each telephone contains a transmitter and receiver. The transmitter converts airwaves into electrical signals easily carried by the wire medium to the receiver. The receiver converts these electrical signals back into airwaves. The messages have to be converted into electrical signals before the telecommunications medium (wire) can carry them. A telecommunications manager must have a basic understanding of how electrical signals are affected by the wire media and the environment surrounding those media.

7.3 CONNECTING THE TELEPHONE TO THE CENTRAL EXCHANGE

The telephones at our residences and any small businesses connect via one pair of wires to a switching system called the *local central office.* In rural areas, the local switching system is found in the center of town. Cables leave the central office and fan out like the spokes of a wheel in all directions. Since the switching system is located at the center of this hub, it is called the *central office, central exchange,* or *central.* In cities, the geographic area of the city is broken up into many smaller geographic areas and a switching system is placed in the center of each smaller area. These smaller areas are called *exchange territories.* Each geographic territory is served by a specific switching system (central exchange) located at the center of the territory (see Figure 7-7).

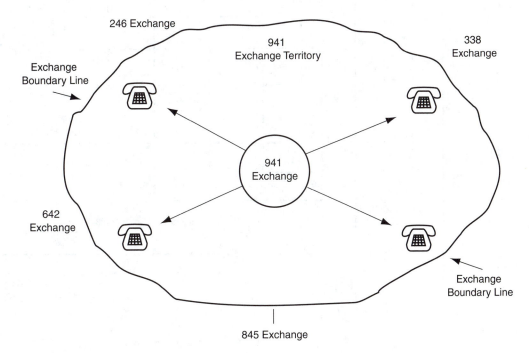

FIGURE 7-7 Central office exchange territory.

The pair of wires that connects the telephone to the central exchange is called the *local loop*. The pair of wires forms a loop that connects the telephone to the local central exchange. One end of the local loop connects to our telephones; the other end connects to the central exchange. At the central exchange, all the wire pairs in all the cables are terminated on connecting blocks mounted on a *main distributing frame (MDF)*. The MDF is simply a large metal frame with connecting blocks mounted on it (see Figures 7-8 and 7-9). The cables that exit the central exchange building to serve telephones are connected to blocks mounted on one side of the

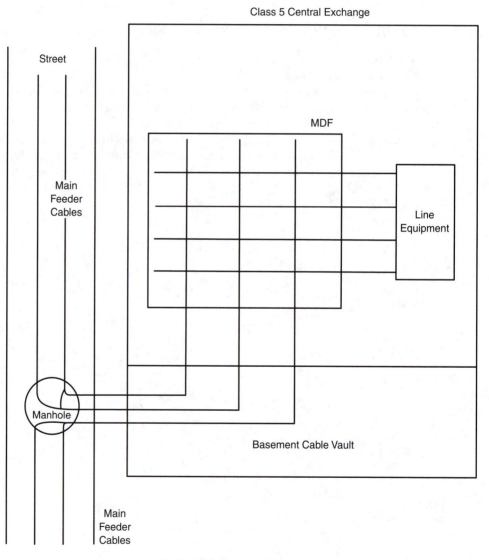

FIGURE 7-8 Main distributing frame.

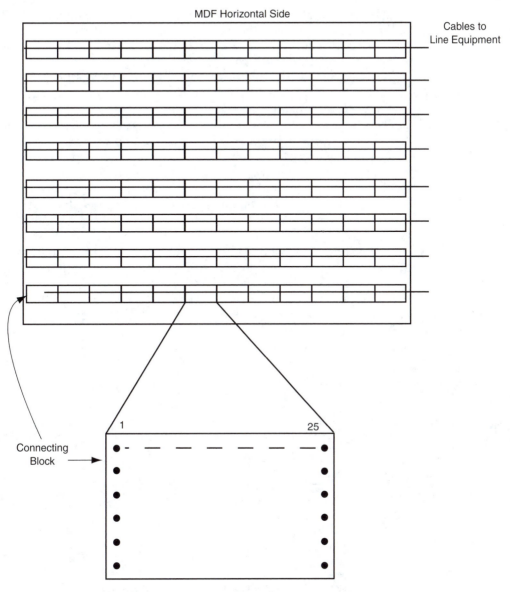

FIGURE 7-9 Cables from main distributing frame to line equipment.

frame. The cables that connect to switching equipment inside the building are connected on the opposite side of the frame.

It does not matter which side of the frame we establish for equipment cable or which side is established for cables that exit the building. Cables that exit the building are referred to as *outside plant cables.* In many central exchanges, the outside plant cables are terminated on blocks mounted vertically on the frame. This side of the frame is called the *vertical* side of the MDF. Blocks on the other side of the MDF are mounted horizontally, and that side of the frame is called the *horizontal* MDF.

The interface device, which interfaces a telephone to a central switching system, is called a *line circuit*. The line circuits of a switching system are cabled to the horizontal MDF. The outside plant cables, which connect to telephones, are wired to the vertical side of the MDF. A short length of twisted-pair copper wire (called a *jumper wire*) is run to connect an outside plant telephone wire pair to a particular line circuit. The MDF arrangement allows us the ability to connect any telephone to any line circuit. If a person moves to another building served by the same exchange, we can simply rewire the jumper to connect the line circuit being used to the new cable pair.

With the development of central switching systems, the local battery supply at the telephone was discontinued. A battery located at the central exchange now powers the telephone transmitter. The telephone connects via the local loop to a line circuit at the switching center. One wire of the local loop passes through components of the line circuit and is connected to the negative terminal of a 52-V battery. The other wire of the local loop passes through components on the line circuit and is connected to the positive side of the 52-V battery. The positive terminal of the central office battery is connected to earth ground. Thus, the negative terminal is negative 52 V with reference to ground. This will result in a negative 52-V battery attached to one wire of the local loop and ground attached to the other. The wire with a negative 52-V battery on it is called the *Ring lead*. The wire with ground on it is called the *Tip lead*. (see Figure 7-10).

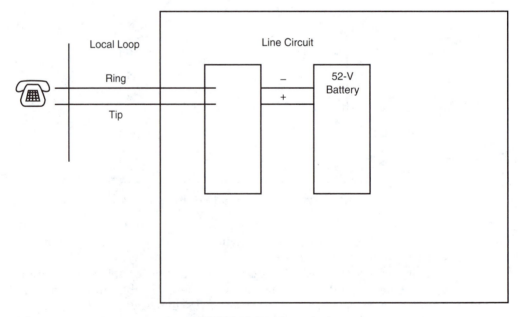

FIGURE 7-10 Telephone circuit.

7.4 ANALOG SIGNAL IN THE LOCAL LOOP

The telephone receives its power from the central exchange via the line circuit in the exchange. When the telephone is on hook, the circuit between the phone and the central exchange is open. In this open condition no electric current will flow. When the telephone is lifted out of its cradle, the hookswitch operates to connect the transmitter, receiver, and dial to the local loop. The operation of the hookswitch closes the circuit between the telephone and the line circuit. With a closed loop, electric current flows from the negative terminal of the central exchange battery, through the line circuit components, over the Ring wire of the local loop, through the transmitter of the telephone, back over the Tip lead wire of the local loop, through components on the line circuit, to the positive (grounded) side of the battery (see Figure 7-10).

When a telephone is taken off hook, electric current will flow. The amount of current flowing depends on how far the telephone is located from the central exchange. The farther from a central exchange it is placed, the less current will be flowing because the resistance of the wire in the local loop increases with distance. The higher the resistance in a loop, the less electric current flowing (Ohm's law, $I = E/R$). The telephone has within it a device called a *varistor* (variable resistor). The resistance of this varistor will vary depending on the voltage applied to the telephone. On short loops, the varistor exhibits a high resistance; on long loops, it has a low resistance. The varistor limits the amount of current flowing on short loops to 60 milliamps (see Figure 7-11).

The transmitter of a telephone and the electronic chip that provides the tones for a touchtone dial require about 8 V of battery power to function properly. If the resistance of a telephone is 400Ω, we can develop 8 V across the phone with 0.020 amp flowing through the phone ($V = IR$; $= 0.020 \times 400$; $= 8$ V). Thus, if the circuit that connects the telephone to the battery in a central exchange has at least 20 milliamps of current flowing in it, the transmitter and dual-tone multifrequency chip receive enough power to operate properly. The varistor in the circuit limits current flow to a maximum of 60 milliamps because a current of more than 60-milliamps contributes to the possibility of *crosstalk*.

The higher the current in a circuit, the more powerful the electromagnetic field surrounding the wire. Loop currents of more than 60 milliamps create a magnetic field strong enough to induce signals into adjacent wires. Crosstalk occurs when the voice signal on one circuit (a pair of wires) is so strong that the electromagnetic field it creates induces a voice signal into adjacent wires strong enough to be heard by users on that circuit. The voice signal is not coupled by physical connection but by the electromagnetic field properties from one wire to a second wire, just as if the windings of a transformer were coupled.

The signal carried by one wire of a circuit will be induced in the other wire. The signal carried by a Tip wire is induced in the Ring wire and vice versa. The two wires used for one voice circuit are twisted around one another to eliminate this induction. The rate of twist for each wire around the other is controlled to so many

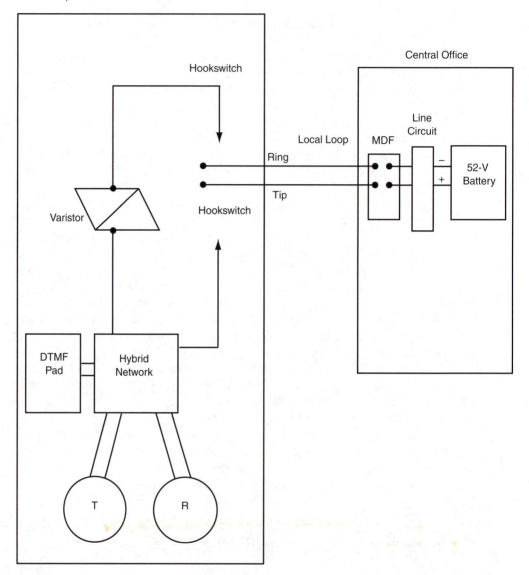

FIGURE 7-11 Varistor of telephone.

twists per inch. The twist results in electromagnetic induction occurring first in one direction and then in the opposite direction. With reversal of induction occurring again and again thousands of times in a wire pair, the net result is zero induction (see Figure 7-12). This twisting of the wire pairs also greatly reduces the possibility of induction (crosstalk) into other wire pairs in the cable.

FIGURE 7-12 Twisted-wire pair. Twisting the wires that serve one telephone around each other eliminates crosstalk. The tighter the twist, the higher-frequency signal it can carry. Data grade (CAT-5) cable has many more twists per inch than voice grade (CAT-3) cable.

7.5 COUPLING ANALOG SIGNALS FROM ONE CIRCUIT TO ANOTHER

When the transmitter of the telephone converts a voice signal into an analog electrical signal, the analog signal is a continuously varying electrical signal. Earlier in this chapter, we covered loop current to a telephone. This loop current will always be somewhere between 20 and 60 milliamps. The exact amount of loop current depends on the length of the local loop. Let's make an assumption that our telephone is located far enough away from the central exchange that 40 milliamps of electric current will flow when the phone is taken off hook. The current will be a steady 40 milliamps until we talk into the transmitter. As the resistance of the transmitter varies in unison with the frequency of our voice signal, it causes the loop current to vary above and below 40 milliamps (see Figure 7-13).

The analog electrical signal is a varying DC signal. Current flows in one direction only. Although electric current flows in one direction, the signal also looks like an AC signal that has a center point of 40 milliamps. Because it looks like an AC signal, it will behave like an AC signal. We use this knowledge to transfer or couple voice signals from one circuit to another. We can use transformers or capacitors to couple voice signals from one circuit to another while isolating the DC voltages of these circuits from each other (see Figures 7-14 and 7-15).

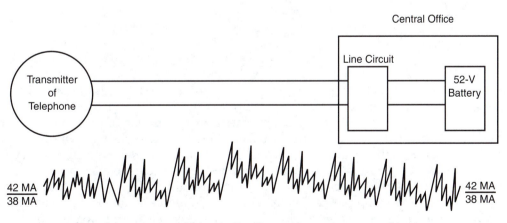

If the current established is 40 milliamps when a phone goes off hook, the voice signal results in fluctuation above and below 40 milliamps.

FIGURE 7-13 Voice signal in local loop.

Analog and Digital Signals

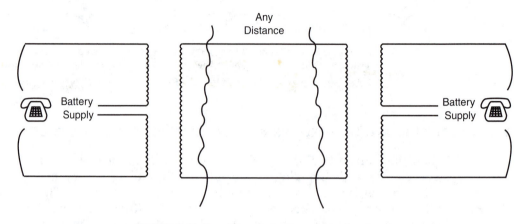

FIGURE 7-14 Transformers used to couple voice signals.

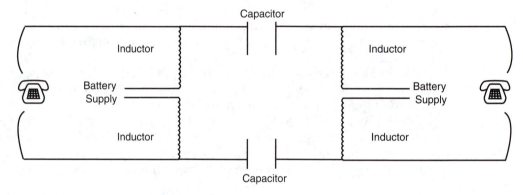

FIGURE 7-15 Capacitor coupling voice signals.

In the transformer coupling circuit illustrated by Figure 7-14, the 40 milliamps of current through the primary winding sets up a magnetic field of a certain strength. Once this strength of field is established, the secondary winding and circuit have no current flowing in them. When the local loop circuit transports an analog electrical voice signal to the primary winding of the transformer, the analog signal causes the magnetic field established by the primary winding to vary. Variations in the magnetic field cause an analog signal to be induced into the secondary winding and into the circuit connected to the secondary winding.

In the capacitor-coupled circuit of Figure 7-15, the 40 milliamps of current in the local loop causes the capacitor to charge to a certain value. When the local loop circuit transports an analog electrical voice signal, the analog signal causes the electric charge on the capacitor to vary in unison with the changes of the analog signal. This changing charge on the capacitor is coupled to the next circuit.

Transformer coupling of signals was used in the long distance (toll) network; capacitor coupling was used in local switching systems. The diagram in Figure 7-16

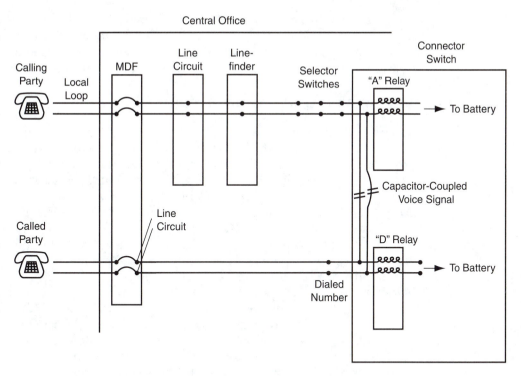

FIGURE 7-16 Strowger connector switch.

illustrates how the voice signal of one telephone and local loop were coupled to the local loop of a second telephone. The device that made the connection between two telephones in a Strowger step-by-step switching system was appropriately called a *connector switch*.

In a connector switch, the calling party was attached to the *A relay* and the called party was connected to the *D relay.* Notice how the capacitors between the A and D relays effectively couple the analog voice signal from one local loop circuit to another local loop while effectively isolating the DC voltages and currents from each other. The inductance in the windings of the A and D relays offers a high opposition to AC currents (and analog voice signals). This inductance prevents the voice signals from being shorted via the common battery supply to both circuits. This type of battery feed to a local loop was called an *inductive-capacitive battery feed circuit.*

Today, we do not use either the transformer or inductive-capacitive battery feed circuits for coupling voice signals. The line circuit that interfaces a local loop to the central exchange includes a codec chip and hybrid network in the circuit. The codec chip converts all analog signals received from the local loop into digital signals. Since the analog voice signal is converted into a digital signal, we cannot use the same techniques to couple the signal from one circuit to another. The technique used to couple digital signals from one circuit to another is to gate them using *silicon controlled rectifiers (SCRs).* These electronic gates are placed between two circuits and are turned on when we wish to connect signals from one circuit to another. Voice signals at the telephone are converted into analog electrical signals at the telephone and then into digital signals at the central exchange. These digital signals are connected via the PSTN switching network to a receiver for decoding.

7.6 CONVERSION OF VOICE INTO DIGITAL SIGNALS

Several standard techniques are used to convert analog signals into digital signals. The standard used in the public switched telephone network (PSTN) to convert analog voice signals into digital signals is *pulse code modulation (PCM)*. The codec chip in a line circuit uses the PCM process to convert analog signals into 64,000-bps digital signals. Other processes are available, such as *adaptive differential pulse code modulation (ADPCM)* and *predictive pulse code modulation.* These processes can code the human voice using as few as 4000 bps but unfortunately they were not perfected in time to use as the standard for the PSTN. They are used in PCS 1900 and voice mail applications, however. These processes are available now because of the tremendous advances made in semiconductor technology.

When the conversion of analog voice signals into digital was first accomplished, the technology available to perform this function used the PCM process. If we code a voice using one protocol, we must decode it using the same protocol. It is a standard rule in communication that for communication to take place, the transmitter and receiver must both use the same protocol. We can interface PCS 1900 mobile telephones and voice mail systems to the PSTN because these signals pass through a protocol converter to convert their 4000-, 8000-, or 16,000-bps signals into the proper 64,000-bps signal used by the PSTN.

The basic process used to convert an analog signal into a digital signal using PCM is to take samples of the analog signal every 1/8000 of a second. These analog samples are then assigned an 8-bit digital code. Using an 8-bit code provides a total of 256 different codes ($2^8 = 256$). Since an analog signal can be almost any value at a given point in time when a sample is taken, the sample taken can assume one of an unlimited number of voltages. Since the PCM process only provides for 256 different voltages to be coded, the sample taken is adjusted to the nearest voltage level for which a code exists by a process called *quantizing.* This results in some distortion of the received signal. This distortion is known as *quantizing noise.*

Once the PCM process in a codec chip converts a sample into an 8-bit digital code, that code must be transmitted out of the codec within 1/8000 of a second. A new sample will be taken of the analog signal in 1/8000 of a second, and the code for the new sample will overwrite the contents in the 8-bit store. Thus, the codec is gated to a transmitter every 1/8000 of a second, and the contents within the store are transmitted over some medium to a receiver. The receiver contains a codec that decodes the 8 bits received into a voltage output. The output voltages are used to reconstruct the original analog signal; they are connected together much like the dots in a child's dot-to-dot coloring book. The connected voltages form an analog signal. The analog waveform is then transmitted to a receiver in the called telephone, and the called party hears the calling party's voice in their telephone receiver.

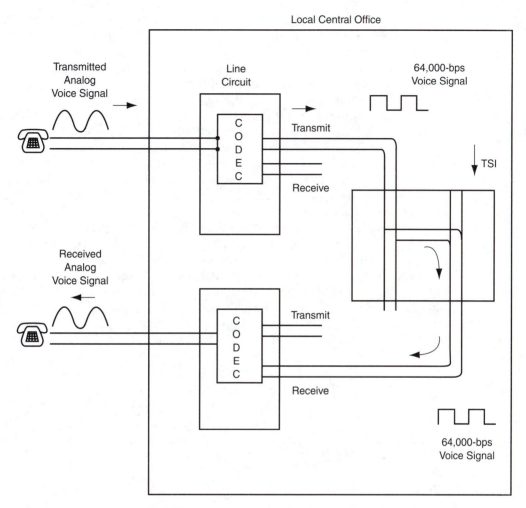

FIGURE 7-17 Coupling voice signals via codecs.

The digital voice signals are connected from one point to another by connecting the coder portion of one codec via a transmission medium to the decoder portion of another codec. There is no need for transformers or capacitors to couple signals from one circuit to another. The elimination of these devices greatly simplifies the design of the PSTN (see Figure 7-17).

7.7 CONVERSION OF THE PSTN INTO A DIGITAL NETWORK

Using digital signals to represent voice or data is much more efficient than using analog signals. Digital signals are easier to interface to each other, the electronic circuit chips are cheaper for digital circuitry, and digital signals are less susceptible to noise. Noise is basically any unwanted signal that has sufficient amplitude that it interferes with the communication process (we will discuss noise in more detail later). Figure 7-18 shows how noise affects an analog signal. If an unwanted signal is induced into the medium carrying our desired signal, the unwanted signal combines with our desired signal. If the combined signal passes through an amplifier, the unwanted signal is amplified along with our desired signal.

Analog signals can only be carried so far by a transmission medium before the signal gets so weak that it must be amplified. Most local telephone loops do not require amplifiers, but the older analog long distance networks used many amplifiers to boost weakened signals. If noise was picked up along the route of a facility that contained an amplifier, the noise was also amplified. A lot of time, money, and maintenance effort were required to keep transmission facilities noise free. Let's contrast this to using digital signals.

When digital signals are used to convey voice, data, or video, the signal is continually varying from one discrete voltage state to another. For example, the signal may be varying from +3 to 0 V many millions of times per second. The two states are used to represent a digit 1 or a digit 0. We send our digital signal by sending the appropriate voltage down the transmission wire. Notice in Figure 7-18 how noise affects this signal. We do not amplify digital signals; we regenerate them. Regenerators are called *repeaters*. Repeaters are placed every so often in the transmission facility. The distance between repeaters depends on the signal carried and the medium

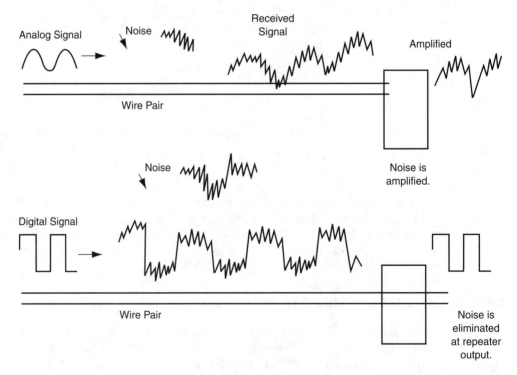

FIGURE 7-18 Effects of noise.

used. For our theoretical discussion, we will place repeaters on our transmission wire every 6000 ft.

When our theoretical digital signal (+3- and 0-V signal) picks up noise, the noise will change the values of the signal. Perhaps the +3-V signal is changed by the addition or subtraction of noise voltages into a +3.12, a 3.09, a 3.0224. a 2.996, a 2.894, or any other voltage between 2.5 and 3.5 V. Our 0 voltage level may be changed by noise into any voltage between -0.5 and +0.5 V. For an analog signal these changes would be disastrous, but for digital signals we can remove the noise without affecting the intelligence of the signal. Notice in Figure 7-18 how the repeater is simply looking to detect whether the voltage received is less than 1 V or greater than 2 V. If it detects a voltage less than 1 V, it regenerates a 0 voltage toward the next repeater. If it detects a signal greater than 2 V, it generates a crisp, clean +3-V signal toward the next repeater. Remember that my +3 V and 0 V signal is only theoretical. As you will discover later, different protocols use different voltage levels to represent a 1 or 0.

Digital signal regenerators strip all noise out of a signal by regenerating crisp, clean new 1s and 0s. The decoder at the receiving end of a circuit uses an 8-bit pattern of 1s and 0s to reconstitute the original analog signal coded by a codec at the transmitting end of a circuit. Because digital signals contain codes not affected by amplitude changes in the signal, "Digital circuits are so quiet you can hear a pin drop." The PSTN has evolved into an all-digital network; very few vestiges of analog components remain in the circuitry between central exchanges.

Although the circuitry between central exchanges is almost 100% digital, the circuitry that connects our telephone to the central exchange is mostly analog. Thus, the PSTN is a digital network, but the on ramps and off ramps to connect to the PSTN are mostly analog. These on ramps and off ramps are mostly analog because the telephone generates analog electrical signals. The medium to connect our telephone to the central exchange must be designed to handle the analog signal generated at the telephone. Twisted-pair copper wire does an excellent job of handling these analog electrical signals. Approximately 60% of a local telephone company's investment is in the outside plant used to connect telephones to the central exchange. It is not in a position to replace twisted-pair copper wire with another technology because of the cost involved.

Some people suggest that the telephone company should replace twisted-pair copper wire with fiber optic cables, but this will be impossible without replacing the current telephone with a much more expensive device. The telephone receives its electrical power from the central exchange via the twisted-pair copper wire that connects it to the exchange. Copper is an excellent conductor of electricity. Glass does not conduct electricity. If we use fiber optic cable composed of glass strands to connect our telephone to the central exchange, we need to buy a different device for a telephone. It will need to receive its power from a source on the customer's premises much like the very first hand-crank telephones did. Our new telephone device would also need to convert our voice signals into infrared light signals, because that is the type of signal the glass fiber is designed to transport efficiently.

Running a glass fiber to everyone's house would be too expensive and impracticable. As discussed in Chapter 5, the most common use of fiber in the local loop is in serving all the houses or businesses in a particular area. Fiber cable is run from the central exchange to a central point in the area to be served. Interface devices are used to strip information from the fiber and place it on twisted-pair copper wire to the home or business. The interface device also contains a power converter to convert AC power received from the power company into DC voltage. This DC voltage is placed on the copper wire to power the telephone.

We will see our connection to the central exchange (our on/off ramp) remaining as twisted-pair copper wire for the foreseeable future. This connection is referred to as the *local loop* and as the *last mile* of the telecommunications network. Another reason this last mile and the telephone will remain analog for the foreseeable future is because all digital central exchanges contain a codec on the line interface circuit. The device expects to see analog signals from the telephone over the local loop. The device contains a coder (analog-to-digital converter) built to change analog signals into digital codes. It does not expect to see a digital signal coming from the telephone and will disregard any digital signal received via the local loop. The existence of codecs on the line interface circuit is the primary reason why modems must be used to transfer digital data over a local telephone circuit.

7.8 DIGITAL DATA OVER THE LOCAL LOOP

As we have seen, the telecommunications industry has introduced a service called **Integrated Services Digital Network (ISDN),** which will be explained in detail in a later chapter. I am briefly introducing ISDN here because it is referred to as an end-to-end digital network. ISDN provides the ability to place digital data directly into the ISDN equipment on each end of the circuit. ISDN lines use twisted-pair copper wire to connect equipment on the customer's premises to the local central exchange. ISDN lines do not connect to regular line circuits at the central exchange; they connect to special line interface circuits called *ISDN line circuits.* The ISDN line circuit does not contain a codec.

If an ISDN line is to be used for the transmission of a voice signal, the ISDN terminal equipment on the customer's premises contains a codec, which converts the analog signal into a 64,000-bps digital signal. The codec in ISDN terminal equipment used for a voice application employs the PCM protocol to convert the analog signal into a digital signal. ISDN terminal equipment will also accept the digital output of a computer. Thus, this equipment uses processes to ensure that voice and data will be represented as digital signals. As you will learn in the chapter on ISDN, these digital signals are transmitted over the local loop so that they look like analog signals.

Another high-speed digital service is *asymmetrical digital subscriber line (ADSL).* Again, this service is classified as a digital service but in fact uses a modem, and the digital data on the customer's premises will be used to modulate an analog signal transmitted to the central exchange. Like ISDN, this ADSL service cannot be interfaced to the exchange using a regular line circuit. ADSL lines are connected at the central exchange to another ADSL modem. Just as with a regular 33.6 modem or any other modem, they must be used in pairs with a modem on each end of the transmission facility. The ADSL modem in a central exchange is part of a device called a *digital subscriber line access multiplexer (DSLAM).* ADSL uses high-frequency analog signals, which are modulated by the digital data to be carried.

Because ISDN and ADSL devices on a customer's premises are transmitting analog signals that carry digital data within the signal, they are carried well by the twisted-pair copper wire designed to carry analog signals. The ISDN and ADSL signals carried by the local loop are at much higher frequencies than the loop was originally designed for, and these signals encounter a much higher loss in signal power than a voice signal does. Because the ISDN and ADSL terminal devices contain vastly improved semiconductor signal logic processor chips, they can recover signals that have experienced a high loss of power over the transmission facility.

Because of the high-frequency signals used, the ISDN and ADSL terminal equipment on a customer's premises must be within 3 mi of the central exchange. It is possible to extend these lines to customers located more than 3 mi from the exchange, but the lines must receive special treatment. The use of special treatment on lines to extend the distance at which a service can operate significantly increases the cost of providing the service.

7.9 DIGITAL DATA CODING TECHNIQUES (AMI, NRZ-L, NRI, MANCHESTER, AND DIFFERENTIAL MANCHESTER)

It is possible to place digital data directly on a twisted wire pair. The first device to use digital signals in the PSTN is a transmission system called a *T1 carrier system.* It carries 24 voice signals in digital format and uses twisted wire pair as its transmission medium. This T1 system uses a protocol that results in a digital signal of 1,544,000 bps placed on the wire. This wire pair must have a regenerator placed on the pair at 6000-ft intervals to regenerate the signal. The signal uses a special protocol called *alternate mark inversion (AMI).* With AMI, the digit 1 is represented by a voltage that is either +3 V or –3 V, and the digit 0 is represented by 0 V. AMI will alternate between +3 V and –3 V to represent successive 1s (see Figure 7-19).

Notice how these uses of alternating voltages to represent 1s make the signal look like an analog signal. AMI also makes it look like the frequency of the signal has been cut in half. This effectively doubles the distance that a signal can travel before

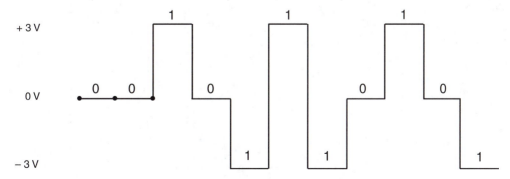

FIGURE 7-19 Alternate mark inversion signal. Each mark (binary digit 1) is represented by alternate polarity.

regeneration is needed. AMI is the predominant protocol used for signals placed over the local loop, but there are other protocols for placing digital signals on copper wire to make the signal look like an analog signal. These other protocols are mainly used in the PC environment. Some of these protocols are **nonreturn to zero (NRZ)**, **nonreturn to zero invert (NRI)**, **Manchester**, and **differential Manchester**. Nonreturn to zero protocol does not use the 0 V as a signal level but uses a voltage level such as +3 V and –3 V (or some other two-state voltage such as +10 V and –10 V, + 15 V and –15 V, and so on). The standard NRZ protocol is also referred to as *nonreturn to zero-level* (NRZ-L) to differentiate it from NRI (see Figure 7-20).

In NRZ-L the signal never returns to zero voltage, and the value during a bit time is a level discrete voltage. Typically, a negative voltage such as –5 V is used to represent a binary digit 1 and a positive voltage such as a +5 V is used to represent a binary digit 0. NRZ-L is the protocol used by the EIA-232 interface of a PC. The EIA-232 interface is used to connect the PC to devices such as an external modem or serial printer. The protocol for an EIA-232 interface specifically states that any voltage level at or more negative than –3 V is to be interpreted as a binary digit 1 and any voltage level at or more positive than +3 V is to be interpreted as a binary digit 0.

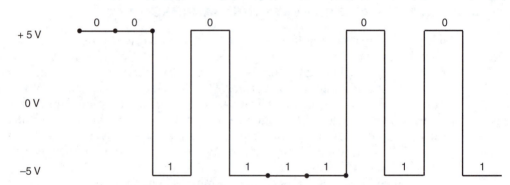

FIGURE 7-20a NRZ-L signal. Nonreturn to zero-level (NRZ-L) +5 V = 0 and –5 V = 1. NRZ-L is used by serial ports.

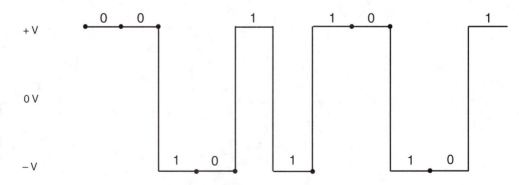

FIGURE 7-20b NRI signal. If there is no transition in the signal at the beginning of a bit, it is a 0. If a transition occurs at the beginning of a bit, it is a 1.

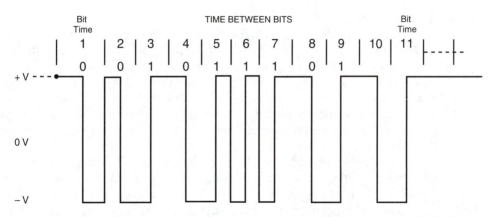

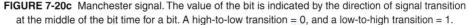

FIGURE 7-20c Manchester signal. The value of the bit is indicated by the direction of signal transition at the middle of the bit time for a bit. A high-to-low transition = 0, and a low-to-high transition = 1.

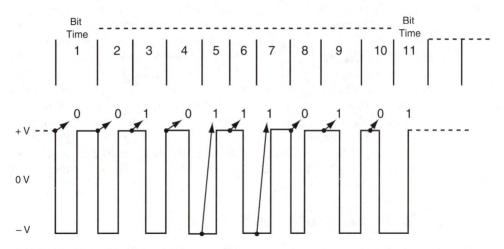

FIGURE 7-20d Differential Manchester. A transition occurs at every bit during the middle of the bit time for synchronization. The transition or lack of transition at the beginning of bit time represents the value of the bit. No transition = 1. A transition (high to low or low to high) = 0.

NRI also uses two discrete voltage levels for its signal, but the value of the voltage present does not represent 1s and 0s. Data is coded as the presence or absence of a transition from one voltage state to the other. It does not matter whether the transition is from the higher voltage state to the lower (such as from +3 V to –3 V) or from the lower voltage state to the higher (such as from –3 V to +3 V). If a transition occurs at the beginning of a bit time in either direction, it is coded as a 1. If no transition in the signal occurs at the beginning of a bit time, it is coded as a 0.

Coding a signal based on the presence or absence of a transition in the voltage level from one bit time to the next is *differential encoding*. It is not necessary to detect an absolute value of the voltage level for each bit. This means that noise should have less impact on the signal and that even if the polarity of the system were to get reversed, it can still be decoded. Both forms of NRZ require that the receiver be able to determine the beginning of each bit. This can be difficult when the NRZ-L signal contains a long string of 0s one after the other. It will also be difficult with a NRI signal when it contains a long string of 1s or 0s. With digital signals, it is imperative that the receiver remain synchronized with the transmitter. Systems that use NRZ coding use a protocol referred to as *asynchronous communication.* This protocol—to be explained in Chapter 8—provides start and stop bits around each 8 data bits to be used as synchronization bits.

Manchester and differential Manchester use a technique known as *biphase coding.* This type of coding requires at least one transition of the signal during every bit time, usually at the beginning or in the middle of the bit time. Where the transition occurs depends on the specific coding protocol being used. Some protocols will use two transitions per bit time to double the data rate capability of the signal. The advantage of a biphase coding system over NRZ pertains to the fact that a transition must occur during every bit. The receiver can maintain synchronization from these transitions and can also use the absence of an expected transition to detect errors in the signal.

For Manchester coding, the transition occurs at the middle of each bit time. A high-to-low transition represents the digit 1, and a low-to-high transition represents the digit 0. As stated earlier, this midbit transition is also used to provide receiver synchronization. Manchester coding is used on local area networks (LANs) that use the *Ethernet* signaling protocol. Differential Manchester coding contains two transitions per bit time. The midbit transition is used only for receiver synchronization, and the absence or presence of a transition at the beginning of a bit time is used for the coding/encoding of data. The absence of a transition at the beginning of a bit time is used to represent the digit 1. The presence of a transition is used to represent the digit 0. Differential Manchester is both biphase and differential and is used for *token-ring* LANs.

7.10 BANDWIDTH VS. POWER LOSS

As mentioned in Section 7.1, an analog signal may be one single frequency or it may be a signal composed of many different frequencies combined together. The analog voice signal is a signal composed of many different frequencies. In telecommunications, the local loop and the old long distance network were designed to handle the frequencies between 300 and 3300 Hz efficiently. The frequencies in this range effectively captured the intelligence contained in a voice signal. We refer to a voice signal in today's digital PSTN as having a spectrum from 200 to 3800 Hz. We also refer to this signal as having a bandwidth of 3600 Hz. The bandwidth of a signal is found by subtracting the lowest-frequency signal from the highest-frequency signal contained in the composite signal. The early digital PSTN had a spectrum from 200 to 3400 Hz, with a bandwidth of 3200 Hz. In the older analog PSTN, the spectrum was from 300 to 3300 Hz with a bandwidth of 3000 Hz.

People often refer to the bandwidth of an analog voice signal as 4000 Hz. To provide for some separation of signals in the old analog long distance network, the analog voice signals were passed through a filter to effectively limit the frequencies passed by the filter to those between 300 and 3300 Hz. The filter was described as having a passband of 300 to 3300 Hz. The frequencies between 0 and 300 Hz, as well as those between 3300 Hz and 4000 Hz, were used as guardbands between signals that had been multiplexed together. Suffice it to say here that the analog voice signal on the local loop can be considered as containing frequencies between 0 and 4000 Hz and thus a bandwidth of 4000 Hz, while the bandwidth of these signals in the toll network has been limited to 3000 Hz. As mentioned, the design of filters at the channel units has led to the reduction in guard bandwidth and increased the bandwidth available to the signal.

If a signal contains a single frequency, its bandwidth will be equal to the frequency of the signal. *Bandwidth* describes the range of frequencies found within a band. Between 0 and infinity we find only one frequency within the band when a signal consists of one frequency. Thus a 2000-Hz signal has a bandwidth of 2000 Hz, while a signal containing a low frequency of 2000 Hz and a high frequency of 2500 Hz has a bandwidth of 500 Hz. The bandwidth of an analog signal is stated in Hertz (Hz), while the bandwidth of a digital signal is stated in bits per second (bps).

The bandwidth of a signal determines the information carrying capacity of the signal. The greater the bandwidth, the more information a signal can carry. For an analog signal we can convey data by changing the amplitude, frequency, or phase of the signal. Each of these occurs during one cycle of a signal. Suppose we change the amplitude during each cycle of a signal and each time we change the amplitude, that change represents a data bit. With a 3000-Hz signal we can change the amplitude of the signal 3000 times per second. Thus, we could transfer information (in this example, data bits) at 3000 bps. With a 1000-Hz signal, we could transfer data at 1000 bps. The higher the bandwidth of a signal is, the higher the information carrying capacity of the signal.

When we wish to transfer information over the local-loop twisted pair, we need high-frequency signals to transfer higher data rates. Although the local loop is purposely designed to handle frequencies between 0 and 4000 Hz, that does not mean we cannot put higher-frequency signals on the local loop. We transmit high-frequency signals over the local loop, but the higher the frequency transmitted, the greater the power loss incurred. This is due to the distributed capacitance that exists between the two wires of the local loop and the inductance in the wire itself. Distributed capacitance adds up as the length of a wire pair is increased and starts to negatively impact signals when the length becomes greater than 1000 ft. Since the connection between computers on a LAN will be less than 1000 ft, twisted-pair copper wire can be used to connect PCs to a LAN without much adverse impact on the 10–100-Mbps signals used.

Refer to Appendix C, Section C.9. The capacitive reactance at 1000 Hz is approximately 1917Ω for 1 mi of a wire pair. If the wire pair is extended to 3 mi, the impedance due to capacitance between the wires will be 639Ω (1917/3). If a 5000-Hz signal is placed on a wire pair 3 mi long, the impedance drops to 128Ω. Since the impedance between the two wires is less at higher frequencies, higher-frequency signals will lose more energy (power) than lower-frequency signals. At higher frequencies, more of the energy will take the low-impedance path offered by the distributed capacitance between the wires and less power will be available to the receiver located at the end of the wire. As indicated in Appendix B, the amount of power lost between the transmitter and a receiver via the telecommunications medium is expressed in decibels.

In addition to power loss of a signal, we are also concerned with the noise that may be inflicted on a signal. When a signal is carried by twisted-pair copper wire, it is especially susceptible to interference (noise) from signals in adjacent wire pairs. It is important to maintain a high signal-to-noise ratio. The wider the bandwidth of an analog signal, the more susceptible the signal is to noise and the more the signal will be impacted by it. Suppose we are using a modulation technique to encode data into an analog signal. Compare the impact of noise on a high-frequency signal carrying 1 million bps to the effect of noise on a low-frequency signal carrying 2400 bps. We can see that a noise spike of 1 sec will cause the loss of 1 million bits in the former but only 2400 bits in the later.

7.11 SUMMARY

Telecommunications requires a transmitter, medium, and receiver. To ensure accurate transmission and reception of signals, the transmitter and receiver must use the same protocols. Protocols specify the rules and procedures that must be followed to set up and maintain accurate, reliable communication. The signals used in telecommunications are either analog or digital. An analog signal is a signal with continuously varying amplitude; it can assume an unlimited number of amplitudes or voltage levels. A digital signal assumes one of a number of finite or discrete voltage levels. When a digital signal uses two voltage states, it is also known as a binary signal. One voltage state is used to represent the binary digit 1 and the second voltage level is used to represent the binary digit 0.

The transmitter of a telephone creates analog electrical signals. The local loop was designed to handle these signals efficiently. Almost all the central exchanges used in the PSTN are digital switching systems. The line interface to these switching systems contains a codec to convert the analog voice signal into a 64,000-bps digital signal using the PCM protocol. Today voice, data, and video are converted into digital signals. The technology required to handle digital signals is much cheaper than the technology required for analog signals, and digital signals are less susceptible to noise interference.

The local loop will remain analog for the foreseeable future due to the enormous cost required to convert it to a digital facility. The capabilities of the local loop have been extended by the improvements made in signal logic processors. These superprocessors make it possible to transmit high-frequency analog signals over the local loop because they can detect the difference between a received data signal and noise. Remember that data does not have to be transmitted as a digital signal. Modems and other devices transmit data by using the data to modulate an analog signal. Three attributes of an analog signal can be changed to represent data: amplitude, frequency, and phase.

The wider the bandwidth of an analog signal, the more information it is capable of carrying in a given time frame. But the use of high-bandwidth signals also makes the data more susceptible to interference from noise. Higher-frequency signals are needed to provide wider bandwidths. Higher-frequency signals encounter higher power losses when transmitted over twisted-pair copper wire.

CHAPTER REVIEW QUESTIONS

1. What type of signal is used to convey information over short distances?
2. What types of signals are used to convey information over long distances?
3. What are the three components of a communication system?
4. What are the differences between simplex, half duplex, and full duplex?
5. What is an analog signal?
6. What is a digital signal?
7. What are the characteristics of an analog signal that can be changed to convey information?
8. What characteristic of an analog signal contains the intelligence of the signal?
9. Why must the local loop be twisted-pair copper wire?
10. What devices were used in analog central exchanges to couple analog signals from one circuit to another?
11. Why do digital voice signals provide quieter communication?
12. Why does a T1 carrier system use AMI coding?
13. What type of coding protocol is included as part of the EIA-232 protocol?
14. How does bandwidth correlate with information carrying capacity?
15. What is the difference between NRZ-L and NRI?
16. Which transition is used for coding data in differential Manchester coding?
17. Which type of coding does Ethernet use?
18. Why do high-frequency signals encounter more loss than low-frequency signals when transmitted over the twisted-pair local loop?
19. If twisted-pair copper wire as used in the local loop causes such high power losses in high-frequency signals, how can twisted-pair copper wire be used for 10-Mbps transmission on local area networks?

PROJECTS AND CASE STUDIES:

Case Study #1

The term ADSL indicates that the signal is a digital signal, but the system uses analog tones to convey signals between the ADSL field terminal and the ADSL central office terminal. So is the signal a digital or analog signal? Fiber optics uses analog signals (light waves), so why do we refer to the signals as digital signals?

GLOSSARY

Analog Signal   A signal analogous (similar) to a voice signal. It can assume an infinite number of values for amplitude, frequency, current, and voltage.

Asymmetrical Digital Subscriber Line (ADSL) A digital subscriber-line technology that can be used to deliver T1 speeds over the local loop.

Crosstalk Occurs when undesired signals appear on a circuit—signals strong enough to be heard on the circuit.

Differential Manchester A digital coding technique. Differential Manchester coding contains two transitions per bit time. The midbit transition is used only for receiver synchronization, and the absence or presence of a transition at the beginning of a bit time is used for the coding/encoding of data. The absence of a transition at the beginning of a bit time is used to represent the digit 1. The presence of a transition is used to represent the digit 0. Differential Manchester is both biphase and differential and is used for *token-ring* LANs.

Digital Signal A signal that can assume one of several discrete states. Most digital signals take on one of two discrete states. The prefix *bi* means two, and a two-state signal is also called a *binary signal.* Each state is represented by a different electrical signal. Two different electrical states can be used to represent the two binary digits (1 and 0). Because this two-state signal is used to represent 1 or 0, it is also called a *digital signal.*

Integrated Services Digital Network (ISDN) Provides the ability to place digital data directly into the ISDN equipment on each end of the circuit. ISDN lines use twisted-pair copper wire to connect equipment on the customer's premises to the local central exchange.

Manchester Coding Used on local area networks (LANs) that employ the Ethernet signaling protocol. For Manchester coding, a transition occurs at the middle of each bit time. A high-to-low transition represents the digit 1, and a low-to-high transition represents the digit 0. The midbit transition is also used to provide receiver synchronization.

Nonreturn to Zero (NRZ) Protocol that does not use the 0 V as a signal level but that uses a voltage level such as +3 V and –3 V (or some other two-state voltage such as +10 V and –10 V, + 15 V and –15 V, and so on). The standard NRZ protocol is also referred to as *nonreturn to zero-level* (NRZ-L) to differentiate it from NRI. In NRZ-L the signal never returns to zero voltage, and the value during a bit time is a level discrete voltage. Typically, a negative voltage such as –5 V is used to represent a binary digit 1 and a positive voltage such as a +5 V is used to represent a binary digit 0. NRZ-L is the protocol used by the EIA-232 interface of a PC. The EIA-232 interface is used to connect the PC to devices such as an external modem or serial printer.

Nonreturn to Zero Invert (NRI) Uses two discrete voltage levels for its signal, but the value of the voltage present does not represent 1s and 0s. Data is coded as the presence or absence of a transition from one voltage state to the other. It does not matter whether the transition is from the higher voltage state to the lower (such as from +3 V to –3 V) or from the lower voltage state to the higher (such as from –3 V to +3 V). If a transition occurs at the beginning of a bit time in either direction, it is coded as a 1. If no transition in the signal occurs at the beginning of a bit time, it is coded as a 0.

Protocol The rules of communication used by devices to communicate with one another.

8

Data Communication

KEY TERMS

ACK
American Standard Code for
    Information Interchange
    (ASCII)
Analog Signal
Asymmetric Digital Subscriber
    Line (ADSL)
Asynchronous Transmission
Automatic Retransmission
    Request (ARQ)
Baud
Binary Signal
Channel Service Unit (CSU)/
    Data Service Unit (DSU)
Continuous ARQ
Cyclic Redundancy Checking
    (CRC)
Data

Data Communication
    Equipment (DCE)
Data Switching Exchange (DSE)
Data Terminal Equipment (DTE)
Digital Signal
Discrete ARQ
Dual Simultaneous Voice and
    Data (DSVD)
EIA-232 Interface
Flow Control
Full-Duplex Transmission
Half-Duplex Transmission
Header
Integrated Services Digital
    Network (ISDN)
Internet Service Provider (ISP)
Modem
NAK
Null Modem

Packet Assembler/Disassembler
    (PAD)
Packet Network
Parallel Transmission
Protocol
Public Data Network (PDN)
Quadrature Amplitude
    Modulation (QAM)
Recommended Standard 232
    (RS-232)
Selective ARQ
Serial Transmission
Synchronous Transmission
Trailer
Trellis Coded Modulation (TCM)
Universal Asynchronous
    Receiver/Transmitter (UART)
Word
X.25

OBJECTIVES

On completion of this chapter, you should be able to:

1   Explain various ways to connect DTE to DCE.
2   Describe a null-modem cable.
3   Describe ASCII coding and Extended ASCII coding.
4   Explain serial transmission.
5   Explain parallel transmission.
6   Explain what a UART does.
7   Explain asynchronous transmission.
8   Explain synchronous transmission.
9   Explain the error-detection technique used by synchronous transmission.
10 Explain the error-correction techniques used by synchronous transmission.

11 Explain the term sliding window.

12 Explain ACK and NAK.

13 Explain baud rate and how it relates to bit rate.

14 Explain how a modem converts a data signal into modulations of an analog signal.

15 Explain what a CSU/DSU is used for.

16 Explain why a modem must be used to connect a PC to a regular phone line.

17 Explain how a 56-Kbps modem achieves that speed.

18 Explain how the end-to-end speed between two PCs can be 115,000 bps when the speed between two modems is 28,800 bps.

19 Explain the function of SS7.

20 Explain how caller ID works.

The purpose of this book is to cover voice and data applications, the public switched telephone network (PSTN), and the public data network (PDN). The PSTN is used to carry both voice and data communication. A telecommunications manager must understand both voice and data networks, since companies use one manager to manage both voice and data facilities. Data communication is closely related to voice communication. Both communication systems are concerned with transmitting information over some distance.

The primary difference between voice and data lies in the signal used to convey information. Voice is converted by the transmitter of a telephone into an analog electrical signal having many different voltage levels. Data is converted into an electrical signal that has two different voltage levels. *Bi* means "two," and this two-state signal is called a *binary signal.* The two states of a binary signal are often called a 1 and a 0. Since the two binary digits, 1 and 0, are used to represent the two different states of a data signal, a binary signal is also called a *digital signal.* The digital signal consists of a series of binary digits (1s and 0s). Binary digits are also called *bits.*

The PSTN started life as an analog network designed to handle analog voice signals. With the introduction of integrated circuit chips, digital communication circuitry has become cheaper and more reliable than analog circuitry. The PSTN has gradually been converted from an analog to a digital network. The telephone, the local loop, and the line circuit are the only parts of the PSTN that remain analog. The first portions of the PSTN to be converted from analog to digital were the multiplexing facilities carrying interoffice trunk circuits. The T1 carrier system developed by Bell Labs multiplexed 24 trunk circuits onto two cable pairs using digital technology. The T1 system contained 24 channel units with a codec in each unit. The codec was an interface between the *analog signal* from the trunk circuit of an analog switch and a TDM digital highway connecting the switch to another switch.

When digital technology was integrated into the switching matrix of class 5 SPC switching systems, the codec was moved from the channel unit of the T1 carrier system to the line circuit of the switch. The interface point between analog and digital now occurs at the line circuit of a class 5 central office. By moving the codec from the line circuit of the switch out to the telephone, we can convert the remaining analog components of the PSTN (the telephone, local loop, and line circuit) to digital.

Several digital technologies have been developed for voice communication that move the codec from the line circuit to the customer's premises. Chapter 9 expands on two of these technologies: *Integrated Services Digital Network (ISDN)* and *Asymmetric Digital Subscriber Line (ADSL).* These two technologies allow us to achieve a complete end-to-end, all-digital PSTN where voice and data are both transmitted using digital signals. The demand for a digital line is low because of its high cost. Most people are still being served by an analog line circuit, local loop, and analog telephone. Perhaps more people will switch to digital telephones when the price of a digital line comes down.

Most signals are two-state digital signals. Each state can be represented by a discrete voltage level. The digital signal consists of a series of bits where each bit is either a 1 or a 0. There is no set standard for all devices as to what the voltage levels for a 1 or a 0 will be. Different devices use different voltage levels to represent a 1 or 0. One device may represent a 1 by using a +3-V signal and represent a 0 by a 0-V signal. Another device may represent a 1 by a –6-V signal and a 0 by a +6-V signal. For a digital communication circuit to work, the devices on each end of the circuit must be using the same protocol for voltage conversion.

T1 carrier systems use alternating +3-V and –3-V signals to represent a 1, and use 0 V to represent a 0. RAM uses a very low voltage to store a 1 and no voltage to store a 0. The digital signal passed between a PC and a modem typically uses –12 V to represent a 1 and +12 V to represent a 0. This signal is called a *bipolar signal* because it has a positive and a negative state. It is also called a *nonreturn-to-zero signal* because the signal does not return to 0 V after sending either a –12-V (1) or +12-V (0) signal. The signal remains at the last voltage state until a new signal is received.

The PC belongs to the category of equipment called *data terminal equipment (DTE).* This equipment is used to transmit and receive data signals. The equipment used to connect a DTE to the PSTN is called *data communication equipment (DCE).* Some people also interpret DCE as *data circuit terminating equipment.* A DCE is used to convert the output of a DTE into a form suitable for the transmission medium. The most common form of DCE is a *modem. The modem converts a digital signal into an output suitable for transmission over the analog local loop.*

The most common connection used to connect DTE to DCE is an *EIA-232 interface* (see Figure 8-1). This interface is also called a *Recommended Standard 232 (RS-232).* It has been approved as CCITT Standard V.24. The EIA-232 interface is the standard for serial transmission of data between two devices. The EIA-232 interface is limited to a cable length of 50 ft between the DTE and DCE. The EIA-232 interface specifies that a signal between –3 and –15 V DC can be used to represent a logic 1, and that a signal between +3 and +15-V DC can be used to represent a logic 0.

The EIA-232 interface cable comes in a 25-conductor and a 9-conductor cable. The 25-conductor cable has a DB-25 connector attached to each end of the cable. The 9-conductor cable uses DB-9 connectors. When the RS-232 standard was developed, it was developed with 25 leads in the interface between DTE and DCE. Personal computers do not use all 25 leads; they only use 9 of them. IBM developed the DB-9 connector for its AT computer. If you look on the back of an IBM PC, you will see a male 9-pin connector. This is the EIA-232 connector of the serial port of an IBM PC. *The serial port on all DTE devices is a male DB-25 or DB-9 connector. The serial port on a DCE device will be a female connector.* It can also be either a DB-9 or DB-25 connector. A cable designed to connect DTE to DCE will have a female connector on one end and a male connector on the other.

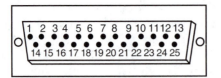

DB-25 Connector

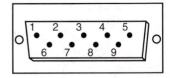

DB-9 Connector

Pin Number	Signal Designation	Pin Number	Signal Designation
1	Protective ground	1	Carrier detect
2	Transmit data	2	Receive data RD
3	Receive data	3	Transmit data TD
4	Request to send	4	Data terminal ready DTR
5	Clear to send	5	Protective ground
6	Data set relay	6	Data set ready DSR
7	Signal ground	7	Request to send RTS
8	Carrier detect	8	Clear to send CTS
9	Positive DC test voltage	9	Ring indicator
10	Negative DC test voltage		
11	Unassigned		
12	Secondary carrier detect		
13	Secondary clear to send		
14	Secondary transmit data		
15	Transmit clock (DCE)		
16	Secondary receive data		
17	Receiver clock		
18	Receiver dibit clock		
19	Secondary request to send		
20	Data terminal ready		
21	Signal quality detector		
22	Ring indicator		
23	Data signal rate selector		
24	Transmit clock (DTE)		
25	Busy		

FIGURE 8-1 EIA-232 interface.

The transmit lead and receive lead in the cable get their designation from the DTE end of the cable. A PC must transmit data on the transmit lead. This lead has to connect to the receive circuit of the DCE. A PC must receive data on the receive lead. Therefore, the DCE must transmit its data over the receive lead of the cable to the PC. A DCE device has its transmit and receive leads flipped inside the DCE. The transmit circuitry of the modem connects to the EIA-232 cable receive lead so that the data from the modem arrives at the receive pin of the serial connector on the PC. The receive circuitry of the modem connects to the transmit lead from the PC. For a DB-25 connector on a PC (DTE), pin 2 is transmit and pin 3 is receive. Since the transmitter of the PC must talk to the receiver of the DCE, pin 2 of the DCE is receive. Similarly, the transmitter of the DCE connects to pin 3 of the EIA-232 so it can talk to the receiver in the PC.

A communication circuit will not work if the transmitter is talking to another transmitter. This is why two PCs cannot be connected using a standard EIA-232 cable. The two computers must be connected using modems or a null-modem cable. *Null* means "none." **Null modem** means "no modem." DTEs have male connectors; therefore, a null-modem cable will have female connectors on both ends of the cable. A null-modem cable is wired so that pin 3 of the female connector at one end of the cable is wired to pin 2 of the female connector at the other end. Thus, the null-modem cable reverses the transmit and receive leads between the two computers. With this reversal, the transmitter of each computer is talking to the receiver of the other computer. Null-modem cables also reverse other leads (see Figure 8-2).

As mentioned earlier, the telephone and local loop are analog devices. They are interfaced to the digital PSTN by a codec on the line circuit at the SPC switch. The local line circuit will only accept analog signals because of the codec in its circuitry. Thus, if the local analog telephone line is to be used for data, the data must be converted into an analog signal. The most common type of interface circuitry for interfacing data to the local loop is a modem. The word *modem* is an acronym for

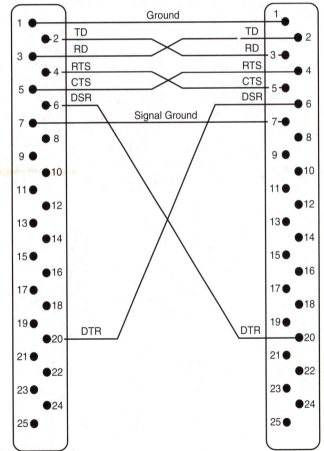

FIGURE 8-2 RS-232 null-modem cable.

modulator/demodulator. When a modem is used to send data over the PSTN, another modem is required at the receiving end. The modems will continuously send an analog signal to each other.

When data from a computer is fed into a modem, it modulates (changes) the analog signal being sent to the other modem. The PSTN handles the modulated signal extremely well because the signal is in the voice bandwidth of the PSTN. At the receiving modem, the signal is demodulated or changed back into a replica of the original data signal and fed into the receiving computer. Since each modem has its own modulator and demodulator, the modem can send a signal and receive a signal at the same time. A modem is a *full-duplex* device but with some protocols is capable of operating in a *half-duplex* mode.

8.1 DATA AND THE PUBLIC SWITCHED TELEPHONE NETWORK

Because the PSTN has been evolving from an analog into a digital network, it is possible to integrate data directly into the digital PSTN if the analog line circuit of a SPC switch is bypassed. *It is impossible to send data over the local telephone line without using a modem because the line circuit in today's SPC switches contains a codec.* The coder portion of a codec changes analog into digital signals. The coder expects to see an analog signal, not a digital signal. The codec does a great job of converting our analog voice signal into a digital DS0 signal, but it cannot do a digital-to-digital conversion. Because the local loop and line circuit are designed to handle analog signals, a modem is required to transmit data when using a local telephone number (see Figure 8-3).

If a facility is to be used only for data and not for voice, a *channel service unit (CSU)/ data service unit (DSU)* is used to interface the data circuit directly into the PSTN. A CSU/DSU is a DCE device. A modem is a DCE device used to interface a digital circuit to an analog circuit, and a *CSU/DSU is a DCE device used to interface a digital circuit to another digital circuit.* The CSU/DSU is designed to do a digital-to-digital conversion and will interface the digital speed of your computer to a different digital

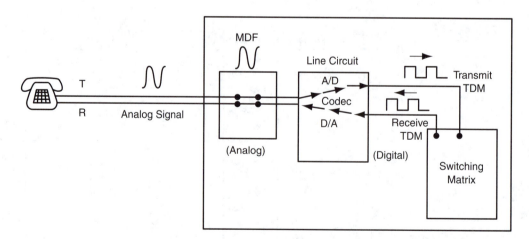

FIGURE 8-3a Voice circuit through codec.

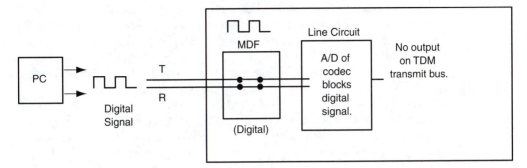

FIGURE 8-3b Data blocked by codec.

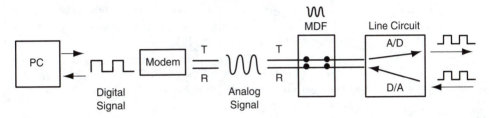

FIGURE 8-3c Modem signal passed by codec.

speed highway, such as 56 Kbps, 64 Kbps, or a 1.544-Mbps PSTN transmission facility. It is now possible to use these digital-to-digital interface devices on the customer's premises and in a local central office because the PSTN network between central exchanges has been converted into a digital network (see Figure 8-4).

It is also possible to transmit data directly over the local loop by connecting the customer's premises to the central exchange, with a digital subscriber line circuit

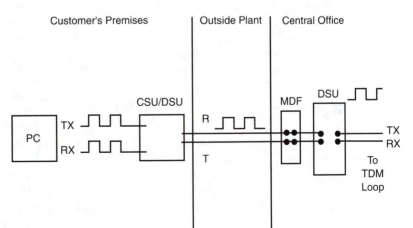

FIGURE 8-4 CSU/DSU interface between customer's premises and central office.

and ISDN rather than an analog line circuit. However, an ISDN line circuit costs $100 a month from the LEC, versus $20 a month for an analog line circuit. Most people prefer their analog local service line over ISDN service, because it is much cheaper than a digital ISDN line. People are also slow to convert to ISDN because services they use are not on an ISDN line circuit. It does not do me any good to convert my line to ISDN if people I transfer data to do not convert to ISDN.

The protocols used to transfer data are not the same for ISDN and a modem. *Protocol* refers to the rules of communication. Each established protocol is a definition of how data is to be communicated from one device to another. The transmitting and receiving ends must use the same communication protocol. Modems can only talk to modems utilizing the same protocol; they cannot communicate with ISDN because the communication protocols are not the same. Remember that a receiving modem is looking for an analog signal from the PSTN. With ISDN, there is no analog signal. ISDN is an end-to-end digital network. If I wish to send data to someone using a modem, I must also use a modem. If I want to use ISDN to transfer data, an ISDN line circuit must be on the other end of the transmission facility. I am able to use ISDN to communicate with other people via e-mail or the Internet because my *Internet service provider (ISP)* has ISDN lines. The ISP serves as an interface and protocol converter between my ISDN line and the lines of other people who are using modems to connect to the ISP.

8.2 DATA COMMUNICATION AND CODING

Data communication is the transmission of data from one location to another. The *data* transmitted can be numeric or alphabetical. When data is transmitted, some code is used to represent the various characters or numbers sent over the system. As we have seen, Morse used a code that was named for him to send messages over his invention (the telegraph). The telegraph was our first means of long distance communication; it was a data communication network. Messages were converted into Morse code. The code was sent as electrical signals over the telegraph lines. The receiving end would decode the signal back into the original message and deliver it to the intended recipient. The telephone replaced the telegraph, and the telephone will be replaced by a computer terminal within the next ten years. As the use of computers has skyrocketed, a data communication network has become essential.

8.3 ASCII CODING

The standard binary code developed for transmission of data is the *American Standard Code for Information Interchange (ASCII). ASCII is the coding scheme used by a PC to store information.* It uses 7 bits, each of which can be either a 1 or a 0. That is, each of the 7 bits can be one of two states (1 or 0). According to mathematical statistics, if we have *seven different* things (bits) that can assume *one of two* states (a 1 or 0), the *total number of different combinations* we can have is equal to 2^7. Since 2^7 is 128, ASCII provides 128 different codes from 0000000 to 1111111. These codes can be converted into decimal numbers by adding the value of each bit in a code. The value of each bit corresponds to the location of the bit in the code. The codes for decimal digits 1, 3, 65, 97, and 127 are:

2^6	2^5	2^4	2^3	2^2	2^1	2^0	=	
0	0	0	0	0	0	0	=	0
0	0	0	0	0	0	1	=	1
0	0	0	0	0	1	1	=	3
1	0	0	0	0	0	1	=	65
1	1	0	0	0	0	1	=	97
1	1	1	1	1	1	1	=	127

We can use a seven-level binary code to represent the decimal numbers from 0 to 127, or we can use the seven-level code to represent other things. ASCII uses the 128 codes that we get with the seven-level binary code to represent the letters of the alphabet and the numerals 0 to 9. See Table 8-1 for the ASCII code chart.

Look in Table 8-1 for the letter *a*. Lowercase *a* is represented by the code 1100001. We have previously shown that if this were converted to a decimal number, it would be 97. But ASCII does not use the code to convert binary to decimal numbers; it uses the code to convert binary numbers to characters and the decimal numbers 1 to 9. When the lowercase letter *a* is pressed on the keyboard, it is stored as an ASCII code 1100001. Therefore when the code is read back, it is converted back into the letter *a*. One note is in order for anyone who connects an oscilloscope to read the signal output of an RS-232 serial port: Although we read an ASCII code from highest- to lowest-order bit (left to right), the bits are actually sent lowest- to highest-order.

PCs contain a hardware device between the data bus of the PC and the RS-232 serial port. This device is called a ***universal asynchronous receiver/transmitter (UART).*** The UART will send a start bit (a 0, which is +3 V or greater). It will then send the least significant data bit and each higher-order data bit until the most significant data bit is sent. It then sends the parity bit, then a stop bit (a 1, which is –3 V or less). Remember our lowercase *a*. It is 1100001 in ASCII code. When the UART sends this code out, it is sent lowest-order bit first, which is sent as 1000011. Suppose we are using even parity; when we put the start, even parity, and stop bits in place, the UART sends 0100001111. It may help to simplify what is happening by stating that although we read ASCII from left to right, the UART sends the code out from right to left. Of course, the receiving UART uses the same protocol and reconstructs the character code starting with the lowest-order bit first.

8.4 EXTENDED ASCII CODING

It was not long before we needed to represent more than 128 different characters. To represent more than this number of characters, an eight-level coding system was developed called *Extended ASCII.* IBM used an eight-level coding system called *Extended Binary Coded Decimal Interchange Code (EBCDIC)* in its mainframes. EBCDIC is

Table 8-1 ASCII Code Chart

Back Five Bits Column #

Column #	Bit Position ($2^4\,2^3\,2^2\,2^1\,2^0$)	Control Codes (Row 0)	Numbers, Punctuation (row 1)	Uppercase (row 2)	Lowercase (row 1)
0	0 0 0 0 0	NUL	sp	@	`
1	0 0 0 0 1	SOH	!	A	a
2	0 0 0 1 0	STX	"	B	b
3	0 0 0 1 1	ETX	#	C	c
4	0 0 1 0 0	EOT	$	D	d
5	0 0 1 0 1	ENQ	%	E	e
6	0 0 1 1 0	ACK	&	F	f
7	0 0 1 1 1	BEL	'	G	g
8	0 1 0 0 0	BS	(	H	h
9	0 1 0 0 1	HT	)	I	i
10	0 1 0 1 0	LF	*	J	j
11	0 1 0 1 1	VT	+	K	k
12	0 1 1 0 0	FF	,	L	l
13	0 1 1 0 1	CR	-	M	m
14	0 1 1 1 0	SO	.	N	n
15	0 1 1 1 1	SI	/	O	o
16	1 0 0 0 0	DLE	0	P	p
17	1 0 0 0 1	DC1	1	Q	q
18	1 0 0 1 0	DC2	2	R	r
19	1 0 0 1 1	DC3	3	S	s
20	1 0 1 0 0	DC4	4	T	t
21	1 0 1 0 1	NAK	5	U	u
22	1 0 1 1 0	SYN	6	V	v
23	1 0 1 1 1	ETB	7	W	w
24	1 1 0 0 0	CAN	8	X	x
25	1 1 0 0 1	EM	9	Y	y
26	1 1 0 1 0	SUB	:	Z	z
27	1 1 0 1 1	ESC	;	[	{
28	1 1 1 0 0	FS	<	\	|
29	1 1 1 0 1	GS	=	]	}
30	1 1 1 1 0	RS	>	^	~
31	1 1 1 1 1	US	?	_	RUBOUT

Most Significant Front 2 Bits ($2^6\ 2^5$):
- 0 0 — Control Codes (Row 0)
- 0 1 — Numbers, Punctuation (row 1)
- 1 0 — Uppercase (row 2)
- 1 1 — Lowercase (row 1)

not the same coding as Extended ASCII. The advantage of using Extended ASCII is that the first 128 codes for it are the same as they are for regular seven-level ASCII. The codes from 128 to 255 are used to represent additional characters (see Table 8-2).

8.5 SERIAL VS. PARALLEL TRANSMISSION OF DATA

Data can be transmitted one bit or several bits at a time. When it is transmitted one bit at a time, it is transmitted over one wire. Each bit is transmitted one after the other onto the same wire. This type of transmission is called *serial transmission*. When data is transmitted several bits at one time, each bit is transmitted over its own wire. This type of transmission is called *parallel transmission* because the bits are transmitted over wires that are parallel. The most common form of parallel transmission is to transmit 8 bits at one time. With seven-level ASCII coding, an eighth bit is used as a parity bit to check the validity of the code. Thus, each character or number is represented by an 8-bit code (7 ASCII + 1 parity bit). By transmitting 8 bits at one time, we can transmit a character at a time instead of a bit at a time. Eight bits make up a *byte*.

In computers, an 8-bit byte is referred to as a *word.* But it is not really a word; it is a character. Each character or number is represented by a unique 8-bit byte. Data inside a computer is transferred over a data bus that is a parallel bus. The first PC had eight wires for the data bus and could transfer data 8 bits (1 byte) at a time. A 486 computer used 32 parallel leads for the data bus and could transfer data 32 bits at a time, or four 8-bit words at a time. A Pentium computer has a data bus with 64 leads and transfers data 64 bits at a time, or eight 8-bit words at a time. PCs define a word as the number of adjacent bits that can be manipulated or stored as one unit. This depends on the size of the data bus, and in a PC the size of the data bus defines a word. In a Pentium PC a word is 64 bits, not 8 bits.

Most printers that connect to a PC are parallel printers transferring 8 data bits at a time. Some older printers are serial printers that connect to an RS-232 port and transfer data 1 bit at a time. It is readily apparent that parallel transmission is much faster than serial transmission, but serial transmission is most often used in data communication, because only one circuit is needed for the transmission of data. Parallel transmission would require establishing eight or more circuits between the two data devices. Parallel data transmission is used when the transmission distance is less than 25 ft. When data is sent over the PSTN, it is sent using serial transmission. It is obvious that we only have one communication circuit over the PSTN. If we only have one circuit for the transmission of data, serial transmission must be used.

8.6 ASYNCHRONOUS VS. SYNCHRONOUS TRANSMISSION

There are two ways to transmit data using serial transmission. One method is to use *asynchronous transmission,* *where each character has its own synchronizing information.* The beginning of a character is designated by a bit called the *start bit*. The start bit is a bit whose value is 0. A bit that has a value of 0 is also called a *space bit*. The end of a character is designated by a bit known as the *stop bit*, whose value is 1. A bit

TABLE 8-2 Extended ASCII Code Chart

Decimal Code	Character	Decimal Code	Character	Decimal Code	Character	Decimal Code	Character	Decimal Code	Character
0	NULL	52	4	104	h	156	£	208	–
1	J SOH	53	5	105	i	157	¥	209	—
2	l STX	54	6	106	j	158	û	210	"
3	§ ETX	55	7	107	k	159	ƒ	211	"
4	© EOT	56	8	108	l	160	á	212	'
5	® ENQ	57	9	109	m	161	í	213	'
6	´ ACK	58	:	110	n	162	ó	214	÷
7	% BEL	59	;	111	o	163	ú	215	◊
8	BS	60	<	112	p	164	ñ	216	ÿ
9	HT	61	=	113	q	165	Ñ	217	Ÿ
10	LF	62	>	114	r	166	ª	218	/
11	VT	63	?	115	s	167	º	219	
12	FF	64	@	116	t	168	¿	220	‹
13	CR	65	A	117	u	169	©	221	›
14	SO	66	B	118	v	170	™	222	fi
15	} SI	67	C	119	w	171	‒	223	fl
16	ÿ DLE	68	D	120	x	172	‒	224	‡
17	◊ DC1	69	E	121	y	173	¡	225	·
18	◊ DC2	70	F	122	z	174	«	226	,
19	!! DC3	71	G	123	{	175	»	227	„
20	DC4	72	H	124	|	176	∞	228	‰
21	NAK	73	I	125	}	177	±	229	Â
22	SYN	74	J	126	~	178	≤	230	Ê
23	ETB	75	K	127	DEL	179	≥	231	Á
24	È CAN	76	L	128	Ç	180	¥	232	Ë
25	Í EM	77	M	129	ü	181	µ	233	È
26	Á SUB	78	N	130	é	182	∂	234	Í
27	Ë ESC	79	O	131	â	183	Σ	235	Î
28	FS	80	P	132	ä	184	∏	236	Ï
29	Ô GS	81	Q	133	à	185	π	237	Ì
30	Ò RS	82	R	134	å	186	∫	238	Ó
31	Ú US	83	S	135	ç	187	ª	239	Ô
32	SP	84	T	136	ê	188	º	240	
33	!	85	U	137	ë	189	Ω	241	Ò
34	"	86	V	138	è	190	æ	242	Ú
35	#	87	W	139	ï	191	ø	243	Û
36	$	88	X	140	î	192	¿	244	Ù
37	%	89	Y	141	ì	193	¡	245	ı
38	&	90	Z	142	Ä	194	¬	246	^
39	'	91	[	143	Å	195	√	247	~
40	(	92	\	144	É	196	ƒ	248	‾
41	)	93	]	145	æ	197	≈	249	˘
42	*	94	^	146	Æ	198	∆	250	·
43	+	95	_	147	ô	199	«	251	°
44	,	96	`	148	ö	200	»	252	¸
45	-	97	a	149	ò	201	…	253	˝
46	.	98	b	150	û	202		254	
47	/	99	c	151	ù	203	À	255	ˇ
48	0	100	d	152	ÿ	204	Ã		
49	1	101	e	153	ö	205	Õ		
50	2	102	f	154	Ü	206	Œ		
51	3	103	g	155	¢	207	œ		

that has a value of 1 is also called a *mark bit.* Asynchronous transmission is also termed *start-and-stop transmission.* The UART discussed earlier uses asynchronous communication. Therefore, the communication between the serial port of a PC and an external modem is asynchronous communication.

When mainframes and minicomputers were the only computers available, most interfaces to these computers were via dumb terminals. The terminals were connected to the computer using asynchronous communication. No one can type at the speed of a computer's processor. There was no need to have the processor dedicated to one user. By employing asynchronous communication protocol, many users could be connected to a central computer. When a terminal had information for the computer, each piece of data was preceded by a start bit. The use of asynchronous (start-and-stop) transmission was an efficient way to connect many devices to a central computer.

The second method of serial transmission is **synchronous transmission.** *This type of transmission sends data as a block of characters at a time.* A data block can be almost any size, but a typical data block contains 128, 256, 512, or 1024 characters. When synchronous data transmission is used, the start and stop bits are not used. A *header* is placed in front of the data sent, and a trailer is placed after the data. The header will contain the destination address of the message, a synchronization signal, and control information. The trailer contains parity-checking information and the address of the sender. The header usually contains about 32 bits; the trailer typically has 8 to 16 bits.

packets

Each character sent using asynchronous transmission requires the transmission of 10 bits (start, character, stop). The start and stop bits are called *overhead bits.* If 256 characters are sent using asynchronous transmission, 2560 bits are required. If synchronous transmission is used, we must transmit the header, the 256 characters, and the trailer. Thus, 256 characters times 8 bits per character is 2048 bits. If we add 64 bits of overhead for the header and trailer, the total number of bits required is 2112. Synchronous data transmission is obviously faster than asynchronous transmission because fewer bits are needed to send the same data. Synchronous transmission is more complex than asynchronous transmission and provides more sophisticated error-detection and error-correction capabilities.

Synchronous transmission is typically used on a dedicated circuit connecting two high-speed computers. Asynchronous transmission is used between the PC (or DTE) and the modem (DCE) on dial-up connections through the PSTN. When we wish to transfer files of data between two PCs, we use a file transfer protocol such as X Modem, Y Modem, Z Modem, or Kermit to transfer the file. These protocols transfer a file between two PCs or between a DTE and a DCE device using blocks of data. For example, X Modem transfers a file 128 bytes at a time. Although the data is transferred as blocks of data, it still is an asynchronous data communication protocol. Each byte in the block of data is surrounded by a start and stop bit. The UART in the serial port dictates that the transmission of data between two serial ports will be asynchronous.

One point that should be cleared up here is that a modem has two interfaces, and each interface has its own protocol. The modem has a serial EIA-232 interface to the PC and another interface to the PSTN. Remember that a modem is simply a protocol converter. It converts the binary signal received from a PC into the modulation of an analog signal to the local loop. It also converts the modulation of the analog signal received from the local loop into a digital signal to the PC. Today's high-speed modems also make another conversion. They convert the asynchronous signal from the PC into a synchronous signal to the PSTN. The modem at the other end of the circuit converts the synchronous signal from the PSTN back into an asynchronous signal to its PC.

Some modems are designed to accept only asynchronous transmission from a DTE device (Bell 103, Bell 202, V.21, and V.23). Some modems will only accept synchronous transmission from a DTE device (Bell 201, V.26, V.27, V.29, and V.33). Several modems are designed to allow the user to select either asynchronous or synchronous transmission between the DTE and DCE devices (Bell 212, V.22, V.24, V.32, and V.34). The EIA-232 interface from the DTE to the DCE device allows the use of either asynchronous or synchronous transmission of data.

When we talk about whether a signal between two DTE devices is asynchronous or synchronous, we are talking about the signal leaving the DTE device. The signal over the PSTN between the two modems (DCE devices) is a synchronous signal when high-speed modems are used. The PC will send bits in serial asynchronous transmission to the external modem. The modem contains memory and software to arrange incoming bits from the PC into a block of data for synchronous transmission over the PSTN. Since a PC is the most common form of DTE device that we encounter, I will use the term *PC* instead of *DTE device* in the remainder of this chapter.

8.7 ASYNCHRONOUS TRANSMISSION ERROR CHECKING

The simplest error-checking method is parity checking. This technique is used for asynchronous transmission. ASCII uses 7 bits for coding and does not use the 8th bit, allowing the use of the 8th bit for parity checking. The transmission system will count the number of 1s sent for each character and place a 1 or 0 in the 8th-bit location as required. There are five types of parity: odd, even, mark, space, and no parity. When *odd parity* is used, the count of 1s for every 8 bits sent will be odd. Suppose we transmit the character *a* (1100001). This character has an odd number of 1s. For odd parity, we place a 0 in the 8th-bit location. This results in an odd number of 1s for all 8 bits. The character capital *A* is 1000001. This is an even number of 1s. When *A* is transmitted, we must send a 1 as the 8th bit to achieve an odd number of 1s for the 8 bits. With odd parity, all transmitted bytes result in an odd number of 1s. The receiver counts the 1s for each byte to make sure they are all odd.

Even parity is similar to odd parity; this error-checking protocol ensures each transmitted byte has an even number of 1s. The 8th bit is made a 1 or 0 to ensure that each byte has an even number of 1s. *Mark parity* is a protocol that makes every 8th bit a 1. *Space parity* makes every 8th bit a 0. As you can see, the 8th bit is used for the parity bit. When an eight-level code such as Extended ASCII is used for coding, the 8th bit is used as part of the code and *no parity* can be used. When a PC transmits data over a serial port, the protocol of the port must be set to match the protocol of the receiving device. The protocol of a serial port consists of data speed in bits per second, the number of data bits per character, parity, and the number of stop bits. The number of stop bits is usually specified as 1 bit. The parity being used by both ends of the data transmission circuit must be specified.

8.8 SYNCHRONOUS ERROR DETECTION AND CORRECTION

Parity checking provides an end-to-end check (PC to PC) on each byte transmitted between two PCs. A modem does not perform any error checking on data it receives from the transmitting PC. The distance between the PC and modem is so short that the possibility of errors is remote. But modems do perform error checking between themselves to ensure that the PSTN did not corrupt the data. The use of synchronous transmission between the two modems allows the use of sophisticated error-detection and control techniques. Modems use a synchronous error-detection technique called *cyclic redundancy checking (CRC).*

There are two types of CRC: CRC-16 and CRC-32. CRC treats a block of data as representing a large number. If the block of data contains 128 characters, it contains 1024 bits ($128 \times 8 = 1024$). Clearly, 2^{1024} is an extremely large number. If all 1024 bits were a 1, the value of this number would be 1 less than 2^{1025}. The block of data is divided by a 17-bit divisor (for CRC-16) or a 33-bit divisor (for CRC-32). The divisor is a prime number, divisible only by itself and 1. Dividing the number represented by the block of data using the prime number divisor results in a 16-bit remainder (for CRC-16), or a 32-bit remainder (for CRC-32).

The CRC calculation is done at the transmitting modem, and the remainder is placed in the trailer behind the block of data transmitted. The receiving modem receives the block of data and trailer. It then calculates a remainder using the same CRC protocol used by the transmitting modem, comparing the remainder calculated to the remainder sent in the trailer. If they compare, the data received is okay. If they do not match, the data was corrupted. After the receiving modem has determined whether errors have occurred, it will notify the transmitting modem. Error conditions are handled by requesting that the originating modem retransmit the data. This technique is called *automatic retransmission request (ARQ).* There are two basic types of error correction: discrete ARQ and continuous ARQ.

Discrete ARQ is also called *stop-and-wait* ARQ because after sending a block of data, the transmitting end will wait for a signal from the receiving modem before sending another block of data. The signal from the receiving modem to the transmitting modem will be either a positive acknowledgment *(ACK)* that the data was good, or a negative acknowledgment *(NAK)* that it was corrupted. For every block of data sent, an acknowledgment is received. The acknowledgment will be either an ACK or a NAK. On receipt of an ACK, the next block of data is sent. If a NAK is received, the modem retransmits the last block of data from its memory.

With the use of discrete ARQ (ACK/NAK), the transmitting modem spends a lot of time waiting for the acknowledgment message. Today's PSTN transmits most data without errors. Thus, most of the received messages are ACKs; few NAKs are received. The modem spends a lot of time waiting just to be told the data was good. *Continuous ARQ* eliminates the need for the transmitting modem to wait for an ACK or NAK. The transmitting modem identifies each block of data with a block number. The receiving modem sends an ACK or NAK to the transmitting modem for each block of data received. The transmitting modem keeps sending data without waiting, as long as it is receiving ACKs every so often. If it fails to receive an ACK after a certain interval of time, the modem stops transmitting.

As long as a transmitting modem continues to receive ACKs from the receiving modem, it continues to transmit data. When the receiving modem detects an error, it sends a NAK, along with the block number affected, to the transmitting modem. Some modems use a protocol that retransmits all blocks from the NAK block forward. Some use a *selective ARQ* protocol. With selective ARQ, only the affected block is retransmitted. CCITT Standard V.42 is an error-correction standard that allows modems to determine during handshaking whether to use MNP4 or LAP-M error-correction protocol. *Link Access Protocol for Modems (LAP-M)* provides a selective ARQ protocol.

8.9. FLOW CONTROL

Modems are continually storing and retrieving data from their memory. Some type of *flow control* is necessary to prevent these memory buffers from overflowing and still keep enough data in memory to allow for retransmission on receipt of a NAK. The flow control management software in the modem keeps track of data flowing into and out of the buffer memory. This software keeps track of the amount of available free memory and instructs the transmitting PC to stop sending when the buffer is almost full.

There are two basic types of flow control. They are software flow control and hardware flow control. Although both types use software programs to control the flow of data across a connection between two devices, we refer to the type based on what activates or terminates flow control software operation. Activation and deactivation can be done through specific messages sent between the two devices. The messages are special characters, and we refer to this as X-on/X-off software flow control. Hardware flow control uses a physical lead or wire between devices to activate flow control. Between a PC and modem, two wires are used to provide a physical path for signals to the flow control software. These two wires are Request-to-Send (RTS) and Clear-to-Send (CTS).

The modem uses hardware flow control to instruct the transmitting PC to stop sending. The EIA-232 interface connecting the PC to the DCE (modem) contains a request-to-send (RTS) lead on pin 4 (DB-25 connector) and a clear-to-send (CTS) lead on pin 5 (DB-25 connector). The RTS and CTS on a DB-9 connector are pins 7 and 8 respectively. When a modem has approached a full-buffer condition, it will remove the CTS signal. On losing the CTS signal, the transmitting PC stops sending. After the modem has sent enough data to free up additional buffer memory, the CTS signal will be sent to the PC and it will start sending data again. The use of RTS and CTS is called *hardware flow control*. When the PC has data to send, it sends an RTS signal to the modem. The modem returns a CTS signal if it is ready to accept data, and an RTS signal is present (see Figure 8-5).

8.10 SERIAL PORTS AND UARTS

The data highway inside a PC is a parallel data bus. The connection to most external devices is via a serial port. The UART between the parallel data bus and the EIA-232 connector converts parallel transmitted data from the computer into serial transmission. It also converts the received serial data from the EIA-232 interface

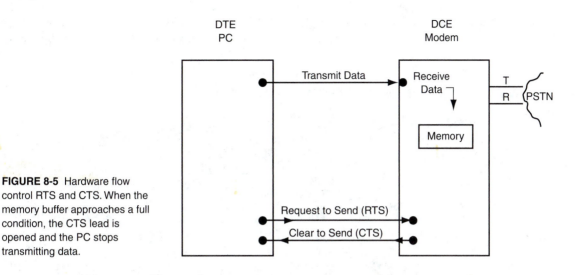

FIGURE 8-5 Hardware flow control RTS and CTS. When the memory buffer approaches a full condition, the CTS lead is opened and the PC stops transmitting data.

into parallel transmission for the computer. Thus, a UART is a parallel-to-serial and a serial-to-parallel converter. The maximum speed of a serial port is determined by the UART and software drivers for that port.

Microsoft Windows 3.11 allowed setting a serial port to a maximum of 19,200 bps. This is because DOS 6.20 only supports 19,200 bps, which is a speed supported by older 8250 UARTs. Many PCs are still manufactured using a 16550 AF UART. This UART can support speeds as high as 115,200 bps. With a 16550 AF UART, the speed of data transmission between the PC and modem can be 115,200 bps. To take advantage of this higher speed, a communication software package will be needed that allows you to set the serial port of the PC to a higher speed than the DOS command "mode" allows. Windows 95 provided the software support necessary to allow setting a serial port to 115,200 bps (see Figure 8-6).

Windows 95 allowed setting a serial port speed as high as 921,600 bps, but the 16550 UART restricts the speed to 115,200 bps. Internal modem cards are now being

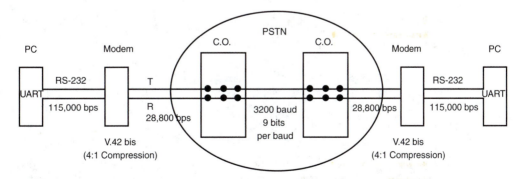

FIGURE 8-6 Personal computer-to-modem speed and modem-to-modem speed. Physical speed between modems is 28,800 bps; end-to-end speed between PCs is 115,000 bps.

sold that use a 16C650 UART. This UART chip achieves 230,000 bps. Eventually the speed of serial ports will rise as PCs, modems, and other devices equipped with this chip replace older devices.

8.11 BAUD RATE AND MODULATION OF THE CARRIER

Modems communicate with each other via audio frequency waves modulated by the digital input from a PC. The audio signal between two modems is called the *carrier signal*. The modulation (or change) to the audio signal can be accomplished by changing the amplitude, frequency, or phase of the audio signal. How fast these signal changes occur is called the *baud rate*. **Baud** refers to the number of changes occurring in a signal.

Suppose we transmit an 1800-Hz signal and change the frequency from 1800 to 1900 Hz. If we change the frequency 2400 times per second, this is 2400 baud. If each change in frequency represents 1 bit, we also have 2400 bps. Today's modems achieve higher speeds because each change in a signal represents more than 1 bit.

Baud rate is limited by the bandwidth of the PSTN voice channel. This bandwidth in the old analog PSTN was 3000 Hz (300 to 3300 Hz). By using 2400-baud signals, part of the bandwidth could be used for guardbands between signals. The bandwidth of the *digital* PSTN is 3600 Hz, and since guardbands are not necessary, modems can transmit at speeds up to 3600 baud over the digital PSTN.

The overall speed of a modem depends on the baud rate it can use and on how many bits can be represented by each signal change (baud). If we wish to use different amplitudes of a signal to represent a *2-bit* combination, we need four different amplitudes. Two bits can assume any of four combinations: 00; 01; 10; and 11. Two bits (called a *dibit*) can be represented by four different amplitudes when amplitude modulation is used:

$$\text{Amplitude } 1 = 00$$
$$\text{Amplitude } 2 = 01$$
$$\text{Amplitude } 3 = 11$$
$$\text{Amplitude } 4 = 10$$

Modems can also use *phase modulation (PM)* to modulate a carrier signal between two modems. The phase of a signal is used to describe the relationship of one signal to another. Suppose we use an 1800-Hz signal for the carrier signal of a modem. Each cycle of an 1800-Hz signal takes 0.0005555555 . . . sec (time $= 1/f$). Each quarter of a cycle takes 0.0001388888. . . sec. Each quarter of a cycle represents 90° of an AC signal. If the timing of one 1800-Hz signal is changed so that it occurs 0.0001388888 . . . sec later than another (reference) signal, it is 90° out of phase with the reference signal. If a signal occurs 0.00277777 . . . sec after the reference signal, it is 180° out of phase with the 1800-Hz reference signal.

There are three types of phase modulation: *two-phase, relative phase*, and *quadrature phase*. A two-phase signal uses 0° to represent a 1 and 180° to represent a 0. A relative phase signal uses the relationship of signals to code a 1 or 0. If the phase received is the same as the last phase received, it represents a 1. If the phase is different, it represents a 0. Thus, to transmit several 0s, the phase is changing back and forth from one phase to another. To represent a string of 1s, the phase remains the same.

A quadrature phase system uses four different phase relationships to code 2 bits at a time. If we use four different phase relationships, we can send 2 bits at a time by using the four different phases to represent the four different states a dibit can have:

Phase 1 of 0° = 00

Phase 2 of 90° = 01

Phase 3 of 180° = 11

Phase 4 of 270° = 10

It is also possible to send 3 bits at a time by using eight different amplitudes or eight different phase shifts. Thus, $2^3 = 8$ different combinations that a *tribit* can have. It is also possible to send 4 bits *(quadbit)* at a time by using 16 different amplitudes or 16 different phase shifts. A V.29 modem achieves 9600 bps, by transmitting 4 bits at a time using 8 different phases of a signal, where each phase can be one of 2 amplitudes. There is a limit on how many different amplitudes or phase shifts a modem can use. That limit is imposed by the detecting capabilities of the receivers and the noise on the transmission facility. The more amplitude levels or phase shifts used to transmit a code, the better the receiver must be at detecting small discrete changes in the signal. The requirement to maintain a high signal-to-noise ratio also limits how many different amplitudes and phase shifts a system can use. The digital PSTN is a much quieter transmission facility than the old analog PSTN, and the digital PSTN allows for more discrete signal changes.

Because of the requirement to maintain a high signal-to-noise ratio and the need to have a change detectable by the receiver, the maximum speed at which a modem can operate often seems to have been achieved. However, chip manufacturers continue to develop new improved Signal Logic Processor chips and continue to push the speed envelope even higher. One point to keep in mind is that although modems such as the V.34 can achieve speeds of 28,800 bps, they can only achieve this speed with an optimal PSTN path. Most of the time, these modems will fall back to lower speeds due to an imperfect PSTN path. Modems capable of 56,000 bps are now on the market. How often this speed can be attained over the PSTN remains to be seen. It is also questionable if the technology they use will be able to transmit at high speed over the PSTN, or will they fall back to less than 28,800 bps, as V.34 modems do?

The newest modem standard on the market is V.90. This standard takes advantage of the fact that most ISPs connect to the local central office using digital trunk lines such as a T1, T3, or primary rate ISDN facility. They have scrapped their old analog line access to the central office. With digital access, they no longer need an analog modem. They still need to appear as if they have a modem attached because users have an analog modem attached at their end of the communication circuit.

ISPs use a device called a *remote access server (RAS)*. The RAS will contain the microprocessor power, programming, and memory necessary to communicate with the distant analog modem.

The connection of digital trunk lines from the PSTN directly into the RAS at an ISP location means that the RAS must do a digital-to-digital conversion. This is not difficult to do. Let's look at the process before the ISP had digital trunks into it. Previously the ISP used analog lines and had an analog modem connected to each line. When a user transmitted data out of a PC, it went into the modem and modulated an analog carrier signal. The analog carrier signal was converted by the codec of the line circuit at the originating central exchange into 8-bit digital codes. These digital codes were transported by the digital facilities of the PSTN and arrived at the line circuit of the terminating central exchange serving the ISP. The line circuit converted the digital codes back into analog carrier signals. The modem at the ISP received the carrier signals and demodulated them to recapture the digital data passed to the ISP's computer.

In the process just described, we made four different signal conversions. On the originating end, we used the digital data from a PC to modulate an analog signal sent to the originating central exchange. At this exchange, the analog signal was converted into a digital signal. At the receiving exchange, we convert the digital signals into analog signals and the modem demodulates the analog modulations into digital bit streams. Let's look at the receiving process at the central exchange on the ISP end of the circuit. If we can take a certain digital bit pattern from the PSTN and convert that to a certain analog signal that the modem then converts to a certain digital bit stream, we can eliminate one conversion step.

If we know that $A + B + C + D = X$ and $X = Y$, why do both conversions? We can cut to the end of the problem by stating that $A + B + C + D = Y$. This is what was done on the ISP end of the circuits. There is no need to do a digital-to-analog and then analog-to-digital conversion; a digital-to-digital conversion is enough. By streamlining the conversion process, the sources of error always present in modulation and demodulation techniques are reduced. But the significant speed gains are made in the direction from the ISPs to their customers. Since they have a direct digital trunk connection to the PSTN, they do not need to use an analog modem. They use special *digital signaling processors (DSPs)* that convert data into 7-bit codes and transmit 8000 of these 7-bit codes per second. $2^7 = 128$; therefore, there are 128 codes to represent data sent 7 bits at a time.

The 7-bit codes fit nicely inside the transport media used by the PSTN, and when they arrive at the line circuit of the connected user, the codec converts them into a certain specific discrete voltage level. This voltage level is transmitted over the local loop to the customer's 56K modem. The voltage level is converted into the appropriate 7-bit code. The voltage level transmitted by the decoder to the local loop depends on the 7-bit code it received over the PSTN from the 56K modem or DSP (RAS) at their Internet service provider. The decoder at the central office line circuit of the user converts the 7-bit code received into one of 128 voltage levels or "voltage codes" and sends a voltage for each 7-bit code out over the local loop to the receiving 56K modem. The receiving modem at the user's location converts each received voltage level back into a 7-bit code. Since the receiving modem receives 8000 of these voltage codes per second and converts each one into a 7-bit code, it receives 56,000 bps. The spread between adjacent voltage levels is small, and a little noise will distort the signal. For a 56K modem this will cause a serious loss of data. Thus, 56K modems can only be used on circuits with low circuit noise, which limits their use to customers residing within 2 or 3 miles of a central exchange.

So this is what makes a 56K modem run fast: we are actually sending digital data as one of 128 different discrete signals between the ISP and the customer's modem. The digital data is changing rapidly and fools the cable pair into thinking it is an analog signal; sending 8000 of these different states per second makes the digital signal look like an analog signal.

A modem also uses compression algorithms that allow it to receive data from a PC faster than it sends data over the PSTN. As data is received from a PC, the modem stores the data in memory. The compression algorithm software looks at the 1s and 0s that came from a PC and compresses the data by eliminating repeating 1s and 0s. In essence, the modem looks at several bytes of data and converts the ASCII code into a new code. The new code is basically a shorthand version of the original ASCII code. The modem transmits this shorthand code to the receiving modem. The receiving modem converts the shorthand code back into ASCII code and sends ASCII to the receiving PC. Most compression algorithms can reduce the bits required to represent data by a factor of 4. By using a 4-to-1 compression ratio (V.42 bis), the end-to-end speed of a 57,600-bps connection can be as high as 230,400 bps. But this speed cannot be realized unless the modems are interfaced to the PC via dual UARTs, or a special accelerator board, in the PC.

Quadrature amplitude modulation (QAM) is a technique that uses 8 different phase angles and 2 different amplitudes for each of those phases to achieve 16 detectable events. With 16 possible signals, QAM can transmit 4 bits at a time by using each of the 16 signals to represent one of the 16 combinations a quadbit can have ($2^4 = 16$ combinations). If we choose 1800 Hz as our signal and change that signal 2400 times per second (2400 baud), we can transmit 9600 bps by transmitting 4 bits per baud.

Baud refers to the number of times a carrier signal changes. Older modems use 2400 baud to meet the requirements of the old analog PSTN. The older analog network used an AM/FDM carrier system to multiplex many voice channels onto one facility. This technology caused some frequency distortion as we modulated and demodulated the signal between groups, supergroups, and mastergroups. These problems limited the frequency changes or baud rate to 2400 baud.

The digital carrier transmission systems used in major toll routes of the PSTN can handle frequency changes of 3600 baud. Modems designed to take advantage of the new digital PSTN are designed to V.34 standards. V.34 modems will use 2400, 2743, 2800, 3000, 3200, or 3429 baud. The baud used depends on the quality of the end-to-end PSTN circuit. If the PSTN path is a digital path and the local loop provides low loss, 3429 baud will be used. V.34 modems continually monitor the circuit and will automatically downshift to a lower baud rate if degradation of transmission occurs. V.34 modems can select from 50 different combinations of baud rate and modulation techniques. The maximum speed of 28,800 bps can be achieved by using 3200 baud and 9-bit *trellis coded modulation (TCM)*. With 9-bit TCM, we can send 9 databits per baud or 28,800 bps ($9 \times 3200 = 28,800$).

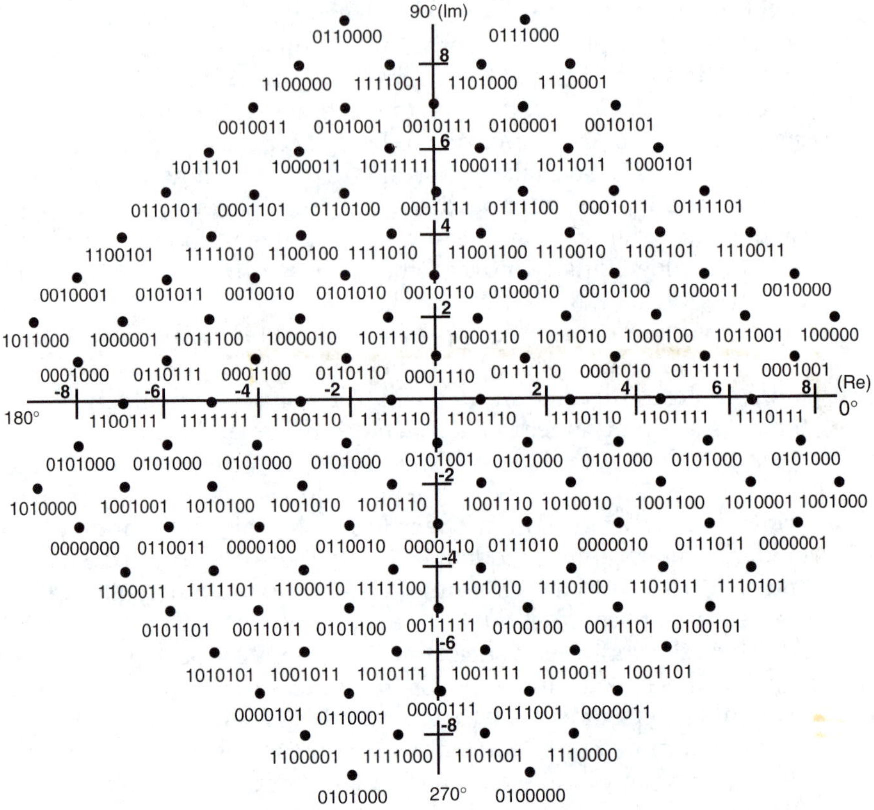

FIGURE 8-7 V.32 bis constellation pattern.

TCM is a technique that expands on QAM. The original TCM was a 5-bit code. It used a 4-bit QAM and added to it a calculated bit to help the receiving modem decode the quadbit. Transmitting 5 bits at a time requires 32 detectable signals. The V.32 standard is the standard for TCM transmitting at 9600 bps. The 32-point constellation (or detectable events of Figure 8-7) is accomplished by using 4 different amplitudes on 8 different phases. V.32 bis was a standard used to achieve 14,400 bps. V.32 bis transmits 6 data bits plus 1 TCM bit per baud. Transmitting 7 bits per baud requires 128 detectable events. Transmitting 9 bits per baud (signal change) requires 512 detectable events, and transmitting 10 bits per signal change requires 1024 detectable events. The ITU specifications for V.34 state that the symbol rate (baud rate) for 28,800 bps is 3200. With a 3200 symbol rate, each symbol must represent 9 data bits. To transmit 9 data bits at one time requires 512 different signals ($2^9 = 512$). V.34 uses a 960-point constellation to provide for the 512 data bits, plus a redundant bit for trellis coding (see Figure 8-8).

I called some major modem manufacturers in an attempt to nail down how they are achieving 960 different signals. How many phase shifts and amplitude levels are being used to achieve a 960-point constellation? Nobody could tell me. It is done by

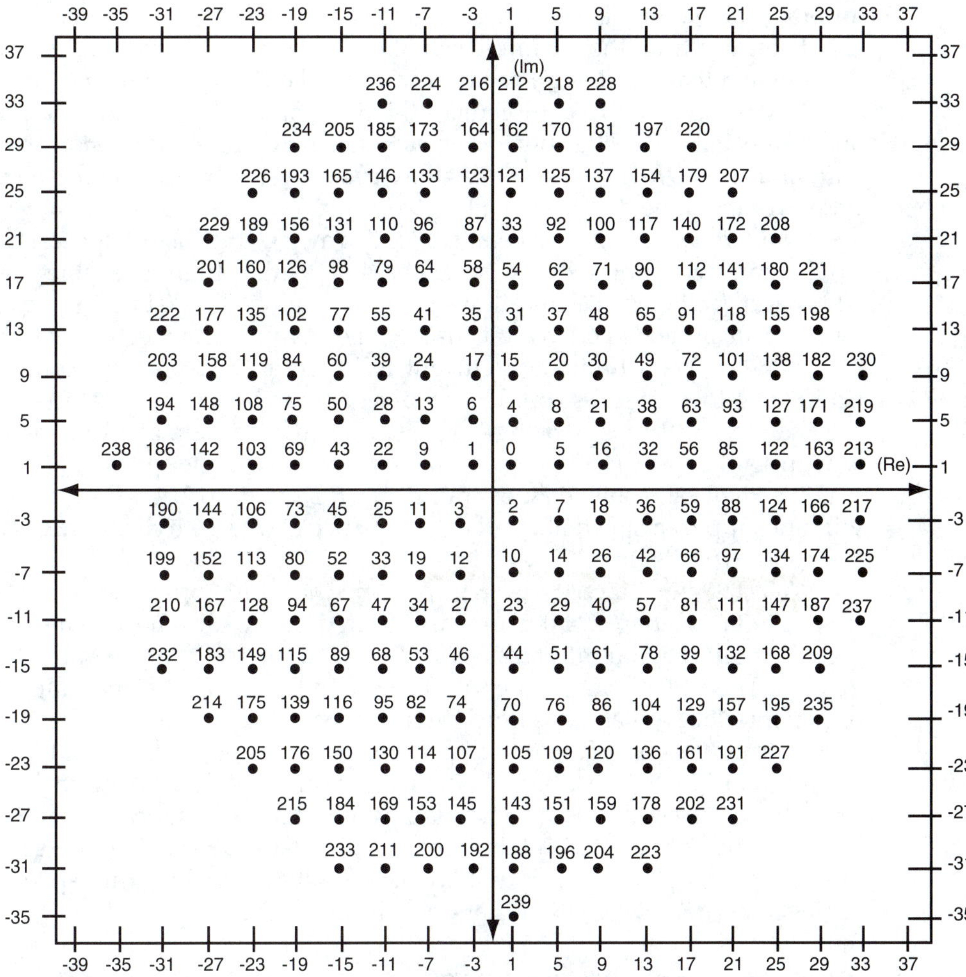

FIGURE 8-8 V.34 quarter-superconstellation with 240 signal points. The full superconstellation is obtained by rotating these points by 0°, 90°, 180°, and 270° (G. Davis Forney, "The V.34 High-Speed Modem Standard." *IEEE Communications Magazine*, Dec. 1996).

the chip; they have no idea what goes on inside the chip. Neither do I! This is the essence of chip design. You know what goes in and what comes out, but not how it achieves what it does. A 33.6-bps speed requires the use of a 1664-point constellation.

Modems contain a microprocessor, ROM, and RAM. The ROM contains software that makes the modem intelligent. Advances in microprocessor technology have allowed modems to achieve the V.34 and higher speeds. When a dial-up circuit is established, the modems send signals back and forth to set up the best possible connection between the two modems. This initial communication between modems is called *handshaking*. The V.32 bis standard established a fast retrain capability. With fast retrain, modems continually monitor the quality of the transmission facility and automatically adjust the speed of transmission up or down. The receiving modem can determine from the incoming signal if it is receiving a data call, fax, or voice call. The modem that handles all types of calls is called a voice, data, fax modem. When the PC uses software to complement the modem, the PC can automatically answer and handle all calls. The newer voice-capable modems are *dual simultaneous voice and data (DSVD).* These modems allow transmission of voice and data at the same time over the same dial-up circuit.

Modem technology has come a long way, but the fastest speed is still limited by the bandwidth of a PSTN voice channel. With a highly reliable digital PSTN, 3200 baud can be used. This allows us to transmit at 28,800 bps. Higher end-to-end speeds can be achieved by the use of file compression techniques. We are approaching a speed limit with current technology. V.42 bis data compression can achieve a 4-to-1 compression. This means data can be sent at four times the speed of the channel. At 9600 bps using V.42 bis, we can send data at 38,400 bps. At 28,800, we can send data at 115,200 bps. Modems now have a line speed of 56,000 bps. With V.42 bis, they can achieve an overall PC-to-PC speed of 224,000 bps, but the UART will not support a speed above 115,200. A special interface card is needed in the PC to support speeds up to 230,400 bps.

A 56,000-bps line speed is approaching the 64,000-bps speed of a DS0 channel. If speeds faster than 224,000 bps are desired, a digital interface higher than DS0 must be used. Digital interfaces (CSU/DSU) to the PSTN are capable of any speed desired. The CSU/DSU can be interfaced to a DS1 or DS3 signal rate. Therefore, a digital interface can achieve much higher speeds than a modem. The elimination of modems and the use of *digital subscriber lines (DSL)* make it possible to achieve rates higher than 224,000 bps. (ISDN is discussed in Chapter 9.) A basic rate ISDN line provides two 64,000-bps DS0 channels to the user. The user can combine these two channels to achieve a 128,000-bps channel. By using a 4-to-1 compression, a PC-to-PC speed of 712,000 bps is obtainable. As stated earlier, this speed cannot be achieved at present due to limitations within the PC.

8.12 CONNECTIONS TO THE INTERNET

Most connections to the Internet are made through a local ISP. The ISP provides the customer with a local telephone number for access. If the ISP is a local business, the local telephone number connects the caller to a modem at the ISP location. The modem connects the caller to a computer at the ISP. The ISP computer connects the caller to the Internet highway. The Internet highway uses synchronous transmission. The communication from our PC to the modem is asynchronous. The communication between our modem and the modem at the ISP is synchronous. The communication between the modem at the ISP and its computer is asynchronous (If the ISP is using analog line circuits and modems; but if the ISP uses digital modems via a remote access server, this connection will also be synchronous.) The computer at the ISP converts the incoming asynchronous signal from its modem to the synchronous signal needed by the Internet. The interface standard for interface to the Internet is the X.25 synchronous data protocol.

When your ISP is not a local company, connections to the ISP are made via (1) a local telephone number that connects via a modem to an X.25 packet network, or (2) an 800 telephone number connected to a modem on the ISP's premises. When ISPs provide a local number for access, the local telephone number connects to a *value-added network (VAN)* company such as Telenet. The VAN provides a packet network connection between the local telephone number and the ISP. The local telephone line connects to a modem at the VAN, and the output of the modem connects to a *packet assembler/disassembler (PAD)*. The PAD converts the asynchronous signal from a customer to the synchronous X.25 signal required by the *packet network.* The customer's data is sent over the packet data network to a PAD at the ISP location. The PAD will connect the customer to the ISP's computer. The ISP's computer will connect the customer to the Internet.

8.13 PACKET SWITCHING NETWORKS

Earlier we covered circuit switching. In a circuit switched network, we establish a circuit path through the PSTN for a call. The circuit established can be used for a voice or data call. The switched circuit is dedicated to connection of one call. The call remains on this circuit for the duration of the call. The PSTN uses circuit switching because it provides a high-quality circuit for a call.

The *public data network (PDN)* uses a *packet switched network (PSN)*. In the PSN, a call does not have to remain on the same path through the network. The PAD assembles data from customers into packets. These packets of data contain a header with the address of the intended destination and a *trailer* containing the address of the sender. The PAD sends the packet to a *data switching exchange (DSE)*. The DSE checks the address of the intended receiver and determines which route to take through the PDN. The route taken through the PDN can vary from one packet to the next. This is possible because each packet contains the destination address. Each DSE reads the address and forwards the packet to an appropriate DSE or PAD. This type of switching is called a *connectionless network* because a connection is not maintained for the duration of a call. A packet is routed through the PDN but no end-to-end connection is maintained.

It is also possible to establish connection-oriented packet circuits. Switched virtual circuits and permanent virtual circuits can be established in the PDN to achieve the equivalent of a circuit switched facility. A virtual circuit allows all packets to use the same route. The connection from a sender to a receiver is made using call setup packets. All packets will use this same setup packet. The VAN carrier determines how the data will be handled. The use of a connectionless network or a connection-oriented network is invisible to the end user. Some people are turning to the Internet to handle voice calls. The quality of voice will be better if a connection-oriented virtual circuit is used.

X.25 is the CCITT standard that defines the interface between a DTE or DCE device, and the PDN. The X.25 standard uses an EIA-232 interface to the DTE or DCE device. A PAD using the X.25 standard will accept serial asynchronous data from the DTE or DCE device and organize the data into structured frames of data for synchronous transmission over the PDN. The advantage of synchronous proto-cols is their extended error-detection and error-correction capabilities. The X.25 packet switching networks check for errors at every DSE and at the receiving PAD. Each receiver of data returns a positive acknowledgment (ACK) if the data is good and returns a negative acknowledgment (NAK) if the data contains errors. Each switch stores the data packet it sent so that it can be resent if a NAK is received. This is why the PDN is also called a *store and forward switching network*.

8.14 SIGNALING SYSTEM 7 NETWORK

The PSTN uses a packet data network for sending messages between switching centers. This network is called the *Signaling System 7 (SS7) network*. All toll switches and many class 5 local switching centers connect to the SS7 network. SS7 can only be used by computer-controlled switching systems. The CPU in one switch can send a signal to another switch simply by sending a message over the SS7 network to the CPU of the destination switch. If a switch has an originating customer making a call to a customer on another switch, the CPU of the originating switch can send a message over the SS7 network to the CPU of the destination switch in-forming it of the number being called. The distant CPU will test the line to see if it is idle. If the line is busy, it will return a message over SS7 informing the originating switch that the line is busy. The originating switch will attach a busy signal to the originating line. If the called line is idle, the terminating switch will inform the orig-inating office that the line is idle. The originating office will issue a message to switching systems in the PSTN to reserve a voice path in the PSTN for the call. The terminating office will ring the called number. The originating office will connect ring-back tone to the originating line. When the called party answers, a message is transmitted over SS7 to activate the reserved path in the PSTN.

The SS7 network is necessary for ISDN and Custom Local Area Signaling Ser-vices (CLASS) such as caller ID, selective call forwarding, and selective call rejec-tion. If a customer has caller ID and receives a call, the CPU of the receiving switch will receive the originating telephone number over the SS7 network. The CPU will then cause this number to be sent over the local loop to the receiving telephone number. Without SS7, local telephone exchanges do not receive originating tele-phone numbers from other exchanges. SS7 is an X.25 data network developed to complement computer-controlled switching systems. Computers can talk to each other by sending messages back and forth between them. SS7 provides a message link that connects the computers. It eliminates the need for the old analog signaling used with all electromechanical switching systems.

8.15 CALLER IDENTIFICATION

Caller identification (caller ID) is provided to a local class 5 central office by the SS7 network. If the line being called has subscribed to caller ID, the line translations database will contain an entry indicating this line has caller ID. When the stored pro-gram control (SPC) switching system receives an inbound call message over SS7, it looks in the database to translate the dialed telephone number into a line circuit location and is informed that the called line circuit has the caller ID feature. The SS7 network passes the originating telephone number to the receiving SPC exchange. A sender is attached to the called line circuit and transmits the caller ID information. The transmission from the sender to the line circuit is in digital format over the time division switching matrix of the SPC switch.

The codec of the line circuit will convert the digital code from the sender into an audio signal and transmit this signal over the two-wire local loop. Caller ID is simplex transmission and no response is required or expected from the customer's location. The customer provides a CPE device that contains a microprocessor and a Bell 202-compatible modem. The ITU V.23 standard defines a modem that transmits/receives 1200 bps, asymmetrical (1200/75 bps), frequency shift keying (FSK) methodology. This is compatible with the Bell 202 modem. Most manufacturers of today's high-speed modems include backward compatibility, including V.23 capability. Therefore, many modems used in a PC support caller ID. If your PC is equipped with software to handle caller ID, you can have caller ID as part of your PC communication capabilities.

Most customers provide a caller ID display unit at their residence to receive the caller ID signals. This display contains a microprocessor and a Bell 202-compatible modem chip. When the modem chip detects a ring, it does not answer the call but does receive the caller ID signal from the central office. The caller ID signal is sent between the first and second ring signal from the central office. The time frame between ring signals is 4 sec. The caller ID signal will be sent between the time frame of 1/2 sec after the first ring, to 1/2 sec before the second ring. Thus, caller ID has a 3-sec window for transmission of the caller ID message to the customer's CPE.

As discussed previously, the signals from a sender, located in the SPC exchange, are converted by the codec on the line card into audio signals that follow the Bell 202 protocol. That protocol results in the following signal being received by the CPE at the customer's location: for each 1 transmitted, a 1200-Hz signal is received, and for each 0 transmitted, a 2200-Hz signal is received. The signals are transmitted at 1200 baud with 1 bit per baud. This results in 1200 bps.

The caller ID protocol between the caller ID display and the class 5 central office uses asynchronous transmission and 8-bit data words with no parity check bit. Extended ASCII coding is used for messages and a checksum is transmitted for error-detection purposes. No provision is made for retransmission, but the checksum is used by the receiver to determine the validity of the received message. Caller ID is transmitted in the following format: (1) channel seizure signal, (2) carrier signal, (3) message type word, (4) message length word, (5) data words, and (6) checksum word.

1. The *channel seizure signal* is 30 continuous bytes of 01010101 (this is the equivalent of 85 in decimal or 55 in hexadecimal). Thirty bytes of code 85 provides a detectable alternating function to the CPE Bell-compatible 202 modem (the caller ID display).
2. The *carrier signal* consists of 130 ms of 1200-Hz tone (mark signal) to condition the receiver for data. The 130 ms of time results in 156 bits being sent.

3. The *message type word* indicates the service and capability associated with the data message. The message type word for caller ID is 04 in hexadecimal or 04 in decimal; in binary code it is 00000100.

4. The *message length word* specifies the total number of data words to follow.

5. The *data words* are encoded in ASCII and provide the following information:

   a. The first two words represent the month.

   b. The next two words represent the day of the month.

   c. The next two words represent the hour in local military time.

   d. The next two words represent the minute after the hour.

   e. The calling party's telephone number is represented by the remaining words in the data word field. If the calling party's telephone number is not available to the terminating central office, the data word field contains an ASCII0. If the calling party invokes the privacy capability, the data word field contains an ASCII *P*.

6. The *checksum word* contains the 2s complement of the module 256 sum of the other words in the data message (that is, message type, message length, and data words). The receiving equipment may calculate the module 256 sum of the received words and add this sum to the received checksum word. A result of 0 generally indicates that the message was correctly received. Message retransmission is not supported.

The caller ID message is composed of binary bits that represent some hexadecimal and some ASCII codes. Rather than show the actual binary bits transmitted to the caller ID display, the ASCII code is represented by a two-digit code. The ASCII code is represented in the following example by a number containing a tens position and a units position. The three most significant bits of the ASCII code (bits 7, 6, and 5) are represented by the tens digit. The lower 4 bits of the ASCII code (bits 4, 3, 2, and 1) are represented by the units digit. For example, the decimal digit 2 in ASCII is 0110010. The bits 7, 6, and 5 are 011 (binary code for 3), and the bits 4, 3, 2, and 1 are 0010 (binary code for 2). Thus, the decimal number 2 is ASCII code 32. Decimal 0 is ASCII code 30. The decimal numbers 0 to 9 are represented by ASCII codes 30 to 39 respectively. An example of a received caller ID message, beginning with the message type word, follows:

$$04\ 12\ 31\ 30\ 32\ 30\ 31\ 32\ 34\ 30\ 39\ 31\ 33\ 38\ 39\ 37\ 35\ 37\ 30\ 31\ 51$$

As noted earlier, some of the two-digit codes are to be converted to hexadecimal and some to decimal. Those to be converted to hexadecimal will be noted with the letter *h*.

04h = calling number delivery information code (message type word)

12h = 18 decimal; number of data words (date, time, and calling number words). Notice that after the words 04 and 12, there are 19 words. The 19th word is the error-checking word (51). Therefore, there are 18 data words in the example given.

ASCII 31, 30 = 10 = the month of October

ASCII 32, 30 = 20 = the 20th day

ASCII 31, 32 = 12 = the 12th hour = 12:00 P.M.

ASCII 34, 30 = 40 = 40 minutes after the hour (that is, 12:40 P.M.)

ASCII 39, 31, 33, 38, 39, 37, 35, 37, 30, 31 = (913) 897-5701; this is the calling party's telephone number.

51h = checksum word

8.16 SUMMARY

Data communication is the transmission of characters, numbers, graphics, and symbols using digital signals. A digital signal is a signal composed of the binary digits 1 and 0. The binary digits are represented by two different voltages. The voltage used will vary from one device to another. The EIA-232 interface standard for a serial port specifies that a 1 is represented by a voltage of –3 to –15 V, while a 0 is represented by +3 to +15 V. DTE such as a PC is used to transmit and receive digital data. DCE is used to interface a DTE device to the PSTN. When a DTE device is interfaced to a regular analog telephone line, the DCE device is a modem. When the DTE is interfaced to a digital facility, the DCE device is a CSU/DSU. Data is transferred within a PC using parallel transmission. A Pentium data bus contains 64 data leads and data is transferred 64 bits at a time. Data is transferred between a PC and an external modem using serial transmission. Serial transmission occurs over one transmission path, and bits are sent one behind the other over this serial path. The device in a PC that interfaces the parallel data bus of the PC to the serial path of an EIA-232 interface is a UART. Current UARTs can support speeds up to 115,000 bps. The UART uses asynchronous transmission.

The transmission between a PC and a modem is asynchronous transmission, where a start bit precedes each character and a stop bit follows each character. Each character is represented by either a seven-level ASCII code and a parity bit, or an eight-level ASCII code and no parity bit. Thus, asynchronous transmission requires the transmission of 10 bits for each character transmitted. The start bit provides synchronization for each byte transmitted. Synchronous transmission transmits blocks of data with no pauses between the bytes inside the block. This allows the use of a synchronization signal at the beginning of the block of data. Since the data is continuous, no additional sync signals are needed within the block. This eliminates the need for start and stop signals and speeds up the transfer of data. Communication between high-speed modems is done using synchronous transmission.

Modems contain sophisticated error-detection and error-correction techniques. Synchronous transmission allows the use of CRC-16 or CRC-32 protocols. The receiving modem notifies the transmitting modem that the blocks of data being received are okay by transmitting ACKs. When a bad block is received, a NAK along with the block number affected is sent to the transmitting modem. The transmitting modem will resend the affected block. Modems also contain a compression algorithm that converts ASCII from the PC into a shorthand code and sends the shorthand code to the distant modem, where it is converted back to ASCII. With a 4-to-1 compression, PCs can send data to the modem four times faster than the modems transmit between themselves. Using 28,800-bps modems and 4-to-1 compression allows a PC to send data to the serial port at 115,200 bps. This is the maximum speed supported by most UARTs in a PC's serial port. PCs being manufactured today contain dual UARTs in the serial port to support a speed of 230,400 bps.

Baud rate is defined as the number of times a signal changes. Baud rate is one of the components determining the physical line speed of a modem. The other component is the number of bits represented by one signal change. To transmit 9600 bps using a signal that changes 2400 times per second (2400 baud), 4 data bits must be transmitted by each baud. TCM requires an additional bit for forward error correction. To transmit 5 bits per baud requires 32 detectable signals ($2^5 = 32$). The V.34 modem requires the use of 960 different signals to transmit 9 data bits per baud. Using a baud rate of 3200, the physical data speed is 28,800 bps (9×3200). For the receiving modem to tell the difference between the 960 different signals requires a sophisticated signal logic processor in the modems, the use of sophisticated error-correction techniques, and a noise-free connection over the PSTN.

The PDN is a network of data switches that allows the public to transmit data in the form of packets. A device called the PAD is employed to interface data from a user to the packet switched network. Each packet contains the destination address in its header and the address of the sender in its trailer. The interface between the PAD and a packet network is an X.25 interface. The Internet is a packet network. An Internet service provider (ISP) will accept data from a user and reformat it into packets of data for transmission over the Internet highway.

CHAPTER REVIEW QUESTIONS

1. What is the difference between data communication and voice communication?
2. What type(s) of signals are used in data communication?
3. What is the most common connection between a DTE and a DCE device?
4. Why do we need a modem if our telephone line connects to a digital central office?
5. What are the two primary types of DCE?
6. What type of coding is used on most data communication?
7. What is serial transmission?
8. What is the difference between asynchronous and synchronous transmission?
9. What type of transmission is used between two high-speed modems?
10. What is the simplest error-detection technique?

11. What type of error detection is used between two high-speed modems?
12. What is the difference between discrete ARQ and continuous ARQ?
13. What type of error correction is provided by LAP-M?
14. What is the maximum speed of the latest UART (16C650 AFN)?
15. How many different detectable events does QAM provide?
16. How many bits are represented by one QAM signal?
17. What is baud?
18. What is the baud rate of older modems?
19. What baud rate must be used in a V.34 modem to achieve 28,800 bps?
20. How do modems achieve an overall data speed of 115,200 bps if the baud rate is only 3600 baud?
21. What is a PAD?
22. What does the X.25 standard define?
23. What facility is used by digital central offices to send signals to other digital central offices?
24. What type of modem chip is inside the CPE for caller ID?
25. What type of signaling is used between the central office and the CPE for caller ID?

PROJECTS AND CASE STUDIES:

Case Study #1

Your sister is thinking about getting a 56K modem. She has asked you for advice. Can you recommend what she should buy? When 56K modems were first introduced, there were two standards. Most people refer to these as X2 and K56. What is the difference between the two? Can we make them work with one another? If I have a 56K modem, how do I know if it will work with your 56K modem? Are 56K modems delivering as promised? How do I know if my telephone line will allow a 56K modem to work at 56 Kbps? Can I use 56K modems on a SLC-96 subscriber carrier system? Will they work over a loaded cable pair?

GLOSSARY

ACK   A positive acknowledgment (ACK) that the data was good. The receiving modem sends an ACK to the transmitting modem to indicate that the data is being received okay.

American Standard Code for Information Interchange (ASCII)   The use of a seven-level binary code to represent the letters of the alphabet and the numerals 0 to 9.

Analog Signal   An electrical signal analogous (similar) to a voice signal. The signal continuously varies in amplitude and frequency. An analog signal has an infinite number of values for voltage, current, and frequency.

Asymmetric Digital Subscriber Line (ADSL)   A digital subscriber line technology that can be used to deliver T1 speeds over the local loop. An ADSL has a high bit rate in one direction and a low bit rate in the opposite direction.

Asynchronous Transmission   Also called *start-and-stop transmission.* The transmitting device sends a start bit prior to each character and sends a stop bit after each character. The receiving device will synchronize from the received start bits. Thus, synchronization occurs at the beginning of each character. Data is sent between two devices as a serial bit stream.

Automatic Retransmission Request (ARQ)   The method of error detection used between high-speed modems. ARQ is either discrete ARQ or continuous ARQ. The receiving device returns a positive acknowledgment (ACK) when it receives a good block of data, and returns a negative acknowledgment (NAK) when the block of data received contains an error.

Baud   The number of times a signal changes its state. If the amplitude of a signal changes 2400 times a second, the signal changes states 2400 times per second or at 2400 baud. If a signal changes back and forth between a frequency of 1800 Hz and 2200 Hz 3200 times per second, the baud rate is 3200. If a signal changes phase 2400 times a second, the baud rate is 2400 baud. Many people confuse baud rate and bit rate. Even some terminal emulation programs state that they will provide an option to change baud rates from 9200 to 19,200. This is incorrect. They should state

that you can change the bit rate from 9200 to 19,200 bps. The selection of baud rate is done automatically by modems.

Binary Signal   See *Digital Signal.*

Channel Service Unit (CSU)/Data Service Unit (DSU)   Also called *customer service unit/data service unit.* The CSU/DSU is a DCE device used to interface a computer to a digital leased line. It is a digital-to-digital interface. It can connect a low-speed digital device to a high-speed digital highway.

Continuous ARQ   Also known as *sliding window ARQ.* Continuous ARQ eliminates the need for a transmitting device to wait for ACKs after each block of data. The device continuously transmits blocks of data and sends a block number with each block. The receiving device checks the blocks for errors and continuously returns positive ACKs to the transmitter. If an error in data occurs, the receiving device returns a NAK along with the block number affected. On receipt of a NAK, the transmitting device will retransmit the bad block of data.

Cyclic Redundancy Check (CRC)   A form of error checking used between modems. A transmitting modem treats the block of data transmitted as representing a large binary number. This number is divided by a 17-bit divisor (CRC-16) or a 33-bit divisor (CRC-32). The remainder is attached in the trailer behind the block of data. The receiving modem performs the same division on the block of data received and compares its calculated remainder to the remainder sent in the trailer.

Data   Raw facts, characters, numbers, and so on that have little or no meaning in themselves. When data is processed, it becomes information. In telecommunications when we speak of data, we are referring to information represented by digital codes.

Data Communication Equipment (DCE) Also called data circuit termination equipment. A device that interfaces data terminal equipment (DTE) to the PSTN. A modem is a DCE used to interface a DTE to an analog line circuit on the PSTN. A CSU/DSU is a DCE device used to interface a DTE device to a leased digital line in the PSTN.

Data Switching Exchange (DSE) Also known as *packet switching exchange*. These are the switches of a packet switched network. The DSE routes packets of information based on the packet address information found in each packet header.

Data Terminal Equipment (DTE) A DTE device is used to transmit and receive data in the form of digital signals. The personal computer is the most common form of DTE.

Digital Signal An electrical signal that has two states. The two states may be represented by voltage or current. For example, the presence of voltage could represent a digital logic of 1, while the absence of voltage could represent a digital logic of 0.

Discrete ARQ Also called *stop-and-wait ARQ*. An error control protocol that requires an acknowledgment from the receiver after each block of data sent. The transmitting modem sends a block of data and then waits for an acknowledgment before sending the next block of data.

Dual Simultaneous Voice and Data (DSVD) A modem that can handle voice and data at the same time.

EIA-232 Interface Also called *RS-232*. The most common interface standard for data communication. CCITT standard V.24 is the same as EIA-232. EIA-232 defines the voltage levels needed for the various signal leads of EIA-232.

Flow Control Controlling the flow of data from one device to another, usually via hardware flow control (RTS/CTS) or by software flow control (XON/XOFF). A modem contains a memory buffer to allow it to compress data before transmitting and to allow it to convert asynchronous data to synchronous data. It must be able to stop the transmitting PC when this memory buffer approaches a near-full condition. Some modems contain a very large memory buffer, which negates the need for flow control.

Full-Duplex Transmission If a transmission system allows signals to be transmitted in both directions at the same time, the system is called a *full-duplex transmission system*.

Half-Duplex Transmission A type of transmission where transmitters on each end of a medium take turns sending over the same medium.

Header A term describing the placement of control information in front of a block of data transmitted using synchronous transmission. The header contains a beginning flag (or sync signal), the destination address, and a control field.

Integrated Services Digital Network (ISDN) The use of digital line circuits to provide end-to-end digital service. The *basic rate interface (BRI)* provides the user with two DS0 channels. The *primary rate interface (PRI)* provides the user with 23 DS0 channels.

Internet Service Provider (ISP) A company that provides individuals and companies with access to the Internet.

Modem A DCE device that interfaces the digital signal from a DTE device to the analog local loop and line circuit of the PSTN, by converting the digital signals into modulations of an analog signal.

NAK   A negative acknowledgment transmitted by the receiving modem when it detects that an error has occurred in a block of transmitted data. If a NAK is received, the modem retransmits the last block of data from its memory.

Null Modem   No modem. A null-modem cable is used to connect two PCs via their serial ports, when they are connected directly without using a modem.

Packet Assembler/Disassembler (PAD)   The device that interfaces data to a packet network. The PAD accepts data from a user and arranges the data into packets that can be processed by the packet switching network.

Packet Network   A data network that transmits packets of data

Parallel Transmission   The use of several transmission leads to allow the simultaneous transmission of several bits at one time. A parallel data bus with 8 leads can process data 8 bits at a time. Or 64 leads on the data bus allows information to be processed 64 bits at a time. Parallel transmission is much faster than serial transmission but requires many more transmission leads.

Protocol   The rules of communication. Each protocol defines a formal procedure for how data is to be transmitted and received using that protocol.

Public Data Network (PDN)   The packet data network accessible to the general public for the transmission of packet data.

Quadrature Amplitude Modulation (QAM)   The transmission of 4 bits per baud. A V.29 modem transmits at 2400 baud with 4 bits

per baud to achieve 9600 bps transmission. QAM uses 2 different amplitudes for each of 8 different phases of a 1700-Hz signal to achieve the 16 detectable events necessary to code and decode 4 bits at a time.

Recommended Standard 232 (RS-232)   The most common interface standard for data communication. CCITT standard V.24 is the same as RS-232. RS-232 defines the voltage levels needed for the various signal leads of RS-232.

Selective ARQ   With selective ARQ, only frames that have errors are retransmitted. When a transmitting modem is using selective ARQ and receives a NAK, only the frame for which the NAK was received is retransmitted.

Serial Transmission   The transmission of bits, one behind the other, over one transmission medium.

Synchronous Transmission   The transmission of data as blocks of bytes. Synchronization of the receiver occurs from a special bit pattern called a *sync signal* placed in front of the block of data information.

Trailer   A trailer is used in synchronous communication; it is data placed behind the block of information transmitted. The trailer contains parity checking information and the address of the sender.

Trellis Coded Modulation (TCM)   TCM is a forward error-correction technique. An extra bit is added to the bits of data transmitted to help the receiver decode the data more reliably. QAM transmits 4 bits at a time. TCM added a 5th bit to the 4-bit code for error-correction purposes.

Universal Asynchronous Receiver/Transmitter (UART)   A piece of hardware (an integrated circuit chip) whose purpose is to interface a device using parallel transmission to a device using serial transmission. Every serial port contains a UART between the parallel data bus of the PC (or modem) and the serial port.

Word In a PC, the number of adjacent bits that can be manipulated or processed. This depends on the number of leads comprising the data bus, which is in turn dependent on the number of registers attached to the data bus. A word can be 8 bits or 1 byte in an 8088 microprocessor environment. A word can be 32 bits or 4 bytes in a 80486-based PC, or it can be 64 bits (8 bytes) in a Pentium-based PC.

X.25 The interface standard to a packet data network.

9

ISDN and ADSL

KEY TERMS

Asymmetric Digital Subscriber
   Line (ADSL)
Basic Rate Interface (BRI)
B-Channel
Broadband ISDN (B-ISDN)
D-Channel
Digital Subscriber Line Circuit
   (DSL)
Hypertext Markup Language
   (HTML)

Integrated Services Digital
   Network (ISDN)
Internet Service Provider (ISP)
Link Access Procedure on the D-
   Channel (LAPD)
Narrowband ISDN
Network Termination 1 (NT1)
Network Termination 2 (NT2)
Network Termination 12 (NT12)

Primary Rate Interface (PRI)
Reference Point
Service Profile Identifier (SPID)
Terminal Adapter (TA)
Terminal Equipment Identifier
   (TEI)
Terminal Equipment 1 (TE1)
Terminal Equipment 2 (TE2)
2B1Q

OBJECTIVES

On completion of this chapter, you should be able to:

1   Explain what ISDN is.
2   Explain what ADSL is.
3   Explain the 2B1Q signaling used by basic rate ISDN (BRI).
4   Describe the various interfaces in a BRI environment.
5   Describe what a terminal equipment identifier is used for.
6   Describe what the service profile identifier is.
7   Describe link access procedure on the D-channel (LAP-D).
8   Discuss the two different signaling protocols used by ADSL modems.
9   Define a SPID and SAPI.
10   Compare and contrast BRI, PRI, and ADSL.

Integrated Services Digital Network (ISDN) is a network providing end-to-end digital connectivity; it supports a wide range of services such as voice, data, and video over one facility. Prior to the introduction of ISDN, different interfaces were needed for each of these services, so that each interface was tailored to provide one specific service. The public switched telephone network (PSTN) was developed to transmit analog electrical signals from one phone to another. The phone converted analog sound waves into analog electrical signals for transmission over the PSTN. In fact, anything to be transmitted over the PSTN had to be in an analog signal form. The local loop was designed for efficient transmission of analog signals between 300 and 3300 Hz. Voice signals fall in this range. The PSTN was designed to handle voice signals, but the network has gradually evolved into a digital circuit network. Analog voice signals are converted into digital signals by a codec on the line card and carried by the digital PSTN.

9.1 ANALOG VIDEO

Video signals are also analog in nature. They are higher-frequency signals than voice and occupy a much greater bandwidth. Video signals used to be transmitted from a television studio to the telephone exchange over local-loop cable pairs that were specially treated. These cable pairs were not loaded. Amplifiers and equalizers were placed at intervals along the cable route to keep the analog signal at a proper power level. At the central exchange, the cable pair from the studio was connected to a cable pair that went to the transmitter and antenna site. This pair also had amplifiers and equalizers placed at intervals along its route. Television networks such as NBC, CBS, and ABC would provide programming to local stations via the frequency division multiplexing (FDM) systems in the old analog PSTN. Special wide-bandwidth channels on the FDM systems were used to get these signals from the network (ABC, CBS, NBC, and so on) to the class 4 toll office serving the local affiliate television station. The signal was connected over cable pairs, from the class 4 office tc the local class 5 exchange, for connection to cable pairs serving the local affiliate TV control center.

Now that the PSTN has been converted into a digital network, video signals are converted into digital signals and transmitted digitally over the digital PSTN. Digitized video signals are converted back into analog signals at the transmitter site and are used to frequency modulate the carrier frequency of the transmitter. Many TV networks and TV stations no longer use the PSTN or local telephone company for transmission of their signals. Network affiliate stations receive signals from the networks via communication satellites. Some mobile units use satellite dishes to beam signals to a satellite. The signal is sent by the satellite to a receiving dish at the TV studio. Local stations also have mobile units that use microwave transmitters and dishes for local remotes. Most local stations transmit signals from their studio to their transmitter site using a microwave dish.

9.2 THE OLD ANALOG PSTN AND DATA

When the PSTN was an analog network, digital data could not be transmitted over the PSTN without first being converted into analog signals. Modems were developed to convert the digital data signals into analog signals that were within the 300- to 3300-Hz range. This conversion process resulted in an analog signal that could be handled by the analog PSTN. The older FDM facilities used in the long distance network portion of the PSTN had a 4000-Hz bandwidth for each voice channel, but some of that channel bandwidth was used for guardbands and signaling. The effective channel bandwidth for the voice signal was 300 to 3300 Hz. As you know, modems are low-speed devices that allow for low-speed dial-up connections between computers. The speed of dial-up modems has improved tremendously and allows for speeds of up to 56,000 bps, but the local loop, or the circuit path in the PSTN, will often limit the speed to less than 28,800 bps. Modems automatically test the circuit being used when the circuit is established and if it cannot support 56,000 bps, the modems will automatically select a lower speed that the circuit can handle.

Using an analog signal and modem for high-speed data requires a high-bandwidth and a high-frequency analog signal. When analog FDM was used to carry the signal over the PSTN, multiple FDM channels were bonded together to support the high-frequency signal. High-speed data being carried by a high-frequency analog signal cannot be carried over a regular dial-up line. This data requires that an end-to-end circuit be leased from a common carrier. These leased circuits are called *private lines* because they cannot be accessed by the general public. The private line is intended for use by a few individuals, concerns, or companies. A private-line facility is required for high-speed data because high-frequency signals require special treatment of the local loop. Special conditioning of the local loop is referred to as *C conditioning*. There are five grades of C conditioning: C1, C2, C3, C4, and C5. The grade of conditioning to use depends on how high a frequency the loop needs to handle. Special conditioning of the local loop to reduce noise is called *D conditioning*.

Conditioning the local loop improves its ability to handle high-frequency signals and reduces the power loss to a signal placed over the conditioned loop. When data is carried by an analog signal, the higher the required data speed, the higher the analog frequency has to be. In some cases the local-loop facility needs treatment with amplifiers and equalizers similar to the treatment used for analog video signals. Today, high-speed data is carried as a digital signal by using multiple DS0 or DS1 facilities in the digital PSTN. If customers need to transmit high volumes of data from one location to another, they usually decide to transmit the data as a digital rather than an analog signal and will lease a T1 carrier facility from the LEC.

If a company requires a high-speed data line, it needs to lease a private data line over the PSTN. In the past, if an insurance company had to transmit high volumes of data between its headquarters in Chicago and a regional office in Atlanta, it had to lease special wide-bandwidth channels on the older FDM PSTN between Chicago and Atlanta from AT&T. Additionally, the LEC had to provide C1 or C2 conditioned local loops in each city for connecting the insurance computers to the AT&T toll exchanges. Now that the toll network is digital, modems are not needed. Low-speed data can connect to a digital local loop using a *customer service unit/data service unit (CSU/DSU)* instead of a modem. The loop is conditioned to handle digital data, and the data is interfaced to the PSTN using the CSU/DSU. High-speed data lines can be connected via a CSU/DSU to a T1 TDM system on the customer's premises. This T1 system can connect directly to leased T1 systems in the PSTN. This type of facility handles only data, and many companies use this arrangement for their data lines. Most residential customers need to use their line for both voice and data. Therefore, they order an analog voice line from the LEC. When they wish to use this line for data, they use a modem.

9.3 CONVERSION OF THE PSTN TO A DIGITAL NETWORK

In the past, modems were needed for conversion of digital signals to analog signals because the PSTN was an analog network. The PSTN is no longer an analog network, but a digital network. However, modems must still be used by regular phone lines because the phone lines are still analog. Engineers developed ways to convert analog signals into digital codes and found that transmitting digitized voice was more efficient and resulted in a higher-quality network than the AM/FDM network. Initially, the analog-to-digital conversion was done at the channel units of TDM carrier systems using a codec in the channel unit. Manufacturers of switching systems caught on to TDM technology and moved the codec from the trunk circuit (channel unit) to the line circuit.

Although a few analog switching systems still exist, the toll switches are 100% digital, and most local switching is also digital. Most IEC networks use digital transmission systems. The only part of the PSTN that remains analog is the local loop and the line circuit. The line circuit in a digital central office has a codec to convert analog signals into digital signals at 64 Kbps (DS0). This line circuit is designed as an interface to an analog local loop. It expects to see an analog signal. The codec on the analog line circuit does not expect to see a digital signal and cannot properly code a digital input. A different line circuit (called a *digital subscriber line circuit*—for example, ISDN) is needed to accept a digital signal from the local loop.

Since the standard line circuit interface of a switch expects to see an analog signal input from a local loop, when a local line circuit is used for switched data, a modem is still required. A modem on the customer's premises converts the digital signal of a computer into an analog signal that can be handled by the local loop *and the codec of the line circuit.* The line circuit codec converts the analog signal back into a digital signal. To prevent all this conversion and reconversion, it would be necessary to move the codec from the line circuit to the telephone (or other station apparatus), condition the local loop to handle the digital signal, and use a line card with a digital interface to the local loop. ISDN does exactly that. The codec resides in the ISDN telephone and converts voice signals into digital signals. The telephone also provides an EIA-232 digital interface. The digital output from a PC can be connected to this port. The ISDN telephone can be used on a voice call while the PC is connected on a data call. Both calls will be handled by the single digital line card and local loop.

As stated earlier, the local loop can receive special treatment that will allow it to handle digital signals. TDM T1 carrier systems use a digital signal (DS1), and cable pairs serve as the transmission medium for the DS1 signal. The DS1 signal used by T1 is a digital signal consisting of a serial data bit stream at 1.544 Mbps. To handle this signal, the cable pair is unloaded and signal regenerators are placed at 6000-ft intervals along the cable route. Each end of the cable pair has a TDM multiplexer/demultiplexer attached to the cable pair. Cable pairs over 18,000 ft can be treated to handle either a digital signal or an analog signal but not both. Cable pairs under 18,000 ft are not loaded. Therefore, cable pairs under 18,000 ft can be used as an analog voice line or a digital subscriber line without special treatment (see Chapter 6). The majority (more than 75%) of all local loops are under 18,000 ft, and these cable pairs can be used as an analog line or a digital line.

Many people are buying ISDN modems for their PCs. They make arrangements with their *Internet service provider (ISP)* for connections with the ISP via ISDN. These customers purchase ISDN service from the LEC and are provided service via an ISDN line circuit in the central office. The ISP is also connected to ISDN line circuits at its serving central office. When an ISDN connection is provided between the

ISDN modems at the customer's PC and the ISP's computer, data can be transferred at 128,000 bps between the two ISDN modems. The ISDN modem in a PC also has several RJ-11 jacks to allow for the connection of telephones or fax machines. Because ISDN is a digital service, the local loop must be an unloaded cable pair and no more than 18,000 ft in length. As discussed in Chapter 4, ISDN is provided to remote areas of an exchange by using subscriber carrier (SLC-96) to establish a carrier service area that can support DS0 services such as ISDN.

9.4 IS ISDN A NECESSARY TECHNOLOGY?

The PSTN has evolved from an analog network into a digital network. Data communication professionals are not very professional when they criticize the PSTN as a voice-only network. The only portion of the PSTN that remains analog is the local loop, and it has been the exchange carriers that have been in the forefront of ISDN development. Many people have criticized the telecommunications industry as pushing ISDN, simply to extend the evolution of the PSTN into an all-digital network. They imply that ISDN was creating a solution where none was needed and that the PSTN is excellent as is, with no need to convert the local loop to digital: "If it ain't broke, don't fix it." ISDN has been called: "*I Still Don't kNow what it's for*," "*It Still Does Nothing*," and so on. If the PSTN is to handle voice only, these critics would be correct. But clearly everyone is aware of the explosion that has occurred in the data world. The PSTN must be designed to handle voice, video, and data. ISDN does offer a solution for this concept.

Opponents of ISDN point to advances in modem design that have provided fast data speed capabilities as a reason ISDN is not needed. The speeds of new modem designs are approaching the speed ISDN offers. Most computers are now equipped with 28,800-bps modems, and many people are deploying new modems with speeds of up to 56,000 bps. These modems provide speeds that allow for fast transfers of files between the ISP and a PC. With these modems, the local loop could remain analog, as long as users wished to use the facility only for voice and connection of PCs to the network. One reason ISDN has not been accepted widely in the United States is that most people have not needed the features it provides. The Internet explosion will change that. Many people are using the Internet, and in a few years everyone will want Internet access. Internet is a packet data network. When sites on the Internet are accessed, large graphic files are transferred from the Internet site to your PC. The PC uses these graphic files to paint the screen. Fast data speeds are needed to minimize wait time between screen changes.

ISDN provides high-speed access to the Internet. This is an excellent application for ISDN, but many ISPs do not offer ISDN access lines. Here is another example of where ISDN is a good solution, but deployment of this application is hindered by high cost and the lack of support from the data community. ISDN provides faster transfer of graphic files than modems do. Opponents of ISDN will cite the explosion that is going to occur in cable modems. With deregulation of local exchange services, cable TV companies are entering the telecommunications business and Internet access market. The wide bandwidth of their coaxial cables allows them to carry high-speed signals. Cable modems will attach to the cable TV facilities and will provide up to 27-Mbps download capabilities and 1-Mbps upload capabilities. This is about 60 times faster than ISDN and 300 times faster than a 28.8-Kbps modem. Cable modem service will cost about $40 a month. BRI ISDN costs about $100 a month.

The deployment of ISDN has been badly handled by everyone involved. Few applications were developed for use of ISDN and it was not widely deployed. It does not do me any good to have ISDN if people I communicate with directly cannot (or will not) get ISDN. As mentioned above, for ISDN to work, I must connect to another ISDN line. When I use the Internet, my ISP interfaces my ISDN line to the X.25 Internet highway and takes care of the protocol conversions needed. If I were to use a regular analog phone line and modem, the ISP would also do a protocol conversion for me. I can communicate with anyone on the Internet and use e-mail because the ISPs take care of protocol conversions. The speed with which *Hypertext Markup Language (HTML)* files are transferred to me depends on the speed of the slowest link. The speed of the on/off ramps to the Internet govern the speed of data transfer.

The window of opportunity for ISDN technology may have come and gone. The LECs should hope this is not the case. I predict that the general public will be demanding digital connectivity by 2005. ISDN is an international standard and appears to be the only affordable technology that LECs can use to meet this demand. The Bell LECs know the future of the local loop depends on its conversion to digital. The first step in the conversion of the local loop to digital will be to replace feeder and distribution cables with fiber.

Many local switching systems have been converted from in-band to out-of-band signaling. This conversion is necessary for three reasons: (1) SS7 is needed to offer Custom Local Area Signaling Services (CLASS) such as caller ID, (2) SS7 is necessary to offer ISDN, and (3) in-band signaling restricts the DS0 channel to 56-Kbps data. Out-of-band signaling is needed to eliminate the bit-robbing signaling technique of in-band signaling and return the DS0 to a full 64-Kbps clear channel for data. In addition to installing SS7 connection to all class 5 exchanges, many LECs have plans to convert from wire, as the local loop, to fiber and coaxial cable. Pacific Bell has anticipated that it will have provided fiber or coaxial cable to 6.5 million homes and businesses in the near future.

9.5 DEVELOPMENT OF STANDARDS

ISDN was initially offered by the LECs in 1986, but because few applications had been developed that could take advantage of the technology, it did not take off. Another reason for the failure of ISDN was a lack of standardization. AT&T, Northern Telecom, Siemens, and other switching manufacturers had their own unique proprietary standard for ISDN. As a result, ISDN station equipment and CPE had to be purchased from the switching manufacturer. An ISDN device designed to work on an AT&T switch would not work on a Northern Telecom switch. A standard ISDN architecture was developed by CCITT in 1984, updated in 1988, and updated again in 1992. These standards are based on the OSI model. The standards are not for the network. They are *interface* standards to the ISDN network.

When a network interface standard was implemented, all switching systems were forced to use the same protocols regardless of manufacturer. Customer services are delivered over two standard interfaces, the **basic rate interface (BRI)**, for basic services, and **primary rate interface (PRI)**, to provide higher transmission capabilities. BRI is the lowest-level ISDN interface. BRI provides four logical two-way digital circuits over one physical cable pair between the customer's location and the central exchange. The physical cable pair is a *digital subscriber line (DSL)*. The **DSL circuit** is a two-way digital circuit operating at 160 Kbps. The ISDN line circuit at the central exchange and the terminal device at the customer's location divide this 160-Kbps bit stream into four logical data channels. Three of these channels (144 Kbps) are used by the customer. The fourth channel (16 Kbps) is used by the hardware of the ISDN line circuit as a maintenance (M) channel. Two of the user channels are 64 Kbps each and are called the *bearer (B) channels*. These B-channels can be used for either voice or data. The third channel is called the *delta (D) channel* and is a 16-Kbps channel that operates only in a packet data mode.

Because BRI consists of two bearer channels and one delta channel, it is called *2B+D*. The **D-channel** provides signaling and control for the two B-channels. The D-channel is used as a vehicle for packet data transport. When most people talk about BRI, they refer to it as 144 Kbps. This is the total user rate for the 2B+D user channels. The D-channel can be used to transport user data in a low-speed packet mode, but most often the D-channel is used only for call setup, and for communication messages between the ISDN devices on each end of the circuit. The user is left with 128 Kbps over the B-channels for use as voice or data. The **B-channels** are called *bearer channels* because they carry the customer's voice or data. The fourth channel is a 16-Kbps channel called the *maintenance* or *overhead channel*. The 16 Kbps are used to support performance monitoring, framing, and timing functions between the customer-premise equipment and the equipment at the central exchange. The total bit rate for all four channels is 160 Kbps.

9.6 BRI AND DSL STANDARDS

Prior to standardization, the BRI supplied by AT&T had 48 Kbps overhead and had a 192-Kbps line rate. The new standard digital line rate of 160 Kbps matches the standard developed by ANSI (T1.601-1988) for a DSL. The DSL standard provides customers with digital access to a digital local exchange using a nonloaded, two-wire, local-loop cable pair. The DSL standard provides the 144 Kbps throughput needed for basic ISDN, plus the 16 Kbps needed as overhead for framing, synchronization, and monitoring. In 1986, ANSI chose to base its basic ISDN standard interface to the local loop on echo canceling technology, using the 2B1Q line coding technique. This decision was reached after more than a year of study on the properties of various line coding techniques such as AMI, MDB, 4B3T, 3B2T, and biphase. These were standards being proposed by some of the world's leading research firms, and some of these coding techniques were already in use by different manufacturers as part of their proprietary standards. After exhaustive laboratory experiments and field trials of the various coding techniques, 2B1Q was found to provide the best service on the various loops in a typical telephone exchange territory. With 2B1Q, the line rate is actually 80 Kbaud because 2 bits are transmitted on each baud to achieve 160 Kbps. Prior to 2B1Q, the local exchange interface to the local loop for

ISDN was 192 Kbps using AMI line coding, which resulted in a 96-kilobaud or Kbaud signal. This provided a user rate of 144 Kbps plus 48 Kbps overhead.

9.7 2B1Q LINE CODING

2B1Q coding is the optimum digital subscriber line technology. The line code is based on *pulse amplitude modulation (PAM)* technology, which takes 2 binary bits and converts them into a multilevel analog signal for transmission, using PAM in a *time division multiplex (TDM)* signal, over the analog local loop. 2B1Q is a four-level code; it codes 2 bits at a time into one of four amplitude levels:

1. If the first bit is a 1 and the second bit is a 0, transmit a +2.5-V pulse.
2. If the first bit is a 1 and the second bit is a 1, transmit a +0.833-V pulse.
3. If the first bit is a 0 and the second bit is a 1, transmit a –0.833-V pulse.
4. If the first bit is a 0 and the second bit is a 0, transmit a –2.5-V pulse.

These four codes are called *quaternary symbols* or *quats*. The first bit determines whether the transmitted pulse amplitude is positive or negative. If the first bit is a 1, the pulse has a positive amplitude. If the first bit is 0, the pulse has a negative amplitude. The second bit determines the level (or amplitude) of the pulse. If the second bit is a 0, the level is 2.5 V. If the second bit is a 1, the level of the pulse is 0.833 V. Notice that *2B1Q* means 2 bits = 1 quat (see Figure 9-1).

The use of 2B1Q line coding reduces the frequency of the line signaling from 160 Kbaud (at 1 bit per baud) to 80 Kbaud (with 2 bits per baud). Local loops are considered to act like low-pass filters and were designed to handle low-frequency signals. Attenuation of signals varies directly with the frequency of the signal transmitted. Lower-frequency signals have less attenuation and can work on longer loops. Lower frequencies also reduce crosstalk into adjacent cable pairs. The lower frequency of 2B1Q makes it an excellent choice for DSL.

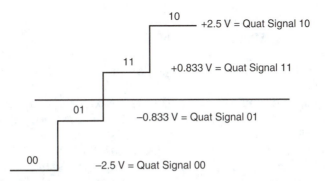

FIGURE 9-1 2B1Q Symbol voltage levels. The diagram indicates how 2B1Q sends 2 bits (a dibit) with each signal voltage. Since four different signal voltage levels are used, it is called *quaternary signaling* and is represented by the letter *Q*. Thus, 2 bits = 1Q (2B1Q). At 80 Kbaud, we transmit 160,000 bps using 2B1Q. The 2B1Q signal is a PAM/TDM signal.

The 2B1Q coding process does not result in the same number of positive and negative pulses on the transmission line, and so 2B1Q includes a scrambling algorithm to achieve the same number of positive and negative pulses on the line. This reduces line current to 0 and allows the use of a longer loop. 2B1Q allows a DSL to work on a cable pair up to 18,000 ft from the central office, if the cable pair is not loaded and has no bridge taps. A bridge tap exists when a feeder cable pair connects to two different distribution cables. The working distance of basic ISDN can be extended beyond 18,000 ft if the DSL is treated by placing a repeater in the line. DSL will work in any cable with regular customer lines, but it will not work properly on a cable pair in close proximity to a cable pair handling an analog subscriber carrier.

The DSL line format is based on the transmission of quats. The basic frame consists of transmitting 120 quats in 1.5 ms. In addition, 80 quats in 1 ms represents 80 Kbaud. At 80 Kbaud (the line rate for DSL) per second, we transmit 120 quats in 1.5 ms. One DSL frame requires 1.5 ms. Each frame contains 240 bits. Since each frame is 1.5 ms long (0.0015 sec), we can achieve 666.66667 frames per second ($F = 1/T$:$F = 1/0.0015 = 666.66667$). With 240 bits in each frame, we achieve 160 Kbps per second (240 bits per frame × 666.66667 frames per second = 160,000 bps).

Each 240-bit DSL frame contains the following bit allocations: bits 1 to 18 for a Sync Word (SW); bits 19 to 234 for information transport of twelve 2B+D channel samples (each B sample is 8 bits and each D sample is 2 bits). The total bits needed for these twelve 2B+D samples is 216 bits ($12 \times (2 \times 8) + 2) = 12 \times 18 = 216$); bits 235 to 240 are used for 6 bits of M overhead channel. Total bits in a frame = 18 + 216 + 6 = 240 bits. Of course, the 240 bits in a frame calculation can also be arrived at by another method. Remember that the basic frame consists of 120 quats. Each quat represents 2 bits. Therefore, $120 \times 2 = 240$ bits per frame (see Figure 9-2).

The Sync Word is used for physical layer synchronization and frame alignment. The M-channel is used for signaling, error detection, and maintenance messages. Since there are eight frames in a superframe, the 6 overhead bits used in each frame for the M-channel provide a total of 48 bits per superframe for the M-channel. Since there are 666.66667 frames per second, 6 bits per frame times 666.66667 frames per second provides 4 Kbps for the M-channel. Notice that the M-channel and Sync Word taken together comprise 24 bits per frame (18 for Sync Word + 6 for M-channel = 24 bits). This provides the 16 Kbps referred to earlier as the maintenance and synchronization channel ($24 \times 666.66667 = 16$ Kbps). Also notice that the 216 bits in each frame for 2B+D = 144 Kbps ($216 \times 666.66667 = 144,000$).

Eight frames are grouped together to form a superframe. Since each frame is 0.0015 sec long, a superframe is 0.012 sec (12 ms) long ($8 \times 0.0015 = 0.012$). We can transmit 83.333 superframes per second. Each superframe contains $240 \times 8 = 1920$ bits, and we transmit 160,000 bps ($1920 \times 83.333 = 160,000$). Superframes are transmitted over one wire of the local loop from the customer to the central office and are superframes no. 1, 2, 3, . . . , 8. Superframes are also transmitted from the central office to the customer over the second wire of local loop and are called superframes A, B, C, . . . , H (see Figure 9-3).

Time:	<- - - - - - - - - - - - - - .0015 sec (or 1.5 ms.) - - - - - - - - - - - - - - - - - ->								
Framing	**Framing**	**2B+D**	**Overhead or Maintenance bits (M1–M6)**						
Quat No.	1 to 9	10 to 117	118s	118m	119s	119m	120s	120m	
Bit Position	1 to 18	19 to 234	M1 235	M2 236	M3 237	M4 238	M5 239	M6 240	
Frame 1	Inverted SW	**2B+D**	EOC	EOC	EOC	ACT	1	1	1.5 ms
Frame 2	SW	**2B+D**	EOC	EOC	EOC	1	1	FEBE	1.5 ms
Frame 3	SW	**2B+D**	EOC	EOC	EOC	1	crc1	crc2	1.5 ms
Frame 4	SW	**2B+D**	EOC	EOC	EOC	1	crc3	crc4	1.5 ms
Frame 5	SW	**2B+D**	EOC	EOC	EOC	1	crc5	crc6	1.5 ms
Frame 6	SW	**2B+D**	EOC	EOC	EOC	1	crc7	crc8	1.5 ms
Frame 7	SW	**2B+D**	EOC	EOC	EOC	1	crc9	crc10	1.5 ms
Frame 8	SW	**2B+D**	EOC	EOC	EOC	1	crc11	crc12	1.5 ms

Total Time for 8 frames (i.e., 1 superframe) = 120.0 ms

One Superframe (There are 8 superframes from the central office ISDN line circuit to the customer's NT1. They are called superframes A, B, C, D, E, F, G, and H. There are also eight superframes from the NT1 back to the ISDN line circuit. These return frames are called superframes 1, 2, 3, 4, 5, 6, 7, and 8.)

NOTE:
SW = sync word; EOC = embedded operations channel; s = space bit; m = mark bit; CRC = cyclic redundancy check for 2B+D and M4; ACT = activation bit; 1 = reserved bit; FEBE = far end block error bit. Note that in 1.5 ms, we acheive the transmission of 120 quats, which is 240 bits.

FIGURE 9-2 2B1Q superframe with detail of M-channel. The chart lays out the composition of a super-frame. Note that there are 240 bit positions in each and every frame from frame 1 to frame 8. Since each of the eight frames contains 240 bits, a superframe (which contains eight frames) contains 8×240 or 1920 bits. Each of the eight frames in a superframe takes 0.0015 sec (1.5 ms) to transmit; therefore the superframe is 8×1.5 ms or 12 ms long. This figure represents one digital subscriber line (DSL) superframe and the bit assignments when the DSL uses 2B1Q protocol. Note that the first 18 bits in each frame (the Sync Word) is used for physical layer synchronization between the customer's NT1 and the central office ISDN line circuit. The last 6 bits in each frame are used to establish a maintenance (M) channel. Bits 19 to 237 inclusive provides 2165 bits used to carry 2B+D information.

9.8 BRI AND NT1

Basic rate ISDN is provided to a customer by a DSL operating at 160 Kbps. For BRI, the customer provides an interface device to the line called a *network termination 1 (NT1)*. The NT1 functions at the physical layer of the reference OSI model and serves as the *demarcation* or *demarc* point. *Demarc* designates the point at which LEC ownership ends and CPE ownership begins. The NT1 provides for termination of the two-wire local loop and termination of the four-wire S/T bus. The NT1 performs impedance matching to the digital local loop and monitors the performance of the DSL. The NT1 performs maintenance functions and ensures accurate timing,

		Group 1	Group 2 to Group 11	Group 12	M Chn	Total
No. of bits:	18	8+8+2 (18)		8+8+2 (18)	6	Number of Bits:
Frame 1	ISW	B1+B2+D	<- - - - - - - - - - - >	B1+B2+D	M	240
Frame 2	SW	B1+B2+D	<- - - - - - - - - - - >	B1+B2+D	M	240
Frame 3	SW	B1+B2+D	<- - - - - - - - - - - >	B1+B2+D	M	240
Frame 4	SW	B1+B2+D	<- - - - - - - - - - - >	B1+B2+D	M	240
Frame 5	SW	B1+B2+D	<- - - - - - - - - - - >	B1+B2+D	M	240
Frame 6	SW	B1+B2+D	<- - - - - - - - - - - >	B1+B2+D	M	240
Frame 7	SW	B1+B2+D	<- - - - - - - - - - - >	B1+B2+D	M	240
Frame 8	SW	B1+B2+D	<- - - - - - - - - - - >	B1+B2+D	M	240
Total Bits	144	144	(10 × 18 × 8) = 1440	144	48	= 1920

FIGURE 9-3 2B1Q superframe without M-channel description. Total bits in each superframe = 1920 total 2B+D bits in each superframe = 1728; each superframe = 12 ms; freq. = 1/t = 1/12 ms = 83.33333 superframes per second; total 2B+D bps = 1728 × 83.33333 = 144,000 bps.

by synchronizing on signals from the ISDN line circuit of the local exchange. The NT1 performs a conversion between the AMI protocol of the four-wire S/T interface and the 2B1Q protocol of the two-wire local loop and vice versa. Of course, it is also the NT1 device that contains an integrated circuit chip that performs the DSL functions and 2B1Q line coding/decoding.

9.9 BRI INTERFACES AND REFERENCE POINTS

The 2B1Q transceiver in the NT1 has two ports. One port connects to customer terminal equipment and is called the *S/T interface*. The S/T interface is defined by CCITT Recommendation I.430. The second port connects to the twisted-wire pair of the local loop through a passive termination hybrid and a line pulse transformer. This interface point to the local loop is called the *U interface* or *U reference point* (ANSI Standard T1.601). Each point where two devices interface with each other is called a *reference point* (see Figure 9-4).

The R reference point is an interface point between a non-ISDN-compatible terminal device that is called *TE2* and an ISDN adapter device called a *terminal adapter (TA)*. The EIA-232 serial output port of a PC is a non-ISDN-compliant device. The PC serial port connects to a TA for conversion of the EIA-232 compliant bit stream into an ISDN-compliant bit stream. The *S reference point* connects an ISDN-compliant device (such as TE1 and TA) to the NT1. The *T reference point* is used to designate the interface between a **network termination 2 (NT2)** and an NT1 when the NT1 connects to a PRI ISDN line to the central exchange.

When the NT1 was owned by the LEC, the T reference point served as the demarc point per CCITT ISDN recommendations. The FCC does not recognize the

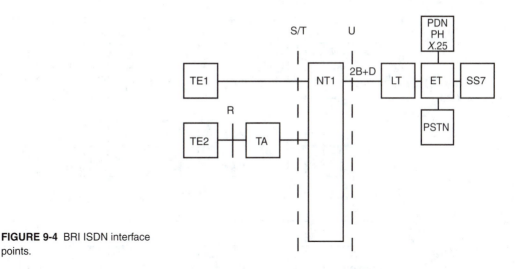

FIGURE 9-4 BRI ISDN interface points.

T reference point as the demarc point; it recognizes the U reference point for this purpose. ANSI has developed the U reference as the point of demarcation. Now that the NT1 is CPE and the demarc is the U reference point, the S and T references are often combined as the S/T bus. For basic rate access, the S and T reference points both have the same electrical specifications. When the ISDN line is a BRI ISDN (23B+D), the NT1 will be connected to an NT2 and the interface between the two is called the *T reference point*. The S and T reference points operate in a TDM bit stream of 192 Kbps. The S/T interface bit stream uses pseudoternary coding for transmission. The NT1 will convert messages received over the S/T bus from the terminal equipment into 2B1Q coding over the U interface to the central office cable pair.

9.10 TERMINAL EQUIPMENT AND THE S/T BUS

Up to eight different terminal devices can be connected to the S/T bus. The *S/T bus* is a four-wire bus that connects the NT1 to ISDN-compliant devices such as terminal equipment (TE), *terminal adapter (TA)*, and NT2. One pair of wires is used for transmit and one pair is used for receive. Each pair is running at a bit rate of 192 Kbps. The user information (2B+D) occupies 144 Kbps. The remaining 48 Kbps are used for control, performance monitoring, and timing. These 48 Kbps of overhead enable the NT1 and TE (or TA) to recover the 2B+D channels from the TDM stream. Both the transmit and receive portion of the S/T bus group 48 bits together to form an I.430 frame, but the frame structures are different for each direction (see Figure 9-5). Both frames use pseudoternary coding for transmission of bits over the S/T bus. This coding is such that a binary 1 is represented by no line signal and a binary 0 is represented by either a positive or negative 750-mV pulse. When 0s are transmitted, each subsequent 0 will have alternate polarity. By alternating the polarity of each 0, the line is DC balanced. This means that over time the net current on the line is 0. This allows the S/T bus to be longer than it would be without using this coding technique.

< ·--------- **Direction of Bits on S/T Bus from NT1 to TE1**

Bit No.	1	2	3	4	5	6	7	8	9	10	11	12	13	14	15	16	17	18	19	20	21
Bit Designation	F	L	B1	B1	B1	B1	B1	B1	B1	B1	E	D	A	Fa	N	B2	B2	B2	B2	B2	B2

Bit No.	25	26	27	28	29	30	31	32	33	34	35	36	37	38	39	40	41	42	43	44	45
Bit Designation	D	M	B1	B1	B1	B1	B1	B1	B1	B1	E	D	S	B2	B2	B2	B2	B2	B2	B2	B2

< ·--------- **Direction of Bits on S/T Bus from TE1 to NT1**

Bit No.	1	2	3	4	5	6	7	8	9	10	11	12	13	14	15	16	17	18	19	20	21
Bit Designation	F	L	B1	B1	B1	B1	B1	B1	B1	B1	L	D	L	Fa	L	B2	B2	B2	B2	B2	B2

Bit No.	25	26	27	28	29	30	31	32	33	34	35	36	37	38	39	40	41	42	43	44	45
Bit Designation	D	L	B1	B1	B1	B1	B1	B1	B1	B1	L	D	L	B2	B2	B2	B2	B2	B2	B2	B2

Note: F = framing bit; L = DC voltage balancing bit; B1 bits for the first B-channel; E = echo of the D bit received from the TE1 by the NT1; D = D-channel bit; A = activation bit; Fa = auxilliary framing bit; N = balancing bit for the auxilliary framing bit; B2 bits for the 2nd B-channel; M = multiframing bit; and S is an unspecified bit.

FIGURE 9-5 I.430 frames. The S/T bus is a four-wire bus (transmit = 2 wires and receive = 2 wires). The NT1 transmits at 144,000 bps over the transmit bus to the TE1. The NT1 will recieve data from the TE1 over the receive bus at 144,000 bps, but the bits will be delayed by 2 bits from the bits on the transmit bus. Thus, when the NT1 transmits bit 3 to the TE1, bit 1 is being received from the TE1.

Terminal equipment (TE or TA) derives its timing from a network clock via the NT1 and S/T bus overhead bits. This timing is used to synchronize the transmitted signals (2B+D). Since station equipment (TE or TA) is being synchronized off the NT, transmission from the TE or TA is delayed by 2 bits. Transmission on the transmit portion of the S/T bus will be offset by 2 bits from transmission on the receive portion of the S/T bus (see Figure 9-5).

9.11 DATA FRAMES ON THE S/T BUS AND INFO SIGNALS

The polarity of the framing bit (F) is always positive, and it marks the beginning of the frame. It is followed by the balancing bit (L), which is always negative. This offsets the positive voltage of the framing bit. Balancing bits are used to keep the current on the S/T bus at zero amps. In the NT to TE direction, the auxiliary framing (Fa) bit is set to 1 in every fifth frame and the multiframing (M) bit is set to 1 in every 20th frame. The Fa and M bits are 0 in all other frames. The Fa and M bits group frames and help keep the line synchronized. The N bit is used to offset the voltage of the Fa bit and serves as the balancing bit for Fa bits. The activation (A) bit is used by the NT to convey to the TE that the interface is active and operational. Remember that with pseudoternary signaling, a 1 bit is recognized as the absence of voltage and a 0 bit is the presence of 750 mV.

When a TE is not in use, no voltage is applied to the line. This is referred to as *INFO 0 signal.* When a TE (or TA) is activated, it transmits a positive 750 mV (a binary 0), a negative 750 mV (a binary 0), and six 1s. These 8 bits (00111111) are transmitted continuously at 192 Kbps from the TE to the NT1 and are known as the *INFO 1 signal.* The INFO 1 signal tells the NT1 that a TE (or TA) wants to activate the line (S/T bus). The NT1 acknowledges by returning 0s in all A, B, D, and E slots. This signal transmitted from the NT1 to the TE consisting of all 0s in the A, B, D, and E slots is known as an *INFO 2 signal.* The TE transceiver quickly synchronizes on the alternating voltages of the stream of 0s. The TE begins sending data over the B- and D-channels. This is known as an *INFO 3 signal.* The TE simply says okay, I have synchronized and here is some data. The NT1 recognizes the incoming data from the TE and sets the A bit to 1 as it transmits back to the TE1. This is called an *INFO 4 signal* and indicates the TE is activated and BRI is operational. Each TE has a unique identification number. When the TE sends the INFO 3 signal, it begins by sending its identification number. As the NT1 receives an ID, it echoes the ID number back to the TE using the E bit slots on the I.430 frame. The TE receives these bits and uses them to verify that it is logically attached to the NT1. Several TE devices can be in contention for connection to the NT1. As they transmit the D-channel INFO 3 signal, the NT1 will accept the information and respond back to one device. It alerts that device that it is attached by sending the ID of the attached device in the E bit locations. All other TE devices will quit sending because they did not receive an E-channel echo of their D-channel bits.

9.12 TERMINAL EQUIPMENT IDENTIFIERS AND SERVICE PROFILE IDENTIFIERS

Because the S/T bus accommodates two bearer channels, only two devices (out of the eight possible devices that can be connected to the S/T bus) can be active at the same time. Each device is interfaced to the NT1 by a TE or TA. To identify which of the eight devices is using one of the B channels, each TE and TA is assigned a ***terminal equipment identifier (TEI).*** The TEI is set on some devices by the use of jumpers, dip switches, or thumbwheels. Some devices have a display and will prompt you to use the keyboard to enter a TEI. Still other devices will have the TEI assigned automatically by the central exchange when the device is used for the first time. The TEI is contained in the second layer's address field. It is 7 bits carried in the second octet of the address field. Seven bits allows for identifying 127 different devices ($2^7 = 128$). Address 0 is not used. Address 127 is used to broadcast a message to all devices.

When a TEI is assigned automatically, it uses messages via LAPD to get a TEI assigned. When the TE1 (or TA) is plugged into the S/T bus, it immediately requests a TEI from the network. The TE sends an unnumbered information frame (I). The contents of the information field indicate this is a TEI request. The address field indicates the TEI is 00 or 127. This indicates to the NT1 and LE that the TE does not have a TEI. The network receives the TEI request and returns a response containing a TEI assignment as part of the information field. TEIs are associated with a *service profile identifier (SPID)* in the database of the LEC central exchange. Automatic initialization cannot take place if the customer has not entered a SPID in the TE. The customer receives the SPID assignment verbally from the LEC customer service center. Only after the customer has manually assigned the SPID to the TE can it initialize automatically. If only one device is attached to the S/T bus in a point-to-point (rather than a multipoint) configuration, a SPID is not needed. The device is identified by inputting the directory number in the TE. With multipoint, each device receives a directory number and SPID. SPIDs have several formats such as: area code + telephone number + a one- or two-digit suffix code + TI—for example, 913 5556666 01 00. The TI will be changed from 00 as it gets an automatic assignment upon initialization.

If only one TE1 were placed on the S/T bus, it could be up to 3300 ft from the NT1. When multiple devices are placed on the S/T bus, the length of the S/T bus from the NT1 is limited to 650 ft if using 150-Ω impedance cable, and each device must be no more than 30 ft from the S/T bus. The D-channel is informed which device is assigned to the bus, and this information is conveyed over the D-channel to the central exchange. The D-channel lets the central exchange know which device has been assigned to the DSL and which B-channel within the DSL has been assigned to that device. Using the SPID for that device, the switch knows which service to provide. A word of caution here. It has been reported that some of Northern Telecom's DMS-100 do not support eight logical devices since they used SPIDs to identify the two B-channels. I am sure that Northern Telecom would have corrected such a problem by now. SPID configurations are dependent on whether a BRI is provisioned as a National ISDN-1 line or as a custom standard. DMS-100 switches now adhere to the National ISDN-1 standard, and AT&T is a custom standard. Additionally, DMS-100 is configured as a multipoint configuration. I believe that the Northern Telecom DMS-100 is suffering from rumors about a past problem because the DMS-100 makes use of the TI to automatically configure TEs and supports multiple terminals. The point to be made here is you should verify ISDN capabilities with your LEC and ISDN vendor before proceeding with ISDN. It may even be possible that the LEC does not offer ISDN in your exchange.

9.13 CONVERSION OF S/T BUS DATA TO U REFERENCE POINT REQUIREMENTS

As mentioned earlier, the DSL, 2B1Q line format over the local loop is using frames that contain 240 bits. Each frame on the S/T bus is 48 bits. The DSL operates at 160 Kbps and the S/T bus is operating at 192 Kbps. The difference in line rates is the difference in overhead. The NT1 has the responsibility of taking the 144 Kbps (2B+D) user information from the S/T frames and placing them in the DSL frames. The 48-bit S/T frame is composed of 16 bits from B1, 16 bits from B2, 4 bits for D, and 12 bits of overhead. Since the S/T interface is running at 192,000 bps, the time for each bit is as follows: $1/192,000 = 0.0000052083$ sec. Thus, 48 bits = 0.000250 sec. Since one frame of 48 bits takes 0.000250 sec, we can transmit 4000 frames per second ($F = 1/T = 1/0.000250 = 4000$). Since each frame contains two 8-bit words for both B-channels, this frame rate satisfies the need to transmit 64,000 bps for each channel—that is, 4000 frames × 16 bits B1 = 64,000. ISDN may be used for analog signals by using a TA to convert the analog output of a phone into a digital bit stream. The TA will contain a codec to perform the analog-to-digital conversion. Remember that the output of a codec is 64,000 bps and the B-channels must operate at 64,000 bps to ensure that every bit generated by a codec is captured.

The DSL operates at 160 Kbps, and the time for each bit input to the 2B1Q coder/decoder is: $1/160,000 = 0.00000625$ sec. Each frame is 0.0015 sec; 0.00000625×240 bits = 0.0015 sec. The DSL will convey 666.66667 frames each second ($1/0.0015$). Each frame contains 12 B1 samples of 8 bits per sample (96 bits), 12 B2 samples of 8 bits per sample (96 bits), and 12 samples of 2 bits per sample for the D-channel (24 bits). With 666.66667 frames per second, the B-channels will have 64,000 bps ($96 \times 666.6667 = 64,000$ bps) and the D-channel will have 16,000 bps ($24 \times 666.66667 = 16,000$ bps). Thus, even though the S/T and U interfaces are running at different speeds and different frame configurations, the NT1 takes care of conversions between the two interfaces to ensure correct passage of 2B+D user information from terminal equipment to the central exchange.

9.14 TERMINAL EQUIPMENT CONNECTIONS TO NT1

The terminal side of an NT1 can connect to an NT2 device via a T interface. When an NT1 interfaces to an NT2, the NT1 has a PRI to the local exchange. When an NT1 only connects to TE1s or TAs, the NT1 has a BRI to the LEC exchange. *Terminal equipment 1* or *TE1s* are terminal devices that are ISDN compliant. TE1 devices include ISDN telephones, PCs equipped with a TA card in one of their expansion slots, and any other device designed to ISDN protocols. TE1s connect directly to the NT1 via the S/T bus. Terminal equipment not in compliance with ISDN protocols (*terminal equipment 2* or *TE2)* must wire to a TA. TE2 devices such as telephones, PCs, dumb terminals, and printers would wire to a TA. The TA will perform a physical interface change as well as doing conversion of analog to digital, protocol conversion, speed conversion, and multiplexing. There is not one do-it-all TA. There are several TA designs to accommodate the variety of different devices you may wish to connect to ISDN. The TA for a telephone will convert analog information coming in over two wires into a DS0 signal sent over the transmit portion of the S/T bus and will take DS0 signals received over the S/T bus, convert them to analog, and send them to the phone. The TA for a PC will accept digital data over the EIA-232 connection from the PC and map these bits into the TDM transmit of the S/T bus. It will map bits received over one of the ISDN channels to the PC. One side of the TA connects to a TE2. The other side of the TA can connect via an S interface to a NT2 or via the T interface to a NT1. Most ISDN service was initially deployed in large businesses through their PABX. When terminal equipment is used behind a PABX, the NT2 is needed as well as a 23B+D PRI NT1 for LEC interface. Why run a BRI from every TA or TE1 to the central exchange via a BRI NT1? Using an NT1 with PRI to the central exchange, the PABX distributes the 23B+D capacity of the NT1 to BRI terminal devices via the NT2. Therefore, most terminals connect via an NT2. The NT2 is placed between the NT1 and TE1, TE2, or TA of Figure 9-4 when the NT1 connects to a 23B+D PRI line.

9.15 S/T BUS DESIGN

In the residential market, there is no NT2. TE1s and TAs connect directly to the NT1 via the S/T bus. Most connections to the bus are done using an RJ-45 connector. This is an eight-pin connector that looks similar to the four-pin RJ-11 connector used on telephones. The four-wire S/T bus connects to pins 3, 4, 5, and 6. Pin 3 is + transmit and pin 6 is –transmit. Pin 4 is + receive and pin 5 is –receive. Pins 7 and 8 are used when a second power source is needed. Pins 1 and 2 are used when a third power source is needed. Power source 1 is provided over leads 3, 4, 5, and 6 from the NT1. These leads provide a phantom circuit arrangement that furnishes 40 V DC. This is a 1-W power source for all TEs. Power source 2 is used for TEs that may require up to 7 W from the NT1. Power source 3 is used to supply power from one of the TEs instead of from the NT1 (see Figure 9-6).

The S/T Bus

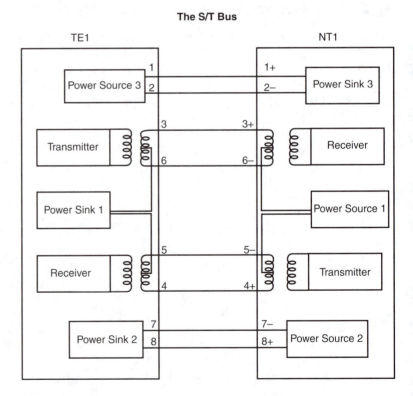

FIGURE 9-6 The S/T bus. Power source 1 of the NT1 supplies power to the TE1 by using a phantom arrangement over the transmit and receive pairs. The TE1 can use its power supply 3 to supply power to the NT1 or other devices. The power sink for a TE1 is pins 7 & 8. Pins 7 & 8 of another TE1 can connect to pins 7 & 8 of the NT1 to receive power from the NT1's power source 2, or pins 7 & 8 of the second TE1 could connect to pins 1 & 2 of the S/T bus to receive its power via source 3 from the first TE1.

9.16 LAYER 2 SIGNALING

Except for our discussion of automatic assignment of TEI, up to this point we have been involved with physical connections, interfaces, and data frame structures. All of these are similar to OSI layer 1 functions. Automatic assignment of TEI is a layer 3 function. Signaling at layer 1 has been to send various patters of on and off voltage conditions as signals between the TE1 and NT1. B-channels are strictly a level 1 layer operation when used in a circuit switched mode. One B-channel is connected over a network to a distant B-channel and they exchange digital signals over a 64-Kbps loop. The B-channels can use any protocol as long as the protocols on both ends are the same. The D-channel is part of the layer 1 physical connection but also provides layer 2 and layer 3 functions.

9.17 LINK ACCESS PROCEDURE ON THE D-CHANNEL

Instead of relying on physical signaling using voltage conditions, the D-channel uses messages for signals. The ISDN D-channel is out-of-band signaling. This signaling takes place between CPU chips and is logical rather than physical. One chip circuit tells another chip what action to take by issuing commands in the message strings between the two chips. Programming is burnt into the chip, telling it what action to take based on the message received. This type of signaling, which uses messages to direct the actions of a CPU, takes place at layer 2 and layer 3. The ISDN data link layer (layer 2) protocol is called *Link Access Procedure on the D-Channel (LAPD).* This protocol (software) defines the logical (not physical) connection between the terminal equipment (TE1 or TA) and the network (NT1 and LE). CCITT Recommendations Q.920 (I.440) and Q.921 (I.441) describe the general principles and operational procedures of LAPD. This procedure is also known as *Digital Subscriber Signaling System No. 1 (DSS1) Data Link Layer* and is a bit-oriented protocol, similar to the *X.25 Link Access Procedure Balanced (LAPB).* The primary function of the data link layer is to provide an error-free communication link between devices and to convey information between layer 3 of these devices using the D-channel. To accomplish this, the data link layer software must do the following:

1. Signal the beginning and end of a data packet by framing the user data
2. Maintain the sequential order of transmitted frames
3. Provide a destination and originating address
4. Return an acknowledgment of received frames to the transmitter
5. Be able to detect bit and frame sequence errors
6. Provide flow control to control the speed of data transfer

One difference between the LAPB of X.25 and the LAPD of ISDN is that LAPB is a point-to-point protocol. LAPD is a point-to-multipoint protocol, which allows its use between an NT1 and multiple TEs. LAPD uses *Carrier Sense Multiple Access/with Collision Resolution (CSMA/CR)* to allow multiple TEs on the S/T bus. I must stress again that LAPD is a protocol (software) operating at layer 2, to carry layer 2 and layer 3 messages. LAPD is a *logical link* between CPUs in the TE and NT1 or LE. The CPUs can send and receive messages framed according to LAPD. A LAPD frame contains the following fields:

1. FLAG Signals the beginning and end of a frame. The flag consists of the bit pattern 01111110.
2. ADDRESS Contains the *service access point identifier (SAPI)* and TEI that is sending or is to receive the message.
3. CONTROL Identifies the type of frame.
4. INFORMATION Contains layer 2 and 3 messages.
5. FRAME CHECK SEQUENCE Uses CRC to check for errors.

The control field indicates the type of frame being transmitted. There are three types of frames in LAPD: information (I), supervisory (S), and unnumbered (U) (see Figure 9-7). Information frames contain layer 2 and layer 3 messages. Supervisory frames control the flow of information frames. Unnumbered frames are used to initiate and terminate a logical link connection.

The address field contains the TEI and the SAPI. The SAPI identifies the service required (voice, switched data, packet data, and so on), and also identifies which B- or D-channel is being used to carry the customer's information. The SAPI is the logical connection between layer 2 and layer 3. Think of the SAPI as a post office box. Messages between layers are placed in the box. Each different service will have a different SAPI or different post office box. This allows each TE to tell the LE what process will take place from the TE and which channel the TE is using.

The address field consists of a TEI, which tells the LE which TE device is active, and a SAPI to tell the LE which process the TE will be using. TEI and SAPI taken together, as one address, are called the *data link control identifier (DLCI)* (see Figure 9-8). Some common SAPI values are 0 for call control over circuit switching, 1 for packet mode communication over the PDN, and 16 for X.25 data over the D-channel. The SAPI address or post office box allows the layer 3 software of the TE to communicate with layer 3 software of the LE in the central exchange to provide the service desired. The layer 3 software is embedded in the information field of the LAPD frame (see Figure 9-7a).

Layer 3 software will vary depending on which SAPI we have used. Signaling is sent from layer 3 of the TE, through the NT1, to layer 3 of the LE. Each signal consists of messages containing information elements. There are about 33 different signaling messages in ISDN, such as SETUP, CONNECT, DISCONNECT, ALERTING, PROGRESS, SUSPEND, RESUME, RELEASE, INFORMATION, and so on.

If a B-channel is to be used for the duration of a call by only one TE, it is set up as a circuit switched connection. A circuit switched connection is started by sending a SETUP message over the D-channel. The TE sends the LE the desired services, the address of the called party, and the B-channel to use. If the central exchange is satisfied that the SETUP message is valid, it returns a CALL PROCEEDING message. A SETUP message is sent to the called party, an ALERTING message received, a CONNECT message sent, and then a CONNECT ACKNOWLEDGMENT sent. Layer 3 on

LAPD Frames

(a)

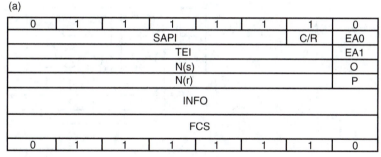

0	1	1	1	1	1	1	0
SAPI						C/R	EA0
TEI							EA1
N(s)							O
N(r)							P
INFO							
FCS							
0	1	1	1	1	1	1	0

Information Frame

(b)

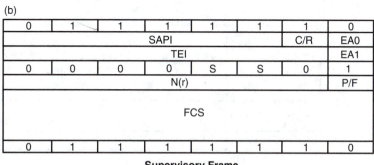

0	1	1	1	1	1	1	0
SAPI						C/R	EA0
TEI							EA1
0	0	0	0	S	S	0	1
N(r)							P/F
FCS							
0	1	1	1	1	1	1	0

Supervisory Frame

(c)

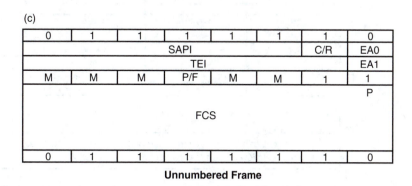

0	1	1	1	1	1	1	0
SAPI						C/R	EA0
TEI							EA1
M	M	M	P/F	M	M	1	1
							P
FCS							
0	1	1	1	1	1	1	0

FIGURE 9-7 LAPD frames.   **Unnumbered Frame**

the D-channel takes care of getting all service requests established (see Figure 9-9). When both ends of the call have received a connect message, the two TE devices can communicate and voice or data can be passed over the B-channel. Signaling messages vary depending on the service desired. Data connections can be on a dial-up basis over the public data network (PDN), packet mode over the B-channel, and packet mode over the D-channel.

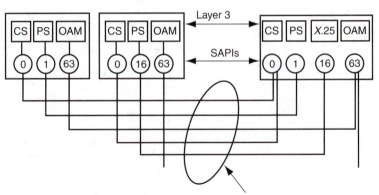

FIGURE 9-8 Layer 3 communication between the TE1 and LE.

The Logical D-Channel

Originating TE	LE Originating Central Office	ISDN Network (PSTN, PDN, and SS7 Networks)	LE Terminating Central Office	Terminated TE
Originating Call Setup:				
SETUP- - - - - >	SETUP- - - - - >	SETUP- - - - - >	SETUP- - - - - >	SETUP - - - - - >
<- - Call Proceeding	<- - Call Proceeding	<- - Call Proceeding	<- - Call Proceeding	<- - Call Proceeding
<- -Alerting	<- - Alerting	<- - Alerting	<- - Alerting	<- - Alerting
<- -Connect	<- - Connect	<- - Connect	<- - Connect	<- - Connect
Connect ACK- ->	Connect ACK - ->	Connect ACK - ->	Connect ACK - ->	Connect ACK- ->
Transfer of voice or data now occurs over the B-channel.				
Disconnect call after conversation is complete or after data has been transferred over the B-channel by issuing the following commands over the D-channel:				
Disconnect - - - >	Disconnect - - - >	Disconnect - - - >	Disconnect - - - >	Disconnect - - - >
< - - - Disconnect	< - - - Disconnect	< - - - Disconnect	< - - - Disconnect	< - - - Disconnect
Release - - - - >	Release - - - - >	Release - - - - >	Release - - - - >	Release - - - - >
< - - - - Release	< - - - - Release	< - - - - Release	< - - - - Release	< - - - - Release

FIGURE 9-9 Signaling message flow over the D-channel.

9.18 CALL SETUP OVER THE D-CHANNEL

The D-channel provides instructions to the network on what to do with the B-channels. It provides TEIs, SAPI, call setup instructions, and call termination instructions. The central exchange knows which services to provide based on the TEI, SPID, and SAPI. For voice calls, the TEI and SPID will instruct the switch to use the PSTN (voice) network. For data calls over one of the B-channels, the TEI and SPID will instruct the central exchange to connect the B-channels to the PSTN. The PSTN will use a 64-Kbps channel path in the PSTN to connect to the called ISDN terminal's ISDN line circuit in the terminated LEC central office. The caller can use both B-channels for data, if desired, by placing a call over each B-channel and bonding the two channels together. This is accomplished by a special command in the ISDN modem's initialization string. If you have devices that transmit packet data, the B-channel will be connected to the PDN.

The D-channel is in charge of getting end-to-end connections established. Once the originating D-channel indicates its wishes to the local switching exchange, the local exchange takes over. The local exchange issues switching instructions over the SS7 network to get the voice or data call established. The far-end terminal will communicate via its D-channel to its local office. Communication between the end offices is handled by the SS7 network. Control information on the D signaling channel is transferred over to SS7 at both end offices.

When on a data call, the B-channel will provide end-to-end 64-Kbps full-duplex data over each B-channel either in a switched mode over the PSTN or in a packet switched mode over the PDN. Since the D-channel of BRI does not require all 16 Kbps for control of the network, low-speed packet data can be sent over the BRI D-channel using $X.25$ protocol. ISDN terminals are assigned a telephone number for voice that is of the following form: area code + exchange code + terminal digits—for example, 913 + 555 + 1212. ISDN terminals may also be assigned a data address (telephone number) of this form: zone code + country code + PDN code + network terminal number—for instance, 3 + 10 + 4 + 5556660000. These 14 digits identify a unique terminal on the PDN.

9.19 LEC CENTRAL OFFICE LE AND TE

The class 5 central exchange provides the line-interface circuitry for either a BRI or PRI line. The local loop interfaces to the line equipment (LE) at the U reference point. The switching system comprises the exchange terminator (ET), packet handler, signaling processor, and switched circuit access to the PSTN and PDN. The interface point between the LE and ET is called the *V reference point*. LE is part of the class 5 switching system, and LE is often considered part of the ET. The V interface is more of a logical interface than a physical interface because the LE may be switched to many different devices, depending on the service desired.

For switched packet data calls over the B-channel, the LE can be connected to a packet handler at the class 5 exchange or it can be connected to an ISDN port of the public switched packet data network (PSPDN or PDN). If connected to a packet handler, the packet handler connects via $X.75$ protocol to the PSPDN. For voice calls and nonpacket data calls, the ET will connect the B-channel to the PSTN. An X.25 TE device can also connect to the D-channel to access the PDN.

The ET at the central office provides access to all services desired by a TE1. Many businesses are using the ISDN D-channel as a way to do credit card validation from point-of-sale terminals (POST). They can purchase D-channel only capability or purchase a complete BRI. With BRI they can use telephones, fax, POST credit card scanners, and alarm systems all over one BRI. The TEs will be connected to the proper service by the ET.

9.20 PRIMARY RATE ISDN AND BROADBAND ISDN

For primary rate ISDN (PRI) (23B+D), the customer must provide an NT1 and possibly an NT2, or a combined NT1 and NT2, which is called a *network termination 12 (NT12)*. NT12s are used in Europe and in some PABXs. NT2s are used to interface PABXs and LANs to ISDN. PRI provides 23B+D service. As with BRI, all signaling is handled by the D-channel. The 23 B-channels can be used for voice or data. PRI appeals to businesses that use PABXs or ACDs. Each B-channel is 64 Kbps, and the D-channel is also 64 Kbps. The basic frame is 193 bits and is the same DS1 frame used for T1. The PRI line rate is 1.544 Kbps (the same rate as T1 carrier). The operational difference between PRI ISDN and T1 is that the ISDN system was designed to handle a full 64-Kbps data rate over each channel by means of out-of-band signaling. T1 uses a bit-robbed signaling technique. ISDN uses a common channel (the 24th channel) for all D-channel signals.

The European version of ISDN uses the E system line rate of 2.048 Mbps. This system provides 30B+D. Conversion of all class 5 trunk signaling to SS7 is necessary for a T1 carrier system to achieve 64-Kbps data capability. PRI B-channels can be added to achieve higher data rates. This technique is used to provide low-resolution video to small businesses and residences. The two B-channels of BRI can be combined to provide 128 Kbps for a video signal. Quite a few businesses are finding this application for ISDN useful. They can feed the signal from surveillance or security cameras to a remote location such as a security company. The LECs also offer other combinations of B-channels over a PRI to achieve higher data rates. Six B-channels can be combined to provide a data rate of 384 Kbps (6 × 64 Kbps = 384 Kbps). This service is called an *H0 channel.*

BRI and PRI services are called *narrowband ISDN* to differentiate these services, provided over wire, from the *broadband ISDN (B-ISDN)* services provided over fiber optic cable. The layer 2 protocol (LAPD) and layer 3 signaling and control protocols of BRI are the same for PRI ISDN and B-ISDN. Narrowband ISDN uses synchronous transmission, but B-ISDN is provided primarily over *asynchronous transfer mode (ATM)* facilities. ATM refers to a switching and transmission service. Data switches in the ATM network provide for the routing of data in the network, and fiber optic cable connects the ATM switches together. ATM is the underlying fabric for *Switched Multimegabit Data Services (SMDS).* SMDS is a network service developed by Bellcore for the RBOCs. The LECs of the RBOCs use SMDS to establish *Metropolitan Area Networks (MANs)* for their large-business customers.

B-ISDN can provide bandwidth-on-demand services using the ATM switches and synchronous optical network or SONET (fiber optic) transmission facilities. SONET is a multiple-level protocol used to transport high-speed signals. It allows multiplexers/demultiplexers to insert or drop out one DS0 signal from the high-speed line without having to demultiplex the whole facility. SONET is the vehicle that makes B-ISDN and the services it can provide possible. These services can be anything from videoconferencing and multimedia computing to medical imaging, broadcast TV, and MAN services.

Perhaps B-ISDN will have more success in the marketplace than narrowband ISDN since B-ISDN offers higher speeds (100 Mbps) and can be used to provide WAN connections or real-time video signals. These high speeds may appeal to business users, but does a residential customer need more than two 64-Kbps channels? For voice calls, the codec output is 64 Kbps. If a PC is making a circuit switched data call, the speed of the PC may be slower than the speed of a B-channel.

The UART that interfaces a PC to its serial port is usually a 8250 or 16550 UART. Although a 16550 UART is capable of 115,200 bps, its speed is limited by DOS and therefore by Windows 3.*xx* to 19,200 bps. Windows 95 can take full advantage of the speed offered by a 16550 UART because Windows 95 has the serial interface drivers required to achieve 115,200 bps and higher. If you are using Windows 3.1x, you will need to purchase a serial accelerator card and associated software to attain speeds higher than 19,200 bps. To take advantage of higher speeds, most people are upgrading their PCs by buying new ones. Most new PCs come with Windows 98 or Windows 2000 and high-speed serial ports.

The TA that interfaces a PC serial port to ISDN must use CCITT Recommendation I.465 to adapt PC speeds to the 64 Kbps of a B-channel. This protocol frames user data so that the receiving terminal knows where the data starts and ends and pays no attention to idle channel time stuffing bits or absence of bits. This protocol is employed when the B-channel is used in a circuit mode. When the B-channel is used in an *X*.25 packet mode, the TA must convert the *X*.25 of the TE to ISDN protocol. A packet handler at the central exchange or at the entrance portal to the PDN will convert the ISDN protocol back to *X*.25 protocol.

I have tried to provide you with some insight into ISDN. You should now have a general view of what ISDN is and how a call is placed over this network. Before implementing ISDN technology as a solution to your communication needs, you need to do research on the different applications available. You should also talk with your ISDN vendors and LEC. Prior to divestiture and deregulation, you could do one-stop shopping at your LEC. The LECs were responsible for the network and all the services they sold. If you purchased a service from them, they would make it work. Today, there is no single source provider, and when trouble with a service occurs, all the providers point a finger of blame at each other.

You can no longer rely on the LEC to provide technical expertise. *You, the telecommunications manager, will be expected to be the expert. You must have a thorough technical knowledge of your telecommunications network and equipment.* The 1996 Telecommunications Reform Act allows LECs to reenter the CPE market without using a subsidiary. This act also allows LECs to manufacture equipment. Hopefully, the LECs will seize this opportunity to become a one-stop shop for all ISDN equipment and services. They have the resources and marketing expertise to design, develop, market, install, service, and provide technical assistance on ISDN products and services.

9.21 ISDN AND THE PERSONAL COMPUTER

The LECs have been slow to deploy BRI. Their most aggressive marketing of BRI was in 1996. Many people have been installing an ISDN line to achieve higher data speeds when connecting their PC to the Internet. As mentioned in Chapter 8, modem technology is developing to a point where a modem can achieve a 56,000-bps data rate and with compression can achieve a 224,000-bps end-to-end connection. This speed approaches the line rate provided by BRI ISDN (128 Kbps), but with 4-to-1 compression, the end speed of ISDN is 512,000 bps. This speed and the fast modem speeds are usually faster than the serial port of a PC can handle. The speed is also faster than most circuit paths in the PSTN can support. Therefore, ISDN and most modem connections will fall back to a lower speed and throughput rate. The 56K modem is actually not designed for an end-to-end dial-up connection between two PCs; it is designed to interface with 56K technology at an ISP. The ISP has digital connections to the PSTN and does not interface via an analog line circuit. The ISP eliminates the analog-to-digital conversion of the codec in a line card, which allows it to run at a faster speed. The 56K modem on one end of a connection must be connected directly to the digital highway of the PSTN to achieve 56K speed.

ISDN can achieve a data rate of 64,000 bps on each channel. By combining both B-channels on a BRI, it is possible to achieve a data rate of 128,000 bps. Thus, even though a modem appears to be almost as fast as ISDN, the actual operating speeds are lower than ISDN. The internal ISDN card for a PC is also equipped with a compression algorithm to speed the transfer of data. With a 4-to-1 data compression, the end-to-end speed for ISDN running both B-channels in a bonded mode is 512,000 bps (4 × 128,000).

It will not be long before someone develops an internal PC ISDN card that will interface to the dynamic memory mapping of a PC in order to achieve a fast data rate transfer between the ISDN card and the data bus of the PC. I think future advances in the ISDN card for a PC will lead to the preference of ISDN modems over analog modems. ISDN will soon be faced with competition from cable modems provided by cable TV companies. The telephone companies are also beginning to deploy *asymmetric digital subscriber line (ADSL)* technology, and this technology may end up replacing ISDN.

9.22 ASYMMETRIC DIGITAL SUBSCRIBER LINE

Pacific Bell issued press releases stating it was going to bring coaxial cable and fiber optic cable to many of its customers by the year 2000. In recent press releases it has seemed to back away from the earlier plans for coaxial and fiber cable installation to most customers. Instead it plans to embrace ADSL technology. According to the president of PAC Bell's Business Communications Services, ADSL was deployed in 1997. He stated that they were still laying fiber, but only to areas with a high concentration of customers.

Every LEC has a large investment in the twisted-pair copper wire local loop. ADSL is a technology that allows the LEC to retain that investment. Pacific Bell plans to use ADSL on copper to deliver T1 speed (1.544 Mbps) to customers. This LEC seems to be in the forefront of delivering ADSL, but other Bell LECs are also conducting field trials of this technology. GTE is currently testing ADSL technology in Texas to access the Internet.

9.23 THE LECS' ANSWER TO CABLE COMPANIES IS ADSL

ADSL was developed by Bellcore Labs of Morristown, New Jersey, for the RBOCs. It was originally created as a vehicle to carry video signals over twisted-pair copper wire. This technology would allow the LECs to provide cable TV services in direct competition with cable TV companies. Asymmetric digital data does not have the same data speed in both directions. ADSL has reached speeds of 6.14 Mbps (that's 6 million bits per second!) in the download direction and 640 Kbps in the upload direction. The download speed is 185 times faster than a 33.3-Kbps modem with no compression. If the modem is a 33.3-Kbps modem and is using 4-to-1 compression (230,000 bps), ADSL is still almost 27 times faster.

Cable modems are capable of 27 Mbps in the download direction and 1 Mbps in the upload direction, but the modem circuits will be shared circuits and overall speed will depend on the number of active users. During periods of heavy use by several users, the cable modem will not provide significantly more than the 6.14-Mbps download capability of ADSL. Thus, ADSL provides the LECs with a vehicle that can go head to head with a cable modem. A basic comparison of the data speeds provided by the various technologies is that to transfer a file of 6 million bits:

1. A 33.3-Kbps modem requires about 6 min for the download.
2. A BRI ISDN will require about 1.5 min for the download.
3. An ADSL will require about 1 sec for the download.
4. A cable modem will also require about 1 sec for the download.

The goal of every Internet user is faster data speeds. The transfer of HTML files every time you select a new address takes time, so you spend much of your time waiting for these file transfers. The LECs, IECs, and cable companies are beginning to compete for the Internet user's business. Some Bell companies have been cutting the prices for ISDN to entice customers to use their service. Cable TV companies have been slow to enter the Internet access market. The cable modems encountered some technical problems, but those appear to have been corrected, and we should see the more aggressive cable TV companies entering this market soon.

ADSL has quickly replaced ISDN in most home user applications as the cost of ADSL equipment has decreased. ADSL equipment is supplied by the local telephone company as part of the service, and the monthly fee is approximately $40 to $60, which is less than ISDN (approximately $100 a month for BRI). One of the drawbacks of ADSL is the need for the telephone company to install and configure the ADSL modem at the customer's home or business. This is being addressed. Companies are trying to develop methods that will allow the customer to install the ADSL modem.

ADSL is 90 times faster than ISDN. This has the advantage of higher speeds but an accompanying disadvantage is that ADSL cannot be placed on a subscriber carrier system such as SLC-96. Since Basic Rate ISDN has an overall bit rate of 160,000 bps, it can be transported using two-and-a-half channels of a subscriber carrier system. The bit rate per voice channel on SLC-96 is 64,000 bps. The total bps for a complete SLC-96 system is 6.312 Mbps. It becomes readily apparent that just one ADSL customer would use all the capacity of the system that is designed to service 96 users. Therefore ISDN is available to areas that are served by subscriber carrier instead of twisted-pair copper wire, but ADSL is not available over subscriber carrier.

To overcome the distance limitation of ADSL (18,000 feet), telephone companies are deploying fiber carrier in the local loop. Fiber cables are being laid underground from the central office to remote switching centers. Many small buildings (remote switching centers) are being placed in strategic locations of the central office's exchange territory. These buildings will house ADSL interface equipment that interfaces the cable pair from the end user's ADSL modem to a fiber optic multiplexer. The fiber optic multiplexer connects to a fiber that goes to the local central exchange. The ADSL signal from a user's modem will come in over a cable pair from the user and at this remote switching station, the signal will be multiplexed onto a fiber for transportation to the Digital Subscriber Line Access Multiplexer (DSLAM) at the central exchange.

The use of remote switching centers and fiber optic connections to the local exchange overcome the 18,000 foot limitation of ADSL over copper and allow the deployment of ADSL to any point in the exchange territory that is within 18,000 feet of a remote switching center. This will allow the telephone companies to compete against cable TV companies that have upgraded their distribution systems from coaxial cable to fiber optic cables in order to provide support for cable modems. ADSL speeds and pricing are comparable to cable modems.

9.24 ADSL AND VIDEOCONFERENCING

Real-time video uses a compression technique called *MPEG2*. This technique can deliver a high-quality video with 6 Mbps. Bell Atlantic ran some trials with ADSL and MPEG2 and discovered that it could only serve about 25% of its market with these technologies due to physical plant problems. Bell Atlantic has turned to wireless services as a method of delivering videoconferencing. Bell Atlantic and NYNEX have invested $100 million in CAI Wireless Cable Company with an option to invest $200,000 million more. Wireless cable (for example, radio wave facilities) companies also have their eye on the Internet access market.

9.25 ANOTHER FIGHT OVER STANDARDS

One reason ISDN languished for so long was that several manufacturers insisted on using their proprietary standards. The same thing is happening to ADSL. There are two competing operating systems for ADSL: (1) *carrierless amplitude and phase modulation (CAP)*, and (2) *discrete multitone technology (DMT)*. In 1996, the American National

Standards Institute (ANSI) adopted an ADSL standard based on the DMT architecture. The CAP camp has been reducing prices of their equipment, and presently the trials of ADSL are using both technologies. The RBOCs have requested that ANSI recognize CAP. The standards war is on again! It remains to be seen how long it will take to settle on one standard.

CAP employs a technology similar to the *trellis coded modulation (TCM)* technology used by high-speed dial-up modems. CAP uses two transmitting frequencies (one in each direction). DMT utilizes 256 different carrier frequencies. Each carrier has a 4-KHz bandwidth. DMT appears to be faster and works better on low-quality circuits. The Paradyne division of AT&T has chosen the CAP technology for manufacture of ADSL terminal equipment. DMT technology is being provided by smaller manufacturing firms. DMT uses Frequency Division Multiplexing (FDM) and Echo Cancellation technologies. The upstream data from a user's ADSL modem to the central office is placed on one of 32 different tones (channels). Data from the central office to the user's ADSL modem are placed on one of 256 channels (tones). Each tone serves as a carrier signal to carry the data placed on it using Quadrature Amplitude Modulation (QAM). QAM is discussed in Chapter 8.

Echo cancellation is used to allow overlapping upstream and downstream channels. The transmitter of the ADSL modem sends a copy of signals it transmits to its own echo cancellation circuitry. The echo cancellation circuitry compares this to signals seen on the cable pair. The echo cancellation circuitry knows that anything seen on the cable pair that matches what it received directly from the transmitter is in fact transmitted data, and it will not pass these signals to the receiver.

Anything on the cable pair that was not sent to the echo cancellation circuitry by its own transmitter must be a signal coming from the distant end, and this signal is passed by the echo cancellation circuitry to the receiver. The use of echo cancellation circuitry allows the use of one cable pair for carrying both transmit and receive data at the same frequencies. Some people mistakenly assume that one wire will carry transmit data and the other wire will carry receive data. This is not true. As we discussed in Chapter 4, electrical signals must have a return path. With the local loop twisted-pair, any signal that appears on the Tip must also appear on the Ring. This premise holds true for all high-speed modems. They transmit and receive modulated tones over the twisted-wire pair where transmit and receive frequencies are the same frequency. They rely on echo cancellation technology at each end to prevent transmitted data at the near end being passed to its own receiver.

The tones used in DMT start at 20 KHz and go up to 1.104 MHz. No tones are used in the lower 0 to 20 KHz band to allow voice signals to exist on the twisted-wire pair. A filter exists in the ADSL Modem to prevent it from receiving 0 to 20 KHz signals. A filter (often referred to as a splitter) is placed between the telephone line and the telephone to prevent it from receiving 20 KHz and upward signals. This use of filters and assignment of frequencies allows the regular telephone line to serve both your telephone and ADSL modem with no interference between the two. You can use your telephone while you surf the Web with your PC and ADSL modem connection.

Each channel in DMT has a bandwidth of 4.3125 KHz as well as a separation of 4.3125 KHz between predecessor or successor channels. The modulation scheme of each cannel using QAM and the number of channels used will vary according to the characteristics of the cable pair. Using two different amplitudes and two different phases on a signal provides four distinct signals. As discussed in Chapter 6, this four-state signal supports transmission of two data bits at a time.

Using two amplitudes and four phases provides an eight-state signal that can carry 3 bits at a time. Each tone can carry from 2 to 15 bits encoded on the tone using QAM. The number of bits to be carried by each tone depends on how many different amplitude and phase shifts we can make to a tone and still be able to recapture this data on the far end. This in turn depends on the transmission characteristics of the cable pair with the most important characteristics being signal-to-noise ratio and attenuation. The modem automatically conducts a signal-to-noise transmission check in order to determine which form of QAM to use. The closer a user is to the central office, the higher the signal-to-noise ratio will be and the higher the data rate.

Which tones to use and which form of QAM to use is determined by the ADSL modem. It selects a combination that will be supported by the transmission facility. The ADSL modem automatically adapts the rate of transmission to fit the capabilities of the cable pair. This is why longer cable pairs result in lower transmission speeds. The longer the cable pairs, the higher the attenuation. The higher the attenuation, the less frequencies we can use on the higher end of the ADSL frequency spectrum. Less frequency for transportation of data lowers the data rate. Rate adaptive ADSL delivers data rates on the downstream channels from 64 Kbps to 8 Mbps. Upstream rates are from 16 Kbps to 768 Kbps. The rate simply depends on the overall transmission characteristics of the cable pair.

The ADSL modem chip is quite a remarkable piece of silicon and engineering. Built on this chip is a spectrum analyzer that continuously monitors the signal-to-noise ratio of each tone. The power level, line attenuation, and noise power of each tone is measured. An equalizer is included to offset the difference in loss between high-frequency and low-frequency tones. Echo cancellation, generation of 256 different tones, and varying implementations of QAM must be part of the circuit. Finally, the ADSL chip must be smart enough to pick out tones that have suffered up to 90 dB of loss and differentiate these tones from noise on the line.

In the first edition, I mentioned that CAP technology would win out in the standards war unless a bigger player jumped on the DMT bandwagon. This has not been the case. Although CAP was used for many early installations of ADSL, most new deployments of ADSL use DMT. The migration to DMT has occurred as trials with DMT have demonstrated its superiority over CAP. The RBOCs are deploying DMT at a rapid pace. In an October 24, 2000 Southwestern Bell news release "Project Pronto" was described as a $6 Billion initiative by its parent company SBC to make DSL available to 77 million people by 2002. This news release from Southwestern Bell contains pricing information as follows:

"Customers interested in ordering DSL service can receive DSL Internet Service for $39.95 per month, with free equipment (after a rebate) and free installation (for customers who choose to self-install) with a one year commitment."

The DSL Internet Service just mentioned provides downstream connection speeds up to 1.5 Mbps and an upstream connection speed of 128 Kbps. Higher speeds are also available. A speed of up to 6 Mbps downstream and 384 Kbps upstream is available from Southwestern Bell as part of their Enhanced Internet Service for as low as $59.95 with a $250 dollar installation charge and a cost of $99 for equipment. Special promotions offer a Compaq Presario computer with Basic DSL Service for $59.95 a month for customers that sign up for 28 months of service.

ADSL modems are available as cards that can be inserted into a PC, but many users opt to go with an external ADSL modem. The ADSL modem can be connected to a hub and serve several computers. I mentioned in Chapter 6 that a bottleneck for regular 33.3 and 56K modems was their serial port interface to the PC. Such is not the case with ADSL. In order to prevent having a bottleneck at the PC connection, ADSL modems connect to a Network Interface Card (NIC) that is inserted in the PC. This provides a LAN connection at 10/100 Mbps between the ADSL modem and the PC.

Most of this discussion has centered on ADSL services to connect to the Internet. Another form of ADSL that is referred to as High Bit-Rate Digital Subscriber Line (HDSL) is not asymmetrical. HDSL is symmetrical (same data rate in both directions). This modem requires two cable pairs. One pair for transmit and one pair for receive. HDSL offers 1.5 Mbps and is an attractive alternative to leasing a T1 line for your business. Newer HDSL modems are available that will work using one cable pair for both transmit and receive. This service is referred to as HDSL II.

9.26 WHERE CAN ADSL BE DEPLOYED?

Remember that the loop limit for BRI ISDN without a repeater is 18,000 ft (approximately 3-mi). Longer loops are possible, but digital regenerators must be on the line. The same is true for all digital signals placed on twisted-pair copper wire. A T1 carrier system transmits 1,544,000 bps. When this signal is placed on copper wire, the signal must be regenerated every mile. ADSL will work on an 18,000-ft (approximately 3-mi) loop without a repeater, if the data rate is below 1.5 Mbps. Higher data rates will limit the length of the local loop. A speed of 6 Mbps will limit the distance from the central office to about 1.5 mi.

9.27 HOW DID THEY DO THAT?

The data speeds achieved by ADSL are staggering when you consider they are using the twisted-pair local loop for a medium. The higher data speeds being achieved by regular 56K modems and by ADSL modems are due to advances in digital signaling processors, codecs, filters, transformers, echo cancellation technology, and compression technique algorithms. Another important factor that allows the attainment of high data rates is the vast improvements that have been made in the local-loop plant. Most outside plant is fairly new and is buried in the ground to protect it from adverse weather conditions.

The semiconductor industry will continue to improve the ADSL chipsets. This will lead to even higher data rates. ADSL utilizes forward error-correction techniques. The microprocessors and software used in today's modems to control their operation has provided them with tremendous capabilities. Whether it is a regular phone-line modem, an ISDN phone-line modem, or an ADSL phone-line modem, the advances in technology over the past several years have been tremendous. We should all thank Hayes for the start of smart modems.

9.28 SUMMARY

ISDN is an end-to-end digital network that expands the digital PSTN to the customer's premises. The final leg of this digital network is the DSL. The DSL uses a 2B1Q protocol to achieve a 160-Kbps transfer rate at 80 Kbaud. Since ISDN is designed to handle digital signals, digital data from a PC can be placed via an ISDN modem (NT1) over the facility. Before voice can be placed over ISDN, it is converted by a codec to a digital signal of 64,000 bps (DS0). The codec chip can reside in a TA or in the input port to an ISDN modem (NT1), which serves a telephone.

Basic rate ISDN (BRI) is provided over the 160-Kbps DSL. Of this rate, the DSL uses 16 Kbps for synchronization and timing. The remaining 144 Kbps is used to provide two bearer channels and one delta channel (2B+D). The D-channel is primarily used to set up calls for the B-channels, to monitor the circuit, and to provide for end-to-end message signaling. The D-channel on each end of an ISDN call is connected to the SS7 network at the exchange terminator. Thus, for ISDN to work, SS7 is mandatory.

BRI has three basic interface or reference points: U, S/T, and R. The U reference point is where the two-wire DSL from the LEC connects to the NT1 of the customer. The U reference point serves as the demarc. The S reference point is where an ISDN-compliant device (TE1) connects to an NT1. The T reference point is where a terminal adapter (TA) connects to an NT1. The R reference point is where a non-ISDN-compliant device connects to a TA. An ISDN device (such as an ISDN modem) will often contain an NT1, a TE1, and a TA as part of its internal circuitry. The only interface we will see is the connection to the ISDN phone line from the LEC (U interface), an RJ-11 connection for a telephone (R interface), and possibly an RS-232 connection for the computer (R interface). The S/T interface will be internal to the circuit card.

When an ISDN modem is installed in a PC, the customer receives a SPID from the LEC. The customer manually assigns the SPID to the software controlling the modem when the modem is installed. When the first ISDN connection to the central office is made, the central exchange checks the SPID database and will automatically assign a TEI to the ISDN modem. The TEI is used to inform the central office, via the D-channel, what service to provide (voice, switched data, packet data, and so on) for the terminal equipment attached to the NT1.

The D-channel of ISDN operates at the lower three layers of the OSI model (physical, data link, and network). The physical layer conforms to I.430 for BRI and to I.431 for PRI. The data link layer conforms to Q.921 (link access procedure on the D-channel or LAPD). The network layer conforms to Q.931 (call setup procedure over the D-channel). The D-channel transmits signaling and information messages from the NT1 to the LE. These messages are used to set up calls for the B-channels. The D-channel provides instructions to the PSTN on what services to provide the B-channel.

BRI and PRI are called narrowband ISDN and use synchronous transmission. Broadband ISDN is asynchronous transfer mode transmission over fiber optic cables. Broadband ISDN provides speeds faster than 100 Mbps and can be used to establish WANs. Another high-speed digital technology being deployed by the LECs is ADSL. ADSL can be deployed to deliver the equivalent of a T1 facility to a customer. It can also be used at higher rates to deliver digital video over the local loop. ADSL has reached speeds of 6.14 Mbps over twisted-pair copper wire. This is 360 times faster than a 33.6 dial-up modem.

CHAPTER REVIEW QUESTIONS

1. How is data transmitted over the PSTN without using ISDN?
2. What part of the PSTN was the first to be converted to digital?
3. Which is faster, a cable modem or ISDN?
4. What is BRI?
5. What type of line is used for BRI?
6. What is the signaling rate on a DSL? How much of this rate is available to the customer?
7. What type of coding is used on the S/T interface?
8. What type of coding is used on the U interface?
9. What is the baud rate of 2B1Q?
10. What is a quat?
11. If the S/T interface is running at 192,000 bps and the U interface is running at 160,000 bps, what happened to the 32,000 extra bits on the S/T interface?
12. What is a TEI?
13. What is a SPID?
14. What is a SAPI?
15. What is the difference between a TE and a TA?
16. How is signaling conveyed in ISDN?
17. What are the frame types in LAPD?
18. What channel carries all signaling and control information?
19. How does a central office know what type of service or connection to provide on an ISDN call?
20. What services are narrowband ISDN?
21. What services can be handled by ADSL?
22. How does the speed of ADSL for the residence compare to ISDN?
23. How does the speed of an ADSL compare to a cable modem?
24. What is the advantage of ADSL technology to the LEC?
25. What is the difference between DSL and ADSL?

PROJECTS AND CASE STUDIES:

Case Study #1

Bill is thinking about getting ADSL lines for his business and home. The business is located five blocks from the central office in your town. His residence is in a suburban area located five miles from the central exchange. Find out if you can get ADSL access at either or both of the locations and what ADSL will cost. If ADSL is not available at either location, what are some alternatives? Are any of these available? What will the alternatives cost? How do they compare with ADSL?

GLOSSARY

Asymmetric Digital Subscriber Line (ADSL)
A digital subscriber line technology that can be used to deliver T1 speeds over the local loop. An ADSL has a high bit rate in one direction and a low bit rate in the opposite direction.

Basic Rate Interface (BRI)   The basic building block of ISDN. A BRI consists of two 64-Kbps bearer channels and one 16-Kbps delta channel (2B+D). The delta (D) channel is used for signaling and sets up calls for the B-channels. The B-channels are used to carry customer voice or data.

B-Channel   The channel that carries the customer's information. This information may be voice, data, or video.

Broadband ISDN (B-ISDN)   Provides speeds faster than 100 Mbps and can be used to establish wide area networks. Broadband ISDN is asynchronous transfer mode transmission over fiber optic cables.

D-Channel   The channel that carries signaling and control information. The D-channel is used to set up calls for the B-channels.

Digital Subscriber Line Circuit (DSL)   Usually refers to a local-loop cable pair that is handling a digital signal using 2B1Q line coding.

Hypertext Markup Language (HTML)   A software language used to create documents, such as home pages, for use on the Internet.

Integrated Services Digital Network (ISDN)
The use of digital line circuits to provide end-to-end digital service. The *basic rate interface (BRI)* provides the user with two DS0 channels. The *primary rate interface (PRI)* provides the user with 23 DS0 channels.

Internet Service Provider (ISP)   A company that provides individuals and companies with access to the Internet.

Link Access Procedure on the D-Channel (LAPD)   The software protocol used at the data link layer (layer 2) of the D-channel. LAPD uses Carrier Sense Multiple Access/with Collision Resolution (CSMA/CR) to allow multiple TEs on the S/T bus. LAPD is a multipoint protocol.

Narrowband ISDN   Designates BRI and PRI, which use synchronous transmission.

Network Termination 1 (NT1)   An interface device provided by a customer to terminate an ISDN line to the customer's premises. The NT1 interfaces the customer's ISDN equipment to the digital subscriber line of the LEC.

Network Termination 2 (NT2)   An interface device that interfaces devices to a PRI NT1. An NT2 is usually used in a PBX environment. NT2s will only be found in a PRI environment.

Network Termination 12 (NT12)   A device combining an NT1 and an NT2.

Primary Rate Interface (PRI)   Consists of twenty-three 64-Kbps B-channels and one 64-Kbps D-channel.

Reference Points   The points at which various devices are connected in an ISDN circuit. The U interface serves as a demarc. It is where the cable pair from the LEC attaches to an NT1. The S/T reference is the interface where a TE1 or TA connects to an NT1. The R reference is the interface where a non-ISDN device attaches to a TA. The V reference is the interface point between the LE and ET in the central exchange.

Service Profile Identifier (SPID)   Contained in a database maintained by the LEC of ISDN line equipment and the services available to each line. This database contains SPIDs. When a terminal is activated, the SPID is used to determine which services a terminal can have access to.

Terminal Adapter (TA)   A device that interfaces a non-ISDN device to an NT1 in BRI and to an NT2 in a PRI environment.

Terminal Equipment Identifier (TEI)   Specific number assigned to each TE1 and TA. The TEI is usually assigned automatically the first time the terminal device is used. LAPD will invoke a query of the SPID database to obtain the TEI. The TEI is used by the LE and ET to determine what service to provide for a B-channel.

Terminal Equipment 1 (TE1)   Equipment designed to ISDN standards and for direct interface to an NT1.

Terminal Equipment 2 (TE2)   Equipment that is non-ISDN compliant. TE2 equipment must be attached to a terminal adapter.

2B1Q   The coding technique used for a digital subscriber line. This coding technique allows 2 bits to be transmitted at one time by using four distinct voltage levels. Each voltage level can represent a particular 2-bit code.

10

Data Networking via LANs

KEY TERMS

Application Layer
Applications Protocol Interface
  (API)
Baseband
Bridge
Broadband
Carrier Sense Multiple
  Access/Collision Detection
  (CSMA/CD)
Cat-5 Cable
Client
Client/Server
Collision Domain

Data Link Layer
DOS-based LAN
Ethernet
Group Ware
Industry Standard Association
  (ISA) Bus
Local Area Network (LAN)
Media Access Control (MAC)
Metropolitan Area Network (MAN)
Network Interface Card (NIC)
Network Layer
Network Operating System
OSI model

Peer-to-Peer
Peripheral Component
  Interconnect (PCI) Bus
Physical Layer
Protocol
Repeater
Server
Session Layer
Token
Token Ring
Transport Layer
Wide Area Network (WAN)
Zero-Slot LAN

OBJECTIVES

On completion of this chapter, you should be able to:

1 Explain what a local area network (LAN) is.
2 Explain what a hub is used for.
3 Describe Carrier Sense Multiple Access/Collision Detection (CSMA/CD).
4 Describe token passing.
5 Describe a zero-slot LAN.
6 Describe how a network interface card (NIC) functions.
7 Define a peer-to-peer LAN.
8 Define a client/server LAN.
9 Discuss how a bridge is used.
10 Discuss how routers are used.
11 Discuss the benefits of using a LAN.
12 Discuss what a network operating system (NOS) is.
13 Discuss the differences between baseband and broadband.
14 Describe the differences between thick-net, thin-net, and twisted-pair media.
15 Describe the OSI model and the functions of each layer.

16 Compare and contrast Ethernet and token-ring LANs.

17 Discuss the function of an applications protocol interface (API).

18 Compare and contrast the ISA bus and the PCI bus.

19 Describe the Ethernet frame.

20 Explain medium access control (MAC) addressing.

Data networking can take place in many different ways. Personal computers can be connected directly to each other in a point-to-point arrangement with a null-modem cable. They can be connected using modems. They can be connected via the public switched telephone network (PSTN) using analog modems. They can be connected by the public data network (PDN) or PSTN using a network terminating unit (NTU) such as a CSU/DSU. These methods were discussed in Chapter 8. Computers can also be connected using *local area networks (LANs)* and *wide area networks (WANs).*

10.1 LOCAL AREA NETWORKS

When only two computers need to be connected, we often use one of the methods discussed in Chapter 8. When we wish to link many computers that are physically close together (such as within one building, or on one floor of a building), we can utilize a LAN. We can then connect two LANs together that are close to one another (such as within the same building) using a device called a *bridge*. We can connect LANs to each other if they are within a larger geographic region (such as a city) using a *metropolitan area network (MAN),* and we can connect LANs separated by any distance using a WAN. The basic first step to getting all computers in a company set up so that they have access to all other computers is to get a LAN established for each group of computers. MANs and/or WANs can then be used to connect the LANs together (see Figure 10-1).

A LAN is a data communication system allowing a number of independent devices such as computers and printers that are usually located within 500 m of each other to communicate directly with each other over a common physical medium. The distance can be more or less than 500 m depending on the type of cable used for the medium. Repeaters can be used to extend the distance, but most LANs are 500 m or less. The use of a LAN allows data, applications, printers, and other resources to be shared efficiently and economically.

This sharing of resources significantly reduces the cost of computing compared to having individual PCs without any interconnection. When PCs first appeared in the business environment, each PC had its own printer and files were shared through "sneaker-net." With sneaker-net, a person copies files onto a floppy disk and then transports the file to a third party by walking (presumably while wearing sneakers) the floppy over to the third party.

As more and more PCs were introduced into the business, it became necessary to implement LAN architecture as a way of reducing costs and improving productivity. The hardware and software for LAN implementation had been developed by 1974 for use in mainframe environments at universities and engineering operations. LAN technologies were developed prior to the explosive

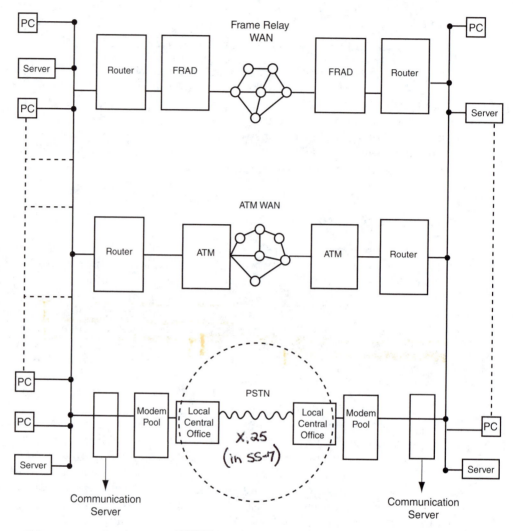

FIGURE 10-1 Wide area networks.

growth of PCs in the workplace. It was a natural evolution to take the matured LAN technology into the business environment during the late 1980s.

As stated earlier, there are many ways to connect computers together. There are also several ways of allowing the sharing of data and printers. Simple and cheap switchboxes can be used to attach many PCs to one printer. PCs can be connected to each other via serial or parallel ports; by using an appropriate software package, files can be downloaded or uploaded from one PC to the other. Applications software packages are available called *zero-slot LANs* that make it possible to connect a few PCs together. They are called *zero-slot* LANs because no hardware is required (no slot is used in the PC for a card).

The next step up from a zero-slot LAN are **DOS-based LANs**. Like zero-slot LANs, DOS-based LANs usually are **peer-to-peer** architecture. All PCs run the same software and can serve as both a client and a server. Zero-slot LANs communicate via serial or parallel ports. DOS-based LANs use hardware inserted into one of the expansion slots of a computer to send and receive data over the communication medium. The use of a network interface card provides a higher speed because the interface card accesses the parallel data bus of the computer. These DOS-based LANs include LANtastic, NetWare Lite, and Windows for Workgroups. These alternatives to full-blown LANs are cheaper, but the cost of LAN hardware and software has been dropping so dramatically that the cost difference is narrowing tremendously. In fact, the continually dropping prices for LAN hardware and software have been one of the major factors in the phenomenal growth of LANs.

People often justify the cost of installing a LAN based solely on the savings that occur by reducing the number of printers needed. High-speed laser printers can be quite costly ($2000 to $5000 each), but I am not so sure that you can fully justify the cost of a LAN on printer savings alone, since switchboxes could be used to allow printer sharing. LAN hardware has dropped so dramatically that the **network interface card (NIC)** for each PC workstation and PC server costs less than $200. The LAN software costs about $10,000 and servers run about $10,000. Spreading the cost of software and servers over many users quickly reduces the price per user to an acceptable level. With only 10 PCs connected to a LAN, the cost is about $2500 per station, including cabling costs. The cost to install a LAN is not insignificant for a small business and it may have to opt for the cheaper alternatives, but surely any business with 20 or more PCs will find that the productivity improvements gained with a LAN can justify its cost.

The use of a LAN allows a business to gain economies of scale on the purchase of applications software. Many vendors offer applications software designed for LANs at a lower cost per user than individual software packages would cost. These LAN packages can be purchased along with a license to allow for *X* number of users. These applications can then be stored on a single computer called a *file server*. The file server is attached to the LAN, and individual PCs called *workstations* or *clients* can access the applications program via the LAN. This arrangement is referred to as a **client/server** LAN. Some PCs are set up as **servers;** they provide services in the way of applications, file sharing, or printer services to the workstations that are called **clients.**

Servers can be minicomputers or microcomputers. With today's high-speed, high-memory, high-storage capacity microcomputers (PCs), the server is most often a high-powered PC. This leads to another benefit of the LAN. A LAN allows for reduced capabilities in the client (workstation PC). The individual user PCs (workstations) connected to a LAN can be equipped with smaller hard disk drives and memory as well as lower-speed processors. This tends to extend the useful life of workstations.

LANs also allow for the installation of **group ware** applications software. These packages provide e-mail and scheduling integration for all users connected to the LAN. Employees can use e-mail to keep each other informed and can attach files to e-mail for quick transfer of information. A LAN often has a public drive on the file server that everyone can access. This public drive is used to transfer files from one user to another. Most LAN software also contains a notification or chat application. This allows users to send instant messages to one another across the LAN. The scheduling integration allows users to post their work and meeting schedules for others to see.

In summary, there are many reasons for installing a LAN, but the key reasons are: (1) they are relatively cheap, (2) they allow sharing of applications software, data files, printers, and modems, and (3) they increase efficiency and productivity of the organization's workforce.

10.2 PUTTING A LAN TOGETHER

A LAN consists of the transmission medium used to connect workstations and servers together, the hardware used by the PCs to connect to the transmission medium, and the LAN-specific applications software used to manage the transmission of data between the PC and the transmission medium of the LAN. To connect workstations, servers, and printers together using a LAN, you first determine the quantity and types of devices that need to be connected to the LAN and what features you want the LAN to provide. You can then decide on the type of network operating systems software that you want to use for your LAN.

If you are planning on connecting two to five computers together, you might simply choose to use Windows for Workgroups and purchase a network card supported by this software. DOS-based LANs can support up to 300 users, but LANs of this size will usually opt for a Novell or Windows NT-based LAN. Both of these *network operating systems* are offered in versions that interface with all the popular operating systems. Many network managers will choose to use UNIX as the operating system for servers on the LAN. The workstations may be using a DOS or Windows 95 operating system. Thus, it is possible to have the Novell network operating system loaded on a server using UNIX as the operating system while the workstations may have a DOS or Windows 95 operating system. The network operating systems software contains an applications program interface (API) that handles the interpretation of messages to allow different PC operating systems to exist on the LAN.

Another decision that needs to be made is whether you want the LAN set up as *Ethernet* or *token ring*. These are the two predominant LAN architectures. About 80% of all LANs are Ethernet. Most of the remaining 20% are token ring. Standards for both LAN architectures are established by the Industry of Electrical and Electronics Engineers as a series of IEEE 802 LAN standards. The standard for an Ethernet LAN is IEEE 802.3; the standard for a token-ring LAN is IEEE 802.5. The original Ethernet employed a bus topology, but most new Ethernet LANs consist of computers wired from a repeater hub in a star arrangement. The repeaters in the hub are connected to a common bus (see Figure 10-2). Token ring employs a ring topology implemented by wiring the computers in a star arrangement. The central point of the star is a wiring hub called a *medium access unit (MAU)*. The MAU links the various stations attached to the MAU so that they are connected in a ring topology (see Figure 10-3).

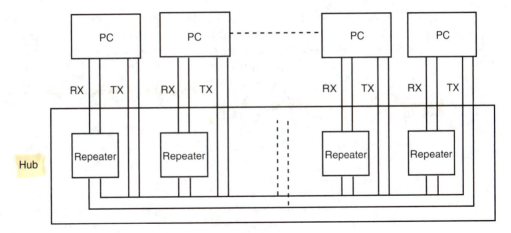

FIGURE 10-2 Twisted-pair Ethernet.

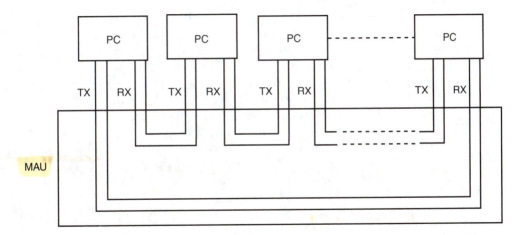

FIGURE 10-3 Token ring.

A third decision that needs to be made when setting up a LAN is the type of transmission medium to use. The choices are thick coaxial cable, thin coaxial cable, twisted-pair copper wire, and fiber optic cable. The first Ethernet-based LAN used thick coaxial cable, and many of these LANs are still in existence. The outer plastic cover of thick coaxial cable is usually a yellow or orange color.

Thick coax required that a device called a *transceiver* (transmitter/receiver) be placed between the PC and the coaxial cable. The transceiver was attached directly to the thick coax, and a nine-conductor cable connected the transceiver to the network interface card (NIC) in the PC. The nine wires included two wires as a transmit pair, two wires as a receive pair, and five wires to allow the NIC control of the transceiver. DB-15 connectors were used to connect the transceiver cable to the NIC. The maximum length for a transceiver cable is 165 ft.

Thick coaxial cable was run in the false ceiling, transceivers were attached to the coaxial cable, and transceiver cables ran to each PC. In this cabling system, the thick coax served as a bus to carry signals for all workstations and servers. The longest distance for this bus without using a repeater was 500 m. The signal speed on the bus could be as high as 10 Mbps. Thick coax is referred to as *10 Base-5*. The first number designates the speed of 10 Mbps; the last digit designates the maximum distance in hundreds of meters.

The base indicates that only one frequency of signal is traveling over the medium and it occupies the entire bandwidth of the medium. The term *baseband* means the signal placed on the LAN medium is at one frequency and occupies the entire bandwidth of the medium. *Broadband* systems are governed by the IEEE 802.7 standard. This standard uses 75-Ω coaxial cable as the medium and allows several applications to run over the medium at the same time using FDM. Broadband allows a system to carry cable TV signals and LAN data using the same medium but requires the use of more complicated electronics. Because broadband LANs require the use of more expensive electronic circuitry, almost all LANs use baseband technology.

The thin coaxial cable used for broadband systems is the same coaxial cable used to connect a VCR to a TV, or to connect the TV to a cable TV jack. Baseband systems use a different type of coaxial cable. Baseband systems use thin coaxial cable that looks similar to the TV cable used by broadband systems, but TV cable cannot be used for baseband systems because this cable has characteristic impedance of 75Ω. The characteristic impedance of both thin coaxial cable and thick coaxial cable used for baseband systems is 50Ω. Thin coaxial cable is cheaper than thick coaxial cable, and a LAN using thin coaxial cable as the communication medium (bus) is referred to as *ThinNet* and *CheaperNet*. Thin coaxial cable could handle signals at 10 Mbps, but the distance was reduced to 200 m. For these reasons, thin coaxial cable is referred to as *10 Base-2*. A NIC designed to connect to thin coaxial cable contains the transceiver, and an external transceiver is not needed.

Today, twisted-pair copper wire is the medium of choice for LANs and is referred to as *10 Base-T*. A NIC designed to connect to twisted pair also contains a transceiver, and an external transceiver is not needed. The standard for 10 Base-T requires that workstation and server PCs be wired in a star configuration to a central wiring center called a *hub* or *concentrator*. The maximum distance for twisted pair is 100 m. As mentioned earlier, the hub contains a repeater for each connection to the hub. The output of all repeaters is connected to a common bus. Each connection to the hub is made via a RJ-45 jack. The NIC in each PC also contains a RJ-45 jack. Twisted-pair copper wire cable is cut to the appropriate length necessary to connect a specific PC to the hub, and RJ-45 connectors are attached to each end of the cable.

Twisted-pair copper wire is rated according to the type of signal it can handle. There are five different grades of twisted-pair copper wire, from Cat-1 to Cat-5. Category 1 to category 3 cable is referred to as *voice grade cable*. Most of the cables used to wire telephones in most office buildings are category 3 cable. Category 3 cable can handle data rates up to 16 Kbps. Category 4 cable can handle data up to 20 Mbps and it was used in most token-ring networks. Category 5 cable can handle data rates of 100 Mbps. The use of Cat-4 cable has diminished because Cat-5 cable

even 1 Gbps

offers higher bandwidth capacity at almost the same cost and can handle higher data rate upgrades of the LAN. *Cat-5 cable* is referred to as *data grade cable.* The difference in grades stems from the construction of the cable. The biggest difference in construction is in the number of twists per inch. Cat-5 cable has about 36 twist per foot, while Cat-3 cable has about 6 twist per foot.

Four wires are used to connect the wiring hub to the NIC. Two conductors are used to connect the transmitter of the NIC to a *repeater* in the hub, and two wires are used to connect the receiver of the NIC to the bus in the hub. The data transmitted by the NIC of a PC is connected through the repeater in the hub to the bus connected to all NIC receivers.

Most LANs use twisted-pair copper wire as the transmission medium of the LAN. By using a wiring hub and repeaters in the hub for each connection, it is possible to achieve speeds of 1 gigabit per second over Cat-5 cable. Fiber optic cable is used in few LANs because the cost for fiber-based NIC and fiber cable is much higher than the cost for twisted-pair copper wire NIC and Cat-5 cable. Fiber will be used where electromagnetic interference (noise) might be a problem. Fiber is also used to connect LANs on different floors or buildings together.

10.3 THE OSI MODEL

As a basic book on telecommunications technologies, this may be the right place to introduce the *OSI model*. Some instructors may choose to skip this topic, since you will be inundated with discussions of the model in later data communication classes. I will provide a high-level overview of the model since we have already discussed the first layer of the model (the physical layer). The open systems interconnection (OSI) model was developed by the International Standards Organization (ISO) to provide a framework for organizing networking technology and protocols.

Protocols are the rules of communication between devices (hardware or software). The path that data travels in going from an applications program to the data bus of a LAN was broken down by the OSI model into seven steps or layers, and the OSI model is referred to as a *seven-layer model* (see Figure 10-4). The nice thing about breaking a process down into seven distinct steps is that it allows working on one part of the model independently of the other parts. Standards are set for how

Layer	Function
7 Application	Involves user applications like Word, Excel, Write, etc.
6 Presentation	Presents data in user's operating systems requirements
5 Session	Enables two applications to communicate
4 Transport	Ensures reliable transmission
3 Network	Sets up the transmission path
2 Data link	Puts data in proper format and controls transmission
1 Physical	Transmits and receives signals in proper format

WAN protocol

HTTP, FTP
TCP
IP

Token ring, Ethernet

LAN protocol

FIGURE 10-4 OSI NIC model

the different layers communicate with each other. This allows a developer to write programs that will perform unique functions in a particular level as long as they provide the proper interface to the level above and below.

We have spent a great deal of time discussing the transmission medium of a LAN. The connection between a LAN and the NIC occurs at the *physical layer*. Most NICs come with two connectors: (1) a BNC connector to allow connection to thin coaxial cable, and (2) an RJ-45 connector to allow connection to twisted-pair cable. Some NICs also include a third connector, which is the DB-15 connector necessary for connecting to the transceiver on a thick coaxial cable.

The physical layer also includes the transceiver, whether it is on the NIC or external on the thick coax. The transmitter will send the appropriate signal at the speed that it is designed for out onto the LAN medium. The receiver will accept signals from the LAN medium and convert them into 1s and 0s. The NIC operates at the physical level as described, but it also performs level 2 functions. The manufacturer of a NIC designs it to work with a specific network technology such as Ethernet or token ring, which are layer 2 and layer 1 functions.

Ethernet uses a *data link layer* that conforms to the IEEE 802.3 standard. This standard specifies that the NIC must monitor the LAN medium to ensure it is not in use before sending data onto the medium. It also specifies that if two devices send data simultaneously, they are to back off, do a random wait, and then retransmit. This 802.3 standard is called *Carrier Sense Multiple Access/Collision Detection (CSMA/CD)*. Token ring conforms to the IEEE 802.5 standard, which specifies that computers cannot transmit unless they hold a *token* that is passed around the ring from one PC to the next. Two completely different methods of controlling access and transmission onto the LAN medium require two different kinds of hardware.

Since the data link layer is part of the NIC card, it is considered part of the hardware, but that does not mean it is only hardware. The NIC card contains its own CPU and also has memory storage of its own. The NIC card handles getting data onto and off of the LAN medium. NICs are programmed at the factory with a specific hardware address that is referred to as the *media access control (MAC)* address. The NIC captures all data packets traveling around the LAN and puts them into a memory buffer on the NIC. It then checks to see if the data packet is addressed to it. If so, it sends an interrupt request via its assigned IRQ to the CPU of the PC. If the data is not addressed to the NIC, it simply discards the data.

To perform all the functions just discussed, the NIC contains programming burned into a ROM chip. This programming is firmware rather than software, since it cannot be changed. The NIC is a combination of hardware and software that performs at the first two layers of the OSI model. The network operating system (NOS) software such as Novell NetWare or Microsoft NT operates at the upper 5 layers of the OSI model. There is much more to the use of an application than the application layer. Basically a NOS can be thought of as an application that allows other applications to run through it.

The NOS controls the NIC through the PC's operating system by using software drivers. This is why any type of PC on a LAN can use any NOS. There are software drivers available to interface each type of NOS to each type of PC and its operating system. The NOS also contains software modules that operate at the *network, transport, session,* and *application layers*. Earlier, I mentioned the *applications protocol interface (API)* that operates at the session level to allow exchange of data between the presentation layer and the transport layer. One of the most common APIs is NetBIOS.

The job done by the session layer is a powerful one. It provides the means by which an Apple PC can be on the LAN and be able to communicate with an IBM PC on the LAN. Data originating via the Apple's operating system is converted by its session layer into the data structure needed by the LAN. An IBM PC on the LAN can receive that data because its session layer will convert the data received from the LAN into the data structure needed by the operating system of the IBM PC. Basically the session layer software performs a function similar to the software driver between the NOS and the operating system. The software driver is working between layers 2 and 3; the session software such as NetBIOS operates between layers 4 and 6.

10.4 CONNECTING TWO LANS TO EACH OTHER

A LAN allows PCs to be connected together so that they can share resources and communicate directly via chat or e-mail. Repeaters can be used to extend LANs. The repeater is a layer 1 device. It simply regenerates new signals based on the signal received. LANs of similar or dissimilar architectures can be connected together by a bridge. The bridge is a device that uses both layers 1 and 2 but is usually described as operating at the layer 2 level. It has the intelligence necessary to be able to determine which addresses reside on which side of the bridge. After going through an initial learning sequence upon attachment to the two LANs, the bridge will only pass frames of data that need to cross from one LAN to the other.

Special bridges known as *remote bridges* can be used on each end of a leased transmission facility to connect LANs that are distant from each other together. Each LAN is connected to a remote bridge, and the remote bridges are connected to a CSU/DSU. The CSU/DSU at each location connects to a leased line such as a 64-Kbps or 1.544-Mbps circuit. Bridges are ordinarily used to connect similar LANs, but bridges do exist that allow the connection of an Ethernet LAN to a token-ring LAN. These bridges are often called *translating bridges, intelligent bridges,* or *encapsulating bridges*.

One nice application for a bridge is to split a LAN up into smaller sub-LANs. This creates two collision domains. If the traffic on an Ethernet LAN increases to a point that collisions are occurring and degrading performance, you can use a bridge to split the LAN into two segments. You would try to split the LAN so that the traffic would be divided into equal loads for each new segment. This effectively reduces the number of collisions that will occur. Sometimes people mistakenly think that increasing the speed of a LAN will increase its capacity. This is not so. If you have collisions at 10 Mbps and increase the speed of the LAN to 100 Mbps, you increase the number of collisions, and when collisions do occur, you have more information to recover.

The use of a bridge creates separate subnetworks that each serve as a *collision domain*. Thus, each segment could be running at 10 Mbps. This effectively doubles the speed of the network. Of course, some traffic will be destined for workstations or servers on the other side of a bridge from the originator, which

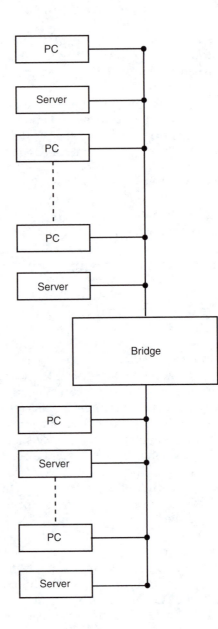

FIGURE 10-5 Bridged LAN.

means we cannot truly double the speed of the LAN. But we can surely increase the performance of high-traffic LANs by using bridges to segregate the traffic into smaller domains. (see Figure 10-5).

10.5 THE PERSONAL COMPUTER

The speed of PCs has continued to evolve over time. Along with this increase in speed has come more memory and larger secondary storage devices. The PCs of today can do more than the mainframe computers of 25 years ago, and they can do it faster. The keys to explosive growth in PCs were standardization and off-the-shelf software. Many of us were introduced to the PC when it first came out, but it was more of a word processor than anything else. If we wanted to make the PC do anything, we had to write programs to perform the desired task. Early PCs even came with a tutorial on how to write programs in BASIC.

The acceptance of DOS as a standard operating system for PCs led to the development of many applications programs based to run on top of DOS. The acceptance of the *Industry Standard Association (ISA) bus* architecture was another big factor in the growth of PCs. The ISA bus specifies the interface leads (the address, data, and control leads) of devices that need to communicate with the central processing unit (CPU). The ISA bus contained 60 leads. This standard led to the fast development of many peripheral devices such as floppy disk drives, hard disk drives, internal modems, game cards, and so on.

Although the ISA standard helped propel the growth in PC use, it had severe limitations. The original ISA bus had 20 address leads and 8 data leads. With the 80286 CPU, the standard was changed by adding additional leads called the *16-bit section* to the ISA connector. Thirty-six extra leads were added to the bus to provide the extra leads needed for 24-bit addressing and 16 data leads. The most obvious flaw for ISA is that the ISA bus only contains 16 data leads and it still operates at a speed of 8.333 MHz. A new bus was developed by Intel to accommodate the higher speeds of today's PCs. It is called the *peripheral component interconnect (PCI) bus.* The PCI bus contains 124 leads and operates at 33 MHz. It contains 32 data leads, but the push is on to upgrade the PCI bus to a 64-data lead bus and to increase the speed to 66 MHz.

Most NICs developed to interface a PC to an Ethernet LAN are designed to plug into an ISA slot. Several manufacturers offer NIC cards that plug into a PCI slot. The ISA or PCI bus provides the CPU on the NIC with access to the IRQ leads, address leads, control leads, and data leads of the PC. This allows the NIC to write and read from the PC's memory. Data is transferred between the NIC and the memory of the PC, 16 bits at a time at a speed of 8.333 MHz (the limit of the ISA bus). A PCI-based NIC can achieve the higher speeds offered by the PCI bus (33 MHz and higher).The NIC is a hardware component that plugs into the ISA or PCI bus of a PC. It is controlled by the PC's *basic input output system (BIOS).* Additional control is provided by higher-level software in the form of software drivers and a NOS. The main CPU of the PC will control the transfer of data between the PC and the NIC. The NIC also contains its own firmware and BIOS to control its interactions with the medium of the LAN.

The open standards development environment for the PC and its peripherals created an atmosphere in which anyone could develop devices to work with the CPU of a PC. The widespread acceptance of DOS provided a stable operating system that a software developer could write applications programs for. As the operating system evolved to include a graphical user interface (GUI), the applications programs became more complex but were easier to use. These higher-powered programs needed more memory and faster CPUs. The CPU chip has continued to evolve into faster and faster chips. Of course, if all you need is a word processor, you do not need much CPU processing power, but many applications in use today require more memory and faster CPUs.

The development of faster CPUs and the growth in number of data leads from 8 to 64 (Pentium motherboard) means that information can be exchanged within the PC at incredibly fast speeds. Many PCs now come with dual Pentium processor chips. The size of programs has continued to climb, and the price of memory has continued to decline. It is now standard to find 32 MB of RAM in most PCs. Most PCs offer CD drives and secondary storage drives of 1 GB or greater. The continued evolution of the PC and applications programs has led to high growth in the number of PCs used in the business environment. This growth has led to an explosive growth in the number of LANs.

One of the main ideas behind the use of a LAN is the ability to share resources between PCs on the LAN and to use less expensive PCs as workstations. Even though users can resort to low-powered PCs as workstations, the price of PCs is so low that they eventually get the latest high-powered PCs anyway. Older PCs do not have a PCI bus, and you may opt to use all newer PCs so that you can utilize PCI-based NICs. You may also want to include larger hard drives on the workstation in order to keep copies of applications software on the local hard drive of a workstation. This can reduce traffic on a LAN. If shared resources can be downloaded to a workstation's local hard drive (C drive), it will eliminate multiple messages over the LAN to retrieve data from a server. Management of resources on a LAN includes determining how those resources will be shared. It may be significantly cheaper to use a 25-user license for an applications program if you think no more than 25 users would be using the software at the same time.

If you have 100 or more workstations attached to your LAN, the cost to have a copy of the applications program on each user's hard drive may be much more than utilizing a 25-user license. The main advantage of a LAN is the ability to share resources and keep costs low, but you need to actually verify what your license agreement states about how the software can be shared. Can the software reside on all local hard drives as long as no more than 25 users are using it at the same time? How much cheaper is a 25-user license versus a 100-user license? Sometimes it is best to actually place a copy of applications on a user's hard drive rather than to have everyone trying to access one copy on a file server. Sharing introduces delay, and the network manager needs to ensure that delays are held to a minimum. The manager of the LAN needs to weigh the cost versus benefits of distributed applications to make the proper choice on where applications should reside.

10.6 THE LAN MEDIUM

The medium of a LAN is a shared medium. Users take turns using the medium, and each user is granted a brief period when they can use it. Thus, the medium is used in a time division multiplexing (TDM) arrangement. Ethernet controls the size of each frame of data that a user can send. Ethernet limits the users' data to 1500 bytes at a time. This ensures that one user cannot hog the medium for extended periods. Since each byte is 8 bits, 1500 bytes is 12,000 bits. With an Ethernet operating at 10 Mbps, this equates to 1.2 ms. Since each user is only granted access to the

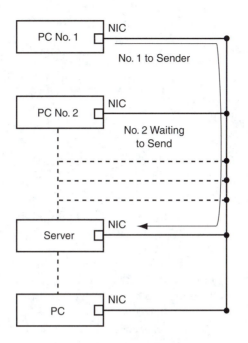

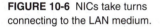

FIGURE 10-6 NICs take turns connecting to the LAN medium.

medium for a maximum time of just over one-thousandth of a second, it appears to each user as if they have instant access to data over the LAN.

The NIC contains a CPU and memory. It is fast enough to interact with the LAN at the speed it was designed for. Thus, to change a LAN from a 10 Mbps LAN to a 100-Mbps LAN involves changing the NIC in each workstation or server. Although the NIC contains its own processor and this processor controls interaction with the LAN, the speed of the CPU in the PC is also important. You do not want the PC's processor to be slower than the processor on the NIC. The faster the PC, and the faster the bus, the faster data can be exchanged between the buffers on the NIC and the internal memory of the PC.

The medium of the LAN is like a high-speed interstate highway. Each PC has a NIC to control the flow of traffic out onto the high-speed highway. Only one user's data can be on the highway at one time, but that data is traveling at near the speed of light for a very short distance. The frame of data that a user's NIC places on the LAN will contain a destination and source address. The NIC that matches the destination address will present the received frame of data to its host PC. Each user is taking turns using the highway (medium) of the LAN (see Figure 10-6).

10.7 THE NETWORK INTERFACE CARD

The network interface card (NIC) is a printed circuitboard (PCB) containing special hardware components, a CPU, memory, and firmware designed to interface to a particular LAN technology such as Ethernet or token ring. The method used to control access to a LAN and the method used to encode and transmit data on a LAN are

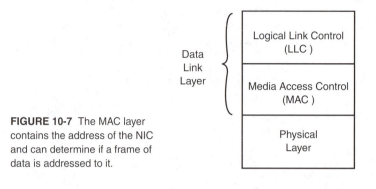

FIGURE 10-7 The MAC layer contains the address of the NIC and can determine if a frame of data is addressed to it.

different for each technology. Thus, NICs have to be designed differently for each of the LAN technologies employed. Ethernet is employed in 80% of all LANs, so we restrict our discussion to these LANs.

If you look at the OSI model, the first layer is the physical layer. This layer defines how bits are to be transmitted. The second layer is the data link layer. This layer controls how data is to be packaged and placed on the physical medium. Thus the first two layers of a LAN protocol are determined by the hardware. An Ethernet card contains the first two layers of the Ethernet protocol within the hardware on the NIC.

The second layer of the OSI model is the data link layer. The IEEE committee split this second layer into two sublayers. The layer next to the physical layer is called the MAC layer. The layer above the MAC sublayer is the *logical link control* sublayer, and it interfaces the data link layer to the network layer. The purpose of the MAC layer is to control access to the LAN. IEEE 802.5 specifies the access methodology used by token ring, and IEEE 802.3 specifies the access used by Ethernet. The MAC layer also contains the address of the NIC and can determine if a frame of data is addressed to it (see Figure 10-7).

Technically Ethernet and IEEE 802.3 are not the same, but for all intents and purposes everyone treats them the same. IEEE 802.3 is the standard specification for an access methodology to the LAN medium that is called CSMA/CD. Since users share the media of a LAN, some method must be employed to control who can use the LAN. With CSMA/CD, potential users monitor the LAN for activity. If activity is taking place on a LAN, the carrier signal will be present (electrical energy will be present). Thus, the NIC hardware will monitor for activity by monitoring to see if any carrier signal (electrical energy) is present on the LAN. If a signal is present on the medium, our NIC will not transmit.

If we have data to transmit out onto the LAN medium, our NIC will receive instructions from the CPU of the host where the data resides, and it will be instructed to transmit that data. The NIC will check for a signal on the medium; if none is present it will form the data into an Ethernet frame and transmit that frame onto the medium. If a collision occurs due to another NIC trying to send at the same time, the collision is detected and the NIC will stop transmitting. Each NIC will go into a randomly assigned wait time before it attempts to retransmit. After waiting its random wait time, the NIC will transmit the Ethernet frame of data onto the medium.

Once an Ethernet frame of data is placed on a LAN medium, it is received by all NICs attached to the LAN. Data is broadcast out onto the LAN medium. Each NIC is programmed at the factory with a 6-byte (48-bit) address. The first 3 bytes identify the manufacturer; the last 3 bytes are the serial number of the card. Each NIC knows its own address. When it receives a frame of data, it checks to see if the frame is addressed to it. If not, the frame is discarded. If the frame is addressed to the NIC, it notifies the CPU of the host that it has data for an application running on the PC by sending an interrupt request (IRQ) to the main CPU. The ISA bus contains IRQ leads and one of these will have been assigned to the NIC on its installation in the PC.

10.8 PACKETS AND FRAMES

When data is transferred from one PC to another over a LAN or WAN, it is transferred according to the protocol for the transmission medium being used. Each service such as Ethernet, token ring, X.25, frame relay, ATM, and so on has its own distinct **protocol**. Because each service is different, it must have its own unique protocol, but there is one common aspect they all use. Data to be transmitted is formed together in a group of 1s and 0s called a *packet*. The different protocols use different-size packets, but they all arrange individual bits and bytes into a larger block of data called a *data packet*. This data packet is then placed inside a larger container called a *frame*.

The frame contains additional information used by the various protocols. Some protocols use frames that contain the address of the destination, the address of the sender, error-checking and error-correction information, and possibly control information. Some protocols contain a header in front of the data being transmitted and a trailer following the data. Some protocols contain only a header before the data being transmitted and do not contain trailers.

When several layers of protocols are involved in transmitting a message from one point to another, each higher-level frame of data will be encapsulated within the lower-level protocol's frame of data. This is usually demonstrated by the use of many sizes of envelopes. Applications data may be thought of as data written on a sheet of paper. This applications data can be put inside a regular business-size envelope that represents the presentation layer. The applications data is encapsulated within the presentation envelope. The presentation envelope contains information that will be used by the presentation layer at the receiving end to determine the proper application layer to invoke. The presentation envelope is now placed inside a larger envelope called the *session layer.* As we continue through each layer in a layered protocol stack, previous-layer frames of data are encapsulated into the frame for the next lower layer.

At the receiving station, the lower-layer protocol looks at its frame of data and determines what to do with the higher-layer frame encapsulated within its frame. The opening of envelopes (lower-layer frames) continues until all we have left is the sheet of data (the original data packet) that is delivered to the application. Regardless of the protocols being used, data is formed into packets and encapsulated

Ethernet Frame

8 bytes	6 bytes	6 bytes	2 bytes	46–1500 bytes	4 bytes
Preamble	Destination Address	Source Address	Type	DATA	CRC

IEEE 802.3 Frame

7 Bytes	1 byte	6 bytes	6 bytes	2 bytes	46–1500 bytes	4 Bytes
Preamble	Start of Frame	Destination Address	Source Address	Length	DATA	CRC

FIGURE 10-8 Ethernet frame and IEEE 802.3 frame.

within frames for transmission to the next protocol. At the physical layer, the frame is transmitted out onto the medium for delivery to the distant PC.

The size of data that a frame can contain is limited to a certain size so that one application or PC cannot monopolize the medium. All data networks that transmit data as frames or packets are using TDM to allow sharing of a common medium by many devices. The more data contained within a frame, the more time the frame occupies the common transmission medium. The maximum size for a data packet in an Ethernet frame is 1500 bytes. (see Figure 10-8).

By now you know that 1500 bytes consists of 12,000 bits. Ethernet running at 10 Mbps can transmit 12,000 bits in 1.2 ms. Thus, the maximum amount of continuous time one user can be connected to the LAN medium is a time slot of 1.2 sec. If another user is waiting to access the medium, they will then be allowed a maximum of 1.2 ms. This small slice of time is unnoticeable to a user and response seems instantaneous. If no one is waiting, the first user will be allocated another 1.2-ms time slot and so on.

The Ethernet and IEEE 802.3 frames contain a header of 14 bytes (112 bits) and a trailer of 4 bytes (32 bits). Since the minimum size of a data packet is 46 bytes, the minimum size for an ethernet frame containing the header, data, and trailer is 64 bytes and the maximum size for a frame is 1518 bytes. Note that the 8 bytes of preamble are not considered either a part of the header or part of the frame. The preamble consists of alternating 1s and 0s. This is used to alert the NIC that a frame is being placed on the medium. The NIC uses the 8 bytes (64 bits) to synchronize its receiver to the incoming 1s and 0s before it sees the header, data, and trailer.

Regardless of the type of protocol used, all data transmissions over facilities that use TDM occur as the transmission of data packets. The various protocols used to transmit data use different-size data packets and different control information within a frame. The frames of data for each protocol are different. When a medium using one protocol must connect data to a medium using a different protocol, a translator is included in the connecting device. If a bridge is used to connect a token-ring LAN to an Ethernet LAN, the bridge will contain a translator to translate from one protocol to another and is called a *translating bridge*.

10.9 THE ETHERNET FRAME

As mentioned earlier, Ethernet is not technically the same as IEEE 802.3. They both use baseband transmission and CSMA/CD but the data within a frame is framed a little differently. Figure 10-8 shows the differences between the two frame formats. On receiving the preamble bits, the NIC receiver is synchronized to the stream of 1s and 0s. It then receives the address for the destination of the frame. If this address matches the address of the NIC, it accepts the frame of data so that it can be passed up to the applications running on the PC.

The Ethernet card is programmed at the factory with a specific hardware address. This address is also called the *MAC address* or *physical address.* The source and destination address fields are 6 bytes wide. The first 3 bytes are used to identify the manufacturer of the NIC. Three bytes is 24 bits. If the 24 bits are grouped into groups of 4 bits each, we end up with 6 groups that are 4 bits each. A grouping of 4 bits can represent the hexadecimal numbers from 0 (0000) to F (1111). The NICs see the address as a continuous stream of 1s and 0s, but we write them in hexadecimal format to make it easier for us to understand.

Many manufacturers make NICs for Ethernet. I will discuss a few of them here. Cisco is assigned the manufacturer's number of 00000C (hexadecimal). 3Com is assigned 00608C (hexadecimal) and Western Digital is assigned 0000C0. Using 24 bits (3 bytes) for the manufacturer's number provides for over 16 million numbers. Also, providing 3 bytes for the serial number gives each manufacturer over 16 million serial numbers to use. Figure 10-9 shows an Ethernet frame with destination and source addresses.

Notice that the address fields in Figure 10-9 are written in hexadecimal using 4-bit groupings (12 groups of 4 bits each). The first 6 groups provide the 24 bits or 3 bytes used to identify the manufacturer. In Figure 10-9, the destination NIC is manufactured by Cisco (00 00 0C) and the serial number is 14,720 (00 39 80) ($3 \times 16^3 + 9 \times 16^2 + 8 \times 16^1$). For ease of representation, hexadecimal is used to represent the data contained within headers of data frames.

Notice that the Ethernet frame contains a type field to identify the data being carried by the Ethernet frame. If this frame were carrying an Internet protocol (IP) datagram, the type field would be 08 00 in hexadecimal format. If the type field were 08 06, it would indicate that the datagram being carried inside the Ethernet frame was an address resolution protocol (ARP). The IEEE 802.3 frame format does not include a type field. The type of datagram being carried would be identified within the data field. The first part of the data field under IEEE 802.3 is the logical link control (LLC) frame header. The first 8 bytes of the LLC header will be used to identify the type of datagram being carried within the LLC frame.

8 bytes	6 bytes	6 bytes	2 bytes	46–1500 bytes	4 bytes
	Destination Address	Source Address			
Preamble	00 00 0C 00 39 80	00 60 8C 00 17 42	Type	DATA	CRC

FIGURE 10-9 Ethernet Frame.

Each NIC has a unique address so that it can receive data addressed only to it, but sometimes a NIC may want to broadcast a message to all NICs on the LAN. This comes in handy when a NIC may not know the addresses of other NICs on the LAN. All NICs are programmed at the factory with a unique address, but they are also programmed to accept messages when the destination address field contains all 1s. In the hexadecimal format that we use for address fields, all 1s is ff ff ff ff ff ff. Anytime a NIC sees this number in the destination address field, it accepts the frame.

The broadcast address is used when a new station is added to a LAN. On power-up, the NIC will broadcast a datagram to the LAN with all 1s in the destination field and its address in the source field. All NICs receive the broadcast; seeing a source address not in their table of addresses, they add the new address to the table. All NICs now know about the presence of the new workstation. The NICs all respond back to the originating NIC. On receiving the responses, the NIC adds all the source addresses to its table and it now knows of all NICs attached to the LAN. We will see later that the process just described is done via the *ARP protocol* (also known as the *ARP response.)*

10.10 COMMUNICATION WITHIN THE LAN

For devices to communicate with each other across a LAN, they must know the hardware address (physical or MAC address) of the device they wish to communicate with. This is why the Ethernet frame contains MAC addresses. If a higher-level protocol is being used on a LAN such as TCP/IP, higher-level network addresses will be used by applications that wish to converse across the LAN or across networks. These higher-level network addresses must be converted within the workstation or server to numerical hardware addresses.

If a workstation on a LAN needs to communicate with a server on the same LAN, the higher-level address will be resolved to the MAC address via tables within the workstation or server. If a workstation is requesting services from a server on a different network, it must pass the frame of data to a router that can send the data on a path that will reach the intended network. The router is attached to two or more different networks. One of the networks it is attached to is the LAN of the originating workstation. In this case the NIC in the workstation communicates with the NIC in the router. The higher-level address is retained at the network layer of the protocol stack, and the router can check it to see where the datagram should be routed to. The router and workstation reside on the same network and must communicate with each other using MAC addresses.

Higher-level addresses are necessary to be able to communicate with devices on other networks, but the passage of datagrams from one device to the next requires that the devices use each other's physical address. This is much the same as using a zip code to get a letter to the right town. Using network addresses (zip codes) is a much easier way to get data to the right network. Once we have done that, we must look deeper and use the actual physical address of the recipient to get the data delivered to the proper place.

10.11 SUMMARY

Data can be transmitted from one computer to another using various methods such as modems and null-modem cables. Data can also be exchanged between computers that are part of a local area network (LAN). The growth of PCs in the workplace has led to an explosion of LANs to connect PCs together. This allows users to share data and provides a convenient way for users to communicate with each other (e-mail).

The most prevalent LAN technologies are token-ring (IEEE 802.5) and Ethernet (IEEE 802.3) technologies, with Ethernet having almost 80% of the LAN market. Ethernet uses an access methodology called CSMA/CD to control which station is allowed to transmit on the LAN. The LAN medium is a time-shared medium, and some method must be in place to limit access to the medium and to control how long a user can use it at one time. With Ethernet the maximum amount of data that can be placed on the medium at each access by one user is 1500 bytes. If they have more data than this to transmit, they must wait for additional access privileges to send the remaining data in 1500-byte increments. This limits each access to just over 1 ms on a 10-Mbps LAN.

NICs are programmed at the factory with a unique hardware address. Each NIC will accept frames of data addressed specifically to it and will also accept broadcast frames. Devices on a LAN must communicate with each other using the MAC address. Higher-level network addresses can be used at the network layer of the OSI, but they must be resolved to hardware addresses when the two devices reside on the same network. Network addresses that are not on the same network must be resolved to the MAC address of the NIC in a router that can route the data out toward the network desired.

The destination address and source address fields contain 48 bits, but these fields are usually expressed in written form as hexadecimal to make them easier to read. After the destination and source address fields is a type field in the Ethernet frame to identify what type (protocol) of data the Ethernet frame contains. In the IEEE 802.3 frame the type field is found in the header of the LLC frame encapsulated within the Ethernet frame. Thus, the Ethernet frame can carry many types of higher-level protocols. The type of protocol being carried is identified either in the header of the Ethernet frame or in the header of the LLC frame.

CHAPTER REVIEW QUESTIONS

1. What is a local area network (LAN)?
2. What is a zero-slot LAN?
3. What is a DOS-based LAN?
4. What is Windows for Workgroups?
5. What is a peer-to-peer network?
6. What is a client/server network?

7. What is a network interface card (NIC)?
8. What type of access methodology does Ethernet use?
9. What are the two sublayers of the link layer in IEEE 802.3?
10. What are the differences between an OS and a NOS?
11. Which layer of the OSI model contains the API?
12. What does *base* in *10 Base-T* mean?
13. What is Cat-5?
14. What is a MAC address?
15. Name the layers of the OSI model.

PROJECTS AND CASE STUDIES:

Case Study #1

Residential Real Estate Company has 50 agents located at three offices around town. They want a web page, e-mail capabilities, and ten PCs at each office that agents can share. If you were to install a LAN at one of the locations and the LAN consisted of 10 PCs, Ethernet 100 Mbps NICs, Category 5 cable, and one server, what would it cost to build this LAN? If you do not know how much it would cost, how much would someone in your town charge to provide all equipment installed?

GLOSSARY

Application Layer   The seventh layer of the OSI model. It does not include end-user application programs but is the layer that provides the support for these applications.

Applications Protocol Interface (API)   An interpreter for an application. The API resides at the session layer of the OSI model. The API permits applications to run on different network operating systems.

Baseband   The predominant method of data transmission on a LAN. The medium carries only one signal at a time and does not use FDM.

Bridge   A layer 2 device. It uses the MAC address to determine whether to pass a frame from one side of the bridge to the other. Most bridges join similar LANs. They can be used on Ethernet to break a larger LAN into smaller segments, thereby creating smaller collision domains. A special bridge known as a *translating bridge* can be used to join an Ethernet LAN to a token-ring LAN.

Broadband   A transmission medium using FDM to allow the medium to carry multiple signals at the same time. Broadband requires special electronics and is seldom used for LANs.

Carrier Sense Multiple Access/Collision Detection (CSMA/CD)   The type of access methodology used by Ethernet. IEEE 802.3

specifies that the NIC will monitor the medium to determine if signals are present before allowing the NIC to transmit onto the medium.

Cat-5 cable   Category 5 cable. This cable contains tightly twisted wire pairs with about three twists per inch. Cat-5 cable is data grade cable.

Client   The workstation in a client/server LAN.

Client/Server   An environment for a LAN where the PCs serve as either a workstation (client) or a server.

Collision Domain   A section of a LAN medium where collisions will be detected.

Data Link Layer   The second layer of the OSI model. It controls how data is to be placed on the medium. It controls the size of frames transmitted on the LAN.

DOS-based LANs   LANs that do not require a NOS.

Ethernet   A LAN technology using a bus or star topology and access via CSMA/CD with frames that limit user data to 1500 bytes.

Group Ware   An application that allows users on a LAN to share information with each other via e-mail and scheduling applications.

Industry Standard Association (ISA) Bus   Originally a 60-pin bus that transferred data 8 bits at a time and had 20 address leads. Thirty-six additional pins were added to another socket referred to as the 16-bit bus. This allowed the transfer of data 16 bits at a time and provided an additional 4 address leads.

Local Area Network (LAN)   A method used to attach several computers together so they can share information, files, programs, printers, and other resources.

Media Access Control (MAC)   The lower layer of the data link layer. It controls access to and from the medium of the LAN.

Metropolitan Area Network (MAN)   A wide area network that spans a geographic area the size of a city.

Network Interface Card (NIC)   Designed to work with a specific technology and designed to plug into an ISA or PCI slot in a PC. The NIC contains a CPU, programming, and memory that allow it to control the flow of information between the NIC and the LAN without assistance from the main CPU of the host PC.

Network Layer   The third layer of the OSI model. The network layer allows for the establishment, maintenance, and control of sessions between computers not on the same physical network.

Network Operating System (NOS)   Designed specifically to allow PCs to share resources via a LAN. The NOS occupies the network, transport, session, presentation, and application layers of the OSI model and includes the APIs and software drivers necessary to interface the NOS to the OS and to user applications.

OSI Model   Developed by the International Standards Organization and used as a template to separate defined functions or protocols into seven separate layers. It is more of a conceptual than a practical framework.

Peer-to-Peer   A LAN arrangement where each PC can act as either a server or a client.

Peripheral Component Interconnect (PCI) Bus   Developed by Intel to provide for faster transfer of data between peripherals and the motherboard.

Physical Layer The lowest layer of the seven-layer OSI model. It contains the hardware necessary to make physical connection to a LAN medium and to transmit electrical signals onto the medium.

Protocol Rules that govern how communication is to take place.

Repeater A hardware device that regenerates digital signals.

Server A PC or minicomputer containing resources or access to resources that can be shared via a LAN with clients (workstations) on the LAN. A print server is a PC that controls the access of clients to the various printers attached to a LAN. A file server contains files and applications software that can be shared with clients.

Session Layer The fifth layer of the OSI model. This layer is responsible for maintaining and terminating sessions between user applications programs. In a client/ server environment, the session layer ensures applications can cross between different operating systems on the client and server.

Token The access methodology used by token-ring (IEEE 802.5) technology. A special bit pattern called a *token* is passed from one station to the next, and only the station with the token can access the medium of the LAN and transmit data onto the medium.

Token Ring A LAN configuration where data passes through each station attached to the LAN. The stations are wired in a ring configuration by wiring all stations to a medium access unit (MAU).

Transport Layer The fourth layer of the OSI model, it provides for end-to-end reliability of network connections by providing flow control with error-detection and recovery mechanisms.

Wide Area Networks (WAN) Networks that span a large geographic area such as between cities.

Zero-Slot LANs LANs that do not require a NIC and thus require no slot on the ISA or PCI bus. PCs are connected via the serial or parallel port.

11

Wide Area Networks

KEY TERMS

Asynchronous Transfer Mode
(ATM)
Bit-Oriented Protocol
Committed Information Rate
(CIR)
Connectionless Protocol
Connection-Oriented Network
Data Link Connection Identifier
(DLCI)
Flag
Frame
Frame Relay

Frame Relay
Assembler/Disassembler
(FRAD)
Header
Information Service Provider
International Data Number
(IDN)
Link Access Procedure Balanced
(LAPB)
Logical Channel Number (LCN)
Modulo 8
Modulo 128

Packet Assembler/Disassembler
(PAD)
Packet (or Public) Data Network
(PDN)
Packet Network
Packet Switching Exchange
(PSE)
Permanent Virtual Circuit (PVC)
Switched Virtual Circuit (SVC)
Virtual Circuit
Wide Area Network (WAN)

OBJECTIVES

On completion of this chapter, you should be able to:

1 Explain what a wide area network (WAN) is.
2 Explain what a PAD is used for.
3 Explain what a FRAD is used for.
4 Explain statistical time division multiplexing (STDM).
5 Explain the public data network (PDN).
6 Describe packet switching.
7 Explain what a connection-oriented network is.
8 Explain what a connectionless-oriented network is.
9 Explain how a logical channel number (LCN) is assigned and used.
10 Compare and contrast a switched virtual circuit with a permanent virtual circuit.
11 Describe the call setup procedure for an X.25 packet network.
12 Describe link access procedure balanced (LAPB).
13 Define modulo 8 and modulo 128.
14 Define bit-oriented protocol.
15 Explain what a flag is in X.25.

16 Define committed information rate (CIR).

17 Define data link connection identifier (DLCI).

18 Compare and contrast X.25, frame relay, and ATM.

19 Explain why X.25 does error detection and correction at each hop but frame relay does not.

20 Define the size of information packets handled by X.25, frame relay, and ATM.

Wide area network (WAN) designates transmission facilities or media used to connect computers or LANs together over a wide geographic area. The PSTN, PDN, and Internet can be thought of as WANs. WANs provide the networks that allow a computer on one network to communicate with a computer on another. Telephone calls can be made over the PSTN because it allows any telephone on the network to call any other telephone on the network. Each telephone has a unique telephone number that serves as its address on the network. The same is true for the public data network. Each station on the PDN has a unique address. Instead of circuit switching, the PDN is a packet switched network, as discussed in Section 8.13.

Every *time division multiplexing (TDM)* system in the PSTN is carrying multiple DS0 circuits of 64,000 bps. When a call is placed over the PSTN, the switching systems inform each other via the SS7 network of the circuit they will use for this particular call. (SS7 operations were discussed briefly in Section 8.14.) The SS7 network reserves circuit paths through the PSTN for a call, and when the called telephone has answered, instructions will be issued via the SS7 network to connect the reserved circuits together. A connected call is provided with a 64,000-bps circuit from their originating class 5 exchange to the terminating class 5 exchange. Voice signals are converted by a codec on the line circuit in a class 5 central office into 64,000 bps. Thus, a 64,000-bps path is needed to carry the voice signal across the PSTN.

In Chapter 8, we discussed why modems are used to convert data signals into modulations of an analog signal. We do this so a regular telephone line can carry the signal. It is necessary to convert the data into modulations of an analog signal because the telephone line connects to a central office line circuit containing a codec. This codec expects to see an analog signal; it cannot accept a digital signal. Once the modulated analog signal reaches the codec, it is converted into a 64,000-bps signal and will be carried by the PSTN. The digital PSTN will allow modems to work over its DS0 circuits at baud rates as high as 3600 baud. The PSTN is a circuit switching network. One DS0 circuit can be switched or connected to any other DS0 circuit, and the switched circuit can carry either voice or data. The selected switched circuits remain dedicated to one user for the duration of a call.

Instead of using modems and dial-up switched circuits, a business can lease circuits in the PSTN for its own use. These leased circuits become private circuits, and the business has a dedicated circuit from one of its offices to another office. With a dedicated private line circuit, which is not switched, the business uses a digital interface device called a *customer service unit/data service unit (CSU/DSU)*. Today, everyone refers to the CSU/DSU as a *DSU*. By using a DSU to connect data circuits to the PSTN, higher data speeds are possible. Many business customers use a DSU for transporting data over their leased private-line circuits.

The DSU that connects a data device to private-line circuits comes in many different configurations. The most popular DSUs are configured to connect to the private-line facilities as a 56-Kbps signal or as a 1.544-Mbps signal. The customer will lease the necessary circuits in the PSTN from the LECs and/or IECs. These leased circuits will become the private-line circuit. The leased circuits are no longer part of the public domain and cannot be accessed by the general public; they provide a private network for the business customers leasing them (see Figure 11-1).

The DSU and leased circuit just described will provide the leasing customers with a 56,000-bps or a 1,544,000-bps circuit to use as they wish. Customers can also lease digital circuits with lower speeds. AT&T's original Dataphone Digital Services (DDS) offered services at 2400, 4800, 9600, and 56,000 bps. With the increase in the volume of data that companies transmit, most need at least a 56-Kbps line. Businesses that need to connect point-of-sale terminals to their centralized computer can still get by with lower speeds. The input circuits to the DSU on each end of the circuit determine the speed required of the leased line. With the appropriate DSU configuration, customers can opt to use a DS1 circuit and use some of the DS1 bit stream for data inputs; other inputs can be used for voice circuits. Many customers are beginning to use voice-over IP, configuring their data networks to carry both voice and data. As mentioned in Chapter 1, the competition for this business represents most of the competition between ILECs and CLECs.

Businesses can lease circuits in the PSTN to carry their data in many different formats or protocols. Some typical WAN transmission protocols are:

1. T1
2. Proprietary T1 protocol, which does not use bit-robbed signaling
3. Fractional T1
4. X.25 packet switching
5. Frame relay
6. Narrowband ISDN (BRI) and (PRI)
7. Asymmetrical digital subscriber line (ADSL)
8. Switched multimegabit data services (SMDS)
9. Asynchronous transfer mode (ATM) protocol
10. Broadband ISDN (BISDN)

Detailed discussion of these services will be handled in more advanced data communication publications and courses. They will be presented in this chapter to give you brief insight into the wide array of services available to connect LANs together. Regardless of the service chosen, it has one basic function, and that is to move 1s and 0s from one LAN or computer terminal to another in the most efficient manner possible. X.25 packet switching technology will be discussed in some detail to provide a basic understanding of how packet switched networks operate.

The type of service to select depends on the cost of the service and its ability to handle the applications that will be run between LANs. Making the correct decision on which service to use requires that you have a very good understanding of the

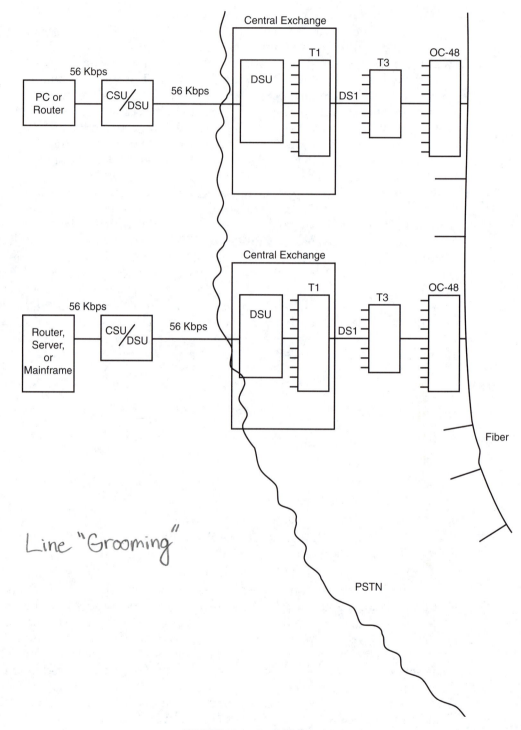

Line "Grooming"

FIGURE 11-1 Private-line network.

needs for specific applications as well as the capabilities of the various services. Each service offering continues to expand its capabilities. What was true yesterday may not be true today. You can stay abreast of new developments by surfing the Internet to such sites as the Frame Relay Forum, ATM Forum, and so on.

11.1 PACKET SWITCHED NETWORKS

All *packet networks* are not the same. Each type has its own distinct protocol, but they all have one thing in common. They all assemble data in packets and then insert the packet into a *frame* designed to the specifications (protocol) of the transport technology used for their specific packet network. Packet networks consist of transmission facilities connected to switches referred to as *packet switching exchanges (PSEs)* or routers. Packet networks also have another thing in common. They take a transport medium and divide it into time slices. They then allow various inputs to have access to the transport medium for brief intervals of time.

The way users are attached to circuits is the main thing that differentiates a circuit switched network such as the PSTN from a packet switched network such as the PDN. In the circuit switched network of the PSTN, each user is allocated to specific time slots within various TDM systems in the PSTN. For voice calls, each user is allocated a DS0 (64 Kbps) channel on various TDM facilities. The switching systems connect these allocated DS0 channels together. This results in the user having a DS0 transmission path from the originating class 5 central office exchange to the terminating class 5 central office exchange. The user is assigned this path for the duration of the call, and no other calls are allowed to use the assigned channels that make up the end-to-end DS0 circuit.

In a packet switched network, the channels or paths within the network are used by more than one user. If the circuit path is a DS0 circuit, it can be used to carry the data of many users by letting each have use of the transmission medium for a brief period of time. This means that the DS0 circuit that is part of a TDM facility is also used by each user in a time allocation (TDM) arrangement. One user may be able to use the DS0 circuit for a brief period, then another user can utilize the DS0 circuit for the transmission of data, and so on. Since each will be sending data to a different destination, a method has to be included to instruct packet switches where the data should be delivered.

When data is grouped into a block of data, it is called a *packet*. User data is placed inside a packet that will contain a *header* preceding the data of the user. This header in the packet of many different protocols will often contain a destination address. Each switch in a packet switched network will look at the destination address in the header and determine how the packet should be routed through the packet switched network to get to its intended destination. Each packet contains information that the switches use to determine how to route the packet (see Figure 11-2). This is different from a voice circuit switched call.

When someone wishes to place a telephone call, they dial the telephone number of the party they desire to talk to. The central office exchange will translate these dialed digits into switching instructions. The class 5 central office will use the SS7 network to get any additional switching instructions needed and will issue

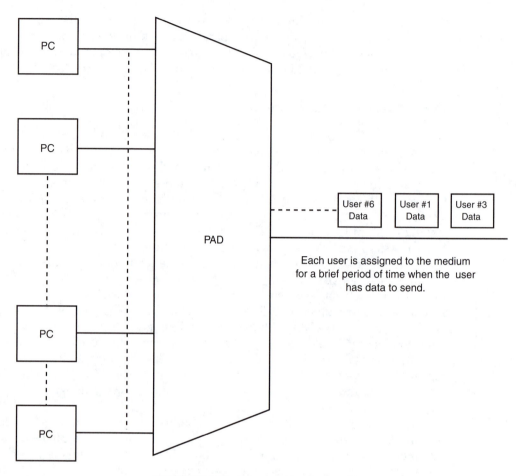

FIGURE 11-2 Packet assembler/disassembler.

those switching instructions over the SS7 network to other switching systems. Once the switching instructions have been issued, circuits in the PSTN remain dedicated to the call for which they were assigned.

Since the voice paths remain connected for the duration of the call, there is no possibility of delay in the voice being received by either party. Since the paths are nailed up at the beginning of conversation, there is no need to provide switching instructions during the call. When the call is completed, switching instructions will be issued over the SS7 network to release the circuits used for the call and make them available for other callers. Because only one call at a time is allowed to use the DS0 circuit path assigned in the PSTN, there is no need to have additional switching instructions during the call. Call setup only has to occur one time (at the beginning of the call) for a call made over the PSTN.

When we take a circuit and use it for more than one user, we must include addressing information with each packet of data transmitted. This adds additional overhead to the data transmitted. Each switch in the packet switched network must look at the destination address contained in the header of each packet to determine how the call should be routed. This introduces delay into the transmission of the message. Voice circuits cannot have delay introduced into the transmission of their signals. This is why circuit switching is preferred over packet switching for voice calls. With circuit switching there is no switching during the call, and messages are delivered instantaneously from one user to the other. The calling and called party remain connected to the assigned circuit, whether they are talking or not. Because the users are permanently connected to the assigned circuit for the duration of the call, there is no delay in the transmission of signals from one end of the circuit to the other.

The fact that a circuit remains connected to two end users in a circuit switched environment, whether they have message to deliver or not, makes circuit switching less efficient than packet switching. With packet switching, many users can use the same transmission path in a time division arrangement. We simply have to include addressing information with each packet of data transmitted to ensure it is delivered to the right recipient. This significantly reduces idle time on the circuit and makes the circuit more efficient. Since a slight delay in data does not violate the integrity of the data, packet technology is a great way to handle data efficiently.

11.2 STATISTICAL TIME DIVISION MULTIPLEXING

Packet data networks are not restricted to using just one DS0 circuit in a time allocation arrangement between users. Many of today's packet networks use DS1 and higher facilities to carry packet data. Let's look at a packet network using a DS1 medium between two DSUs that dynamically allocate the channels within the DS1 bit stream. Recall from Chapter 4 that a device that does dynamic allocation is referred to as a *statistical time division multiplexer (STDM; also called a *stat mux*). STDMs are known as *asynchronous multiplexers* as well to differentiate them from the synchronous time division multiplexers used in the PSTN.

A typical example of a synchronous time division multiplexer used in the PSTN is a T1 multiplexer. It has 24 channels. Input 1 on a T1 multiplexer will be assigned to channel 1 in the TDM facility connecting to the distant T1 multiplexer. Input 2 is assigned to time slot 2 on the TDM facility. Each of the inputs to a multiplexer is assigned to the time slot in the TDM path that corresponds to its channel unit number. Channel 24 input is assigned time slot 24 on the TDM path. Whether or not a channel unit is active, its time slot is reserved in case the channel unit becomes active. Each channel's input is synchronized to its corresponding time slot in the TDM composite signal between multiplexers.

Asynchronous TDMs (STDMs) do not have permanently mapped time slots to channel units. The time slots can be assigned dynamically. Thus, channel unit 1 can be connected to any of the 24 time slots in the DS1 TDM path between STDMs. For one transfer of data, channel 1 may be assigned to TDM slot 5. For the next transfer, it may be assigned time slot 10 (or any other time slot not in use). This is why the header must contain address information. The receiving STDM needs to know where the data should be delivered. The address in the header will tell the receiving multiplexer which output channel the data should be delivered to. If the data originates at one multiplexer on channel 1 and is transported using time slot 10, the address in the header will tell the receiving multiplexer to deliver the data it is receiving on time slot 10 to the channel 1 output.

The STDMs' ability to dynamically assign slots in the TDM path when an input channel has data to transmit makes the system much more efficient than a T1 system. A T1 system can only have 24 DS0 channel units, but a STDM could have more. With a regular T1 system if channels 1, 2, 6, 8, and so on are idle and have no data to transmit, their corresponding time slots in the composite TDM signal are idle. This wastes capacity. We can increase the utilization of the composite TDM circuit by using a STDM and increasing the number of input channels.

Depending on the volume of traffic that each user needs to send at the same time, it may be possible to have 48 DS0 input channels to a STDM that uses a DS1 facility to the distant STDM. Of course, it will be impossible for all 48 inputs to transmit at the same time because there are only 24 time slots available in the DS1 path between STDMs. The STDM contains buffer memory for each of its inputs. It can hold data in the buffers until a time slot is available between STDMs. Because a STDM has more input channels than time slots available in the TDM path, it cannot permanently map time slots to inputs but must allocate them as needed. Because the STDM can handle more inputs than a regular TDM, the idle time in the composite TDM signal is reduced, making more efficient use of the facility.

This is the same principle used when a company leases one DS0 channel for a private data line facility between computer centers. The DS0 facility provides for the transmission of 64,000 bps. A STDM can be attached to each end of the TDM facility and many inputs can be connected to the input of the STDM. For example, you could use a STDM with each of its 48 inputs rated for a maximum speed of 4800 bps. Dumb terminals may be attached to the inputs of a STDM in one location, and the 48 channels at the distant STDM could be connected to the inputs of a mainframe. This arrangement allows 48 remote terminals to interact with a mainframe without having to provide 48 separate transmission facilities between the two sites.

If all terminals transmitted at a steady rate of 4800 bps, the total bit rate would be 230,000 bps. This would exceed the bit rate of the facility and we would have problems. User terminals will not be transmitting data continuously. They transmit data a few bits at a time and have more idle time than transmit time. The PDN takes advantage of these characteristics to multiplex many users onto one transmission path. Because all terminals are not transmitting continuously, we can take advantage of the idle time associated with each input and combine them using STDM technology. We can use a 64,000-bps or even a 56,000-bps facility because the terminals do not transmit a steady stream of data. They transmit bursts of data. The STDM can buffer these bursts of data and then allocate a slice of time on the 64,000-bps facility to transmit the buffered data to the distant STDM. Each user will appear to have instant communication with the mainframe. Because no one can type at 4800 bps, we will have many periods of idle time on each input to the STDM. The STDM can buffer each input and assign it a time slot on the 64-Kbps path between locations.

Each user's input will be buffered by the STDM and will be dynamically allocated a time period on the DS0 channel. The header for each packet of data will inform the STDM on the distant end which output channel unit the data should be delivered to. If terminal 1 is connected to input channel 1, then channel 1 at the distant STDM should receive the data typed by terminal 1. The packet of data will include information in the header that specifies the data is to be delivered to channel 1 at the distant STDM. The packet of data travels across the DS0 link and will be connected to output channel 1. If terminal 10 has data to send, it will be allocated time on the DS0 highway between STDMs. When it arrives at the distant end, the address in the header will tell the STDM to deliver the data to the output of channel 10 on the receiving multiplexer.

11.3 SWITCHING OR ROUTING IN A PACKET SWITCHED NETWORK

Packet networks are also called *packet switched networks* because they must be able to look at the address in the header of a packet and make switching (routing) decisions. The packet switches use the destination address included in the header in the same manner that a STDM does. The STDM uses the address to determine which output channel on the distant STDM data should be delivered to. The packet switch uses the address to determine which network path the data should take to be delivered to the proper recipient. Some packet networks include the address of the intended destination in the header for each packet transmitted. Some packet networks will use a call setup packet to determine the route that a packet should follow through the packet network. On determining the path for packets, a number will be assigned for that path. This circuit path number will then be used as the address in subsequent headers. Regardless of the technique used, the header contains some piece of data used by the switches to route the data across the network.

11.4 PACKET (OR PUBLIC) DATA NETWORK

Let's look at the forerunner of all packet networks: the *packet (or public) data network (PDN)*. The PDN is based on technology developed for the Department of Defense's *Advanced Research Project Agency Network (ARPANET)*. ARPANET was the forerunner of the Internet network. The two predominant service providers offering access to the PDN when it was first developed in the early 1970s were Tymnet and Telenet. Telenet used the ARPANET technologies for the base of its PDN. Tymnet used a proprietary network and developed specialized software for its Tymnet. The two networks could not talk directly to each other.

Through the cooperation of Telenet, Tymnet, and the telecommunications carriers, a standard was developed to allow interconnection between separate PDNs. This standard is the X.25 standard, which defines the interface to a PDN and not what goes on inside the network. Telenet, Tymnet, and any other provider of PDN services can use proprietary protocols and software within their PDN but must provide an X.25 interface to the network. This use of the X.25 standard means many different manufacturers can develop products to access the network. It also means that a user can access any PDN with one device and does not need a proprietary access device for each network.

Companies that provide PDN services are referred to as *value-added network (VAN)* carriers. A VAN leases circuits in the PSTN from the major LECs and IECs. They connect these leased circuits to their packet switching equipment to form a data network. The PSTN is a network composed of many DS0 channels multiplexed together using TDM. Most IECs use OC-48 multiplexers that are connected by fiber cable to carry calls over their network. An OC-48 multiplexer uses TDM technology to multiplex 32,256 DS0 channels together. These channels can be used for voice services or can be leased to a VAN to be used as part of its PDN.

Telenet was started in 1975 and was acquired by General Telephone and Electronics (GTE) in 1979. In 1986 Telenet acquired the Uninet PDN from United Telecommunications. GTE combined resources with Southern Pacific Railway to provide long distance services under the name *Sprint.* United Telecommunications joined the Sprint venture and later became the sole owner of the Sprint long distance services and Telenet. It renamed the Telenet operation *SprintNet.*

Tymnet was started in the late 1960s by Tymshare, Inc. to market the services of Tymshare. It began services as a public data network provider in 1977 and was acquired by McDonnell Douglas Automation in 1983. In 1989 British Telecom acquired the network; in 1993, it sold the U.S. portion of the network to MCI to gain FCC approval for a merger of business interests with MCI. British Telecom had purchased a 20% interest in MCI and was contemplating a merger with it. The joint public data network was called *Concert.* In late 1997, WorldCom made a competing bid for MCI and eventually won approval for the merger with MCI. British Telecom and MCI agreed to continue their partnership in the joint Concert venture. MCI has been upgrading the network with state-of-the-art computers and switching centers. It is increasing capacity between switching centers and increasing the speed of its switches. MCI is also replacing the old legacy software used for Tymnet switching centers with more robust and universally available X.25-compliant software.

AT&T also has a PDN, referred to as *Accunet Packet Service.* AT&T does not provide a PAD access to its network. Thus, it cannot accept dial-up access from asynchronous terminals. The customer must provide a PAD to access Accunet. AT&T announced to the federal government in November 1998 that it would no longer support X.25 services. These services were part of the FTS-2000 contract bid on by the major IECs. The government has stated that X.25 capabilities must be bid on as part of the FTS-2001 contract for telecommunications services to the federal government.

Sprint has indicated it will keep X.25 PDN services and will continue to provide these services to the government. It has stated that it is ready to pick up any X.25 business AT&T does not want. AT&T has indicated that it will provide frame relay services in lieu of X.25 services. As noted earlier, MCI will continue in the X.25 business, because it is actively upgrading the old Tymnet with the latest switching systems and transmission media.

The data network is referred to as a PDN. The PDN can be used to allow anyone access to a data terminal (public); it can also be used to allow companies to construct private data networks. Just as a telephone company can lease circuits that are part of the PSTN facilities to private companies for their own use, a VAN can lease circuits that are part of the PDN so they can be used by companies to form a private data network. The leased circuits in both scenarios are no longer part of the public domain. In the case of the PDN, customers are not leased a physical circuit; they are leased guaranteed slices of time on the PDN circuits.

VAN carriers also lease access facilities from a LEC. The customers of the VAN carrier use these facilities to access the data network. Customers can be provided with dedicated leased lines from their business location to the data network and can use these dedicated access facilities and leased circuits in the PDN to construct a private data network. VANs can also lease local telephone numbers or 800 XXX-XXXX numbers from the LECs and IECs to provide their customers with dial-up access to the PDN.

11.5 PACKET ASSEMBLER/DISASSEMBLER

A business contracts with a VAN to get access to the PDN. Access to a PDN must be done via a device called a *packet assembler/disassembler (PAD)*. The PAD must format data into packets that comply with the X.25 protocol and can be implemented by a software package in the customer's DTE. A separate hardware/software device that is appropriately called a *PAD* can also be used to connect regular asynchronous terminals to the synchronous PDN. Anyone wishing to transmit data over the PDN has to connect to the PDN via a PAD. Customers that use telephone numbers for dial-up access to the PDN will be connected to a PAD provided by the VAN carrier. The PAD emulation software in the DTE or the PAD (hardware device equipped with software) forms the data into packets and transmits them onto the PDN facility using a synchronous protocol. To provide the ability to transmit data between any two PADs, the PADs are connected to switching systems in the PDN. The switching systems are called *packet switching exchanges (PSEs)*. PSEs are connected together in a mesh arrangement (see Figure 11-3).

Several commercial *information service providers* such as America Online (AOL) and Prodigy have used the PDN to connect their customers to their computers. The VAN provides telephone numbers in each major city for AOL customers to dial. When customers dial these numbers, they are connected to modems at the VAN location. The outputs of the modems are connected to a PAD, which in turn connects to the PDN. The PAD is programmed to put an address in the header of the data packet it sends out onto the PDN. AOL customers come in on certain inputs to the PAD. Data received on these inputs is formed into a packet and addressed to the PAD that resides at AOL headquarters. AOL has been replacing the access via VANs with its own access network.

The VAN provides local telephone numbers in many different cities for its customers that use dial-up access. AOL is one of its customers. The local or 800 XXX-XXXX telephone numbers connect to modems attached to a PAD. Each telephone number connects via a modem to a specific port on the PAD. The PAD knows by the port accessed who the customer is. In the case of AOL customers, the telephone number dialed will terminate via a modem into certain specific ports on the PAD. The PAD is programmed to transfer data from these ports to the PDN and instruct it to deliver the data to the PAD at AOL. The PAD at AOL connects to its computer. This computer provides numerous services, including access to the Internet.

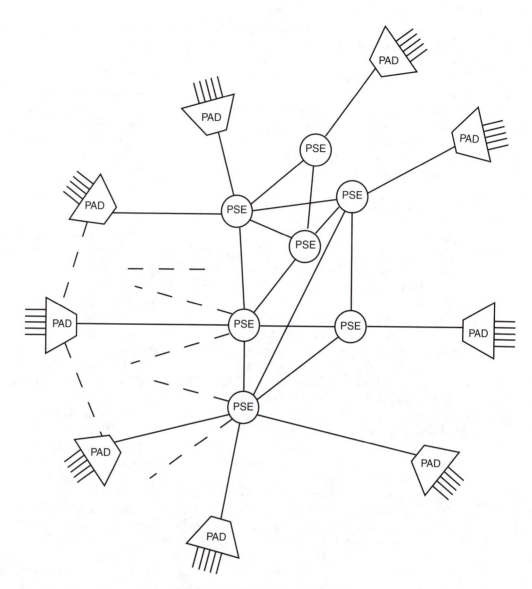

FIGURE 11-3 *X.*25 Public data network.

Other customers of a VAN will have their own access numbers to dial for access to the PAD at the VAN location, or they may have a dedicated facility that connects their terminals to the PAD. Customers can have dial-up access or dedicated access to the PDN. In the AOL scenario just presented, AOL is the customer and the PAD is preprogrammed to insert the address of the AOL computer as the destination address into the header of the packet of data. (see Figure 11-4).

Other customers using dial-up access may want the ability to access any other terminal on the PDN. Once connected to the PAD, they will be able to provide it

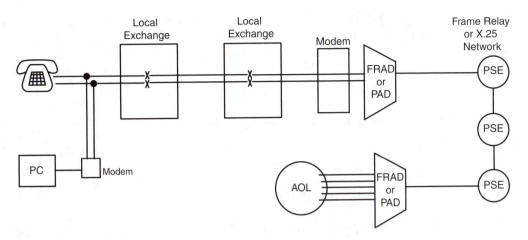

FIGURE 11-4 AOL access via local dial-up numbers connected to the PDN.

with an address for their intended destination. The addresses used in the PDN look like telephone numbers but do not have a fixed length. Telephone numbers are always 10 digits long when the area code is included. PDN addresses or numbers usually have between 12 and 14 digits. Some addresses are 15 digits long. The header contains a field to inform the switches of the PDN how long the address is.

11.6 ADDRESSING USING INTERNATIONAL DATA NUMBERS (X.121)

Each station on the PDN is assigned a unique address consisting of decimal digits. The X.121 ITU standard specifies the format of the numbering system. The address is referred to as an *International Data Number (IDN)*. The IDN varies in length and can be up to 14 decimal digits long. The first four digits of the IDN are referred to as the *data network identification code (DNIC)*. The first three digits of the DNIC specify the country; the fourth specifies the network. The remaining digits are used to identify the specific DTE on the PDN. As a country needs more network addresses, it is provided with additional country codes. The United States has been assigned codes 310 through 316.

Many companies contract with a VAN as a way to provide data services to their customers. A company that offers real-time quotes on stock market activity can use the PDN to provide information to customers. They will provide customers with the telephone numbers that can be used for dial-up access. They will require an authorization ID and password when a customer requests services. Because the customer dialing in is not a direct customer of the VAN, the company providing the stock quotes must agree to pay for the PDN usage. This is done on the initial call setup. The call setup packet will contain customer identification and a request for reversal of charges. If the company that has the contract with the VAN agrees to the reversal of charges, the connection is established.

The PDN uses the X.25 protocol, but its use is not limited to the PDN. Many LANs utilize the X.25 protocol. The SS7 network discussed earlier is an X.25 network.

All X.25-based networks do not use the International Data Numbering format (X.121). X.25 makes it possible to use other addressing schemes instead of the X.121. In the initial call setup packet, a bit of data in the header indicates the addressing scheme used. This bit is called the *A bit*. We will discuss the format of call setup packets in the next section.

11.7 LAYER 3 OF THE X.25 PROTOCOL (THE NETWORK LAYER)

The PAD is the device or software responsible for assembling data in packets and transmitting them onto the PDN. At the receiving end, it is responsible for taking the user's data out of the packet and reassembling it into the original data stream. The PAD is responsible for taking this data and placing it inside an X.25-compliant frame. The X.25 protocol is a three-layer protocol. This protocol existed before the OSI model was introduced, but it maps neatly onto the first three layers of the OSI model.

The top or third layer is the *network layer*. This layer is the packet layer. It is responsible for breaking the user's data down into packets and adding headers to them. It is also responsible for establishing a circuit between the PAD and the PSE and transferring packets between them. The network layer will generate a call setup packet that results in a *logical channel number (LCN)* being assigned between the PAD and the PSE. This LCN is used to identify a logical connection between the PAD and PSE and will be used for subsequent packets from this user. The facility between the PAD and PSE may be a 56-Kbps circuit. Many users will be sending data over this facility to the PSE, and the LCN will be utilized to identify a particular specific user to the PSE.

The X.25 PDN is referred to as a *connection-oriented network*. A connection is established for each user. Since many users are employing the same physical circuits, the user is assigned a LCN. The LCN is utilized to identify user connections over the common physical medium they share. These user connections are referred to as *logical connections* because the user does not remain on the physical medium 100% of the time. The field of data within an X.25 packet that identifies the LCN is 12 bits wide. This allows for the storage of 4096 different numbers. Zero is not used for a LCN; this means a total of 4095 LCNs can be assigned.

The originating PSE will determine from the destination address the best route to take through the PDN. The PSE is physically connected to other PSEs via some medium that is carrying TDM channels. One or more of these channels will provide a connection between the PSEs. These channels are used for multiple users, and each user will be assigned an LCN between the PSEs. The PSE that connects via a PAD to the desired receiving terminal will also assign a LCN between it and the PAD. Thus, throughout the PDN, we have LCNs assigned between all the devices that will be handling the user's data. These LCNs are connected together to provide an end-to-end logical circuit, referred to as a *virtual circuit.* The user appears to have a dedicated physical circuit through the PDN, but it does not. It is sharing the physical circuits with many other users. Because the circuit appears dedicated to the user, it is called a *virtual circuit.*

Most people do not refer to the connections between the PSEs as logical connections (even though they are). They reserve the use of the term *logical channel* to identify the connection on each end of a circuit between the PAD and the PSE. Because many users implement the PAD function in the DTE with software, the LCN is often referred to as the channel between the DTE and the receiving DCE at the PSE. Thus, LCNs often designate the channels assigned on each end of a PDN connection, while *virtual circuit* refers to the overall end-to-end connection.

A particular LCN can be permanently assigned to a particular user. This user will always be guaranteed a connection to the PSE and therefore guaranteed a connection to the PDN. A user with a permanently assigned LCN has a *permanent virtual circuit (PVC)*. This provides a service similar to a leased private-line circuit. These circuits allow companies to build private data networks using PVCs in the PDN. PVCs require no call setup since they are permanently assigned and always available to the user. The connection between the DTE (with PAD software) and the PSE will always be the same LCN for every transmission of packets.

Users that do not have a permanent virtual circuit (PVC) assigned to them will use a *switched virtual circuit (SVC)* for connection to the PDN. If we go back to the scenario of an investment firm providing dial-up access to customers, the customers will access a PAD that uses a SVC for access to the PDN. The originating PAD will issue a call request packet to the PSE. The call request packet will contain a LCN and the address of the DTE the user desires to contact. The PSE will route the call setup packet through the PDN based on the address of the terminating station. The PSE that connects to the terminating PAD will establish a LCN with the PAD. Remember that the PAD is often contained within the terminating DTE when the DTE has direct access to the PDN.

Each end of the circuit established for the transfer of data will have a LCN assigned to identify the logical channel used to connect each end of the PDN to the DTE. The assignment of LCNs on each end of the PDN is done independently. This means the LCNs for each end of the circuit will most often be different LCNs. The connections between PSEs are done according to the way a particular VAN has designed its network. It does not have to use LCNs, within the PDN, between PSEs; however, this is the technique most VANs will use. The preceding discussion pertains to SVCs. Remember that PVCs do not require a call setup packet because the LCNs are permanently assigned for a PVC and every transaction will use the same LCN. SVCs do require call setup packets. They must have a call setup packet to establish the LCNs that will be used.

Once the LCN is assigned via a call setup packet, it will be used as the address in the header of packets that are transferring user data. The PSEs have kept a copy of the DTE address to LCN mapping, and they know based on the LCN where the packet of data should be delivered. If a certain path between PSEs becomes congested, the PSE can use an alternate route to send packets to their final destination, but the LCNs on each end of the circuit will remain the same for every packet transmitted. Thus, within the PDN, dynamic routing can occur; every packet does not have to take the same route but they usually will.

The network layer is responsible for setting up and controlling a logical connection between the originating PAD and originating PSE and for forming the user's data into packets. The network layer at each PSE that will be involved in the transmission of packets will be responsible for the same functions between PSEs.

General Format Identifier	Logical Channel Group Number
Logical Channel Number	
Packet Type Identifier	
Length of Originating DTE Address	Length of Terminating DTE Address
Variable-Length Field Containing Originating and Terminating DTE Addresses	
Length-of-Facilities Field	
Variable-Length Field Used to Set the Quality of Service	
Variable-Length Field Employed for User Data	

FIGURE 11-5 *X*.25 call setup packet.

The network layer at the terminating PSE and the terminating PAD will be responsible for getting the data to the DTE. At the originating PAD, the network layer will send its *X*.25-compliant packet to the next lower layer, the data link layer. It will be the responsibility of the data link layer to transfer the packets out onto the physical link connecting to the originating PAD to the receiving DCE at the PSE. At the originating PAD, the network layer packet is received by the data link layer and will be encapsulated within the packet formed by the data link layer.

As mentioned earlier, call setup packets are not required for PVCs but are required for SVCs. The call setup packets consist of a call request packet, an incoming call packet, a call accepted packet, and a call connection packet. The call request packet will be issued by the originating PAD (within the originating DTE) to the terminating PAD (DTE) via the PSEs in the PDN. If the terminating DTE accepts the call, it issues a call accepted packet and the originating PSE will issue a call connected packet to the originating DTE. The basic format of these packets is shown in Figure 11-5.

The general format identifier is used to identify what type of addressing scheme is used in the address fields, to specify what type of sliding window protocol is used (modulo 8 or modulo 128), and to indicate whether the data in the information field is user data or control data. The logical channel group number and the logical channel number fields are used to assign a logical channel on the link between the PAD (DTE) and PSE (DCE). The call setup packet will be used to establish a LCN for the SVC data session. Once the LCN is established, it will be used in the address fields of the packets containing user data. When the data session ends, a **call clear packet** will be issued to clear the LCN assignments. The packet type identifier will indicate what type of packet the frame contains. Packet types include the following: call request, call accepted, call clear request, clear confirmed, interrupt, interrupt confirmation, receive ready, receive not ready, reject, reset request, reset confirmation, restart request, restart confirmation, registration request, registration confirmation, and data. The format of a network layer packet containing user data is shown in Figure 11-6.

The network layer packet containing user data will contain a general format identifier indicating whether packet acknowledgment is required between the PAD and PSE or required end to end, noting what type of flow control is used (modulo 8 or 128), and verifying that the packet contains user data. The LCN fields

General Format Identifier	Logical Channel Group Number
Logical Channel Number	
Receive Packet Number and More Bit	Transmit Packet Number and User Data Bit
Variable-Length User Data	

FIGURE 11-6 User data packet.

will contain the LCN established by the call request packet to identify the DTE to the PSE logical connection. The receive and transmit packet numbers are used to reassemble data in the proper order and to ensure that all packets have been received. The user data bit is set to 0 for a packet containing user data. The *more* bit is set when large blocks of data are broken down into several packets. Many types of packets are used in *X*.25. Figures 11-5 and 11-6 are used to provide a basic illustration of the *X*.25 network layer packet format. These packets will be encapsulated within the data link layer as the information field.

11.8 LAYER 2 OF THE X.25 PROTOCOL (THE DATA LINK LAYER)

The data link layer of *X*.25 is based on the *high-level data-link control (HDLC)* protocol. There are several versions of HDLC, and each is used for a particular application. *Logical link control (LLC)* is used at the data link layer when *X*.25 is used on a LAN. *Link access procedure on the D-channel (LAPD)* is used by ISDN to control the setup of B-channels in an ISDN environment that allows multiple terminals access to a link. *Link access procedure for modems (LAPM)* is used between modems. *Link access procedure (LAP)* was used when the PDN used analog facilities in the PSTN for its network. Today, the digital circuit-based PDN uses **Link access procedure balanced (LAPB)**. LAPB is a point-to-point protocol; it is used on a link that connects two devices. In our case, the two devices are a PAD and the DCE at the PSE.

The data link layer is responsible for controlling the flow of data across the link between the PAD and the PSE. LAPB is a balanced protocol because both ends of the logical link have equivalent status. Either end of the logical channel can issue messages to the other end of the link without having to request permission. The LAPB software will control the flow of data across the link and perform error detection and error correction. These mechanisms are similar to those used in modems (LAPM) and were discussed in Chapter 8. Error detection consists of using a CRC checksum in each frame transmitted. The receiver performs the same CRC calculation and verifies that its checksum matches the CRC number transmitted within the frame. If they match, the receiver transmits a positive acknowledgment (ACK) for that frame number to the originator. If they do not match, the receiver sends a negative acknowledgment (NAK) for that frame number to the originator.

The data link layer contains information within the header that indicates how many frames can be outstanding without acknowledgments. The older PDN allowed seven frames to be outstanding. This was referred to as *modulo 8*. Later when the PDN started using satellite transmission circuits, it took longer to get data transmitted and acknowledged over the satellite circuits, and the ability to have 127 frames outstanding before getting an acknowledgment was added to the X.25 protocol. This is referred to as *modulo 128*. Thus, flow control depends on the modulo selected. Modulo 128 has a sliding window of 127 frames. As long as ACKs are received within 127 frames of the frame being transmitted, the originator keeps sending additional frames.

Each frame transmitted contains a number for that frame. The receiver sends acknowledgments (ACKs) for each frame received. Using the sliding window protocol, the transmitter keeps sending as long as it is receiving ACKs from the receiver. If the receiver detects an error, it sends a negative acknowledgment (NAK) along with the frame number affected. The originator will then retransmit that frame. In the PDN, this error detection and error correction occurs at every PSE. Each PSE keeps copies of frames it has sent, and each receiving PSE or PAD checks for errors in the frame. Each PSE must keep a copy of frames sent so that it can retransmit the frame if a NAK is received. Doing error checking at each hop that a packet is routed through adds additional overhead to the transmission of data but ensures very reliable error-free transmission. This is what makes the PDN so reliable.

The data link layer is also responsible for synchronizing the receiver to the transmitter. This is what makes the X.25 a synchronous network. The synchronization of the transmitter and receiver is accomplished by transmitting a certain pattern of bits, referred to as a *flag.* The pattern of bits used for a flag is 01111110. This flag is transmitted constantly from the transmitter to the receiver until the transmitter has a frame to transmit. As long as a receiver sees the flag, it will remain synchronized to the transmitter. The flag serves as a signal to establish synchronization between the sender and receiver. The HDLC protocol and therefore LAPB protocol require that flags be sent when the link is not in use. Thus, LAPB is always active on the link between a DTE and the PSE and between PSEs, whether or not they have user data to send. LAPB is a point-to-point full-duplex protocol. It controls the link connecting two points.

Flags are also called *interframe signals;* they are transmitted continuously when frames are not being transmitted. The receiver monitors the bits being received over the link and gets one flag after another until a frame is sent out onto the link. When this happens, the frame starts and ends with a flag. The first 8 bits of the frame will be a flag. The next 8 or 16 bits will be an address. The address will not have the bit pattern of a flag. The receiver detects this and knows that it is receiving a frame of data. When the receiver encounters the flag at the end of the frame, it knows that all the data received belongs to that one frame.

A receiver does not know that it is receiving a frame until it has received an address (which will not look like a flag). Therefore, a receiver must store bits as they are being received to ensure it has them all available when the bits received are part of a frame and not part of a flag. The receiver stores one flag after another. On receiving two flags in a row with no nonflag data between them, it disregards the flags. The continuous flags keep the receiver synchronized to the transmitter. Also, remember that the link between devices such as between the PAD and the PSE is a

FLAG	ADDRESS	CONTROL	INFORMATION	FCS	FLAG

FIGURE 11-7 LAPB frame.

full-duplex channel. The physical circuit connecting a PAD to the DCE at the PSE is part of a full-duplex circuit. The PAD and PSE can be transmitting and receiving at the same time. LAPB keeps the receiver in the PSE synchronized to the transmitter of the PAD and also keeps the receiver of the PAD synchronized to the transmitter in the PSE.

The transmitter ensures that flags only occur at the beginning and end of a frame. If any other field such as the address field, the control field, the data field, or the frame check sequence field contains five 1s in a row, the transmitter inserts a 0. Anytime five 1s are received from the link layer by the transmitter after the opening flag has been detected, the transmitter inserts a 0. The receiver recognizes the beginning of a frame by the non-flag-like data that follows the opening flag. After detecting the start of a frame, the receiver looks for any bit pattern consisting of five consecutive 1s followed by a 0. It will strip out the 0 when it follows five consecutive 1s. HDLC is referred to as a *bit-oriented protocol*. It does not use special codes to control actions of the link; it uses only one special pattern of bits to designate the start and end of a frame. HDLC does not have to look for special codes within the frame to determine what actions to take. Each bit within the frame will be part of a field of data or can be used for control purposes.

Figure 11-7 illustrates a LAPB frame. The information field of the LAPB frame contains the *X*.25 packet formed at the network layer (layer 3). The network layer packet placed inside the information field can vary in size between 16, 32, 64, 128, 256, 512, 1024, 2048, and 4096 octets. The network layer packet contains the addresses of the originating and terminating DTEs (or it contains LCNs) to identify the logical network connections between the DTE (PAD) and PSE (DCE). The address field in the LAPB Frame contains the address of the PAD of the DTE and DCE of the PSE. The PAD address is set to *A* and the DCE address is set to *B*.

11.9 LAYER 1 OF THE X.25 PROTOCOL (THE PHYSICAL LAYER)

The physical layer is concerned with connecting one device to another. The *X*.25 protocol specifies using *X*.21 or *X*.21bis. The *X*.21bis is basically the standard RS-232D cable used to connect many types of DTEs to many types of DCEs. With a dedicated leased line between the user's DTE (equipped with PAD software) and the DCE at the PSE, the user's DTE will connect to the DSU via a standard EIA-232 interface. The DSU will connect to two cable pairs (a transmit and a receive pair) that terminate at the central exchange DCE. The DCE will connect to the PSE via an EIA-232 interface. For a dial-up connection, a user's DTE will connect to an analog modem. This modem will be connected via a dial-up connection over the PSTN to a modem connected to a PAD. The PAD will connect to the PSE via an EIA-232 interface (see Figure 11-8).

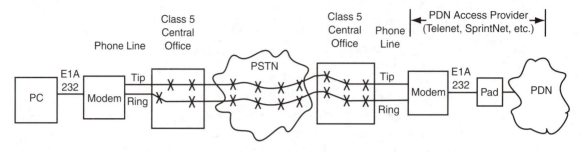

FIGURE 11-8 An EIA-232 interface.

11.10 COMPARISON OF X.25 TO FRAME RELAY

Frame relay is described as a fast packet network. It is faster than X.25 because it does not do the error detection at each hop that X.25 does. Frame relay takes advantage of the improvements made in transmission technology. Today's digital circuits running over fiber optic are much less prone to electromagnetic interference than the older analog network. X.25 was developed when the PDN was built using analog transmission circuits in the PSTN. The facilities used in the PSTN today are digital, and most circuits within the PSTN are carried over fiber facilities. These error-free circuits negate the need to do error checking on every packet at every node. Instead of doing error detection at the link layer, error checking for frame relay is done at a higher layer. In addition, frame relay uses a more streamlined flow control procedure that also leads to higher performance and efficiency.

In an X.25 network a PAD is used to provide a user with access to the network. The device that provides a user with access to the frame relay network is a *frame relay assembler/disassembler (FRAD)*. Nodes in a frame relay network will check frames for errors and will discard those with errors. Frame relay relies on the receiving application to detect a missing frame and request the transmitting DTE to resend that frame. Nodes in the frame relay network do not have to keep a copy of frames because they are not involved in the error correction process. Error correction is done on an end-to-end basis.

Frame relay uses PVCs. The ability to provide SVCs was recently added to the frame relay protocol, but most carriers still offer PVC only. The frame relay protocol is a two-layer protocol consisting only of a data link layer using LAPF protocol and a physical layer. Multiplexing and switching of logical connections takes place at layer 2. Frame relay uses the term *data link connection identifier (DLCI)* to refer to the logical channel established between a FRAD and the network. Frame relay was developed for use over primary rate ISDN (PRI), and like PRI, call control signaling is carried on a separate logical connection from user data.

Frame relay provides users with a *committed information rate (CIR)*. The service provider guarantees the handling of data at the CIR purchased by a user and will handle bursts of traffic above the CIR if possible. User data is encapsulated within a LAPF frame by the FRAD and transmitted across the frame relay network.

Since frame relay serves only as a transport mechanism, it can carry packets that use any protocol. Many companies have been replacing their X.25 network with frame relay. As mentioned earlier, AT&T has announced it will no longer offer X.25 PDN services but will offer frame relay instead.

11.11 ASYNCHRONOUS TRANSFER MODE

As the speed of LANs continues to grow, the speed of the WAN must also increase. X.25 was originally engineered around 56-Kbps and 64-Kbps transmission links. X.25 circuits of 2.048 Mbps are now being used. Since frame relay was developed for use over PRI, its speeds have been limited to the 1.544 Mbps of 24-channel PRI and the 2.048-Mbps speed of the European 32-channel PRI. Another fast packet technology has been developed that uses fixed packet sizes. The packets are all 53-byte packets called *cells.* A cell contains 5 bytes of header information and 48 bytes of user data. This fixed packet technology is referred to as *cell relay* and *asynchronous transfer mode (ATM)* packet switching.

ATM was developed as a transport vehicle for broadband ISDN (BISDN). ATM is even more streamlined than frame relay. Like frame relay, it takes advantage of today's digital technology over fiber to eliminate the need for low-level error correction. Like X.25 and frame relay, ATM is a packet switching technology and multiplexes many logical channels over one physical medium. X.25, frame relay, and ATM are all connection-oriented services. With ATM the connections are referred to as *virtual channels* and *virtual paths,* but ATM is also designed to carry the **connectionless protocol** of *Switched Multimegabit Data Services (SMDS)* used by the Bell Companies to provide WAN service.

SMDS provides speeds from DS1 (1.544 Mbps) to DS3 (45 Mbps). BISDN provides speeds of 155.52 Mbps to 622.08 Mbps. Speeds of 45 Mbps to 622.08 Mbps dictate the use of fiber optic facilities. Companies must connect via a fiber facility to the ATM network. AT&T, Sprint, and MCI are now offering services that are referred to as *on-demand services.* The customer connects to the fiber ring of an ATM-based WAN and can transport voice, video, and data over one access to the WAN. The carrier has intelligent switches (routers) and gateways on the WAN to connect voice, video, and data packets to the most appropriate network (packet switched or circuit switched).

11.12 VOICE-OVER ATM

Voice signals cannot be subjected to much delay in the transmission of signals from the transmitter to the sender. Because X.25 packet switching introduces a significant delay in transmission due to error checking at each hop, it is not suitable for the transmission of voice signals. Frame relay does not do error detection and correction at each hop or switching node, which makes it a more suitable transmission vehicle than X.25 for voice signals. ATM includes a *quality of service (QoS)* field to indicate that voice packets should receive a higher priority than data packets. ATM also uses fixed-size packets that are only 53 bytes long. The use of QoS and small fixed packets makes ATM the preferred transport vehicle for voice that has been packetized.

ATM can be used as a transport vehicle for other protocols. Frame relay, X.25, BISDN, and IP protocols can be encapsulated within an ATM frame. Voice-over IP is quickly becoming an accepted technology. When ATM is used to carry the IP packets, a high-quality voice circuit is achieved. Packet switching still does not offer the excellent quality achieved by a voice call over the circuit switched PSTN but has significantly narrowed the quality gap between circuit switched and packet switched voice signals.

Several router manufacturers are offering the capability to interface packet voice to the circuit switched PSTN. You can buy an IP phone that connects to your Ethernet-based LAN. The IP phone contains a NIC and will notify other NICs of its MAC address on connection to the LAN. The DHCP will receive the MAC address and assign an IP address that stays associated with the IP phone until it is un-plugged from the LAN and moved to another location. The router that interfaces the LAN to the PSTN will contain management software to control IP telephony within the LAN and between the LAN and the PSTN.

Calls between two IP phones connected to the LAN will be connected directly together the same way PCs would be connected together if they were transferring data files between themselves. They simply transfer IP packets that contain voice. The applications software and hardware in the IP phone convert the data within the voice packets into voice signals coupled to the receiver of the telephone. When a call is placed from an IP phone to a local telephone number on the PSTN, the IP packets will be addressed to a NIC in the router. A special gateway card in the router will convert the IP packets into the appropriate protocol for connection to the PSTN.

Each router vendor offers several gateway interface cards. Cards are available that will interface the LAN to either an analog loop start or a ground start line circuit in the local central office. Cards are also available to interface the LAN to T1 or PRI carrier systems facilities to the class 5 central exchange. The use of voice-over IP can be used to replace a PBX, if routers are used to interface the IP packets to the circuit switched PSTN. Voice-over IP can also be routed over a company's WAN to connect to IP phones at a distant LAN equipped with voice management software.

Most LANs are Ethernet-based LANs. The use of voice-over IP may drive many people to convert to ATM-based LANs. IP phones and voice call management software allow setting the priority of IP voice packets higher than data packets. Voice packets are much smaller than data packets and with voice compression will take up less than 5% of the traffic on a LAN. Data-intensive applications may hog the medium of the LAN for extended periods. Even though the largest block of data is 1500 bytes, the sender may beat the IP phone to the punch. There is no QoS setting for MAC frames. ATM will provide a better transport vehicle for voice packets. Companies that want to combine voice and data on their LAN may want to look at implementing an ATM-based LAN.

While ATM-based LANs appear to be the best solution for the transport of voice traffic, there are many Ethernet-based LANs and many businesses using frame relay-based WANs. These businesses can and will use voice-over IP on their existing facilities. The telecommunications manager must be aware of the amount of traffic that voice-over IP will add to these facilities and must ensure that the facilities are sized to prevent delay in the transmission of packets.

11.13 SUMMARY

Wide area networks (WANs) are used to connect computers or LANs together across a wide geographic area. LANs are networks connecting computers together within a building. Metropolitan area networks (MANs) are networks that connect computers or LANs together within a larger geographic area (usually within one city). WANs are networks that extend across the country. The PSTN is a WAN used to connect telephones together. The Internet is a WAN used to connect users to other computers on the Internet. The PDN is a WAN used to connect data terminals or computers to one another.

The PDN is an X.25-based WAN. The X.25 protocol uses LAPB at the data link layer to provide a very reliable transport medium. It does this by performing error detection and error correction at each node that a packet passes through. X.25 is a three-layer protocol. The X.25 packet is formed at the network layer and then inserted into the LAPB frame of the data link layer. User data is placed inside a packet at the network layer. This packet will contain a header that is used by the PSEs to route the packet through the PDN. X.25 uses a call setup packet to establish LCNs at each end of the PDN. The LCNs are connected together by a virtual circuit through the PDN. Access to the PDN is done through a PAD. The PAD may be implemented in software at the DTE, or it may be implemented in a device called a PAD.

The PAD connects to a DCE at the PSE via a circuit leased from the LEC, IEC, or bypass carrier. Many users utilize the circuit (link) between the PAD and DCE in a time-sharing arrangement. Each user's time slice is identified by a LCN. The link is used in a statistical time division multiplexing arrangement. This provides a mechanism to gain efficient use of the link. X.25 was covered in enough detail to provide a good understanding of packet switching. X.25 is being replaced by frame relay and ATM, which are simpler protocols. They can be simpler because the circuits used today for data networks are basically error-free circuits.

Voice-over IP will continue to grow. Telecommunications managers must stay abreast of developments in this area. Mid- to large-sized businesses with private network facilities and WAN connections will probably be able to reduce their telecommunications bill by using voice-over IP, but the technology must be deployed on a selective basis since it does not provide toll-quality voice. When used over frame relay facilities, you should investigate the sizing of the present facilities and the impact that VOIP will have on those facilities. New standards for voice-over frame relay continue to evolve and should soon provide a high-quality voice transport medium.

CHAPTER REVIEW QUESTIONS

1. What is a WAN?
2. What is a VAN?
3. What is a PDN?
4. Why does frame relay not do error correction at each node in the network?
5. What is a LCN?
6. What is a PVC?
7. What is a SVC?
8. What protocol is used at the data link layer by X.25?
9. What protocol is used at the data link layer by frame relay?
10. What is a call setup packet used for?
11. Is a call setup packet required for a PVC?
12. What is the function of the PAD?
13. What device is used to access the frame relay network?
14. Which packet-switched network was designed to support BISDN?
15. Which packet-switched network was designed to support SMDS?

PROJECTS AND CASE STUDIES:

Case Study #1

Residential Real Estate Company has 50 agents located at three offices around town. They want a web page, e-mail capabilities, and ten PCs at each office that agents can share. Bill's Consulting has installed a LAN in each office. What services can they get from local LECs to connect these LANs together in a WAN? What will the WAN facilities cost?

Case Study #2

Can you still get X.25-based services in your city or is Frame Relay the only option? What is the primary reason that Frame Relay is faster than X.25.

GLOSSARY

Asynchronous Transfer Mode (ATM)   A transport and switching technology based on fixed-size packets called *cells*. The fixed-size cells allow the switching systems to process the packets very fast. Fixed-size cells also prevent one user from monopolizing the medium for extended periods of time.

Bit-Oriented Protocol   Includes HDLC and its derivatives such as LAPB. These do not use ASCII characters for control information but use each bit or group of bits for specific functions.

Committed Information Rate (CIR)   The minimum bandwidth or speed guaranteed to a user for the transfer of data across the network.

Connectionless Protocol   Also referred to as a *datagram service*. Rather than establish an end-to-end connection through a network, each frame of data contains the destination address that routers can use to route the packet through the network.

Connection-Oriented Network   A virtual circuit is established for the transfer of data. The virtual circuit provides a connection over which subsequent packets can be delivered.

Data Link Connection Identifier (DLCI)   The logical channel assigned between a FRAD and the frame relay network.

Flag   A special pattern of bits used to flag the beginning and/or end of a frame. For the *X*.25 frame the flag bit pattern is 01111110.

Frame   The term used for a packet formed at the data link layer for transmission out onto the transmission medium connecting to the network.

Frame Relay   A fast packet network that uses a separate logical channel for call setup and control and does end-to-end error correction. It is a two-layer protocol designed to run on PRI facilities.

Frame Relay Assembler/Disassembler (FRAD)   The device used to access the frame relay network.

Header   The term used for the addressing and control information placed in front of

user data when a packet is formed or placed in front of the information field when a frame is formed.

Information Service Providers   The term used for companies like Prodigy and America Online. Any company that contracts with customers to provide them with information is an information service provider. Most of these companies provide information to their customers over the PDN.

International Data Number (IDN)   A number assignment technique used to assign terminal numbers to the DTE (with PAD software) connected to the PDN. The numbers are of the form 31264298997678. The addresses are usually 12 to 15 digits long.

Link Access Procedure Balanced (LAPB)   An HDLC-based protocol used on point-to-point links. It is a full-duplex protocol that allows either end of the link to initiate actions.

Logical Channel Number (LCN)   A number used by the DTE (PAD) and DCE at the PSE to identify the originator and destination of the packet containing the LCN.

Modulo 8   Indicates that the sliding window allows 7 frames to be outstanding without acknowledgment.

Modulo 128   Indicates that the sliding window allows 127 frames to be outstanding without acknowledgment.

Packet Assembler/Disassembler (PAD)   The software in the user's DTE that forms the user's data into packets according to the *X*.25 protocol and then delivers or receives these packets to or from the PDN. The PAD can also be a hardware device (with software) that allows for dial-up connections to the PDN.

Packet (or Public) Data Network (PDN)   A network built by a VAN using leased circuits in the PSTN from a LEC and IEC. The most common PDNs are Telenet, now owned by Sprint and called *SprintNet*, and Tymnet, now owned by MCI and called *Concert*.

Packet Networks   Networks that use STDM techniques to transport data packets from many users over one physical circuit.

Packet Switching Exchanges (PSEs)   Nodes in the PDN used to route packets through the network.

Permanent Virtual Circuit (PVC)   A circuit that has been permanently assigned a path through the data network by virtue of a permanently assigned logical channel number.

Switched Virtual Circuit (SVC)   A circuit that is set up by a call setup packet. The call setup packet at the beginning of a data session establishes a virtual circuit for the subsequent transfer of data. When the data session ends, a call clear packet is issued to release the LCN assignments.

Virtual Circuit   The term used to describe a circuit in a data network where many users share the same physical circuit in a TDM arrangement.

Wide Area Network   A data network that spans a large geographic area from across town to across the country. A WAN is used to connect LANs or computers in different locations together.

Internet Services

KEY TERMS

Address Resolution Protocol (ARP)
Domain Name
File Transfer Protocol (FTP)
Internet Protocol (IP)
IP Address
IP Datagram

Multipurpose Internet Mail Extension (MIME)
National Science Foundation Network (NSFNET)
Simple Mail Transfer Protocol (SMTP)

Transmission Control Protocol (TCP)
Uniform Resource Locator (URL)
User Datagram Protocol (UDP)
Well-Known Port Number

OBJECTIVES

On completion of this chapter, you should be able to:

1 Describe the evolution of the Internet.

2 Explain what transmission control and Internet protocols (TCP/IP) are.

3 Explain how TCP/IP fits the OSI model.

4 Explain how port numbers are used in TCP/IP.

5 Describe how an IP datagram is formed.

6 Explain the differences in addressing between IP version 4 and IP version 6.

7 Describe the network ID and host ID portions of an IP address.

8 Explain how class A, B, C, and D networks are differentiated to a router.

9 Explain how the hosts file is used.

10 Explain how address resolution protocol (ARP) works.

11 Define Internet domain name.

12 Explain what a Uniform Resource Locator (URL) is.

13 Define user datagram protocol (UDP).

14 Explain the differences between TCP and UDP.

The Internet has grown tremendously over the past few years. It has had as much impact on our society as the telegraph, telephone, television, and personal computer. As mentioned in Chapters 1 and 2, access to the Internet is crucial to those who seek information. The U.S. government has funded a program to establish an information superhighway to allow for more effective commerce over the Internet.

This program is called the *National Information Infrastructure (NII)*. As more and more business is conducted over the Internet, the highways of the Internet must grow in their ability to handle the increased traffic. Fortunately, fiber optic and dense wave division multiplexing technologies can provide the capacity needed for the foreseeable future.

12.1 EVOLUTION OF THE INTERNET

The Department of Defense established an agency to fund a project to allow dissimilar computers to communicate with each other via a WAN. This project was started in 1969, and was referred to as *ARPANET*. ARPA awarded grants to researchers at universities, colleges, and private industry to encourage the development of software that would allow the interconnection of dissimilar LANs across a WAN. The first researchers on the project were from the University of California at Los Angeles (UCLA), the University of California at Santa Barbara (UCSB), Stanford Research International (SRI), and the University of Utah at Salt Lake City. This project evolved into what is known as the *Internet*.

AT&T developed UNIX operating system software and was involved with universities in the deployment of UNIX. The University of California at Berkeley released its version of UNIX called *UNIX B* or *Berkeley Software Distribution (BSD)*. The *TCP/IP* protocol developed by the Internet researchers was included with the 4.2 BSD release in 1983. That year, the ARPANET was split into two entities, the Military Network (MILNET) and the Internet. The Internet Activities Board (IAB) was formed to oversee activities regarding the Internet. The Internet switched from network control protocol (NCP) to TCP/IP as its standard. Also during 1983, the University of Wisconsin developed Name Server, which allowed users to find sites without knowing the exact address of a remote system on the Internet.

In 1986, the National Science Foundation established the **National Science Foundation Network (NSFNET)** for research work. The NSFNET started as a 56-Kbps WAN connecting the supercomputers of several universities together. As the number of computers connecting via the network grew, the Internet backbone grew to a 1.544-Mbps WAN in 1988 and a 44.736-Mbps WAN in 1992. Also along the way, the NFSNET became the backbone network for both the NSFNET and the Internet. These backbone services were contracted out to Advanced Network Services and the ANSNET between 1992 and 1996. In 1996, MCI contracted to provide the Internet backbone via 155 Mbps (OC-3) over fiber optic facilities.

To gain approval of its merger with WorldCom in 1998, MCI sold its Internet service provider (ISP) operations to UUNET but retained the Internet backbone facilities. MCI is also under contract with NSF to provide a very high speed (OC-12) backbone for a second generation NSFNET. Universities and colleges are using the NSFNET to develop technologies for the next generation of Internet services. This second-generation NSFNET does not carry any Internet traffic and provides universities with their own private network for doing joint research.

The early ARPANET was an X.25-based WAN that all researchers connected their computers to. Researchers had several computers at their site connected to the ARPANET. They would dedicate some of these to their research work and would use others for general purpose computing. A file server was set up on the network to allow researchers to post their ideas. Researchers would issue a report as a *request for comments (RFC)*. Other researchers would respond to an author's RFC. RFCs are public documents; anyone can access them to gain a more thorough understanding of how the technologies work.

As mentioned in Chapter 1, several sites can be found on the Internet that provide information on RFCs. The IAB has contracted the registration of Internet addresses with Network Solutions, Inc. and has contracted the directory and database services to AT&T. The database containing Internet registration information and RFCs is at **www.internic.net.** You can also find RFCs at several other sites, such as **www.yahoo.com.** At this site you can select **computers** and **Internet,** select **standards,** and then select **RFCs.** RFCs 791, 919, 922, and 950 provide information on IP. TCP is covered in RFC 793, and FTP is covered by RFC 959. The first step in an RFC search is to view the index file for all RFCs.

12.2 TCP/IP PROTOCOL

Universities developed the TCP/IP model from research work on how they could join dissimilar LANs together across a WAN. The Department of Defense (DOD) formed a WAN to connect its computers together. This WAN was the ARPANET. The DOD invited universities to join the ARPANET and to research ways they could connect from their LAN to other universities that were using different LAN protocols. The ARPANET was the forerunner of the Internet, as pointed out earlier. The outcome of this research work was a protocol called transmission control protocol/Internet protocol (TCP/IP).

Many universities use UNIX, and TCP/IP was included as part of the UNIX distribution between universities and colleges. TCP/IP has become the de facto standard for the Internet; it is a five-layer model (see Figures 12-1 and 12-2). There is no official model since the TCP/IP is a de facto standard, but the models in Figures 12-1 and 12-2 provide an overview of the TCP/IP. Some people combine the physical and data link layer into a layer called the *access layer.* This leads to a four-layer TCP/IP model.

The physical layer is concerned with the physical interface to a transmission medium and sending the appropriate signals used by the network the computer is connected to. The data link layer is concerned with forming data link frames that contain the MAC address of the two computers attached to the network. In the case where TCP/IP is being used on a LAN, the MAC addresses could be workstations,

Layer	Function
5 Application	Provides access to applications programs
4 Transport	Provide end-to-end communication between applications
3 Internet	Sets up the transmission path between PCs
2 Data Link	Puts data in proper format and controls transmission
1 Physical	Transmits and receives signals in proper format

FIGURE 12-1 TCP/IP model.

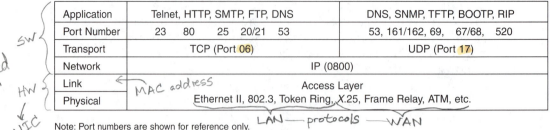

Application	Telnet, HTTP, SMTP, FTP, DNS					DNS, SNMP, TFTP, BOOTP, RIP				
Port Number	23	80	25	20/21	53	53, 161/162, 69,	67/68,	520		
Transport	TCP (Port 06)					UDP (Port 17)				
Network	IP (0800)									
Link	Access Layer									
Physical	Ethernet II, 802.3, Token Ring, X.25, Frame Relay, ATM, etc.									

implemented with: SW, HW, NIC — *MAC address* — *LAN — protocols — WAN*

Note: Port numbers are shown for reference only.

FIGURE 12-2 TCP/IP model.

servers, or routers, such as an Internet server. The Internet layer uses the IP software to provide the routing functions necessary to connect across a WAN to another network.

IP is connection-less *TCP is connection-based*

The IP does not provide for reliable connections. If users desire to ensure that the IP connection is reliable, they will use the TCP on top of IP as the transport layer protocol. TCP software turns the unreliable IP connection into a reliable connection by doing error detection and correction at the transport layer. If a highly reliable connection is not needed for the transport of data, users may decide to employ the *user datagram protocol (UDP)* at the transport layer instead of TCP. It does not do error detection and correction, and thus UDP provides faster transportation of user data.

Many protocols can be transmitted over IP. That is why IP is able to allow the internetworking of various protocols. Approximately 100 protocols can be transmitted via IP. The two major protocols used are TCP and UDP. The header of an IP frame contains much information about the data within the IP packet. The header of a TCP or UDP packet will carry information about the application protocol (port number) that will be using the TCP/IP or UDP/IP transport protocol. Each application within a host must have a unique port number address so that the transport layer knows where to deliver and receive data for each application. These unique addresses are called *port numbers.* Port numbers are like mailboxes; each application has its own mailbox. The applications program and the transport layer use the mailbox (port number) to transfer data between them (see Figures 12-2 and 12-3).

The TCP header is placed in front of the user's data, when the TCP packet is built at the transport layer. The four most significant bits of byte 13 allow us to vary the length of the header, but notice that the header in Figure 12-3 contains 20 bytes. Most TCP headers are 20 bytes long. The TCP header contains information used to detect errors and request retransmission of data. Error detection and error correction are handled by the transport layer (TCP) and not by the network or link layer. TCP turns the unreliable connection of IP into a reliable connection. The TCP packet is passed down to the network layer, where it is encapsulated inside the IP packet. The IP packet contains a header followed by the TCP packet.

One of the pieces of information the IP header contains is the type of protocol it is carrying. The header contains a protocol field. If the IP packet is carrying a TCP packet, the protocol number in this field will be 6. If the IP packet is carrying UDP, the protocol number will be 17. The other 100 protocols also have a number that is used to identify them. TCP protocol is used to provide reliable transport. UDP does

Bytes 1 & 2	Bytes 3 & 4	Bytes 5, 6 ,7, & 8	Bytes 9, 10, 11, & 12
TCP Source Port	TCP Target Port	Source Sequence. No.	ACK Sequence. No.
4 Most Significant Bits of Byte 13	4 Least Significant Bits of Byte 13	Byte 14	Byte 15 & 16
TCP Header Length	Reserved	Session Flags	Sender Window Size
Bytes 17 & 18	Bytes 19 & 20		
TCP Checksum	Urgent Data Size		

FIGURE 12-3 Typical TCP header.

not provide for error detection and control and provides an unreliable transport vehicle. Due to the error-free quality of networks, UDP is used by several applications. UDP will be faster than TCP because of the reduced functions (no error correction).

Once a TCP/IP or UDP/IP connection is established, various applications can flow over the connection. Applications that use TCP are *Simple Mail Transfer Protocol (SMTP), File Transfer Protocol (FTP),* Telnet, *Hypertext Transfer Protocol (HTML),* and *Domain Name Server (DNS).* Applications that use UDP are *Boot Protocol (BOOTP), Simple Network Management Protocol (SNMP), Trivial File Transfer Protocol (TFTP), Domain Name Server (DNS),* and *Routing Information Protocol (RIP).* Each transport protocol (such as TCP and UDP) can serve many different applications. The transport protocol at the receiving PC knows which application to deliver data to because, as noted earlier, each application is assigned a unique identifier—a *port number.* The port number that identifies which application a packet of data is for resides in bytes 3 and 4 of the TCP header.

A port number identifies the application that will flow over TCP/IP or UDP/IP. The Internet Assigned Numbers Authority assigned port numbers to each application that uses TCP or UDP. These port numbers are called *Well-Known Port Numbers.* The port number for FTP is port 20 for the data transfer, and port 21 for the control information on the file being transferred. Telnet is port 23. World Wide Web http is port 80. DNS is port 53. A 2-byte field is used to identify the port number. This provides the capability to identify 65,536 ports (applications). There are presently fewer than 1000 ports assigned to applications. The header of the UDP or TCP packet will carry both a source port address and a destination port address. Thus, when the transport layer (TCP or UDP) is stripped out of the IP packet, the port address tells the transport layer which application needs the data.

As mentioned earlier, many protocols use IP for transmission of data. Two of these are TCP and UDP, but there are many more. More protocols that use IP are being developed every year. As mentioned earlier, we now have *voice-over IP (VOIP).* The IP packet can carry data as large as 65,536 bytes. This far exceeds the packet size of transmission media such as Ethernet, which has a limit of 1500 bytes.

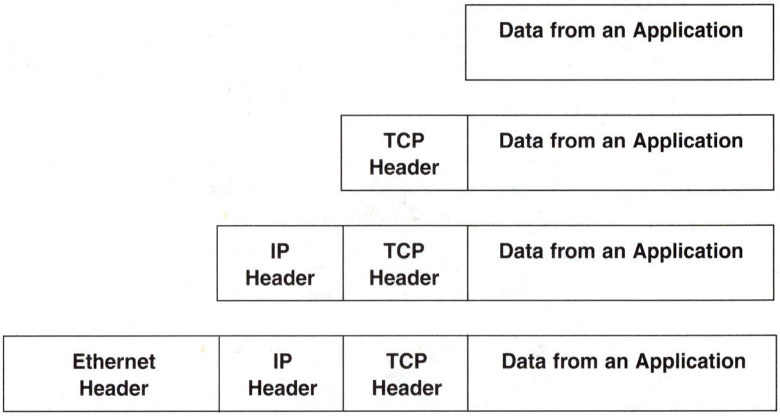

			Data from an Application

		TCP Header	Data from an Application

	IP Header	TCP Header	Data from an Application

Ethernet Header	IP Header	TCP Header	Data from an Application

FIGURE 12-4 Application data inside a TCP/IP/Ethernet frame.

IP can be placed over many different transmission links. IP packets can be placed inside Ethernet frames, frame relay frames, ATM frames, and so on. The data link layer interfaces between the IP layer and the physical layer to segment the IP packets into smaller packets that the transmission link and protocol can handle. TCP and UDP are implemented in software at the transport layer. IP is implemented in software at the network layer. These higher-layer protocols will be sent out onto some transmission link by the link layer that resides in some hardware device such as a network interface card (see Figure 12-4).

12.3 THE IP PACKET

The IP provides for a connectionless network. The IP packets are referred to as datagrams because all of them contain source and destination addresses. Since IP is a connectionless protocol, each and every packet must contain the source and destination address. The advantage of a connectionless protocol is that datagrams are routed independently of each other. This means that a failure of one component in the network, such as a router, will not take down the network. Datagrams will automatically be rerouted. The length of an IP header can vary, but for IP version 4 it is most often 20 bytes long (see Figure 12-5).

The first 4 bits of the first byte inform the receiving device whether IP version 4 or IP version 6 is being used for this *IP datagram.* Bytes 3 and 4 specify the total length of the IP datagram. This 16-bit field can provide a number as high as 65,535. Therefore, that is the maximum-size IP packet allowed. This is larger than some transmission links allow. Ethernet has a maximum size of 1500 byes in a frame. The IP datagram will be segmented into smaller packets to fit inside the frame of the data link and physical layers. Bytes 7 and 8 are used to inform the receiving host on how the datagram was fragmented. Byte 10 informs the target address whether TCP, UDP, or some other protocol is being used. Some numbers encountered in the protocol field are 01 hex (ICMP), 06 hex (TCP), 08 hex (EGP), and 11 hex (UDP). The TCP, UDP, or other protocol's packet will follow the IP header. The last 8 bytes of the header contain the source and target addresses.

4 Most Significant Bits of 1st Byte	4 Least Significant Bits of 1st Byte	3 Most Significant Bits of 2nd Byte	5 Least Significant Bits of 2nd Byte
IP Version	IP Header Length	Precedence	Type of Service
Bytes 3 & 4	Bytes 5 & 6	Bytes 7 & 8	Byte 9
Total IP Length	Datagram ID No.	Fragment Area	Time to Live
Byte 10	Bytes 11 & 12	Bytes 13, 14, 15, & 16	Bytes 17, 18, 19, & 20
Protocol ID	Checksum	Source IP Address	Target IP Address

FIGURE 12-5 Typical IP version 4 header.

No! The Class A, B & B addresses partition this space!

One of the major problems with IP version 4 is the length of the address fields. They are 4 bytes long. This provides 4,294,967,300 different addresses. When IP was developed, no one thought that a network would grow beyond this number of hosts. The Internet has been able to grow that large and was running out of addresses when IP version 6 was developed. It has 16 byte fields for source and target addresses. This provides 2^{128} different addresses.

I will be using IP version 4 addressing for the remainder of our discussion. A typical IP version 4 address is written in dotted decimal notation such as 192.140.173.4, and the numbers between decimals are stored in the 13th, 14th, 15th, and 16th byte of the header for the source address or bytes 17, 18, 19, and 20 for a destination (target) address. Each decimal digit is stored in 1 byte. This provides an address of the following form: byte.byte.byte.byte. In IP version 6, addresses are written in double byte format and are separated by colons. IP version 4 format is much shorter than IP version 6.

12.4 IP ADDRESSING

The IP protocol provides an address field for the address of the originator of an IP packet that is called the *Source Address.* The IP protocol also contains an address field for the destination of an IP packet that is called the *Target Address.* In IP version 4, both of these addresses are 32 bits long. The 32-bit field consists of 4 bytes. Each Byte can contain a number from 1 to 255. The addresses are written down using the number from each byte with a decimal point placed between bytes. Address possibilities with this arrangement range from 0.0.0.0 (all 0s in each byte) to 255.255.255.255 (all 1s in each byte), but all 0s and all 1s are used for special purposes and cannot be used as part of an IP address for a host. All 1s in a host ID is used to broadcast a packet to all hosts on the network. All 0s in the host ID is used to identify a subnetwork.

Instead of specifying a particular user's computer, the **IP address** is used to specify a network and then a host on that network. *Host* is a term used to refer to a workstation, server, or other computer attached to a network. The address field contains a network ID and a host ID. The network that a host is attached to may contain one computer, or it may be a LAN that has many computers attached to it.

This means we can have networks of many different sizes. A network containing many hosts is classified as a class A network. Medium-size networks are class B networks, and a small network is a class C network.

There are not many very large (class A) networks, but there are a lot of small (class C) networks. Since we must use the same 4-byte address field for all types of networks, the address field was designed with flexibility to handle the different-size networks without wasting a lot of address numbers. If there are fewer than 255 large (class A) networks, the first byte of the address field can be used to identify the network. The remaining three bytes of the address field can be used to identify a particular host on the class A network. This is exactly the approach that IP uses for addressing hosts on a class A network.

There are more than 255 medium-size networks. This means we must use more than 1 byte of the address field to identify which class B network a host is on. IP uses the first 2 bytes of the address field to identify which class B network a host is connected to and uses the last 2 bytes of the address field to identify the specific host. There are many small (class C) networks with just a few hosts attached to each network. For a class C network, the first 3 bytes of the address field are used for the network ID and the last byte is used for the host ID.

A method is needed to inform the receiving device (such as a router) how to treat the information in each of the 4 bytes used for an address. Should only the first byte be used to identify which network to connect to, should the first 2 bytes be used, or should the first 3 bytes be used for the network ID? The method IP version 4 uses to determine which bytes of the address field should be utilized for network ID is to use the first bit of the first byte of the address field to identify whether the address is a class A address. The second bit of the address field is used to identify whether the address is a class B address, and the third bit of the address field is used to determine whether the address is a class C address.

If the first bit of the address field is 0, the address is a class A address and only the first byte of the address field should be used for the network ID. The remaining 3 bytes should be employed for host ID. If the first bit of the address field is 1, the address is not a class A address and routers look at the next bit to determine if the address is a class B address. If the second bit is 0, the address field contains a class B address and the first 2 bytes will be used for the network ID, with the last 2 bytes used for the host ID. If the second bit is 1, the address is not a class B address and routers look at the third bit of the address field. If the third bit of the address field is 0, the address is a class C address and the first 3 bytes will be used for the network ID. The fourth byte will be used for the host ID. Finally, if the first 3 bits are 1s and the fourth bit is 0, the address is a class D address, which is used for multicast of a packet to many hosts.

Since the first bit of the first byte is 0 for class A addresses, the first number for a class A network will always be 0NNNNNNN in binary code. This means the first number in decimal will always be less than 128. The first byte will identify the network and the remaining 3 bytes will identify the host. The address 127.0.0.0 is reserved for a loop-back to test TCP/IP for interprocessing on a host. Data sent by a PC to 127.0.0.0 never leaves the machine but is returned back to an application within the host. The first byte of the address field for a class A network will contain

localhost

	Class A Addresses		
0NNNNNN	HHHHHHHH	HHHHHHHH	HHHHHHHH
1 to 126	0 to 255	0 to 255	0 to 255

Note: N = Network ID and H = Host ID. The three host ID fields cannot all be 0s, nor can they all be 255.

FIGURE 12-6 Address field of an IP packet for a class A network.

127 reserved for localhost

24 2

a number between 1 and 126. The last 3 bytes are used for host ID. This provides 2^{124} (16,777,216) possible addresses. Since we are not allowed to use all 0s or all 1s for a host ID, we are left with 16,777,214 host addresses for a class A network (see Figure 12-6).

Class A addresses can run from 1.0.0.1 to 126.255.255.254. This numbering system allows IP addressing to be used for 126 class A networks, and each of these networks can have from 1 to 16,777,214 hosts attached to the network. Needless to say, a class A network address would never be assigned to a company that only has a few hundred hosts attached to its network, because that would waste a class A address. You would never see a class A network with one host on it. Because there are so few class A addresses, they are all used. Many large organizations have to use multiple class B networks to gain enough addresses for all their hosts because no class A addresses were available.

The first bit of an address field for a class A network will be 0; for all other networks the first bit will be 1. For a class B network address (see Figure 12-7), the first bit will be 1 and the second bit will be 0. Thus, the first byte for all class B network IDs will be between 128 and 191. The second byte will be between 0 and 255. The first byte provides for 64 numbers (128 to 191 inclusive = 64). The second byte provides for 256 numbers (0 to 255). These 2 bytes put together provide for 16,384 (64 × 256) different class B networks. Eliminating all 1s and all 0s for a host ID, the last 2 bytes of the address field provide for 65,534 host IDs. A class B network address should be used for large networks.

A class C network address starts with 110 as the first 3 bits of the first byte of the address field. Thus, the first byte will be decimal numbers 192 to 223. This provides 32 numbers (192 to 223 inclusive). The second and third bytes can be between 0 and 255, which provides for 256 possibilities in each byte. Since the first 3 bytes are used for the network ID, this provides for 2,097,152 class C networks. The last byte is used for the host ID. Eliminating all 0s and all 1s as a host ID leaves the numbers 1 to 254 that can be used for a host ID. A class C network address is for small networks that have between 1 and 254 hosts attached to it (see Figure 12-8).

	Class B Addresses		
10NNNNN	NNNNNNNN	HHHHHHHH	HHHHHHHHH
128 to 191	0 to 255	0 to 255	0 to 255

192.168.X.X

FIGURE 12-7 Address field of an IP packet for a class B network.

reserved as a non-forwarding IP addr.

	Class C Addresses		
110NNNNN	NNNNNNNN	NNNNNNNN	HHHHHHHH
192 to 233	0 to 255	0 to 255	0 to 255

FIGURE 12-8 Address field of an IP packet for a class C network.

Class A networks will have network addresses between 1 and 126, class B networks will be between 128 and 191, and class C networks will be between 192 and 223. Class D addresses are used for multicast and will be between 224 and 239.

12.5 GENERATING THE IP ADDRESS

Most people are familiar with the IP addressing used on the Internet. Everyone surfing the Internet encounters the dotted decimal notation addresses. Most people also know that when surfing the Internet, they use names such as **www.kc.devry.edu** instead of a dotted decimal address. There are special computers attached to the Internet designated as *DNS*. The DNS converts addresses using a naming convention into dotted decimal addresses, thus providing the dotted decimal notation to your computer. Your computer then launches a request to the dotted decimal address.

IP addresses are used wherever the IP protocol is used. This means they are used on other networks beside the Internet. The IP protocol is also used on LANs and WANs. This means IP addressing is also used on LANs and WANs. Devices on a LAN must have a hardware address (MAC address) to communicate with each other, as discussed in Chapter 10. If IP is being used, the IP address will be converted to a hardware address. Once the hardware connection is established, the IP connection can be established. When an application is in use on a host and it requests information from another host, the application will check a hosts file on the PC for the IP address of the target host.

[handwritten margin note: PPP uses this! with arrow pointing to "verted to a hardware address"]

The hosts file contains a listing of the aliases used for computers connected to the network. When a user specifies a particular server, workstation, printer, or other host on the network, the host file will resolve the name of the target host to the IP address of the target host. The host's file resides on each computer (host) connected to the network and performs the functions needed on a LAN that a DNS provides for the Internet.

Once the hosts file generates the IP address, the IP address on a LAN will be mapped to the hardware address of the target computer by a table containing IP-to-MAC address entries. When a computer is connected to a LAN for the first time, it issues a broadcast packet to all computers on the LAN to inform them of its presence. These broadcast packets contain the IP and MAC address of the new computer. All computers attached to the LAN will store this new IP-to-MAC address in their IP-to-MAC table. Each computer will then send a response back to the new host and provide it with their IP-to-MAC address. Every computer on a LAN keeps a table in an *address resolution protocol (ARP)* cache that informs it of the IP and MAC address of each computer on the LAN.

12.6 ADDRESS RESOLUTION PROTOCOL (ARP)

The ARP cache stores hardware and IP addresses as well as a timer for each entry. The timer is set to 15 minutes when an entry is stored and gradually counts down to 0 seconds. The ARP cache will be used as long as the counter has not reached zero. When the timer reaches zero, the cache is listed as invalid but will still be present and will still be used. The application will simply be notified that the timer has expired. Every time communication takes place with another host, the ARP cache is updated.

Higher-level network addresses are used when we leave a LAN and go out onto a WAN or the Internet (the world's biggest WAN). We do not need hardware addresses unless the two computers are on the same network. When the two computers are on the same network, we *must* use hardware addresses for the two computers to communicate with each other. Applications use IP addresses and must resolve IP addresses to MAC addresses. This is done by using an ARP request. On an Ethernet-based LAN, an ARP request is placed inside an Ethernet frame and broadcast to all computers on the LAN. Figure 12-9 displays the Ethernet header that precedes an ARP packet.

The hexadecimal code 08 06 in the protocol field of the Ethernet header informs the receiving device that the packet is an ARP packet. The first two bytes of the ARP packet identify the hardware type. For a 10-Mb Ethernet NIC, the hardware type is 1 and is displayed in the 2-byte field as hexadecimal 00 01. The next 2 bytes are used to identify the protocol. For an IP protocol requesting an ARP, this will be 08 00 in hexadecimal. The fifth byte will be used to inform recipients how long the address field is in bytes. Ethernet addresses are 48 bits (6 bytes) long, and the hardware length field on Ethernet will be hexadecimal 06.

The sixth byte of the ARP packet is used to inform recipients how long the protocol address is. For an IP protocol with IP version 4, the length of the IP address field is 4 bytes, and this field will contain 04 in hexadecimal. Bytes 7 and 8 will be used to indicate the ARP operation to be performed. Hexadecimal 00 01 is for an ARP request. Hexadecimal 00 02 is for an ARP reply. The remaining four fields of an ARP packet contain the Source Hardware Address, the Source Protocol Address, the Target Hardware Address (usually all 1s for a broadcast), and the Target Protocol Address. For IP protocol, the Source and Target Protocol Address will be the IP addresses.

When the Target Hardware Address is unknown, we broadcast the ARP packet to all NIC cards, and Figure 12-10 shows all 1s (broadcast) in the target hardware address field. Each NIC card will look at the Target Protocol (IP)

First 6 Bytes	Bytes 7 to 12	Bytes 13 and 14
Target Hardware Address	Source Hardware Address	Protocol Type
ff ff ff ff ff ff	Address of the Host Generating the ARP Packet	08 06
Broadcast Address		ARP

FIGURE 12-9 Ethernet header that precedes an ARP packet.

Bytes 1 & 2	Bytes 3 & 4	Byte 5	Byte 6	Bytes 7 & 8
Hardware Type	Protocol Type	Hardware Length	Protocol Length	Operation
00 01	08 00	06	04	00 01
Ethernet (10 Mb)	IP Protocol	6 Bytes for Hardware Address	IP Protocol Version 4 Uses 4 Bytes	Operation Code 01 for ARP Request
Variable Field	Variable Field	Variable Field	Variable Field
Source Hardware Address	Source Protocol Address	Target Hardware Address	Target Protocol Address
00 60 8C 00 20 34	C2 81 98 72	11 11 11 11 11 11	C2 81 98 50
	194.129.152.114	Broadcast to All NICs	194.129.152.40

FIGURE 12-10 ARP packet in an IP protocol environment.

Address. Only the host, which has been assigned the IP address contained in the Target Protocol Address field, will return an ARP reply packet. The reply packet will contain the IP and MAC addresses or the replier. The host that had generated the ARP request will use the information in the ARP reply packet to update its ARP cache. Figure 12-10 contains addresses used as an example to demonstrate the ARP protocol. Sometimes the Target Hardware Address field will contain the expired ARP cache address instead of all 1s.

12.7 INTERNET DOMAIN NAMES

Internet addresses fall in one of 13 top-level domains, which are **.edu**, **.com**, **.gov**, **.int**, **.mil**, **.org**, **.net**, **.web**, **.arts**, **.info**, **.nom**, **.rec**, and **.store**. The .nom domain is used for personal domains and for personal web pages. Companies that create and host web pages for others use the .web domain. The other *domain names* are self-explanatory. If I type **http://www.kc.devry.edu** into my browser, the browser checks in my hosts file to see if I have already received a resolution of the named site into an IP address. If the hosts file contains a resolution, it used the resolved IP address. If the hosts file does not contain a resolution, a resolution request is launched to a DNS on the Internet. One DNS does not contain the name of all hosts on the Internet. The resolution request will be sent to one DNS after another until the *fully qualified domain name (FQDN)* for the request is found.

Internet names consist of labels separated by periods. Note that in our example, **kc** is followed by a period; **devry** is followed by a period, as well. Each of these is referred to as a *label*. Labels can be up to 63 characters long and can consist of alphanumeric characters. As we saw earlier, domain names are limited to three characters. When a DNS request is launched, the DNS searches for the name from the rear to the front. It starts its search by looking at the highest-level domain name. In our example, it first finds a DNS that contains addresses for **.edu** sites. It will request the DNS to look for a match to **devry.edu**; if that is found, it will ask for **kc.devry.edu**. There is another level above the domain name used to specify the country where a host is located. If it is not specified, it is assumed to be the United

States. Two characters are used to specify country domains. When launching a DNS search, it is not necessary to type the **http//www.** part of the address. If searching for a **.com** address, it is not necessary to type the **.com**.

12.8 BROWSING THE INTERNET

The browser is a software program that interprets special files called *Hypertext Markup Language (HTML)* files. The HTML file contains tags used by the browser as instructions on how to display the file. Browsers are designed to look for the document or web page specified in the *URL* field. In Netscape the URL will be typed into the window labeled **Netsite** in the menu bar. In the Internet Explorer browser, the web site we wish to visit will be typed into the window labeled **address** in the menu bar. We can type the web site address using the fully qualified domain name and let a DNS translate this into a dotted decimal address for us, or we can type in the dotted decimal address to reach a host on the Internet.

The URL consists of six component parts, of which three are optional (port, filename, and variable) (see Figure 12-11). For http, the default port is 80. It does not have to be specified in the URL. The http application will communicate via port 80 with TCP at the transport layer to provide it with data for the TCP packet. The TCP packet will be enclosed in an IP frame and sent out over the Internet. The IP address of the target host will be contained in the IP datagram and will be used to route the TCP packet to the proper server containing the desired resource.

Once the TCP packet is delivered to the proper server, the transport protocol will see that the packet is delivered to port 80. If the filename is not specified, the browser will look for **index.html**. Therefore, every web site begins with **index.html** as its first web page. This page is commonly called the *home page.* If you type **http://www.pearsoncustom.com/authors/** as the address that you wish to get a web page from, you will get an index.html file (or main.html file). It will contain linkages to other pages. If you desire to look at one of these links, simply click that link on the home page. You could also access linked files by typing the location of the file. As an example, if you wanted to access the file for frequently asked questions that exists as a link on the Authors page, you could access it directly by typing **www.pearsoncustom.com/authors/faqs.html** rather than using the link.

The World Wide Web is a client/server network. Users will invoke a request via their PC (client) by entering the URL in the address field provided by the web browser and pressing the Enter key. The resource desired will be located on a server attached to the Internet highway. The server receives the request and will return a response to the requesting client. Various servers on the Internet will cache a copy of web pages that have been requested. These cached copies are often

Service	Domain Name	Port	Pathname	Filename	Variable
http://www	pearsoncustom.com	80	read/authors/	Index.html	

FIGURE 12-11 Example of WWW address.

sent to a requesting client rather than the server being contacted that contains the original source document. This helps speed up the response, but sometimes this causes problems. Sometimes the cached copy is out of date. I have modified my web page, and the modifications were not immediately available over the Internet because an older cached copy was being returned to requesters. Sometimes it took several days for the cached copies to get updated with the changes I made on the original document on my server.

Your PC also caches web pages, and it will often use cached copies instead of going to the host site where the original web page resides. It pays to clean out your cached pages on a regular basis. With a Netscape browser this is done by opening Netscape and then selecting **Edit** from the menu bar. In **Edit**, select **Preferences**, select **Advanced** in the category list, then select **cache**. You can then clear memory cache and disk cache.

When you select **Edit** from the menu bar and then select **Preferences**, the category that you start with is Netscape. You will notice that this allows you to set the page that your browser starts with when you open the browser. You can click the home page button to have Navigator start with a home page. In the field below that you can specify the location of the page you wish to use as the home page. While on this page you can clear **History** if you do not want anyone to be able to see the sites you have visited. You can set the number of days that you wish to keep in **History** as well.

You can also clear the cache and **History** file in Internet Explorer. To clear the cache, select **View** on the menu bar and then select **Options**. For Internet Explorer version 3, select the **Advanced Tab**. For Internet Explorer version 4, select the **General Tab**. After selecting the appropriate **Tab** (**Advanced** or **General**) for your version of Internet Explorer, select **Settings**. You can then choose to view the files or may delete them. I usually view, delete and then view again. Cookie files may not get deleted, and you can get rid of them in the view mode. Select all files (**Edit**, **Select All**) and then select **File**, **Delete**.

You can also delete your **History** file. For Internet Explorer version 3, under **View**, **Options** select the **Navigation Tab**. You can then view or delete the **History**. You can also select the page you wish to use when you start your browser on the **Navigation Tab**. In the **Start** field use the drop-down arrow to select **Start Page**. In the Address field type in the address of the web page you want to use at Startup, such as **http://www.prenhall.com**. For Internet Explorer version 4, the **History** and **Startup** page are on the **General Tab**.

12.9 FILE TRANSFER PROTOCOL

FTP is used to request files from a file server on the Internet. FTP requires two TCP connections for each file transfer. One connection is for the control port (port 21), and the other is for the transfer of the file via port 20. FTP can be used to transfer single or multiple files. Many versions of FTP marketed by different software companies include a graphical user interface (GUI). These interfaces make it relatively easy to use FTP. Older FTP programs do not support the Windows 95/98 extended filenames and extensions. They will truncate *html* to *htm* when the file is transferred from your PC to an FTP server.

FTP Command	Name
OPEN	Open a Connection with FTP Server
CWD	Change Working Directory
DIR	List the Files in Current Directory
LS	List Abbreviated Information on Files in Directory
MKD	Make Directory
RMD	Remove Directory
PORT	Destination Port
USER	User ID
NLST	Name List
PASS	Password
PWD	Print Working Directory
QUIT	Quit Session
RETR	Retrieve File
TYPE	ASCII or Binary File Transfer

FIGURE 12-12 FTP commands.

You may not have to buy a special FTP program unless you want a GUI and other benefits FTP software packages offer. DOS and the DOS command program of Windows 95/98 contain an FTP program that can be activated by typing **ftp**. FTP can be invoked by commands. Using the DOS command interface, type **ftp**. Once you have typed **ftp**, you can type **help** to get a list of the FTP commands. Some of the most frequently used FTP commands are shown in Figure 12-12.

The FTP command will be delivered to the target host via TCP/IP. The TCP packet will contain a TCP source port number that is a random port number. All TCP sessions begin with a random number in the Source Port field. The TCP Target Port field will contain hexadecimal 00 15, which is decimal 21, when the client is trying to establish an FTP session. A typical FTP address might be 208.128.98.1. To use FTP, you first establish a connection to the WAN or Internet (via your ISP). You can then use the OPEN command followed by the address of the FTP server. The server and client will go through a three-step handshake procedure to establish a TCP connection and the server will send a response code 220 to indicate it is ready to do a file transfer.

After receiving response code 220, the client (this is you) will send another packet with 21 in the Target Port field and the packet will contain USER XXXXX, where USER is the command and XXXXX is the user's name. At the **ftp** prompt you simply type the user command, followed by a space, followed by your user name. If the user is allowed access to the file server, the file server will return a packet with 21 in the Target Port field and response code 331 (along with something similar to **password required**). The client (this is you) responds with another packet addressed to port 21 simply by tying your password at the **ftp** prompt. This packet contains the pass command along with the user's password. If the password is correct, the server will return a response code 230. This response code will be accompanied by verbiage such as **User XXXXX logged onto server**.

After users are logged on, they can change the directory they are in, print a working directory, make a new directory, list the directory, and so on. If users wish to transfer a file from the server to their PC, they will first issue a PORT command. This command instructs the server where the file should be delivered. The PORT command is followed by the IP address of the requesting host plus the port number where the file should be delivered. The server will return a response code 200 to the client so that it knows the PORT command was received. FTP will now open port 20 via the same three-step handshake procedure used to open port 21. It will then transfer the file over port 20 to the location specified by the PORT command. You type the command **QUIT** to end the FTP program. You can then type **Exit** to exit DOS and return to Windows. If you do this exercise, I think you will agree that GUI-based programs, such as FTP, are worth the price. They are so much easier to work with.

Many companies provide free software downloads from their web sites. You are allowed to download these files using your web browser. Many companies used to maintain files on an FTP server for access via FTP. Most companies have converted their FTP sites into web sites. In fact, many FTP sites are becoming outdated because they are no longer being maintained and updated. FTP still comes in handy to upload files. The browser does not provide upload capabilities. When I make updates to the copies of home page files that reside on my PC, I use FTP to upload them to my ISP site.

12.10 E-MAIL OVER THE INTERNET

E-mail is one of the most popular services used over the Internet. The protocol used for e-mail is SMTP. *Multipurpose Internet Mail Extension (MIME)* is a protocol that extends the capabilities of SMTP to allow it to transfer graphic files, executable files, and other binary files. SMTP uses TCP as its transport protocol. The TCP packet carrying SMTP will contain hexadecimal 19 (decimal 25) in the TCP Target Port Address field. The port number of 25 informs the receiving server that the user is requesting an SMTP session.

The establishment of all TCP session is quite similar; only the port addresses and commands differ. E-mail is a store-and-forward operation. E-mail is delivered to an e-mail server, where it can be retrieved by a user logging into the mail server. The e-mail address is the name of the person followed by the @ sign followed by the domain name for the mail server, such as **mary_cole@earthlink.net**. A SMTP session starts with the three-step handshake procedure required to establish a TCP connection. Once the TCP connection is established, the user can invoke the SMTP commands. Commands sent to the mail server are answered with a response code returned to the client.

When the client starts a SMTP session, the response to the TCP connection setup with a Target Protocol Address of 25 will be a response code of 220. If users wish to send an e-mail, they will issue a HELO command. The server's response to the HELO command is response code 250. The user then sends command MAIL FROM: **user@host** to identify who is sending the e-mail. The response code from the server to the client will be response code 250 again. Basically, response code 250 means "okay, I received your command."

The next step in sending e-mail is to send the RCPT TO: **user@host** to inform the mail server where the e-mail should be sent. The server responds okay (response code 250). The next command sent by the client is DATA. The server will instruct the client what characters it should send at the end of the e-mail message. The e-mail message and the end-of-message characters will follow the DATA command. The client will then issue a QUIT command to end the e-mail session. A takedown procedure is then implemented to end the TCP connection. If the user wishes to get new messages off an e-mail server, they issue a HELO and MAIL FROM command followed by a TURN command to the mail server. The mail server will then send e-mails being held for that user to the client.

12.11 TELENET

Telenet is an applications program and protocol that allows a PC to connect to a remote computer as a terminal and interact with the remote computer. This application was used by many people to activate search facilities such as Gopher and Archie prior to the development of browsers. Telenet compares to a browser as DOS compares to Windows 95. The browser and search engines provided on the Internet have relegated Telenet, Gopher, and Archie to minor roles. If you try to access Gopher sites today, you find that many of them are no longer accessible. Telenet is still used as a network management tool. LAN managers often have sophisticated LAN management software loaded on the LAN to provide them with management capabilities, but they will often use Telenet to activate a test such as PING across a WAN.

12.12 INTERNET JOB OPPORTUNITIES

The growth of the Internet has led to a tremendous number of job opportunities. Every company needs a presence on the Internet and needs a web master to keep the web server online and keep web pages current. ISPs need people familiar with UNIX, Microsoft NT, web page design, and WAN technology. The IECs have added Internet access offerings as one of their services. This required them to form joint ventures with existing ISPs or to establish their own ISP operations. People are needed to staff the help desk at these ISPs and to manage the network facilities of the ISP.

12.13 SUMMARY

The Internet is the largest WAN in the world and provides people with access to a tremendous amount and variety of information. It allows people to communicate quickly via e-mail and e-phone. The IP protocol that powers the Internet has found its way into LANs and is now being used to provide a transport vehicle for voice calls over the data networks. TCP provides error detection and error correction at the transport layer for IP connections. For applications that do not require error detection and correction, UDP is used as the transport protocol. Port numbers are used inside IP, TCP, and UDP packets to identify which protocol should receive the packet of data.

Addressing of hosts attached to the Internet is done using dotted decimal notation addresses. The addressing schema provides for type A, type B, type C, and type D networks. The network address consists of a network ID and a host ID. Class A addresses run from 1 to 126 in the network ID portion of the address. Class B addresses run from 128 to 191, and class C addresses run from 192 to 233. The address resolution protocol (ARP) is used on LANs to resolve an IP address to a MAC address.

When attaching to a site on the Internet, most people use a naming convention to connect to a computer on the web. The web has domain name servers (DNS) that resolve the name to a network address following the dotted decimal notation format. A user types a name in the URL field and the PC will check its host file to see if it has already been resolved to dotted decimal notation before launching a request to a DNS.

HTTP has been instrumental in the explosive growth of the Internet. The capabilities of browsers continue to expand, and ISPs are offering many services such as search engines that the browser can use to search for data on the Internet. The growth of the Internet has created many job opportunities for telecommunications professionals.

CHAPTER REVIEW QUESTIONS

1. Who provided us with the browser?
2. What level of the OSI model does IP belong to?
3. What level of the OSI model does TCP belong to?
4. What level of the OSI model does UDP belong to?
5. Is IP a connector-oriented protocol?
6. Does TCP provide a connector-oriented protocol?
7. What is the basic function of TCP?
8. Why would someone use UDP instead of TCP?
9. What are port numbers used for?
10. What is SMTP used for?
11. What is FTP used for?
12. What has replaced Gopher?

PROJECTS AND CASE STUDIES:

Case Study #1

Residential Real Estate Company has 50 agents located at three offices around town. They want a web page, e-mail capabilities and ten PCs at each office that agents can share. Bill's Consulting has installed a LAN in each office and connected them together via a WAN. How should the web site for this company be handled? How do you get a name registered for ResidentialRealEstate.com?

GLOSSARY

Address Resolution Protocol (ARP)   The protocol used to resolve IP addresses to hardware addresses (MAC addresses). When the IP protocol is used, computers are referred to by higher-level applications using an IP address. In order for two computers on the same (network) to communicate with each other, they must use MAC addresses. When an IP packet is on the network containing the destination computer, ARP converts the IP address into the MAC address.

Domain Name   The top level in the Internet naming conventions. There are 13 domain names: **.edu**, **.com**, **.gov**, **.int**, **.mil**, **.org**, **.net**, **.web**, **.arts**, **.info**, **.nom**, **.rec**, and **.store**.

File Transfer Protocol (FTP)   Used to request files from a file server on the Internet or any other network. FTP establishes two TCP/IP connections for a file transfer. Port 21 is used as a control port and Port 20 is used for the transfer of data.

Internet Protocol (IP)   A network layer protocol used by many networks in general and by the Internet in particular. It provides the routing functions necessary to route data packets across a WAN and is also used on LANs. IP provides for a connectorless, unreliable transfer of data.

IP Address   The network layer address. It contains both the network and station identity. The IP address for IP version 4 is 4 bytes long. To provide many more addresses, IP version 6 has been released. It has a 16-byte field.

IP Datagram   A packet of data that contains source and destination addresses. Encapsulated within the IP datagram are the TCP or UDP header information and the actual user data from an applications program.

Multipurpose Internet Mail Extension (MIME)   A protocol that extends the capabilities of Simple Mail Transfer Protocol to allow the transfer of binary files such as executable programs and graphics.

National Science Foundation Network (NSFNET)   Network that started out as a 56-Kbps *X*.25 packet-based wide network to connect computers of several universities together for research work. The network has grown in size and speed and has become the network backbone for the Internet. This backbone is currently a 155-Mbps (OC-3) fiber optic facility.

Simple Mail Transfer Protocol (SMTP)   The protocol used for sending e-mail over the Internet. SMTP is identified in the TCP header as port 25.

Transmission Control Protocol (TCP)   A transport layer protocol used in conjunction with IP at the network layer to form a TCP/IP protocol. TCP turns the unreliable connection of IP into a reliable connection by performing error detection and correction. The TCP protocol is identified inside an IP packet by a Protocol field number of hexadecimal 6.

Uniform Resource Locator (URL)   The identifier used to designate where an application or resource resides on the Internet. The URL consists of six parts: Service, Domain Name, Port Number, Pathname, Filename, and variable. Port Number, Filename, and Variable are optional fields. The URL is typed into browser Address or Netsite field when using **http.** The computer then sends a request via the Internet to port 80 on the destination host for an HTML file located on that host.

User Datagram Protocol (UDP)   A transport layer protocol used with IP when error detection and correction are not needed. When data is being transmitted over a highly reliable error-free transmission medium, error detection and correction are not needed and the use of UDP will provide for a faster transfer of data than the TCP protocol. The UDP protocol is identified inside an IP packet by a Protocol field number of hexadecimal 11.

Well-Known Port Numbers Port numbers that have been assigned by the Internet Assigned Numbers Authority to each application using TCP or UDP. These port numbers exist between the upper two levels of the TCP/IP and UDP/IP protocol model. They identify which application should receive data from the transport layer.

13

Mobile Telephones and Personal Communication Systems

KEY TERMS

Advanced Mobile Phone System (AMPS)
Base Station Controller (BSC)
Base Transceiver Station (BTS)
Basic Trading Area (BTA)
Cellular Radio
Code Division Multiple Access (CDMA)
Code Excited Linear Predictive (CELP) Coding

Digital Advanced Mobile Phone System (DAMPS)
Global System for Mobile Communications (GSM) 1900
Improved Mobile Telephone System (IMTS)
Major Trading Area (MTA)
Mobile Telephone Serving Office (MTSO)

Mobile Telephone System (MTS)
Personal Communication System (PCS)
Personal Communication System 1900 (PCS 1900)
Roaming
Time Division Multiple Access (TDMA)

OBJECTIVES

On completion of this chapter, you should be able to:

1 Describe the evolution of mobile telephone technology.
2 Discuss the differences between DAMPS and PCS-1900.
3 Understand TDMA and CDMA.
4 Understand MTA and BTA.
5 Understand the various components that make up a PCS-1900 network.
6 Describe how SS7 is used to support mobile telephone roaming.
7 State the purpose of subscriber module identity.
8 Describe how the BTS connects to the BCS.
9 Describe the operations of a MTSO.

The newest wave of technology to hit telecommunications is the technology behind the *personal communication system (PCS)*. This technology takes advantage of the technology advances made in SPC switching systems, the SS7 network, advanced intelligent networks (connected by SS7), and mobile telephone systems. PCS itself is not so much a technology as a concept. The concept of PCS is to assign someone a *personal telephone number (PTN)*. This PTN is stored in a database on the SS7 network. That database keeps track of where a person can be reached. When a call is placed for that person, the *artificial intelligence network (AIN)* of the SS7 determines where the call should be directed. To provide mobility to the PTN, it was decided that the primary location for most PTNs should be reached by using radio waves rather than copper wire for a transmission medium. The PCS is provided by radio frequencies in the 1900-MHz range, and the service is often referred to as a *personal communication system 1900 (PCS 1900)*. Before discussing PCS, I think it appropriate to briefly review the developments that have occurred in mobile telephone systems.

13.1 THE BEGINNINGS OF MOBILE TELEPHONE SYSTEMS

Mobile telephone systems got underway in 1946 and were called *MTS*s. These systems used the radio frequencies between 35 and 45 MHz. Although MTS usually stands for "mobile telephone system," this acronym also meant "manual telephone system." All calls had to be handled by an operator. These early systems used one frequency for the mobile phone and the base station. The handset had a push-button. The mobile party used a push-to-talk protocol. When the button was depressed, the transmitter of the mobile unit was activated. This would cause a signal to be transmitted to the base station. On receiving the signal, the base station would light a light above a jack at the operator's position. The operator would answer the call by plugging a patch cord from her position into the jack connected to the base station transmitter. She would ask the calling mobile party for information on whom they wished to call. The operator would use the other end of the patch cord to connect the mobile telephone caller to the desired PSTN destination.

The PSTN telephone was connected to the base station's transmitter through the patch cord at the operator's position. A conversation could now take place, with the caller and called party taking turns. The MTS systems usually had only one radio frequency for both the base and mobile phones. Therefore, only one phone call at a time could be placed over the MTS. Because all mobile phones used the same frequency, the MTS was like a party line. All MTS mobile phones could hear any conversation on the system. To place a mobile-to-mobile phone call, the mobile caller would reach the operator by pressing the push-to-talk button. The person would then ask the operator to ring the other mobile phone. Once the operator rang the mobile phone, she was no longer needed. The base station was not needed for a mobile-to-mobile call, because the transmitters and receivers of both phones were tuned to the same frequency.

The early MTS phones were connected to a transmitter/receiver located in the trunk of the car. This transceiver was a tube-type device and placed a heavy drain on a car's battery. The transceiver was about 24 in. long, 18 in. wide, and 6 in. high. The receiver of the MTS mobile unit contained a mechanical decoder device. The tube-type MTS mobile units were replaced by transistor-type mobile units around 1970. Solid-state logic circuits in these units served as the decoder circuit. When the operator placed a call to a mobile phone, the dial pulses from her dial were converted into pulses of a signaling tone. Let's assume that a 40-MHz signal is being used by a MTS. If a 5 was dialed, the base station would modulate the 40-MHz carrier signal 5 times with a tone of 2400 Hz.

All receivers were tuned to accept the 40-MHz signal and would see the 5 pulses of tone. The output of the receiver went to a decoder as well as to the receiver of the handset. The input of the decoder included a filter that would only pass the frequency of the signaling tone. The decoder would count the pulses and store the digit. The decoder contained a chain circuit. This chain circuit was wired at the time of installation as the phone number of the mobile phone. The chain circuit of the decoder for each telephone was wired so that if the first digit dialed matched the first digit of the mobile phone number, the decoder remained active and looked at the second stream of tones sent for the second digit of the phone number. If the second digit matched the wired number, the decoder looked at the next digit, and so on. This process was done for each digit. If at any digit, a match was not made between the received number of tones and the number wired in the decoder for that digit, the decoder released and reset itself. If, however, each digit was matched as it was received, on receiving the last digit, the decoder recognized that the received number matched the number assigned to this mobile unit and would ring the phone. It also sent a signal out on a wire that could be connected to the horn relay of the vehicle, if desired. This would cause the horn to blow one time for about 2 sec. On hearing the ringer or horn, the mobile customer would pick up the handset and press the talk button to answer the call.

MTS mobile phone numbers had no relationship to the telephone numbers of the PSTN; MTS numbers were distinct and separate from PSTN numbers. The Bell Company of each state kept a registry of numbers assigned to the MTS systems in that state. Each MTS service provider would get a number assignment from the Bell Company records department when a new MTS phone was installed. These mobile phone numbers were usually five-digit numbers. They could not be accessed directly over the PSTN and were not part of the PSTN number plan. To reach a MTS mobile, callers had to know what serving area the mobile was in. Callers would have their local operator connect them to the operator of that serving area. They would then ask the operator of that area to ring the desired mobile number.

For the MTS system that I maintained, we had a base station transmitter, receiver, and antenna on top of a large hill. The transmitter had a range of about 50 mi. I cannot remember the exact power output, but most of these systems had a 100- to 200-W output. The transmitters in the mobile units had a much lower level of power output (about 20 W). The mobile transmitter had a range of about 15 mi. To provide coverage for the 50-mi radius from the base transmitter, we had receivers placed in the north, east, south, and west directions, about 20 mi from the base transmitter/receiver site. We had a total of five receiver sites. One of these (the central site) was also the site for the base transmitter. All sites were connected by various land transmission media to the toll office. The toll office had a mobile communication bay, which received the inputs from all receivers and the base transmitter. The equipment in this bay would determine which receiver was providing the strongest signal and would connect the signal from that receiver to a jack at the operator's position. This jack was labeled *MTS*. The operator knew that a call coming in on this jack was a mobile phone call (see Figure 13-1).

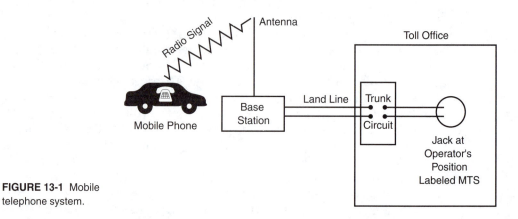

FIGURE 13-1 Mobile
telephone system.

13.2 IMPROVED MOBILE TELEPHONE SYSTEMS

Improved Mobile Telephone System (IMTS) was introduced about 1964 but was not implemented where I worked until around 1973. The major benefit of IMTS was that it used several different frequencies and could support many different conversations at the same time. IMTS was also connected to the local class 5 office instead of the toll office. IMTS was connected to regular telephone numbers and line circuits. Unlike MTS, IMTS was assigned a regular PSTN telephone number. Anyone could reach an IMTS phone by dialing the PSTN number assigned to the mobile phone. This eliminated the need for operators to handle mobile phone calls. When the IMTS mobile unit picked up their handset, they were connected to a line circuit in the central office and received a dial tone. They placed their call just as they would from a regular telephone. The frequency spectrum used by IMTS was 454 to 512 MHz.

When a call was placed to a mobile IMTS phone, the caller dialed a number in the PSTN that was assigned to the mobile telephone. The PSTN telephone number was connected to a mobile radio terminal in the central office. The ringing signal for the called number activated a relay in the radio system's terminal equipment. The terminal equipment in the central office communicated with the controller of the transmitter at the antenna/transmitter site. The control system would automatically select an idle, available transmitting frequency and would modulate that frequency with signaling tones representing the dialed telephone number. The decoder of the mobile IMTS units would decode the signaling tones. If the decoded number matched the number programmed into the mobile unit, it would ring.

13.3 ADVANCED MOBILE PHONE SYSTEMS

MTS and IMTS used base station transmitters with 100 to 200 W of power and mobile transmitters with 5 to 25 W of power. These high power levels allowed the MTS and IMTS systems to cover a wide area using one base transmitter for the area. The next phase in mobile phone systems was the introduction of *cellular radio.* Cellular radio is also known as *advanced mobile phone systems (AMPS).* AT&T had been testing cellular radio technology in the late 1970s and early 1980s. This technology took a novel approach. It uses many low-powered base transmitters (7 W) spread out over a geographic area (multiple base transmitter sites) and employs computer technology to decide which transmitter should be used. All the base station receivers and low-powered base transmitters are connected to a centrally located computer, which controls all operations. When a mobile phone is turned on, it communicates with several base site receivers. The computer determines which receiver is closest to the mobile unit on the strength of the received signal. The mobile unit transmits its identity (its telephone number) to the central computer. The central computer always knows the location of the mobile unit and which receiver/transmitter is closest to the mobile unit. When a call comes in for a mobile unit, the computer will use the transmitter closest to the mobile unit and send an alerting signal to the mobile unit.

On October 13, 1983, Illinois Bell implemented the first cellular network. The large metropolitan area of Chicago was divided into smaller geographic areas called *cells.* Each cell site was equipped with its own transmitting and receiving base station. Each station was assigned several transmitting frequencies and several different receiving frequencies. By using a low-powered transmitter in the base station and mobile units (7 W), the frequencies could be reused in many cells around the city. The frequencies were not reused in adjacent cells but were utilized again in cells several miles away. Basically, cells were arranged in a group of seven cells forming a distinct pattern that kept repeating across the metropolitan area. Frequencies could not be assigned to more than one cell in the pattern.

With seven cells in the pattern, you have seven times the number of frequencies assigned in the cell. If each cell site had 20 frequencies, you would have a total of 140 (7×20) frequencies for the pattern. The 140 frequencies for the pattern of cells would continually repeat as the pattern repeated. No adjacent cells would be using the same frequencies. Thus, this arrangement would result in our theoretical city having a total of 140 different radio frequencies. Different companies use different numbers of cells in the cell pattern and different total numbers of frequencies in each cell. The quantity used is dictated by how many calls the system must handle at the same time. By using low-powered transmitters, we can reuse frequencies many more times than a high-powered transmitter permits. This allows us to provide many more communication channels without having to use many more frequencies (see Figure 13-2).

The example above resulted in 140 frequencies. There are 400 frequencies available for cellular radio. A cellular service provider may use more than 7 cells in a pattern and may use as many as 128 of the 400 frequencies at each cell site. One of the problems with radio communication is that we continually run out of frequency allocations. The cellular technique allows us to use a set of frequencies over and over. If we use 100 high-powered transmitters at different frequencies in Chicago, we could only have 100 mobile conversations at the same time. If we use 100 low-powered transmitters, which can be reassigned 10 times, we can have

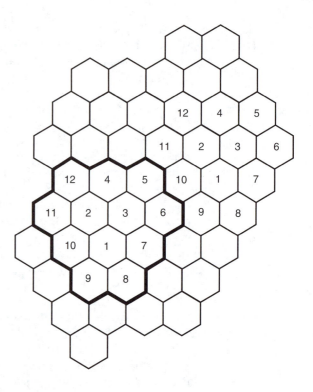

FIGURE 13-2 Twelve-cell reuse pattern.

1000 simultaneous mobile conversations (assuming they are spread out across Chicago). The key to cell technology is the centralized computer at the *mobile telephone serving office (MTSO)* and the constant communication between a mobile unit, the base station cell site, and the computer of the MTSO. The computer will always know which cell site it should use to connect a call over.

Cellular radio uses frequencies from 825 to 845 MHz and from 870 to 890 MHz. The lower frequency band (825 to 845) is assigned as the transmit frequencies of mobile units. Base station receivers are tuned to these frequencies. The upper frequency band (870 to 890) is assigned to the base station transmitters. Mobile receivers are tuned to receive on these frequencies. The use of two frequencies for each call provides a full-duplex communication medium. The FCC controls the allocation of these frequencies and will provide a license for frequencies in a certain geographic area called a *cellular geographic serving area (CGSA)*. The CGSA is designed to fit the borders of a *standard metropolitan statistical area (SMSA)*. A SMSA defines geographic areas used by marketing agencies. The SMSA was also used by the 1984 MFJ to define most *local access transport areas (LATAs)*. Therefore, most CGSAs should correspond to the area covered by a LATA.

The cellular radio telephone system contains many cell sites, a central switching system, controllers at each cell site, and a central controller at the switching system. Voice channels from each cell site are connected to the MTSO. *MTSO* also stands for *mobile telephone switching office*. In the PCS network, the *mobile switching center* is also called the *MSC*. Each cell site controller is connected to the central controller of the MTSO by messaging links. The central controller also has messaging links connected to the central switching system. A cell site can contain up to 128 different transmitters and receivers. The cell site controllers keep track of the signal strength received from a mobile unit and report this information to the central controller. The central controller uses this information to decide which cell site should handle a call. The central controller keeps a list of all active mobile units and their cell site location.

The central controller and switching system for cellular radio is called the *MTSO*. The MTSO can be a standalone switch, owned by a private cellular company, or it can be integrated into a local switch if the LEC is the cellular service provider. The MTSO is similar to a class 4 switching center but is much smaller. It is like a class 4 switching center because it does not have line circuits; it only has trunk circuits. Trunk circuits connect the MTSO to the transmitter site and connect it to a switch in the PSTN. Many MTSOs are connected to the SS7 network. This allows a cellular phone to roam. The mobile telephone is continuously reporting to the closest cell site. As you drive across country and pass from one service provider to another, the central controller of the MTSO owned by other service providers will report your location over the SS7 network to your home base.

Suppose you have cellular service from Southwestern Bell in Kansas City, Missouri, and you are traveling to Atlanta. Anyone can call the telephone number in Kansas City assigned to your mobile phone and the call will be automatically forwarded over the PSTN to the MTSO presently serving your mobile phone. If your mobile phone is turned on, it has been communicating with other MTSOs. These MTSOs have been informing the MTSO in Kansas City of your location using the SS7 network. Suppose you are just outside St. Louis and someone calls your number in Kansas City. The Kansas City MTSO will contact the St. Louis MTSO over the SS7 network and ask it to ring your mobile phone. The Kansas City MTSO will also issue instructions to IEC switches in the PSTN and the MTSO in St. Louis to reserve a voice path between the Kansas City and St. Louis MTSOs. If the mobile phone is answered, the reserved voice path is established and the person that called you will be connected to your mobile phone. They do not know where you are. They probably assume you are in Kansas City, since that is the PSTN location for the number they dialed (see Figure 13-3).

13.4 DIGITAL ADVANCED MOBILE PHONE SYSTEMS

The major difference between AMPS and the *digital advanced mobile phone system (DAMPS)* is that AMPS uses analog radio signals and DAMPS uses digital radio signals. DAMPS is a technique that places multiple calls over one radio frequency using pulse code modulation (PCM) and time division multiplexing (TDM). The PCM signal is not the standard PCM signal found in the PSTN, which converts an analog signal into 64,000 bps. DAMPS technology converts the analog signal into 16,000 bps. DAMPS was introduced in 1992 for the existing AMPS, and DAMPS allows cellular operators to carry four times as many calls as a regular

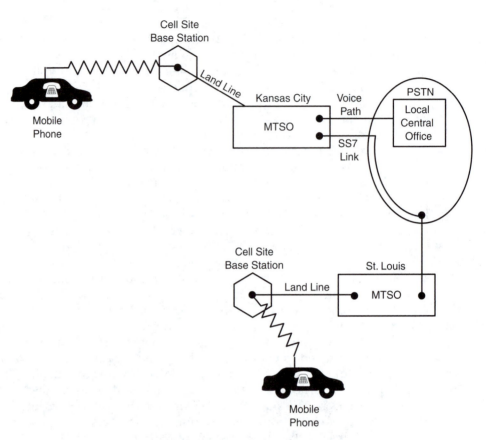

FIGURE 13-3 Advanced mobile telephone system.

AMPS system. DAMPS is backward compatible with analog AMPS. The system will handle either a regular analog call per channel or multiple calls per channel. Manufacturers of mobile telephone sets provide a switch on the set that allows the user to select between AMPS and DAMPS. Cellular service providers could gradually change their systems from AMPS to DAMPS. This technology was used in the 825- to 895-MHz frequency band for AMPS. Cellular service providers will also use DAMPS in the higher-frequency band of 1850 to 1990 MHz to provide PCS.

13.5 PERSONAL COMMUNICATION SYSTEM

The PCS is quite similar to the technique used to provide DAMPS. PCS uses cellular technology; it also uses digital technology (CELP at 7.95 Kbps) and TDM. PCS utilizes lower-wattage transmitters than AMPS in the base and mobile stations. It also employs higher radio frequencies than AMPS, using radio frequencies between 1850 and 1910 MHz for mobile unit transmitters and between 1930 and 1990 MHz for base station transmitters. The channels are numbered from 512 (1850.2 MHz and 1930.2 MHz) to 810 (1909.8 MHz and 1989.8 MHz). This provides 299 carrier slots

(512 to 810 inclusive). Channel 512 is a transmitting frequency of 1850.2 MHz by the mobile unit transmitter. This of course is also the receiving frequency in the base station unit for channel 512. The base station transmits at 1930.2 MHz, and the mobile unit also receives at this frequency for channel 512. Each channel is assigned a frequency for the base transmitter, and a different frequency for the mobile unit, to provide a full-duplex communication circuit.

The base unit is located at a **base transceiver station (BTS).** The BTS site will consist of a small building to house the base station transmitters and receivers, and an antenna tower approximately 150 ft high. The antenna tower will contain 9 antennas arranged in a triangular array. Each side of the triangle will contain one transmitting and two receiving antennas. Looking at one side of this triangular array, the transmitting antenna is located at the center point of the side of the triangle; the receiving antennas are located on each side of the transmitting antenna. The receiving antennas are placed about 3 ft on each side of the transmitting antenna. With a triangle arrangement, we have three separate antenna arrays pointing in three directions. If a circle is drawn around this triangle, you would see that each side of the triangle covers 120° of the total coverage area for the site. The triangle array provides a full 360° of coverage for the site.

13.6 THE FCC AUCTION, MTAS, AND BTAS

The FCC held an auction in 1995 to sell the licenses for PCS frequency allocations. This auction raised $7 billion. The United States was divided into 51 segments. Each segment was termed a **major trading area (MTA),** and each MTA was subdivided into a **basic trading area (BTA).** There are 492 BTAs. The auction sold two licenses per MTA and four per BTA, for a total of 2070 licenses. Several companies were sold licenses on different frequencies in the same BTA. These licenses were for broadband PCS, which provides a bandwidth of 30 KHz. The bandwidth of broadband PCS is necessary for voice communication. Narrowband PCS is a service used for paging service providers. The narrowband frequencies were auctioned off in 1994 and the broadband PCS frequencies the following year. Many of these licenses were bought by IECs and LECs. They intend to use PCS technology to provide local phone service. By using radio waves to provide local service, the IECs can get into the local services market without having to lease local loops from the LECs. Sprint purchased licenses for 29 MTAs. AT&T purchased licenses for 21 MTAs. Prime Communications (NYNEX, Bell Atlantic, US West, and Air Touch) purchased 11 licenses. Bell South, Pacific Telesis, and Southwestern Bell also purchased licenses for BTAs in their area of LEC operations. Existing cellular service providers were not allowed to bid for an MTA license.

13.7 DAMPS VS. GSM VS. CDMA

As stated earlier, many cellular service providers will use DAMPS (IS-136), which is a **time division multiple access (TDMA)** technology, to provide PCS. AT&T Wireless (the old McCaw Cellular) will use DAMPS. This allows it to use its current equipment. It merely changes the frequencies of the transmitters and changes some software in its controllers. Other companies entering the PCS market have chosen to use the *Global System for Mobile Communications (GSM) 1900* standard for PCS. GSM 1900 also uses TDMA technology. The method used to change an analog voice signal into a digital code varies between systems. Early DAMPS systems used a process called *adaptive pulse code modulation* to achieve digital voice at 16 Kbps. The PCS 1900 system uses a process called *code excited linear predictive (CELP) coding* to achieve digital voice with 7.95 Kbps. CELP is also available in a 16-Kbps chip. Another method of providing PCS will use an evolving technology called *code division multiple access (CDMA)* technology.

Development of the GSM standard began within CEPT, the Council of European PTTs, in 1982. The goal of the CEPT was to develop a standard for the digital cellular network that would allow international *roaming.* The first GSM networks entered service in 1992. By the middle of 1995, more than 75 networks were in service in 45 countries. The GSM standard is the most successful digital cellular standard in the world. More than 30 million customers use it, with growth of around 600,000 new customers per month.

GSM has spread far beyond the European community. GSM standards have been adopted by South Africa, Saudi Arabia, Egypt, Australia, China, Singapore, New Zealand, and many other countries. More than 120 operators in 80 countries have signed a Memorandum of Understanding on GSM with plans to implement GSM standard networks. GSM standards were based on the frequency spectrum around 900 MHz. In 1992, a variant on GSM called *DCS 1800* was also specified as a standard by the European Council. DCS 1800 uses the same network technology as GSM but operates at frequencies in the 1800-MHz band. DCS 1800 was used to provide enhanced services to the MTS community. The first DCS 1800 networks were installed in 1993 in France, the United Kingdom, and Germany. GSM 1900 is simply another evolution of the original GSM standard. The frequencies of operation were just changed from 1800 to 1900-MHz.

CDMA has been adopted as IS-95 Standard for PCS Networks. IS-95 Q-CDMA was codified as a standard in 1993. CDMA can provide ten times the capacity of DAMPS. CDMA is not backward compatible with analog AMPS in the same way that DAMPS is. CDMA cannot reuse DAMP technology. To implement CDMA, the service provider must install a new network. This is a problem for existing cellular service providers but not for new entrants to the mobile communication market. Sprint is not a new entrant to cellular communication, but in order to bid on PCS licenses, it was forced to sell its old cellular markets. It spun off the cellular division, which is no longer part of Sprint. Thus, for all intents and purposes, Sprint had to start with no technology in place. This made it easy for it to decide on CDMA technology. With this technology, each active mobile transmitter is assigned a code by the central controller of the MTSO. Each transmission includes the code assigned to a particular call. Many mobile units can be transmitting over the same frequency and at the same time, but each transmission has a unique code appended to it. The receivers will determine which signal is meant for them by checking the code assigned.

In the AMPS, the different mobile phone calls were allocated to a specific radio frequency. Each radio frequency was 30 KHz wide. Thus, AMPS employs *frequency division multiplexing (FDM).* DAMPS also used FDM and the same 30-KHz

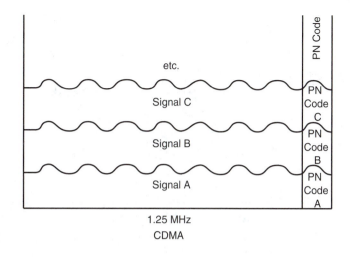

FIGURE 13-4 CDMA showing several signals and PN code. The receiver will receive all signals but will only amplify and decode the signal for the PN code assigned to the signal that matches the PN code assigned to the receiver.

bandwidth per frequency, but each call does not occupy the 30-KHz spectrum for the full duration of time; instead, it occupies a certain portion of time on a specific frequency. CDMA uses a wider bandwidth than AMPS and DAMPS. CDMA is referred to as a *spread spectrum technology* because the signal is spread out over a wider bandwidth. The bandwidth of each frequency is 1.25 MHz. Multiple signals will be placed using the same frequency and all will be transmitted at the same time. As a result, they will interfere with each other, and it is important that transmit power levels be very low. CDMA terminals transmit at less than 1 W. In PCS the terminals transmit at 0.85 W. Each voice signal is assigned a pseudo-random number (PN) code pattern that will only be recognized by the receiver assigned to recognize that PN (see Figure 13-4).

CDMA is another example of the advances made in signal processing technology and Super Very Large Scale Integrated Circuit design. The advances in these technologies have made CDMA possible. The receiver must be smart enough to analyze a captured signal, despread the signal, and identify which signal should be reconstituted and then code that signal into a 16-Kbps code. CDMA technology allows the reuse of transmitter frequencies in adjacent cells and allows the cell site coverage to be larger. The signal processor provides a higher gain for the received signal, and so the larger cell radius is due to receiver improvements, not higher transmitter power. As stated before, CDMA also allows lower transmitter power. In fact, the use of CDMA *requires* lower transmitter power. Since a wider coverage area per cell reduces the number of BTSs needed and lowers the cost of the PCS system, many PCS providers will elect to use CDMA if they are implementing new systems.

13.8 PCS NETWORK

The PCS network is similar to the DAMPS network. In PCS, the MTSO contains a home location register (HLR) database, a visitor location register (VLR) database, an equipment identification registry (EIR) database, an authentication center (AUC), and a short messages services service center (SMS-SC). The MTSO connects to a base

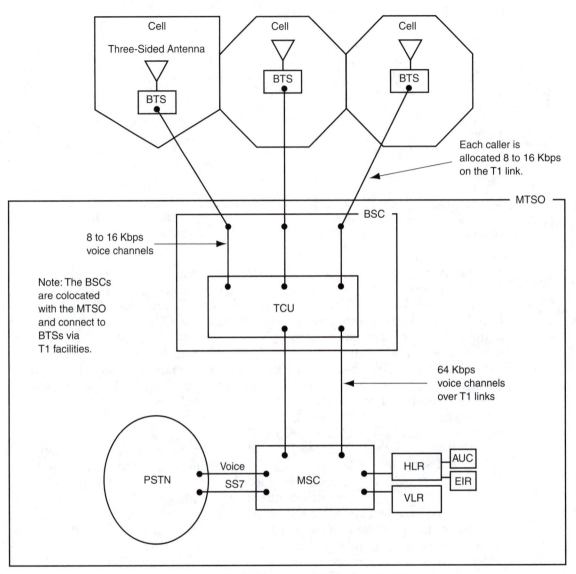

FIGURE 13-5 Block design for PCS 1900.

station subsystem (BSS). The BSS consists of a TCU (located at the MTSO), a *base station controller (BSC),* and several base transceiver stations (BTSs) (see Figure 13-5).

One TCU is required for each PCM link between the MTSO and the BSC. The TCU provides an administrative and maintenance interface to the BSC and performs a digital rate conversion between the BSC and the MTSO. The DS1 link connecting a BSC/TCU to a BTS will contain twenty-four 64-Kbps (DS0) channels. Each 64-Kbps channel contains 4–8 calls. Remember that the mobile units are using coders that convert voice into 8–16 Kbps. The TCU will demultiplex the 8 or 16 Kbps voice channels and convert each voice channel to 64 Kbps. This conversion is necessary because the codecs (coders/decoders) of the PSTN use 64-Kbps coding. We cannot pass the 8 or 16 Kbps of the mobile signal out over the PSTN because when it arrives at the distant decoder, it will not be decoded properly. The decoder at the end of a PSTN facility is looking for 64 Kbps to decode. Likewise the TCU will accept a 64-Kbps voice signal from the PSTN, convert it to 8 or 16 Kbps, and multiplex this signal onto the appropriate DS0 channel to the BTS.

The BSC connects to all the transmitter sites. The transmitter sites are called *BTS sites.* The BTS provides an interface between the mobile phone and the BSC; it is controlled by the BSC. Communication is constantly taking place between the BSC and the BTS. The BSC tells the BTS which radio channel to use for communication with a particular mobile station. The BTS contains the transmitters, receivers, antenna connection equipment, and antennas. The BSC will receive information from all BTSs connected to it. This allows the BSC to make a decision on which BTS should handle a call. The BSC is informed of the signal strength received by each BTS and will assign the call to the BTS receiving the strongest signal. As a mobile moves from one cell to another, the BSC will recognize the changing signal strength and will reassign the call from one BTS to another. One BSC can control up to 32 to 64 BTSs. Large networks will use several BSCs.

13.9 INITIATING A PCS CALL FROM THE MOBILE STATION

When mobile users initiate a call, their equipment will search for a local BTS. Each BTS has at least one of its radio channels assigned to carry control signals in addition to traffic. The BTS will inform the BSC of each call attempt. The BSC will allocate a dedicated bidirectional signaling channel and will set up a route to the MTSO. The BSC is responsible for the management of the radio resource within a region. Its main functions are to allocate and control traffic channels, control frequency hopping, undertake handovers (except to cells outside its region), and provide radio performance measurements. When a mobile unit initiates a call it must provide its *International Mobile Subscriber Identity number (IMSI).* This is a unique number that will allow the PCS system to initiate a process to confirm that the mobile customer is allowed to access it. Each PCS phone has an entry in the HLR of its home MTSO. The HLR contains information about the services the subscriber is allowed to access; it also contains a unique authentication key and associated challenge/response generators. The AUC will verify that the customer is legitimate and allow the call to proceed.

Whenever a mobile station is switched on, and at intervals thereafter, it will register with the system. This allows the location of the caller in the network to be established and its location area to be updated in the HLR. A location area is a geographically defined group of cells. When the first registration attempt occurs, the MTSO will use the IMSI to interrogate the customer's home-base HLR and will add the customer data to its associated VLR. The VLR now contains the address of the customer's HLR, and the authentication request is routed back through the HLR to the customer's AUC. This generates a challenge/response pair used by the local network to challenge the mobile station. In addition, the PCS system will check the mobile equipment against an EIR in order to control stolen, fraudulent, or faulty equipment.

The authentication process is powerful and is based on advanced cryptographic principles. It especially protects the network operators from fraudulent use of their services. The TDMA nature of GSM coupled with its frequency-hopping facility make it difficult for an eavesdropper to lock onto the correct signal and monitor a conversation. CDMA makes it impossible for someone to eavesdrop on your conversation since the voice is a digital signal with a pseudorandom number (PN) code pattern. Customers desiring a secure conversation over facilities using GSM TDSA should use encryption.

Once the user and the user's equipment are accepted by the network, the mobile station must define the type of service it requires (voice, data, supplementary services, and so on) and the destination number. At this point, a traffic channel with the relevant capacity will be allocated and the MTSO will route the call to its desired destination. The MTSO will use the SS7 network to establish call setup over the PSTN. If the called number is busy, the local switching system will return a busy signal.

13.10 RECEIVING A PCS CALL

When a PCS number is dialed, the MTSO will look in its HLR to identify the location of the PCS number. If it is informed that the location is in a remote MTSO, it will launch a query over SS7 to find the mobile phone. The MTSO will send out a paging signal to page the mobile unit. If the mobile unit is in the area, it will respond to the page. The serving MTSO will connect the caller to the mobile unit via its BSS if the PCS phone is still in the local serving area of the MTSO. If the serving MTSO finds the PCS phone in a remote MTSO, the MTSO will connect to the remote MTSO over the PSTN and will instruct the remote MTSO to connect the call via the remote's BSS.

13.11 SMART CARD

The *Smart Card* or *Subscriber Identification Module (SIM)* card contains a microprocessor and a small amount of memory. With a SIM card, a customer can use any PCS phone that has a card reader to make a call. The SIM card—the size of a credit card—is inserted into the PCS phone to activate the phone. In effect, while the card is in the phone, the PCS phone becomes personalized and becomes the user's personal phone. All the customer's personal data, personal identification number (PIN), services subscribed to, authentication key, IMSI, speed dialing lists, and so forth are stored in the SIM card.

13.12 SUMMARY

I am sure that the technology behind PCS and the services it offers will advance rapidly. Check out the Internet and check the web sites of the LECs and IECs to see what is happening currently with PCS. As the technology grows, you will need to keep abreast of the changes. Future revisions of this book will incorporate these advances, but I think I have given you a good start on PCS.

Although many different voice coding techniques are used in wireless communication, you do not need to understand how they work. You simply need to know that they exist so that you will know the type of coding your terminal equipment requires. Your equipment must utilize the same coding technique that the wireless radio system is using. If the wireless system uses CDMA, you must purchase CDMA terminal equipment. If the system is using 16-Kbps CELP, your terminal equipment must also be 16-Kbps CELP. Although this chapter made reference to the CDMA RF signal being changed to 16Kbps, the CDMA systems in use today can also change the signal to 8Kbps. This allows for placing more calls over the T1 facility between the BTS and BSC.

CHAPTER REVIEW QUESTIONS

1. What is PCS?
2. What is PCS 1900?
3. What is MTS?
4. What is AMPS?
5. What is DAMPS?
6. What is TDMA?
7. What is CDMA?
8. What is BSC?
9. What is BTS?
10. What is MTSO?
11. What is HLR?
12. What is MTA?
13. What is BTA?
14. What is the difference between PCM in the PSTN and in DAMPS?
15. How does a company get a license to sell mobile telephone service?

PROJECTS AND CASE STUDIES:

Case Study #1

Residential Real Estate wants to provide all agents with a mobile telephone. What service providers are in your town? What are the service plans available from each? What do each of the plans cost? Which service provider uses TDMA and which uses CDMA? What are the differences between the two?

GLOSSARY

Advanced Mobile Phone System (AMPS)
Also called *cellular radio.* The serving area for AMPS is broken up into cells. Low-powered transmitters are used for AMPS. Radio frequencies can be reused in nonadjacent cells. AMPS uses frequencies from 825 to 845 MHz and from 870 to 890 MHz.

Base Station Controller (BSC)   Connects to all the base transceiver station (BTS) sites serving a BTA. The BTS provides an interface between the mobile phone and the BSC. The BTS is controlled by the BSC. Communication is constantly taking place between the BSC and the BTS. The BSC tells the BTS which radio channel to use for communication with a particular mobile station.

Base Transceiver Station (BTS)   The BTS site for a PCS 1900 system will consist of a small building to house the base station transmitters and receivers and an antenna tower approximately 150 ft high. The tower will contain 9 antennas arranged in a triangular array. Each side of the triangle will contain one transmitting antenna and two receiving antennas.

Basic Trading Area (BTA)   Each of the 51 major trading areas (MTAs) was subdivided into a BTA. There are 492 BTAs. PCS licenses were sold for each BTA.

Cellular Radio Also called *advanced mobile phone system (AMPS)*   The serving area for AMPS is broken up into cells. Low-powered transmitters are used for AMPS. Radio frequencies can be reused in nonadjacent cells. Cellular radio uses frequencies from 825 to 845 MHz and from 870 to 890 MHz.

Code Division Multiple Access (CDMA)
Has been adopted as IS-95 standard for PCS networks. IS-95 Q-CDMA was codified as a standard in 1993. CDMA can provide ten times the capacity of DAMPS. CDMA is not backward compatible with analog AMPS in the same way that DAMPS is. CDMA cannot reuse DAMPS technology. To implement CDMA, the service provider must install a new network. CDMA is spread spectrum technology. Many conversations are multiplexed over one frequency. Each conversation is assigned a code. The receiver strips out conversations individually by using the code.

Code Excited Linear Predictive Coding (CELP)   Also known as *vector-sum excited linear predictive (VSELP) coding.* This speech coding algorithm is EIA standard IS-54. This is the speech coding technique recommended for TDMA cellular radio systems. The bit rate is 7.95 Kbps. VSELP is also available in a 16-Kbps chip. AT&T makes a digital signal processor using this technology; it is called a *DSP-1616.*

Digital Advanced Mobile Phone System (DAMPS)   A technique that places multiple calls over one radio frequency using pulse code modulation (PCM) and time division multiplexing (TDM). The PCM signal is not the standard 64,000-bps PCM signal found in the PSTN but is a 16,000-bps signal. DAMPS was introduced in 1992 for the existing AMPS and is backward compatible with anolog AMPS. The TDM feature allows DAMPS to carry four times as many calls as AMPS.

Global System for Mobile Communications (GSM) 1900   A standard for mobile phone service based on Time Division Multiple Access (TDMA).

Improved Mobile Telephone System (IMTS)
Introduced around 1964. The major benefit of IMTS was that it used several different frequencies and could support

many different conversations at the same time. It was also connected to the local class 5 office instead of the toll office. IMTS was connected to regular telephone numbers and line circuits. An IMTS phone was assigned a regular PSTN telephone number. Anyone could reach an IMTS phone by dialing the PSTN number assigned to the mobile phone. This eliminated the need for operators to handle mobile phone calls.

Major Trading Area (MTA) The United States is divided into 51 areas for mobile telephone licensing purposes. Each of these segments is called a *major trading area (MTA)*. Each MTA was subdivided into a basic trading area (BTA). There are 492 BTAs.

Mobile Telephone Serving Office (MTSO) The central controller and switching system for cellular radio. The MTSO can be a standalone switch owned by a private cellular company, or it can be integrated into a local switch if the LEC is the cellular service provider.

Mobile Telephone System (MTS) Mobile telephone service began in 1946 and was called *MTS*. These systems used the radio frequencies between 35 and 45 MHz. Although *MTS* stands for "mobile telephone service," it also meant "manual telephone system." All calls had to be handled by an operator.

Personal Communication System (PCS) Not so much a technology as a concept. The concept of PCS is to assign someone a personal telephone number (PTN). This PTN is stored in a database on the SS7 network. That database keeps track of where a person can be reached. When a call is placed for that person, the artificial intelligence network (AIN) of the SS7 determines where the call should be directed.

Personal Communication System 1900 (PCS 1900) PCS is provided by radio frequencies in the 1900-MHz range. PCS 1900 is the latest evolution in mobile communication. PCS 1900 can be provided by the time division multiple access (TDMA) technology of DAMPS or can use code division multiple access (CDMA) technology.

Roaming The mobile telephone of AMPS or DAMPS is continuously reporting its location to the closest cell site. As you drive across country and pass from one service provider to another, the central controller of the MTSO owned by other service providers will report your location over the SS7 network to your home base.

Time Division Multiple Access (TDMA) TDMA technology is used by DAMPS. With TDMA, each conversation occupies a time slot on a common transmission medium. In DAMPS, each conversation occupies a time slot on a particular radio frequency.

14

Telecommunications Management

KEY TERMS

Computer Inquiry II Ruling of   Electronic Data Interexchange   Internet
   1981   (EDI)

Telecommunications management is the management of telecommunications. Management is generally thought of as the major functions a manager carries out, such as planning, organizing, directing, controlling, delegating, developing, staffing, and, on occasion, praying. When all of these functions are applied to telecommunications technology and its uses, we are managing telecommunications. The telecommunications manager's job is not easy. To be done properly it requires someone with all the skills a good manager must possess: leadership, communication, planning, negotiation, interpersonal, conceptual, and technical skills.

There are many management positions in telecommunications with many different companies. One manager may manage a network mostly used for voice, another may manage a data network, and another may manage in an Internet or intranet environment. In today's downsizing environment, a telecommunications manager finds himself or herself managing all types of systems and networks. I personally think telecommunications managers can do a better job if they have good technical skills in the technology used to supply telecommunications services to their company, its vendors, and its customers. But good technical skills are not enough. They are simply a good starting point.

14.1 INTERPERSONAL SKILLS

Although telecommunications managers must have good technical skills, they must also excel in interpersonal skills. Like other types of managers, they achieve their goals and objectives via the efforts of other people. Many telecommunications managers are actively involved in the actual work done to install and maintain network components. For many managers, being a technician is part of the job. We have to be careful when we put on our technical hat that we do not get too involved technically. If it is known that a manager gets good results based on technical rather than management ability, it will be difficult to advance within the organization. Everyone will think the manager is the glue holding everything together, and if we promote him or her, things will fall apart.

It is much better to be known for your excellent interpersonal skills and your ability to put together a well-oiled operation. When you are viewed as a manager rather than a technical expert, your chances for advancement are usually greater. There will be some positions where you cannot avoid being required to be a supertechnician, but the sooner you can change your role to that of a manager, the better off you will be. You can start by providing training, development, and guidance to your reports. Nothing motivates a person more than receiving training and a bigger share of the technical work. Train your reports to be supertechnicians, and you will not have to fill that role.

Treat everyone respectfully, fairly, and equitably. Notice that I did not say treat everyone the same. I do not believe in that philosophy, but many managers do. You will have to make your own decision on how you will behave in this area. I believe strongly in the *Situational Leadership* philosophy proposed by Paul Hersey and Kenneth Blanchard. You treat reports according to their maturity level (knowledge) and the difficulty of the task at hand. Should you treat the person who does not carry their fair share of the workload the same way you treat your hardest-working employee? If you do, it will not be long before the hard worker becomes disenchanted with your lack of control. Discipline in the workforce is a necessary part of management. If you set high standards (that are achievable with some effort), people will strive to reach them. Set low standards and that is what you will get.

If you treat people with respect and understanding, you are on the road to getting a well-motivated team put together. Next, get their input on what they would like to do and what training they would like to receive to reach their goals. You have taken the second step. Third, get a training and development schedule put together. Fourth, as your reports become better trained, involve them in a participative management atmosphere. Fifth, gradually delegate more responsibility (along with recognition and higher pay) to each report.

It is also important to establish good rapport and relations with your peers, customers, and vendors. It is much easier to get their help when you need it if they like you. I could spend hours talking about the importance of interpersonal skills, but I think you already know that they will be the key to your success in any endeavor. I mention them here because many managers get so caught up in what they are doing that they forget the big picture and the importance of always paying attention to how they deal with other people.

14.2 PLANNING

A telecommunications manager must determine what telecommunications technologies and services are necessary to support the mission, goals, and objectives of the organization. Long-term plans must be developed that will take into account the future direction of the company. These plans must try to establish which technologies and services the company will need in the future and the extent to which current technologies and services must expand. How long before we must replace our 250-station CBX? Will we be able to use voice-over IP, and if so, when? Will our data transmission needs outgrow our current WANs? What will we need?

Long-range planning involves trying to answer as many questions as possible about the future growth and telecommunications services needed to support that growth.

Once the telecommunications manager has a fairly good handle on the long-range plans of other departments in the company and the telecommunications technologies needed to support those plans, he or she is in a position to convert those plans into operational plans. The manager must stay abreast of changes in technology. What was a good plan yesterday may be out of date today. Telecommunications technologies are continually evolving. You must adapt your operational plans to include consideration of these changes from the perspective of the long-term plan.

Operational plans are short-term plans. They cover what we will be doing in the near term or immediate future. They not only include selection of technology and services, but they also include how we do what we do. We often hear of severe telecommunications outages that have had a drastic impact on a company. Many of these outages could have been prevented with proper planning. Anytime you are going to be working on something that can impact the performance of your company's operations, voice, or data network, you must put together a plan on how the work will be done. This is often referred to as a *method of procedure (MOP)*.

Never do anything to an active working component that is part of your network without putting your planned actions together in a MOP. The MOP should also include checkpoints that allow you to verify that your work is not impacting network performance. It should also include a disaster recovery procedure in case something goes wrong. If you accidentally blow a fuse, do you have a spare immediately available? Do you have the contact number where you can reach a manufacturer's representative if necessary?

Planning can almost never be overdone. It is better to plan five times and implement once rather than the other way around. Good planning will involve your peers and reports in developing the plans. Not only will they help you develop a better plan; they will also buy into your plan through participation in the process. Nothing is worse than having a great plan that everybody else thinks is terrible. A plan will only be successful if it is carried out with due diligence. This will not happen if people who must execute parts of the plan have not bought into it. Once again we see the importance of interpersonal skills and participative management.

14.3 STAFFING

Staffing is important to a manager because you get your work done through other people, but for a good manager, staffing is automatic. It happens as a result of good interpersonal skills and good planning. If you are planning properly, you will accurately forecast your staffing requirements. If you are treating your reports fairly and equitably, providing them with training, recognition, and leadership, you will not lose any of your good people. Leadership includes providing guidance, direction, development, and discipline. This leads to poor performers either improving or leaving your department. You will not have any losers—only winners. A winning team attracts other good people. People are the key to your success. Treat them well and lead by example, and you will always have a winning team wherever you go.

14.4 DIRECTING

A manager provides directions to ensure that everyone understands what is expected of them. Most people want to do a good job. If reports are provided with training and adequate directions to do a job, they will do it well. Directing is best done in a participative environment. Most people respond best when asked to do something rather than being told what to do. You want to try operating as a participative manager as much as possible, but there will simply be times when you have to tell people what to do. If your house is on fire, do you want the fire chief to hold a team meeting to decide what everyone should do, or do you want him or her to take charge and tell them what to do? When your business is on fire, you often have to tell people what you want done. After you have put out the fire, you can start down the road to creating a warm and fuzzy atmosphere.

Please do not get me wrong. I firmly believe in participative management. I also believe that there are occasions when the person in charge must not abdicate the position of authority and must take charge of a situation to bring it under control quickly. A good manager knows when and how to take control without alienating the people involved. A good technique for participative management is the use of the joint goal-setting sessions, procedures, and reviews that are known as *management by objective (MBO)*. The MBO process requires joint setting of goals by managers and their reports. I have found that several companies violate this joint goal-setting process. They tell the report what their goals should be. This is management by directive, not management by objective.

A good manager uses the MBO process with frequent reviews of an employee's performance. These review sessions should be used to build a good bond with the employee. One way that can be accomplished is to make sure you spend more time telling the employee what they did right than what they did wrong. You should also spend a good deal of time reviewing the employee's concerns and desires, and go over the training plan for the employee. When the employee senses that you are genuinely concerned about them and their development, they will follow your directions willingly. They will view you as someone they want leading their department. Excellent leaders provide excellent directions.

14.5 TELECOMMUNICATIONS JOB OPPORTUNITIES

Telecommunications offers tremendous job opportunities. As mentioned earlier, telecommunications used to mean voice communication over a distance using electrical signals—in other words, services provided by the local telephone company. The telephone company offered excellent job opportunities and good pay. As the telecommunications infrastructure has been replaced with state-of-the-art equipment, the number of job opportunities at the local telephone company has plummeted. Computer-controlled switching systems and buried cables require little maintenance. The deployment of fiber optics technology has drastically reduced the amount of equipment that needs to be maintained. Telephone companies have been continually downsizing their workforce and management employees during the past decade or more.

Not only has the local telephone company workforce been hard hit by technological advances; deregulation of the industry has also reduced job opportunities. The *Computer Inquiry II ruling of 1981* by the FCC deregulated telephone station equipment (telephones, keysystems, and PBXs). Telephone companies were no longer allowed to provide station equipment to new customers; customers had to buy their telephones, keysystems, and PBXs from private companies. (Chapter 2 provided further discussion of the impact of this legislation on the telecommunications industry.)

By now you are probably asking why I started this section by saying that there are tremendous job opportunities in telecommunications and then went on to say the telephone company has been downsizing since the 1980s. Well, the telephone company is no longer the only game in town. The job opportunities at the local telephone company have been shrinking, but the explosion of voice communication, data communication, and the Internet in the private business sector has led to explosive growth in job opportunities in this area.

14.5.1 CLECs and ILECs

Telecommunications deregulation legislation has produced many competitors to the local telephone company and AT&T. You need only look in the local phone book's yellow pages to find numerous companies offering telecommunications services. You will find many competing long distance service providers. In larger cities, you will find several *competitive local exchange carriers (CLECs)* competing with the *incumbent local exchange carrier (ILEC)* to be the telecommunications service provider for local businesses. You will find many companies that specialize in the sale, installation, and service of keysystems and PBXs. There are usually several paging and mobile telephone service providers that you can select from. All of these businesses did not exist prior to deregulation—and they all need managers and telecommunications specialists.

The growth in numbers of CLECs vying to provide telephone service and take business away from the established ILECs reverses the merger mania/downsize cycle. When established ILECs merge and consolidate operations, they always end up laying people off. The CLEC needs the same personnel that an ILEC does. They duplicate the services and workforce of an ILEC. This creates more jobs. There can be instability in this area if the FCC allows ILECs to purchase CLECs. If this were to happen, the ILEC would lay-off some employees, as they have with past mergers.

The jobs available in the CLEC environment relate to marketing and sales of telecommunications services to the business community. CLECs also need technicians to install, monitor, and maintain the telecommunications network. In addition, they need telecommunications managers to oversee these operations. The growth of

CLECs also provides more work for companies that manufacture telecommunications technologies. CLECs must purchase switching systems, routers, fiber cable, and multiplexers for their network. They also have to hire contractors to install these network components. All this activity creates jobs for telecommunications managers and technicians.

14.5.2 Managing Data Networks

Today's explosive growth in data communication, LANs, WANs, and the Internet has also created a tremendous growth in job opportunities. The enormous growth of the computer industry and the growth in the telecommunications infrastructure required to connect these computers together have led to the creation of millions of jobs. We have moved from an era where agricultural and manufacturing jobs were the driving force in our economy to an era dominated by jobs in the service industry. We have also moved into an era called the *information age*. These changes have led to the prolific growth in the use of computers and the growth within telecommunications.

It used to be that the telephone company met all of a company's telecommunications needs. They had people within the marketing department that determined what services you needed based on their assessment of your needs. With deregulation, businesses had to carry out their own needs assessment. This led to the creation of many telecommunications management jobs in the business sector. With the growth of data and the burgeoning needs of data communication, each business has to have someone to manage their data networks. The data requirements of a business continue to grow. As this happens, the data network must grow. Thus this network is in a constant state of flux. The number of managers required to handle the telecommunications needs for a business will continue to increase as the economy grows.

14.5.3 Managing Mobile Telephony

The growth in cellular radio and paging has also created many job opportunities. Firms in these areas need personnel to work in their network operations centers. They also need personnel who can determine the number of circuits required between their mobile telephone switching center (MTSO) and the PSTN office it connects to. This work is referred to as *provisioning*. An engineer determines how many circuits are needed and coordinates efforts with the LEC to provide (provision) the number of circuits required. People who do provisioning are called *network engineers, network analysts,* or *network managers*. The CLECs also have to do provisioning and also hire people to do this work.

14.5.4 Consulting

Some people do not want to work for a large corporation. These individuals start their own companies or work for companies that specialize in providing contract personnel to other companies. Telecommunications companies are so short of qualified personnel that they are constantly hiring contractors and/or consultants. Several of my students have worked for contracting firms while completing their studies. Several contracting firms contact me on a monthly basis to ask if I know of anyone that would be interested in a particular job. Several of my colleagues do consulting and contract work. There are plenty of opportunities should you desire to start your own company.

The positions that a telecommunications student can move into after graduation or after they perform a brief stint as a technician include the following: network operator center (NOC) technician, NOC manager, network analyst, network engineer, LAN manager, network manager, network design engineer, network design manager, central office maintenance supervisor, central office engineer, telecommunications consultant, data communication manager, voice communication manager, special services manager, associate director of telecommunications, Internet consultant, product manager, project manager, technical writer, sales engineer, and customer service engineer.

Remember that job opportunities in telecommunications are not limited to the traditional telecommunications carriers such as Bell, AT&T, Sprint, MCI World-Com, GTE, ILECs, CLECs, and competitive access providers. Job opportunities are also not limited to the manufacturers such as Lucent, Nortel, Siemens-Stromberg Carlson, Erickson, Fujitsu, Cisco, 3-Com, and Motorola. Every business needs a telecommunications manager and every ISP needs a network manager.

14.6 THE TELECOMMUNICATIONS PART

So far I have spent a great deal of time talking about management and little about telecommunications. It is important that you understand that the interpersonal part of our business is much more important than the technical end. If you have good management skills, you will be surrounded by good people who will take care of the technical end of the business for you.

Having just made the above statement, I now switch positions. I think it is important for a telecommunications manager to know the technology well. If we can have both an excellent technician and an excellent manager wrapped up in one person, that's what we want. A manager that understands the technology stands a better chance of choosing the right technology for the right business applications.

In some companies, the telecommunications manager is the technician, the clerk, the janitor, and just about everything else required to keep a company's telecommunications systems humming. To make the right telecommunications choices, you must continue your course of study. How will you know if frame relay or ATM is the best WAN for your particular business application unless you are well versed in the technologies. Should you use a CTI-based ACD or a standalone ACD system? You can get the opinions of others to make a decision, or you can learn as much as possible about the technologies available before making a choice.

You must evaluate all technologies against the requirements of the application to which the technology will be applied. Once you have made a technically feasibility analysis, you should do a financial feasibility analysis. A benefit versus cost analysis will provide an indication of which technology will fill the needs of the application at the lowest cost. As a manager, you will be expected to put budgets together for the long-range and operational plans. You can do a better job of forecasting the cost of a technology when you understand the technology.

Continue with your study of telecommunications so that you can learn the technologies in sufficient depth to make informed decisions. I would also encourage you to read as much as you can over the next few years on ways to improve interpersonal, communication, and negotiating skills. An excellent telecommunications technician is not as hard to find as an excellent telecommunications manager that has excellent technical skills. Since your goal is to be a manager of telecommunications, you need to work on developing management and technical skills to the highest degree possible. Do not devote all your development time to one at the expense of the other.

14.7 MANAGING TELECOMMUNICATIONS GROWTH

Telecommunications technology continues to grow and change. To effectively manage it, a telecommunications manager must maintain a high level of technical knowledge. Managers are not usually technical experts, but they should be technically savvy. They should have a good understanding of the telecommunications technology being used by their business and should be able to evaluate how new technologies can be used at their company.

This technical knowledge can be gained by education. You can attend a university, college, or technical institute that offers a degree in telecommunications. This course of study will provide you with technical training on various technologies used to provide voice and data communication networks as well as providing you with training in the management and accounting skills needed by a manager. The latter skills can always be enhanced by more in-depth study, but management and accounting disciplines are fairly stable over time. Such is not the case with telecommunications technology. It changes overnight.

The only way to stay abreast of telecommunications technology is to constantly read. Surf the net, read books, attend trade shows, and get additional training via seminars. I do all of this and still find it almost impossible to keep up with the changes occurring. To make proper technology decisions, a manager must listen to their technicians and may have to call in a consultant. It is better to admit that you need help in making the right decision on the various technologies available to your company than to select the wrong options.

As stated earlier, we are in the information age. Everyone in the company needs access to more and more information. You can also use the sharing of information with vendors and customers as a strategic measure. Many companies use a network referred to as the *Electronic Data Interexchange (EDI)* to exchange data with their vendors and customers. EDI can be done as a private network, or it can be done via companies that serve as an EDI service center. EDI is used by many companies to support just-in-time inventory. EDI allows immediate contact with a vendor to order materials from them. It also provides support for paperless ordering, confirmation, shipping notification, receiving notification, and payment to the vendor.

The need to provide access to information requires control. The right people must have access to the right information (data) and must be restricted to being able to access only the information they need. Methods and procedures must be in place to limit and protect sensitive data from unwarranted access. The same is true for voice services. The right people must be able to access various parts of your voice network such as voice mail and long distance capabilities from remote access. This remote access must have tight security measures in place to protect from unauthorized use. To select the right technologies for your voice services, your data services, and your security needs, and still provide access to the information required by employees and business partners, you must understand both the technologies available and the business applications that can benefit from using telecommunications technology.

Information is shared today via telecommunications networks and is often referred to as *telecommunications-based information technology (TBIT)*. It is left to the telecommunications manager to make sure the right telecommunications technology, services, and networks are in place to support the information technology needs of the business. As these needs grow, so must the telecommunications technology. It is impossible to eliminate the telecommunications manager's job. Someone will always be needed to manage the telecommunications technology due to the constantly changing IT needs of the business.

14.8 MAKING TELECOMMUNICATIONS DECISIONS

We have been discussing the telecommunications industry and telecommunications technology from an introductory point of view. I am sure that by now you realize there is much to learn before you can properly select the technologies that would be needed by your company. Each company, business, and industry has unique telecommunications requirements. This book is intended to acquaint you with some of the technologies and processes used in the telecommunications networks.

After reading this book you will have a good understanding of telecommunications technologies, but an in-depth knowledge of these technologies will come only after you have completed more advanced courses and done additional reading. Each chapter in this book is covered in much greater detail by books devoted to specific topics. For example, in Chapter 9, I discussed Integrated Services Digital Network (ISDN) and Asymmetrical Digital Subscriber Line (ADSL) technologies. Books and manuals are available that discuss only ISDN and cover it in great detail. In Chapters 10 and 11, I discussed LANs and WANs. If you are a student using this book as an introduction to telecommunications, in later terms you will have a LAN class along with a book that discusses only LANs and LAN technologies at length. You will also have a class on WANs, in which you will use a book that covers only WANs in excruciating detail. Only after completing all the technology courses in your telecommunications management degree program will you be able to properly diagnose an operational or IT need and be able to prescribe just the right telecommunications solution for it.

14.9 AGE OF COMMUNICATIONS

Our economic growth is tied directly to developments in the area of transportation: transportation of people, transportation of goods, and transportation of knowledge. The transportation of knowledge is critical to our decision-making capabilities and intellectual growth. The amount of information available to us, and the speed with which it can be accessed, are directly correlated with progress in a country's telecommunications infrastructure. The ability to learn has always been directly related to the access a person has to reference materials and instructors. The growth of telecommunications has expanded the base of intellectual resources beyond what is available from local intellectuals and the local library. In some countries, the synergy of intellectual resources has been less than optimal due to a poor or nonexistent telecommunications infrastructure. Other countries have been able to achieve a high level of education for their citizens despite the lack of such an infrastructure. These countries have been favored by certain factors, including in some cases small size—which may bring intellectuals close together and facilitate the exchange of information.

Instant access to news of world events is possible due to the advances in telecommunications. For example, news coverage of the Gulf War provided us with an immediate presence on the war scene. This event dramatically demonstrated the advances that had been made in satellite and radio communications, with live pictures of air raids and antiaircraft fire that were captured in Baghdad and beamed immediately into our living rooms. Radio-controlled and laser-guided missiles, along with the use of "spy satellites" to capture the actions and movements of enemy troops, played a crucial role in the swiftness with which this action was brought to a conclusion. Intelligence gathering has been made much easier with advances in telecommunications technology.

The benefits we enjoy from this technology are not limited to military uses. It is true that most successful military ventures belong to those who have the "intelligence edge." It is equally true, in the business world, that those with the best, most accurate, latest, and fastest information sources win most of the battles. Providing your company with the best telecommunications technology available not only gives it a competitive edge; this strategy is necessary to survive.

Business associates, friends, and family can now easily talk with each other over extended distances on a daily basis due to advances in our telecommunications infrastructure. Telecommunications progress allows immediate access to numerous sources of information and enables individuals to make more informed and quicker decisions. The explosion of the Internet is the latest development to make more resources available for learning and research.

The technologies mentioned in Sections 3.12 to 3.14 have improved tremendously over the past 20 years and will continue to improve. These technologies provide a vehicle allowing us to reach out and access any information we need. In this information age, we can get on the *Internet* highway and access many libraries. As we move into the future, the developments in deregulation, PC hardware, software, and telecommunications technology will facilitate even greater accomplishments.

LECs, IECs, and other common carriers will strive to become our one-stop communication center. They will offer us local telephone service, long distance service, television service, and Internet access service. Since they will be able to offer all these services, they will seek to provide bundled services at discount prices. Can we do without these services? Of course we can. We do not need television. We can drive to the library and get books. We can call our stockbroker to get financial information. We can also do without a car. We can walk everywhere we want to go. Just as a car allows us to be more productive, the enhanced services available via telecommunications allow us to accomplish more.

In short, access to information will be critical in the years ahead. This access must be available in our homes. Many children will be denied equal educational opportunities because their family cannot afford a PC and Internet access. Still other children (and adults) will not take advantage of these technologies. I am not advocating that everyone become an Internet junkie. I am merely saying that people can take advantage of the good Internet sites to broaden their horizons and expand their intellectual database. In the 1940s, everyone did not have a car, nor did everyone have a radio. In the 1950s, not everyone had a television set, nor did everyone have a telephone. Internet access will be the next service to fall into this category. Within the next ten years, more and more people will take advantage of the emerging telecommunications technologies. As growing numbers of people install these technologies, think of the new business opportunities that will be created. I see great applications in the educational field. The data capacity of a CD-ROM provides a great vehicle for interactive education on any subject, at any age level, from prekindergarten to graduate programs.

Having many libraries and live meeting room forums available over the Internet can also enhance the learning process. Many meeting rooms are used today for people with mutual interests. Some companies now set up meeting rooms over Internet or intranet facilities. Many companies make use of customer knowledge by providing technical rooms. A person can access these rooms via the Internet and ask if anyone present can help solve a problem. Since many people come and go in these rooms, the problem can be posted for access by later attendees. The company has technicians assigned to monitor the bulletin board of the meeting room. If the technician has an answer to posted problems, they will provide a reply. Many times, one customer's problem can be solved by another customer who has experienced the same problem.

Establishing meeting rooms can provide a company with many intellectual resources. Many software companies are now charging a flat fee of $25 to provide support on their products. They would enjoy better customer relations if they provided an address on the Internet for a technical room. I also see opportunities for an entrepreneur to set up pay-per-question sites as the quest for knowledge becomes greater. You could establish sites that require a user identification number. The user pays a flat fee for enrollment. They could then access the site for information over a certain time frame and ask unlimited questions. I could offer a course in telecommunications! Everyone would not have to start school on the same date. They could progress through modules at their own pace.

It takes some time for technologies to be accepted. We still use checks and the mail to pay our bills. Many companies have set up direct deposit of checks and many employees use direct deposit. Most of us have a problem setting up direct payment from our checking account to pay bills. We are hesitant to allow companies access to our accounts, so we do not do direct payments. I do not have any direct payment accounts set up, but I would be receptive to paying by phone or Internet if companies would provide such a setup. Companies are reluctant to accept this method of payment due to the possibility of fraud. In the near future, I expect that someone will provide a secure method for making transactions via telephone or the Internet. Instead of a bankbook, we will have a software program and ledger. We will be able to use dial-up access to check on our account balance at the bank and view all deposits and withdrawals. Many banks provide this service over voice mail. It will not be that difficult to provide the same information as a file transfer. This information could be downloaded directly into a software program at our PC. We can enter bills we wish to pay and the amount for each. The software program can automatically launch calls over the telephone network or Internet to provide an amount and account number, as well as to authorize transfer of funds for payment of bills.

There are many ways to use the technologies coming online. The extent to which they will be used is limited more by our acceptance of technology than by the lack of technology. ISDN has been around for 15 years with almost no acceptance by the average consumer. The future of the information age, and how much we benefit from emerging technologies, will only be limited by our vision and active involvement in the use of technologies developed. When PCs were first introduced, many of us did not accept the technology because you had to be a technical expert to use a computer. You had to build your own system and write your own software. The development of standards brought many manufacturers into the business. Some people started software companies and developed operating systems software and applications software. People who knew nothing about hardware or software are now able to use the technology developed by others and so became computer users.

14.10 SUMMARY

Since this is the last chapter of the book, its time to thank you for the many hours you have spent reading it. If you read this book as part of a degree program to prepare for a management position in telecommunications, I hope it has served you well. I trust it has laid a solid foundation on which you can continue to build. If you are a telecommunications manager and read the book to broaden your knowledge base on the varied technologies presented, I hope it has fulfilled your needs. I have attempted to cover all the major technical points but have tried to avoid presenting the material in the format of another dry, technical, jargon-filled book. I hope my writing style made the book an easy read for you.

The telecommunications industry has a bright future and offers many challenging opportunities. I wish you much success in your future within that industry. I encourage everyone to drop me an e-mail at **mary_cole@earthlink.net.**

All suggestions on improvements for the next edition of the book are most welcome. Let me know what you think of content and style. Thanks again for spending time with me, and I hope you have a happy, successful career.

GLOSSARY

Computer Inquiry II Ruling of 1981 A ruling by the FCC that deregulated station equipment and stated that this equipment must be owned by the customer or some non-telephone company entity.

Electronic Data Interexchange (EDI) A network that allows companies, their vendors, and customers to exchange data. EDI uses a data network such as an X.25 network.

Internet A network of computers that can be accessed by other members of the Internet community. Most individuals become members of the community by purchasing access to the Internet from a local Internet service provider or commercial online information access provider such as Prodigy or America Online (AOL).

Appendix A

Basic Electricity

KEY TERMS

Acid
Ampere
Ampere-Hour
Atom
Battery
Cell
Compound
Conductor
Direct Current (DC)

Direct Current (DC) Voltage
Dry Cell
Electric Current
Electromotive Force (EMF)
Insulator
Ion
Material
Matter

Molecule
Multimeter
Ohmmeter
Resistance
Static Electric Charge
Voltmeter
Volt, Voltage
Wet Cell

The world around us is filled with devices that use electricity. These devices depend on and cannot operate without electricity, which has been here since the beginning of time. Lightning bolts contain electrical energy. Once we learned what electricity was and how to control it, we were able to use it to help provide a better standard of living. Once we learned how to generate electricity, it became a source of power for other devices. One of the first practical devices to use electricity as a power source was the electric motor. Electric motors were used to automate factory operations, and they paved the path for the industrial age.

Motors, electric heaters, and air-conditioners consume most of the electrical energy used today, but countless devices have been developed that use low levels of electricity. These devices include communication systems, computers, calculators, watches, test equipment, medical equipment, cameras, radios, compact disc players, televisions, musical instruments, smoke detectors, alarm systems, and pacemakers. Almost every day another device is invented that uses electricity to make our life better. *Electronic devices* and components have been used to make the computers and communication devices that have led us into the information age. Devices composed of semiconductor materials on small chips and using small amounts of electric current to perform some function are called *electronic components*. The *Pentium* chip is an electronic component. The chips used for modems and signal logic processors are electronic devices.

This book covers communication systems that use electronic devices and components to convey messages over distances as short as one-billionth of an inch or as long as billions of miles. *Telecommunications* means to communicate over a distance, as we have seen. Prior to the introduction of electricity and electronic devices, smoke signals or drums were used to communicate over distances of more than a few miles. The use of electrical signals is a much better way to communicate over a distance, and the distance covered can be far greater with electrical signals. Today, we can communicate over any distance using electronic devices and electrical signals.

Since human beings switched from smoke signals to electric signals, the devices used to convey these signals in the electronic communication systems have continued to improve. Many advances have been made over the last 25 years in telecommunications and information processing technologies. Today's communication systems fall into two basic categories: voice communication and data communication. Telecommunications is communication of either voice or data. Since telecommunications uses electronic devices and electrical signals for the communication of messages, a telecommunications manager should understand how electricity works.

A.1 STATIC ELECTRICITY

Lightning is a phenomenon caused by *static electric charges*. These charges can be developed when two dissimilar materials are rubbed against one another. When you walk across a carpet in a dry environment, you can develop a static charge in your body. When you touch another object or person, the charge will jump from you to the other object or person in the form of a spark. When particles of water (a cloud) move through the atmosphere, a static charge is developed in the cloud. When the charge becomes very high, a spark occurs between the cloud and earth in the form of a lightning bolt. Static charges are developed when electrons are added or removed from an object. Electrons exist in all materials. Electrons have a negative charge. When the material has more electrons than normal, it has a negative charge. If the material has lost some of its electrons, it has a positive charge. Electrons can be transferred from one material to another in several ways. One way is via friction, as when the two materials are rubbed against each other. Some materials transfer electrons between them easily via friction. Other materials resist the transfer of electrons regardless of how hard they are rubbed together.

A.2 MATTER, MOLECULES, ELEMENTS, AND ATOMS

Anything that occupies space or has weight is called *matter.* All materials are matter. Examples of matter are metal, water, air, skin, earth, plants, and so on. All matter is composed of small particles called *molecules.* A molecule is made up of smaller particles called *atoms* or *elements.* Elements are the building blocks of the universe. All matter in the universe is composed of substances called elements. There are 103 known elements in the universe.

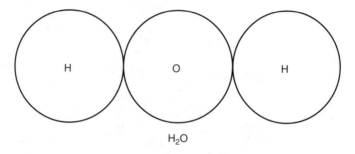

H_2O

FIGURE A-1 One molcule of water.

Some matter consists of a single element. The smallest unit that an element can be broken down into and still retain the characteristics of the element is an atom. Some *materials* are made by combining two or more elements. A material composed of two or more elements is called a ***compound.*** The smallest unit that a compound material can be broken down into and still retain the characteristics of the compound is a molecule. For example, water is a compound composed of two elements: hydrogen and oxygen. Water can be broken down or subdivided into one molecule and still be water. If the water is broken down any further, it will no longer be water but will be atoms of the elements that make up water. The chemical symbol for hydrogen is H and the chemical symbol for oxygen is O. It takes two atoms of hydrogen (H_2) and one atom of oxygen (O) to make one molecule of water (H_2O) (see Figure A-1).

When a material is made strictly from one element, it is called a *pure material.* Copper is a pure material. It is composed solely of atoms of the copper element; it can be broken down into a molecule of copper, which consists of one atom of the copper element. Other examples of some of the 103 known elements include: hydrogen, helium, oxygen, nitrogen, radon, chlorine, calcium, manganese, iron, sulfur, aluminum, carbon, silicon, nickel, and zinc.

A.3 PROTONS, NEUTRONS, AND ELECTRONS

Atoms are made up of protons, neutrons, and electrons. Protons and neutrons exist at the core or center of an atom; electrons exist in the outer part of an atom (see Figure A-2). An atom looks like a small solar system. The center of an atom is a nucleus composed of protons and neutrons, and electrons revolve around this nucleus much like planets revolve around the sun. Protons and neutrons are the same size and are much smaller than electrons. If electrons were magnified to the size of a baseball field, the protons and neutrons would be the size of a baseball.

Although electrons are much bigger, they are lighter than protons and neutrons. Since protons and neutrons weigh more than an electron, they are said to have more mass than an electron. Protons and neutrons weigh about 1840 times as much as an electron. Protons have a positive electric charge, neutrons have no electric charge, and electrons have a negative electric charge. Protons and electrons

One Atom of Copper (CU)

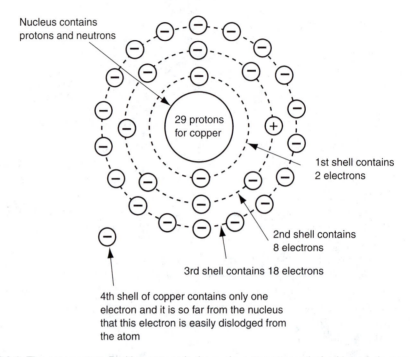

Nucleus contains protons and neutrons

29 protons for copper

1st shell contains 2 electrons

2nd shell contains 8 electrons

3rd shell contains 18 electrons

4th shell of copper contains only one electron and it is so far from the nucleus that this electron is easily dislodged from the atom

FIGURE A-2 The copper atom. Positive protons in the nucleus attract negatively charged electrons to keep them from flying away. The electrons circle the nucleus at high speed, much like planets circling the sun.

exist in all elements. Neutrons exist in all elements except hydrogen. The protons and neutrons exist at the core (or nucleus) of an atom, and the electrons revolve around this core.

A.4 OPPOSITES ATTRACT

Because of the existence of protons in the nucleus, the nucleus has a positive charge. Circling around the nucleus are negatively charged electrons. An atom contains as many protons as electrons; therefore, the atom has no electric charge. When an atom loses an electron, it has a positive charge. An atom with a positive charge is called a *positive ion*. When an atom has picked up an extra electron, it has a negative charge. An atom that has a negative charge is called a *negative ion*.

It has been known for ages that if a glass rod is rubbed with silk, the glass rod becomes electrified or charged. It is also known that if a stick of sealing wax is rubbed with wool, it will be charged with an electric potential. The process of rubbing the glass rod with silk removes some of its electrons from the glass rod and deposits them on the silk. The glass rod becomes positively charged and the silk becomes negatively charged. The process of rubbing a stick of sealing wax with wool removes electrons from the wool and deposits them on the sealing wax. The sealing wax becomes negatively charged and the wool becomes positively charged.

If two charged glass rods are suspended close to each other, they repel each other because they both have a positive charge. If two charged sealing sticks are suspended close to each other, they repel each other because they both have a negative charge. If a charged glass rod is suspended close to a charged sealing stick, they will attract each other. This experiment was used to prove the theorem that like charges repel and unlike charges attract. Objects can be electrically charged by friction or by applying an external electrical source.

A.5 ELECTRON SHELLS AND ATOMIC NUMBERS

Electrons circle the nucleus at various distances from it. Each electron has a particular orbit about the nucleus. Unlike the situation in our solar system, there is more than one orbit at the same distance from the nucleus. These orbits are combined into a graphical representation called the *shell*. Electrons closest to the nucleus are in the first shell. The first shell is also called the *K shell*. Electrons farther away are in the second shell (*L shell*), then the third shell (*M shell*), the fourth shell (*N shell*), and so on. The maximum number of electrons that can be in a first shell is 2. The maximum number of electrons that can be in the other shells is as follows: 8 in the second, 18 in the third, 32 in the fourth, 18 in the fifth, 18 in the sixth, and 2 in the seventh (last) shell.

Elements are listed in a chemical table called the *periodic table*. The periodic table lists elements according to their atomic number, designating the number of protons in the nucleus of an atom. Since an atom is electrically neutral and has as many electrons as protons, the atomic number is also the number of electrons revolving around the nucleus. The element that has only one proton (and also one electron) is hydrogen, and it is the first element listed in the periodic table.

A.6 CONDUCTORS

Copper (Cu) is the 29th element listed in the periodic table because it has 29 protons (and 29 electrons). Two electrons can be in the first shell, eight in the second, and eighteen in the third shell. This leaves only one electron in the fourth shell. The positive charge of the nucleus attracts the negatively charged electrons. The attraction is greatest for electrons closest to the nucleus. The speed of the electrons in their orbits keeps them from being pulled into the nucleus, and the pull from the nucleus keeps the electrons from flying out of their orbits. Since copper has only one electron in the last shell, it is not held tightly in place and is easy to dislodge. The last shell of an atom is called the *valence shell*, and the electrons in it are referred to as *valence electrons*. Therefore, copper has one valence electron.

Materials or elements that conduct electricity have electrons that are held loosely in place. These materials are called **conductors**. Copper is a material (an element) that conducts electricity well because it has one electron held loosely in place (see Figure A-2). The loosely held valence electron in the outer part of an atom is called a *free electron*. Since the pulling force of the protons in the nucleus on the electron in the last shell is weak, it is easy to dislodge these electrons with an external power source.

If a negative voltage source is attached to a conductor, the negative voltage will dislodge or repel the loosely attached electrons of many atoms of copper on the end of the conductor attached to the voltage source. The negative voltage supply forces these electrons to move down the conductor away from the negative applied external voltage. If a positive voltage source is attached to the other end, it attracts the loosely held electrons from many atoms of the copper conductor into the positive source. The push of the negative voltage source on one end of a conductor, along with the pull of the positive voltage source on the other end, causes an orderly flow of electrons from the negative to the positive end of the conductor.

Electrons flow out from the negative voltage source (such as a battery) into the conductor to replace the electrons that have moved down the conductor away from the negative battery terminal, and the process continues. This orderly flow of electrons is called *electric current*. Thus, to make electric current flow we need a voltage source and a conductor attached between the negative terminal and the positive terminal of the voltage source.

A.7 THE BATTERY

Before the invention of the battery, scientists and inventors were trying to communicate using static electric charges as signals. Static electricity provides a short burst of current flow in the form of a spark. Once the spark occurs, the static charge is lost. Italian physician-physiologist Luigi Galvani (1737–1798) accidentally discovered electricity that had a steady current flow, when he crossed two different pieces of metal on a dead frog's leg and the leg twitched.

Galvani's discovery was passed on to a fellow Italian physicist, Count Alessandro Volta (1745–1827). He took the frog out of the equation by proving that any porous material that conducts electricity could be substituted for the frog's leg. These discoveries led to the development of the *chemical cell*. A chemical cell uses a chemical reaction to provide a steady voltage source, and inventors seized on this source to make rapid advances in telecommunications technology.

In 1800, Volta made a chemical cell composed of alternate pieces of copper and zinc separated by vinegar-soaked cloth. This device was called the *voltic pile,* and it produced a steady flow of electric current for a brief time. The chemical reaction that takes place between the vinegar and the dissimilar metals causes electrons to be removed from the copper and deposited on the zinc. Thus, the zinc plate will be negatively charged and the copper plate will be positively charged. The essential constituent of vinegar is acetic acid (CH_3COOH). Later versions of the voltic pile used hydrochloric acid (HCL) as the electrolyte (see Figure A-3). The voltic pile was the first device that could produce a continuous electric current. It is also referred to as the *voltic cell.*

Volta was a pioneer in the study of electricity. He was educated at the Jesuit College in Como, Italy. By age 24, he had mastered all that was known about electricity, and he published his first paper on the subject in 1769. His invention of a

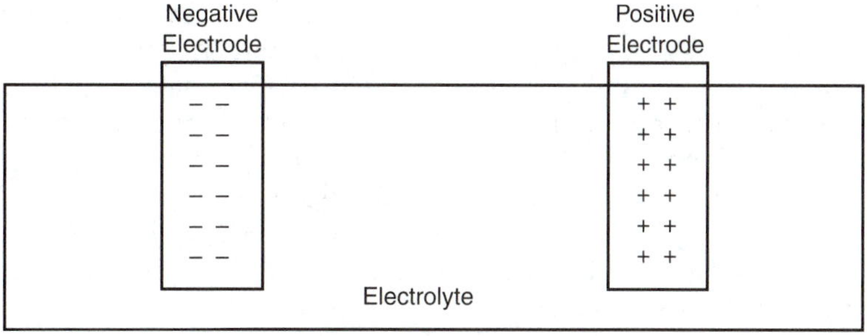

Negative Electrode   Positive Electrode

Electrolyte

FIGURE A-3 Voltic Cell (chemical cell).

simple electrostatic device called the *electrophorus* in 1775 received widespread acclaim. He taught for four years in Como and was then appointed professor of experimental physics at Pavia in 1779. He retired from academic life at age 74. In honor of the pioneering work done by Volta, the unit that measures electromotive force is named for him. ***Electromotive force (EMF)*** is the force that causes electric current to flow. This force is measured in ***volts***. EMF is sometimes represented by the symbol E. Many people prefer to use the term *volts* instead of *EMF*, and most formulas use the symbol V for volts. Throughout the remainder of this chapter, I will use V to represent volts in formulas, the voltage across a device, and the voltage of a cell.

The chemical cell invented by Volta generates electricity by a chemical reaction within the cell. The ***cell*** is a device that transforms chemical energy into electric energy. A ***battery*** is usually an assembly of two or more cells, connected in series, but the term is also used for a single-cell battery. The common flashlight battery is a single-cell battery. If a chemical cell generates 1.5 V of pressure, we can generate 3 V of pressure by using two 1.5-V cells in series to form a 3-V battery.

Direct current (DC) voltage is produced by the chemical reaction within a cell. DC voltage is a voltage that causes electrons to flow in one direction at a rate directly correlated with the voltage. The higher the voltage applied to a circuit, the higher the current. If the voltage is doubled, the current will double. Current is measured as the number of electrons flowing in a circuit. Therefore, the higher the voltage, the higher the number of electrons flowing in the circuit. Electrons will flow out from the negative terminal of a cell into a conductor and will return to the positive terminal of the cell. Therefore, electron flow in the conductor is from the negative terminal of the cell to the positive terminal of the cell.

A chemical cell has three basic components: (1) a negative plate, (2) a positive plate, and (3) an electrolyte (Figure A-3). An electrolyte is a conducting medium in which the flow of current is accompanied by the movement of ***ions***. Atoms of a liquid, or gas, that have become electrically charged are called positive or negative ions and will flow like electrons. The electrolyte reacts chemically with the positive plate and negative plate of a cell to transfer electrons from the positive plate to the negative plate. Most electrolytes are ***acids*** or are acidic in nature.

The negative terminal or plate of an electrolytic cell is called the *cathode* and the positive terminal or plate is called the *anode*. In Volta's time, it was assumed that electric current flowed from the positive to the negative terminal of a cell. This is called the *current flow theory.* Today, we believe that current flow is due to the flow of electrons and we use the *electron theory,* which states that electron flow outside a cell is from the negative to the positive terminal. It really makes no difference whether you state that electric current flows from positive to negative or electrons flow from negative to positive, but most people prefer the electron theory.

A chemical cell produces a voltage that results in a steady flow of electric current when an external device is attached to the cell. Electric current is the orderly flow of electrons in a circuit. A steady flow of electrons in one direction at a constant rate is called **direct current.** Direct current is electric current that flows from the negative terminal of a battery out over a wire connecting some device to the battery, through the device and returning over a second wire from the device, to the positive terminal of the battery. The electric current flow is at a constant rate and moves continuously in one direction. The amount of electric current flowing is measured in **amperes** and is directly proportional to the voltage produced by the battery. Since a battery produces a voltage that causes a direct current flow, the voltage produced is called a *direct current voltage* and is simply stated as a *DC voltage.*

DC voltage is a measure of how much potential difference or electrical pressure exists between the negative and positive terminals of a battery. The chemical cell used in a car battery (and in the battery of a telephone central exchange) is constructed using lead for the cathode (negative terminal), lead-peroxide for the anode (positive terminal), and diluted sulfuric acid for the electrolyte (see Figure A-4). The chemical reaction within the cell removes electrons from the lead-peroxide plate and deposits them on the lead plate. The chemical reaction results in millions of electrons being removed from the lead-peroxide plate. This leaves it with a positive charge. The point at which an external connection is made to the lead-peroxide plate is called the *positive terminal.*

All the electrons removed from the lead-peroxide plate are deposited on the lead plate, and it becomes negatively charged. The point at which an external connection is made to the lead plate is called the *negative terminal.* A negative lead plate has many excess electrons and is an area with a high concentration of electrons. The

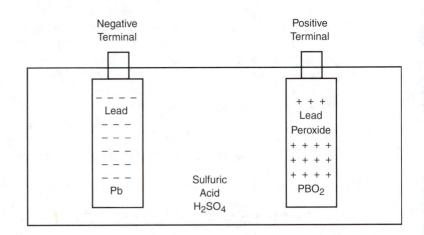

FIGURE A-4 Lead-acid wet cell.

positive lead-peroxide plate has a deficiency of electrons and is an area with a low concentration of electrons. The difference between the amount of electrons on the negative plate and the positive plate is potential difference or electrical pressure. For the lead-acid cell described, the chemical reaction will result in a difference of electrical pressure—that is, 2.17 V.

The difference in electrical pressure is what causes electric current flow. As noted, this pressure is called *electromotive force* (electron moving force) and is usually abbreviated *EMF*. The unit of measure for electrical pressure (EMF) is volts. In our universe, many different occurrences are the result of things moving from an area of high pressure to an area of low pressure, or from an area of high concentration to an area of low concentration. Flows of this nature are responsible for weather changes, and for the osmosis process by which plants take in food and water for survival. The movement of electric current is the movement of electrons from an area of high concentration (the negative plate) to an area of low concentration (the positive plate). The higher the difference between the area with a high concentration of electrons and the area with a low concentration of electrons, the higher the potential difference between the two and the higher the pressure (voltage) developed between the two plates.

The amount of voltage that a cell produces depends on the materials used in the construction of the cell. There are many different chemical cells, constructed from many different materials. Each different construction produces a different voltage. The standard flashlight battery is a single cell composed of a carbon rod (positive terminal) physically in contact with a mixture of manganese dioxide and carbon. This is the anode or positive plate. An electrolyte composed of ammonium chloride and zinc chloride mixed with flour, water, and starch to form a paste surrounds the anode (see Figure A-5). The electrolyte is surrounded by an outer case composed of zinc. This zinc serves as the cathode (it also serves as the negative terminal). The cell is then packaged inside a steel case. This cell is called a *carbon-zinc cell* and develops 1.5 V between the positive and negative terminals.

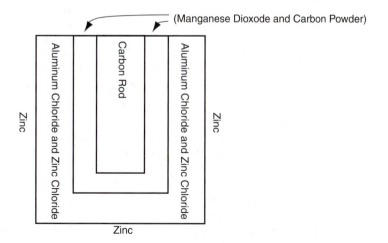

FIGURE A-5 Typical dry cell (flashlight battery): carbon-zinc or Leclanché cell, 1.5 V.

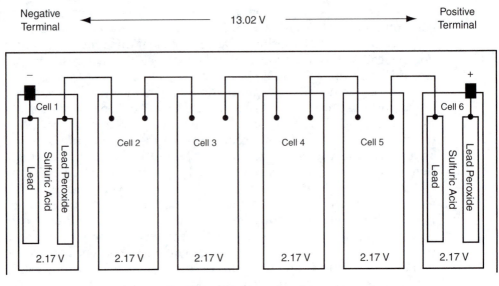

FIGURE A-6 Twelve-volt car battery.

The carbon-zinc cell is sometimes referred to as a *Leclanché cell* in honor of its inventor. To increase the amount of voltage supplied by a chemical cell, several cells are connected in series to form a battery. A car battery is composed of six lead-acid cells connected in series (see Figure A-6). Each of the cells produces 2.17 V. The six-cell car battery produces 13.02 V (6 × 2.17 V). The voltage of a battery is defined as the voltage of one of its cells multiplied by the number of cells connected in series to form the battery. Most people refer to the battery in a car as a 12-V battery because they use 2 V per cell in their calculations instead of 2.17 V per cell.

A.8 DRY CELLS AND WET CELLS

There are two types of cells: primary and secondary. Most *primary cells* are constructed with the electrolyte in a paste form and are referred to as **dry cells** (see Figure A-5.) *Secondary cells* and a few primary cells are constructed with the electrolyte in a liquid form and are referred to as **wet cells.** (see Figure A-4). Primary cells cannot be recharged because they consume their active materials during operation. Secondary cells do not consume their active materials during operation. The materials are altered chemically. A secondary cell can be recharged by reversing the flow of electricity through the cell. The reversal of current flow through a secondary cell restores the original chemical condition of a charged cell.

Cells are rated according to how much current they can provide and how long they can provide current using a measure called **ampere-hours**. The number of amperes supplied to a circuit times the number of hours that those amperes can be delivered is stated as ampere-hours. The standard 11/2-in. diameter dry cell flashlight battery has a 3 ampere-hour capacity or rating. The maximum current that this cell can deliver is 150 milliamps. This means it can provide 150 milliamps for 20 hr (0.150

A × 20 = 3 AH), or 30 milliamps for 100 hr, and so on. This standard flashlight battery is referred to as a size *D* cell. The smaller *AA* and *AAA* cells have lower ampere-hour capacity ratings and were designed for use in low-current applications. Although they are rare, there are primary chemical cells that deliver high current. One type of primary cell used in railway signaling applications is a wet cell using a caustic-soda electrolyte with a copper anode and a zinc cathode. This cell is rated as high as a 1000 ampere-hours and can provide currents as high as 15 amps. Most high-current applications either use commercial AC electricity or secondary cells as a power source. Commercial AC electricity is the electricity supplied to our homes and businesses by the local power company. AC is vastly different from DC and is discussed later.

A.9 CELL SIZE VS. AMPERE-HOURS AND VOLTS

The size of a cell has no bearing on the EMF developed between the positive and negative plates. All wet cells composed of lead and lead peroxide develop approximately 2.17 V per cell regardless of size. The size of the cell determines how long a cell can deliver electricity before it discharges. The larger the cell, the more active material it contains, and the longer it will supply electricity before it reaches a discharged condition. For this reason, the batteries in a telephone central exchange are made using large wet cells. Each cell is about 2 ft high, 1 ft wide, and 1 ft long.

An engineer determines the size of a cell to use by calculating how much electric current is needed for 8 hr. Electric current is measured in amperes. If a central office uses an average of 200 amps each hour, then in 8 hr it will use 1600 amps. Since cells are rated in ampere-hours, the engineer would need to furnish a 1600 ampere-hour battery in this case. This rating for each cell in the battery ensures that the battery can deliver 200 amps for 8 hr. A battery charger connected to the commercial AC power continuously charges the batteries in a central office. If the AC power fails, the charger quits working. The engineer selects cells for the battery that allows it to provide power for 8 hr with the battery charger off. This provides an 8-hr time frame for the restoration of commercial AC power or the attachment of an emergency generator to the battery charger's input.

The wet cells used in car batteries and telephone central exchanges are constructed using a layered process. Instead of one large lead plate and one large lead-peroxide plate being used to achieve a high ampere-hour capacity, many smaller plates are utilized. The cell is constructed with several lead plates and several lead-peroxide plates placed in alternate layers and separated by an insulator such as glass, plastic, or rubber. The lead plates in one cell are all connected together and attached to the negative terminal for the cell. The lead-peroxide plates are connected together and attached to the positive terminal for the cell. Higher ampere-hour ratings can be achieved by either increasing the size of individual plates or increasing the number of plates in the cell.

A central exchange is equipped with 24 cells connected in series. This provides approximately 52 V to the equipment (24 × 2.17 = 52.08 V). Most people refer to the central exchange battery as a 48-V battery. They use 2 V per cell in their calculations (24 × 2 V = 48 V). Because some people use 2 V per cell in their calculations and others use 2.1 V or 2.17 V per cell, the central exchange battery may be called a 48-V supply, a 50-V supply or a 52-V supply. In the older electromechanical switching systems, the battery consisted of 23 cells and was called a 48-V battery even though it was actually producing 49.9 V ($23 \times 2.17 = 49.91$ V). Remember that the cell's size has nothing to do with its voltage. A lead-acid wet cell produces 2.17 V per cell regardless of how large the cell is. The voltage of a battery depends on the number of cells we connect in series. Older electromechanical telephone switching centers used a 23-cell battery. Today's stored program, electronic switching centers use a 24-cell battery.

A.10 FLASHLIGHT CELLS IN ACTION

The 4.5-V flashlight battery used in Figure A-7 is composed of three 1.5-V dry cells connected in series. The electrical diagram for a circuit is called a *schematic diagram*. The schematic diagram for Figure A-7 is Figure A-8. In this figure, an external device (a lightbulb and switch) is connected between the material having an excess of electrons (the negative terminal) and the area with a deficiency of electrons (the positive terminal). The electrons will flow through the external device as they move from the area of high concentration (high-pressure area) to the area of low concentration (low-pressure area). The flow of electrons in the circuit will cause the external device to operate (in this case the lightbulb will light). As long as the

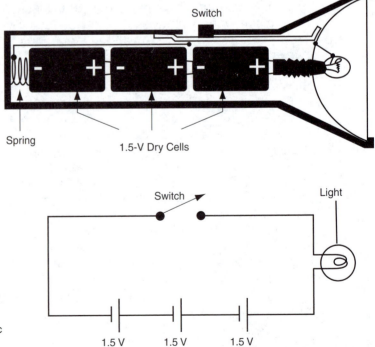

FIGURE A-7 A 4.5-V battery composed of three 1.5-V cells. Three dry cells connected in series equals a 4.5-V battery. When the switch is moved forward, the negative end of the battery is connected to the lightbulb, and it will light.

FIGURE A-8 Schematic diagram for Figure A-7.

switch is on, electron flow will continue until most of the excess electrons on the negatively charged material have flowed through the external circuit and reached the positive material.

As electrons leave the negatively charged material, it slowly loses its negative charge. As electrons flow into the positively charged material, it slowly loses its positive charge. If the chemical reaction in the cell stopped, both materials would slowly lose their electrical charge as current flowed from the negative to the positive terminal. The chemical reaction within the cell continues to maintain the electrical charge on each material. Eventually, the chemical reaction will stop as materials are changed by the chemical reaction. In the case of a dry cell flashlight cell, the carbon anode is actually consumed by the chemical process. The voltage difference between the two terminals is slowly reduced until the difference in potential is so low that current will not flow. At this point, we say the battery is discharged.

The cell for a flashlight uses a carbon rod located in the center of the cell as the positive terminal. This carbon rod is surrounded with a manganese dioxide-carbon mixture as one of its active materials. This mixture is surrounded by and enclosed within an acid paste and an outer shell made of zinc (see Figure A-5). The acid paste chemically removes electrons from atoms of carbon and deposits them on the zinc material. The manganese dioxide-carbon mixture is positive because it has lost many electrons. With many additional negative electrons added to it, the zinc shell has a negative charge. The initial chemical reaction continues until the pressure difference created between the carbon and zinc reaches a point where the chemical action stops. At this point the zinc material has accepted all the extra electrons it can store and will repel any attempt to add more electrons to it. At this stage, the cell is fully charged.

The cell remains fully charged and the chemical reaction ceases until the cell becomes partially discharged. If an external device is attached to the cell, electron flow starts and the chemical reaction begins to recharge the cell. The point at which a cell composed of carbon and zinc becomes fully charged results in a pressure difference (EMF) of 1.5 V per cell. This type of cell is called a *dry cell* because no liquid electrolyte is used. The electrolyte is in a paste. If two of these cells are connected in series, they make a 3-V battery. If three are connected in series, they make a 4.5-V battery.

Another common dry cell is the *alkaline cell*. The anode is composed of a mixture of manganese dioxide and carbon, while the cathode is composed of powdered zinc. The electrolyte is potassium hydroxide. The alkaline cell develops 1.5 V per cell. Whereas a carbon-zinc cell has the carbon rod/anode encased by an outer shell of zinc, the alkaline cell has the powdered zinc cathode inside the anode, and a brass strip inside the powdered zinc connects to the bottom of the outside steel shell.

A.11 ELECTRON FLOW, CONDUCTORS, AND INSULATORS

When a battery is connected to a device using copper wire, the wire serves as a conductor to carry electrons from the battery to the device. Electrons from the negative terminal of the battery will repel loose or free electrons in the copper, forcing them to move away from the negative battery terminal and through the device on their way to the positive terminal. The chain reaction flow of electrons in a circuit is called *electric current. Electric current is the flow of electrons.* How much current flows depends on the pressure (voltage) applied to the conductor and on how much opposition or *resistance* the conductor and device have to the current flow.

All conductors have resistance or opposition to current flow. The amount of resistance depends on how easy it is to dislodge electrons from their atoms. *Different elements have different resistance.* Some elements have their electrons held so tightly to the atom that it is very hard to dislodge them, and they have a very high opposition or resistance to electric current flow. The elements or materials that have a high resistance to electric current flow are called *insulators.* Materials having a low opposition to electric current are called *conductors.*

Electricity will only flow if there is a complete loop of conducting materials between the negative battery terminal and the positive battery terminal. If the loop is opened, electron flow ceases. The purpose of a switch in a circuit is to provide a means whereby the loop can be opened or closed. When the switch is on, the loop is closed and current flows. When the switch is off, the loop is opened and current flow ceases (see Figure A-7).

The electron flow process in a conductor is similar to what would happen if you had a tube filled with tennis balls (or baseballs or billiard balls) and added a ball to one end of the tube. Assume the tube represents a conductor and the balls represent electrons. If you were to force a ball into one end of the tube, all the balls would move down the tube in a chain reaction until the ball at the far end popped out of the tube. This is how electricity flows. The electrons bump each other down the conductor at almost the speed of light. If you had a tube from New York to California filled with balls and forced a ball into the tube in New York, a ball would pop out of the California end almost immediately. The speed of this chain reaction is near the speed at which light travels. Each electron travels a short distance, but when one electron enters the end of a conductor attached to a negative battery terminal, the bumping process results in another electron leaving the positive end of the conductor and flowing into the positive terminal of the battery.

A.12 VOLT AND AMPERE DEFINED

We stated earlier that a volt is the measure used to indicate how much electrical pressure difference there is between the negative and positive terminals of a cell. So, exactly how is a volt defined? A volt really defines the stress (pressure) that exists between the two terminals of a battery. We have all those electrons piled up on the negative plate and they want so much to get back to the positive plate. This creates pressure or stress. Just as pounds per square inch measures how much pressure is trying to force water through a pipe, voltage measures how much pressure is trying to force electrons to flow through a conductor or other device.

We can begin delving into what a volt is by first understanding what an electrical charge is. The basic unit of electrical charge for a static charge (an electrical charge that is not moving) is the *coulomb,* named in honor of French scientist Charles Augustin de Coulomb (1736–1806). He discovered Coulomb's law: "The force of attraction or repulsion between two static electric charges is directly proportional to their product and inversely proportional to the square of their distances." A *coulomb* is defined as a unit of electrical charge equal to 6.25×10^{18} electrons. That's 6250 trillion electrons. So, if the positive plate of a cell has lost half of this number (3125 trillion) and the negative plate has gained the same amount (3125 trillion), the difference between the two plates is 6250 trillion electrons or 1 coulomb. The formula for a coulomb is: Q (charge stated in coulombs) equals the number of electrons divided by 6250 trillion electrons per coulomb.

$$Q = \text{number of electrons}/(6.25 \times 10^{18} \text{ electrons})$$

An electric current flow of 1 amp is defined as the flow of 6250 trillion electrons (1 coulomb) past a given point in 1 sec. Thus, the coulomb is used to express a static or stationary electric charge, but when that charge is moving or being transported, the term *amperes* is used. Amperes is coulombs in motion. If 1 coulomb moves past a given point in 1 sec, the current flow is 1 amp. If 12 coulombs move past a point in 1 sec, that is 12 amps. If 12 coulombs move past a point in 3 sec, that is 4 amps. The formula for amperes is:

$$I = Q/t$$

Current is measured by the number of electrons that flow past a point in a unit of time, where I is the symbol for current in amperes, Q is the symbol for the charge of the electrons stated in coulombs, and t is the time in seconds. Thus, electric current flow is the number of coulombs per second that flow in a circuit. This symbol I comes from the French word for *intensity* and stands for the intensity of current flow. This symbol is used in many formulas to designate the amount of current flow.

Now back to that volt. The volt is a measure of the pressure that makes current flow (it is a measure of the energy that makes current flow). If two charged plates that were not part of a cell were connected together by a conductor, they would rapidly lose their charges and then current would cease to flow. But when the charged plates are part of a cell, they will receive new charges to replace charges that flow out of the cell. The chemical reaction in a cell attempts to maintain the charge on the plates. The chemical reaction creates the pressure. When a cell is not being used, the plates charge to a certain potential and the chemical reaction stops. At this point the cell is fully charged.

The chemical reaction stops when a cell is fully charged because the negative plate has received all the electrons it can hold and the positive plate has given up all the electrons it can give up. The charged negative plate repels any attempt to add more electrons and the positive plate will not respond to the chemical action trying to pull off additional electrons. The pressure between the plates equals the pressure from the chemical reaction. With no pressure differential, ions and electrons stop flowing—that is, the chemical action stops. When the cell is being used, pressure between the plates starts to decrease because they lose some of their charge as electrons flow in the outer circuit from the negative to the positive terminal. The chemical reaction starts up to replace lost charges on the positive and negative plates.

Since cells can be constructed from different materials, each construction has its own unique chemical reaction. Different chemical reactions generate different electrical pressures or energy levels. Therefore, cells of different construction develop different voltages.

The volt measures the amount of energy stored per coulomb of charge. A *volt* is defined as the potential difference necessary to obtain 1 joule of work when 1 coulomb of charge flows. A *joule* is the amount of heat energy produced by a 1-Ω resistor when a current of 1 amp flows through it for 1 sec. A joule is equal to 0.7376 of a foot-pound. A foot-pound is the amount of energy required to raise (or lower) 1 lb a distance of 1 ft. Thus, a joule is the amount of energy required to raise (or lower) 0.7376 lb a distance of 1 ft. Why are we talking about work? Work is required in the form of a chemical reaction to separate electrical charges. These separated electrical charges, or potential difference, represent the potential or capacity to do work. By connecting a device to the terminals of a battery, the device receives energy and can do work. As stated earlier: One volt is defined as the potential difference necessary to obtain 1 joule of work when 1 coulomb of charge flows; this is stated in a formula as follows:

$$V = W/Q$$

Voltage = energy (W) in joules divided by charge in coulombs (Q). Voltage is the pressure, force, or energy exerted on electrons that try to make them move. The unit of voltage is the volt, symbolized by V. As noted, 1 V is the difference in potential between two points when 1 joule of energy is used to move 1 coulomb of charge from one point to the other. The chemical reaction of a cell determines the amount of push being exerted on electrons. The higher the push developed, the higher the voltage. Remember that the size of the cell and its components have no bearing on the voltage developed. The size simply determines how long current can be supplied before the cell is in a discharged state.

A.13 CHEMICAL ACTION IN A TYPICAL LEAD-ACID WET CELL

The most common type of secondary cell uses: (1) plates made of spongy lead (Pb) for the negative electrode (cathode), (2) plates made of lead peroxide (PbO_2) for the positive terminal (anode), and (3) an electrolyte of diluted sulfuric acid (H_2SO_4). Combining water with sulfur dioxide makes sulfuric acid.

$$H_2O + SO_2 \longrightarrow H_2SO_4$$

An electrolyte conducts electricity; therefore, insulators must be placed between the plates or the cell would discharge internally. The spongy lead and lead peroxide are placed in a firm metal grid for support. This metal support grid is usually made from alloy of lead and antimony. The cell is then called a *lead-antimony cell*.

When an external circuit is connected to a cell, the electrons start to flow. This causes some of the sulfuric acid molecules to break up into two positively charged hydrogen ions (2H) and a negatively charged sulfate ion (SO_4) (see Figure A-9).

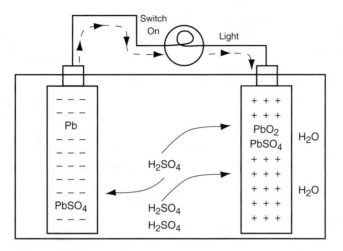

Electrons leave negative terminal and flow through switch and light to positive terminal.

FIGURE A-9 Wet cell action during discharge.

Each SO_4 ion near the lead plate combines with one atom of lead to form one molecule of lead sulfate ($PbSO_4$), and the ion gives up two electrons to the lead plate. The two hydrogen ions (2H) are attracted to the lead-peroxide plate, take two electrons from the lead peroxide, and then unite with one atom of oxygen from that plate to form one molecule of water. The chemical action results in adding two electrons to the lead plate, making it negative, and removing two electrons from the lead-peroxide plate, making it positive. This establishes a difference of potential (voltage) between the two plates.

A second molecule of H_2SO_4 will break up into two positive hydrogen ions (2H), which are also attracted to the lead-peroxide plate to take away two electrons and combine with an oxygen atom to form another molecule of H_2O. The SO_4 ion from this second molecule of H_2SO_4 is attracted to the lead-peroxide plate and combines with a lead atom to form a molecule of lead sulfate ($PbSO_4$). These chemical reactions are taking place simultaneously for many molecules of sulfuric acid, lead, and lead peroxide. The end result of the chemical reaction while a cell is furnishing electrons to an external circuit is to convert the lead and lead plates into lead sulfate, convert the sulfuric acid into water, and gradually discharge the battery. The chemical actions are expressed by the following chemical equation:

$$Pb + PbO_2 + 2H_2SO_4 \longrightarrow 2PbSO_4 + 2H_2O$$

As the water created by the chemical action is added to the acid, it dilutes the acid. Sulfuric acid is denser than water. The denser a liquid is, the easier it is for an item to float on the liquid. *Specific gravity* is the density of another liquid relative to water. For this reason, specific gravity is also called *relative density.* The specific gravity of water is 1 because the density of water relative to water is 1. A liquid denser than water has a specific gravity greater than 1. A device called a *hydrometer* can be used to check the specific gravity of the electrolyte in a wet cell.

The specific gravity of sulfuric acid in a fully charged cell is 1.280. A cell that is three-fourths charged has a specific gravity of 1.250, a one-half charged cell 1.220, a one-fourth charged cell 1.190, and a completely discharged cell 1.130. Thus, as a cell discharges, the chemical reaction creates more and more water that dilutes the sulfuric acid, lowering its specific gravity. The hydrometer can be used to check the charged condition of a cell.

A high-discharge test can also be used to test a cell or battery. A high-discharge test consists of placing a low-resistance connection between the negative and positive terminals, while also placing a voltmeter across the terminals. If the voltmeter indicates that the voltage remains above 1.8 V under high-discharge current, the cell is good.

Passing an electrical current backward through the cell charges a wet cell. This will occur if a voltage source higher than the cell or battery is connected to the battery terminals. In the case of a car battery, a generator or alternator is connected to the battery. When the car is running, these devices put out about 14 V DC. This is a higher voltage than the battery voltage, and current will flow backward through the battery. This reverse current causes each molecule of water in the electrolyte to break down into two positive hydrogen ions and one negative oxygen ion. Two hydrogen (2H) ions near the positive lead-peroxide plate combine with an SO_4 ion from the lead sulfate to form one molecule of sulfuric acid (H_2SO_4). The oxygen ions from two molecules of water combine with the lead ion portion of the lead sulfate on the positive plate to form lead peroxide (PbO_2). The pair of hydrogen ions (2H) from the second molecule of water are attracted to the negative lead plate, where they combine with an SO_4 ion from the lead sulfate on that plate to form another molecule of sulfuric acid (H_2SO_4) (see Figure A-10). Thus, during charging,

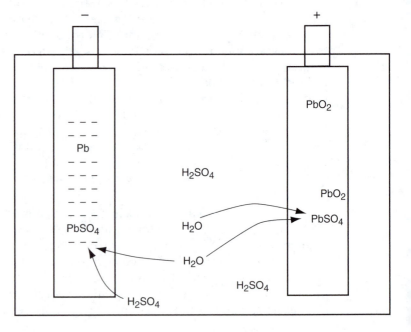

FIGURE A-10 Ion flow during charging of wet cell.

the electric current breaks the water down and its charged atoms (ions) combine with the lead sulfate to form spongy lead at the negative plate and lead peroxide at the positive plate. The formula for this chemical action is:

$$2PbSO_4 + 2H_2O \longrightarrow PbO_2 + Pb + 2H_2SO_4$$

The charging rate of a battery can be high when the battery is in a highly discharged condition. Most of the plate area will contain sulfates, and a high current can be used because there are many sulfates for the hydrogen ions to combine with. As a battery reaches a fully charged condition, the charging rate must be reduced. If the charging current is too high through a fully charged battery, the current will heat the water and cause it to decompose into hydrogen and oxygen, which merely gas off into the atmosphere. If high charging currents are kept on a battery too long, the heat produced will warp the plates. For this reason, devices that charge a battery contain circuitry to monitor the voltage of the battery and the value of the charging current. As a battery reaches a charged condition, the charging device lowers the charging current by lowering the output voltage of the charging device. When the battery is fully charged, the voltage of the battery will equal the voltage of the charging device and no current will flow through the battery.

When the battery in a car is fully charged, its voltage equals the voltage output of the alternator and no current flows into or out of the battery. The alternator supplies all the current demanded by the car (engine, radio, lights, and so on), and no current is drawn from the battery. Thus, the battery is only needed to start your car. Once the car is running, the alternator is working and supplies all electrical needs.

A.14 NICAD BATTERIES

Another popular secondary cell is the nickel-cadmium (NiCad) cell. The positive plate is made of cadmium and the negative plate of nickel hydroxide; the electrolyte is potassium hydroxide. This cell develops 1.2 V. Five of these cells are connected in series and put inside one package to form the 6-V battery used in camcorders, cellular phones, and other portable electronic devices. The electrolyte is in the form of gel.

A.15 RESISTANCE

Every element and compound contains atoms with electrons. The protons in the nucleus attract the electrons with a force that keeps the electrons within the atom. As mentioned earlier, elements that have electrons held loosely in the outer shell are good conductors, but even the best conductor has some opposition or resistance to current flow because it still takes some degree of effort or energy to force an electron out of the atom. When an atom loses an electron, it becomes positive and quickly attracts an electron from a nearby atom. Electrons in a conductor are always moving about between atoms. They move in all different directions, jumping back and forth between adjacent atoms. When we attach an external electrical source to the conductor, we force all free electrons to move in one direction. The movement of electrons in one direction constitutes current flow. Elements that are good conductors are silver, copper, aluminum, zinc, iron, and carbon. Brass, which is a compound made of copper and zinc, is also a good conductor.

Elements and compounds that have electrons in their outer shell held tightly to the atom by a strong pull from protons in the nucleus are not very good conductors. We call these elements and compounds *insulators*. An insulator has a high opposition or resistance to current flow because electrons are bound so tightly to their atoms that it takes a very high voltage to dislodge them. Every element and compound has some resistance; those with low resistance to electron movement are called *conductors* and those with high resistance are called *insulators*. Insulators are just as important as conductors. They allow us to contain electron movement within a desired path. Good insulators are glass, plastic, mica, ceramics, and dry air. Insulators are also called *dielectrics*. Just as every conductor has some resistance to electron flow, every insulator can be forced to allow electron flow if the voltage is high enough. Air is a good insulator, but lightning travels through air when the voltage becomes very high. Think of insulators as poor conductors.

A.16 VOLTAGE/CURRENT/RESISTANCE RELATIONSHIP

The more voltage supplied to a device, the more current will be flowing through the device. The more opposition or resistance the device offers to current flow, the less current will be flowing through the device. Voltage and current are directly related. Double the voltage applied to a device and the current flow will be doubled. Resistance and current are inversely related. If resistance goes up, current goes down. If resistance is doubled, the current flow will be cut in half. Ohm's law states these relationships between voltage, current, and resistance:

$$\text{Ohm's Law}$$
$$I = V/R$$

In other words, current in amperes = voltage/resistance. This formula can be rearranged to find the voltage that will be dropped across a device with current flowing through it: voltage dropped equals current times resistance.

$$V = IR$$

This relates to what we discussed earlier: A volt is defined as the potential difference necessary to obtain 1 joule of work when 1 coulomb of charge flows. A joule is the amount of heat energy produced by a 1-Ω resistor when a current of 1 amp flows through it for 1 sec. When 1 amp (I) flows through 1 Ω of resistance (R), then $IR = 1 \times 1 = 1$ V per Ohm's law. The formula can also be rearranged to find the resistance of a device when we know the voltage dropped across it and the current flowing through it: resistance equals voltage divided by current.

$$R = V/I$$

Current is measured in amperes and *resistance is measured ohms*. Ohm's law formulas are used to determine how much voltage drop, resistance, and current we have in any device or circuit. Keep in mind that the voltage formula is used to calculate voltage dropped across a component in a circuit. It cannot and is not used to calculate voltage supplied by a battery except by extrapolation. As will be noted shortly, the amount of voltage applied to a closed loop equals the sum of all voltage drops in the loop. Thus, if we calculate the voltage across each device in a series circuit and add them together, their sum equals the voltage of the supply. More about this shortly.

A.17 THE COMPLETE SYSTEM

A battery can be thought of as being similar to a water tower. A water tower is filled with water, and the weight of the water causes it to exert pressure downward. Energy is stored and ready to do work. When a pipe connected to the water tower is opened, the gravitational pull results in water flowing out of the pipe. In today's water distribution system, additional pumps are placed at intervals along the main water distribution pipe to increase the pressure behind the water in the pipe. When we open the faucet at our sink, the pressure behind the water causes water to flow out of the pipe. If there were no way to get additional water back into the water tower, we would soon run the tower dry. But water coming out of our facets finds its way back into streams and rivers and is pumped back up into the water tower. Our water system has a way to supply us with water and also has a way to return water to the tower.

Batteries are like water towers because a supply of electrons is stored in a battery under pressure. Batteries can also be thought of in terms of a water pump. The water pump places a great deal of pressure behind a supply of water. Batteries are electron pumps; they place considerable pressure behind a supply of electrons. The battery supplies electrons to our devices and circuits when a circuit is completed between the negative and positive terminals. These circuits must have a return path to the positive side of the battery to return the electrons we do not use to the battery. If a device is connected only to the negative terminal of a battery, electric current will not flow. The excess electrons on the negatively charged plates are trying to get back to the positive terminal. If the positive terminal is not attached to the circuit, the electrons refuse to flow. When both terminals are hooked to a device, electrons will leave the negative terminal and flow through the device in their journey back to the positive terminal.

Resistance can be thought of in terms of water pipes. Small water pipes have a higher resistance to the flow of water than larger pipes. The higher the resistance, the less water current we will have. The same is true in electric circuits. The higher the resistance in the circuit, the less electrical current we will have. Large-diameter conductors have low resistance and small-diameter conductors have higher resistance.

Now that we can generate electricity using a battery and we know current will flow through any device attached between the negative and positive battery terminals, we can put switches in the circuit to turn the device on and off. In Figure A-7, the flashlight is turned on and off by operating the switch. When the switch is closed, the flashlight is turned on. Electrons will flow from the negative terminal, through the spring at the bottom of the cells, through the metal case to the switch, through the closed switch, to the lightbulb, through the lightbulb, and return to the positive terminal. This flashlight has three cells connected in series. The negative terminal of one cell connects to the positive terminal of another cell.

The overall voltage of any battery is equal to the voltage of each cell, connected in series, added together. Since each cell produces 1.5 V, the voltage for this battery (three cells in a series-connected arrangement) is 4.5 V. If a water main had three pumps in series along its route, the total pressure would be found by adding the pressure developed by each pump. Each cell in a battery is an electron pump, and when they are connected in a series arrangement, the total voltage equals the sum of the individual cell voltages. A device called a *voltmeter* can measure voltage. Often the voltmeter is part of a device called a *multimeter.* This multimeter can be used as a voltmeter to read voltage, as an ammeter to read current, or as an *ohmmeter* to read resistance.

A.18 SERIES RESISTANCE IN A CIRCUIT

When components in a circuit are connected in series, adding the value of all resistance together gives the total resistance for the complete circuit. A series circuit is a circuit in which the output of one device connects to the input of the next device. In a series circuit, the electron current flow will flow through one device, then the next, then the next, and so on. You can tell when a circuit is a series circuit because the electron flow can only take one path. The resistance of each device is shown in a schematic circuit (see Figure A-11).

If we substitute resistors for the resistances of the devices in Figure A-11, we achieve an electrically equivalent circuit. Therefore, I will use resistors to demonstrate circuit operations. We can tell that the resistors of Figure A-11 are wired in a series arrangement because the electric current will flow from the negative terminal of V_s and will first flow through R_1, then R_2, then R_3, then R_4, then R_5, before returning to the V_s positive terminal. Also note that the output of the negative

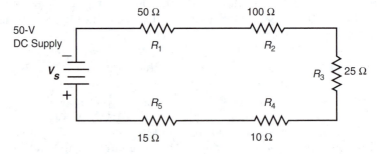

FIGURE A-11 Series circuit.

terminal of V_s is wired to the input lead of R_1, the output of R_1 is wired to the input of R_2, the output of R_2 is wired to the input of R_3, and so on. The total resistance (RT) of a series circuit is equal to the sum of the individual resistances:

$$RT = R_1 + R_2 + R_3 + R_4 + \ldots$$

The R Total for Figure A-11 $= R_1 + R_2 + R_3 + R_4 + R_5$
$$= 50 + 100 + 25 + 10 + 15 = 200 \ \Omega$$

The first step in trying to determine how much current flows in a circuit is to determine the total voltage applied to the circuit as well as the total resistance of the circuit. Ohm's law can then be applied to determine the current flow. In Figure A-11, let us assume that the supply voltage (V_s) is 50 V. Ohm's law states that current in ampere (I) = EMF in volts (V) divided by resistance in ohms (R). For Figure A-11: $I = V/R = 50/200 = 0.25$ amp. In a series circuit the current flow is the same at any point in the circuit. The amount of current leaving the battery is 0.25 amp, and the amount of current returning at the positive terminal of the battery is also 0.25 amp. The current through each of the resistors is 0.25 amp as well.

Many people find this point counterintuitive. They think each resistor will eat up or use current and less is sent to the next component in line. Each component will use up energy, but the energy it uses is what creates the resistance of the component. The resistance of the component has compensated for its use of electricity. This is much the same as trying to pump water through several pipes in series, with each pipe having a different size. The smallest pipe (largest resistance) is the major component limiting current flow, but the resistance of all pipes (total resistance) limits the overall water flow. The same amount of water flows through each pipe and the current flow is the same everywhere in the pipeline. In an electrical circuit, total resistance limits the current flow, and the current flow is the same at any point in the circuit.

Another factor to consider in a series circuit is the voltage supplied to each resistor. This can be determined by using Ohm's law, which states that the voltage dropped across any component or device is as follows: voltage (V) = current (I) × resistance (R). In Figure A-11:

Voltage across R_1 = 0.25 amp × 50 Ω = 12.5 V
Voltage across R_2 = 0.25 amp × 100 Ω = 25 V
Voltage across R_3 = 0.25 amp × 25 Ω = 6.25 V
Voltage across R_4 = 0.25 amp × 10 Ω = 2.5 V
Voltage across R_5 = 0.25 amp × 15 Ω = 3.75 V
Total = 50 V

Notice how the sum of the voltages across each resistor is equal to the applied voltage. The voltages across each resistor are called *voltage drops*. The person who first observed this was the 19th-century German physicist Gustav Robert Kirchhoff

(1824–1887), and his observation is known as *Kirchhoff's voltage law.* This law states that the sum of all the voltage drops around a single closed loop in a circuit is equal to the total source voltage in that loop.

Kirchhoff's Voltage Law

Total voltage = $VR_1 + VR_2 + VR_3 + \ldots$

Now you can see that it is not current that drops in a circuit, but voltage. The current is the same everywhere in a series circuit, but the voltage is not. We start out with a supply voltage of 50 V. If we use a voltmeter to measure between the + and – terminals of the battery in Figure A-11, we read 50 V. If we measure between the + terminal of the battery and the junction between R_1 and R_2, we will read 37.5 V because 12.5 V was dropped across R_1. Each resistor will cause the voltage to drop until at the output of R_5, the voltage is zero. Again, this is similar to water flow in pipes. The largest pressure drop will occur across the pipe with the highest resistance, and the total pressure drop in the system will equal the total applied pressure from all pumps in the system. Also note that the voltage dropped across R_1 is directly proportional to how much of the total resistance it represents. The total resistance for our example series circuit is 200 Ω. R_1 is 50 Ω. This is one-fourth of the total resistance (50/200 = 1/4). The total applied voltage is 50 V and 1/4 of 50 is 12.5 V.

A.19 PARALLEL RESISTANCE IN A CIRCUIT

When components are connected such that all their inputs are connected together and all the outputs of each are connected together, they are wired in parallel. This is similar to a multilane highway. The lanes are in parallel; traffic can flow over any lane. Figure A-12 is a parallel circuit. Notice how point A is connected to one end of all resistors and point B is connected to the other end of all resistors. Each resistor has its input lead connected to point A and output lead connected to point B. If current is flowing from point A, when it gets to resistor R_1 it has three paths to choose from. Some of the current will flow through R_1 and some will flow toward R_2 and R_3. Thus, in a parallel circuit current *is not* the same at every point in the circuit. If a 50-V battery were connected across points A and B, the 50 V would appear across each and every resistor in parallel. Thus the voltage across R_1 is 50 V, the voltage across R_2 is 50 V, and the voltage across R_3 is also 50 V. The current in each resistor equals the applied voltage divided by the value of the resistor.

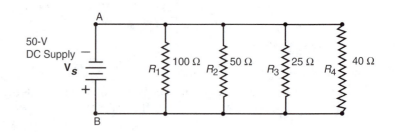

FIGURE A-12 Parallel circuit.

Because a parallel path provides multiple paths for current flow, when resistors are placed in parallel, the total circuit resistance decreases. The formula used for calculating total resistance of a parallel circuit is:

Parallel Resistance Formula

$$1/RT = 1/R_1 + 1/R_2 + 1/R_3 + 1/R_4 + \ldots$$

This formula can also be stated as follows: The reciprocal of the total resistance is equal to the sum of the reciprocals of each resistance. In Figure A-12, the reciprocal of $R_1 = 1/R_1 = 1/100 = 0.01$ Ω. The reciprocal of $R_2 = 1/R_2 = 1/50 = 0.02$, and the reciprocal of $R_3 = 1/R_3 = 1/25 = 0.04$. The reciprocal of $R_4 = 1/R_4 = 1/40 = 0.025$. The next step is to get the sum of these reciprocals: $0.01 + 0.02 + 0.04 + 0.025 = 0.095$. The reciprocal of the total resistance equals the sum of the reciprocals: $1/RT = $ sum of the reciprocals. In Figure A-12: $1/RT = 0.095$. Therefore, to get total resistance we take the reciprocal of both sides of the equation. The reciprocal of $1/RT = 1/1/RT = RT$. The reciprocal of the sum of the reciprocals: $1/0.095 = 10.526316$ Ω. Therefore $RT = 10.526316$ Ω. In a parallel circuit RT will always be less than the lowest resistance, and we meet this test (10.526316 is less than 15). Now that we have RT, we can use Ohm's law to find total current.

Total current (I total) = total voltage (50 V)/total resistance (10.526316) = 4.75 amps. The current in each resistor can also be found using Ohm's law. IR_1 (current in R_1) = VR_1 (voltage across R_1)/R_1 (resistance of R_1) = $50/100 = 0.5$ amp. Likewise, $IR_2 = VR_2/R_2 = 50/50 = 1$ amp and $IR_3 = 50/25 = 2$ amps. $IR_4 = 50/40 = 1.25$ amps. The total current for all four resistors equals $IR_1 + IR_2 + IR_3 + IR_4 = 0.5 + 1 + 2 + 1.25 = 4.75$ amps. This equals the I total found above. This proves another theorem of Kirchhoff. The second law of Kirchhoff is called *Kirchhoff's current law:*

Kirchhoff's Current Law

Total current approaching a point = Total current leaving a point

The total amount of current approaching any point in a circuit is equal to the total current leaving that point. The total current at point A = 4.75 amps; 0.5 amps of that current flows through R_1, 1 amp of that current flows through R_2, 2 amps of that total current flows through R_3, and 1.25 amps of that current flows through R_4. These individual currents flowing through R_1, R_2, R_3, and R_4 will combine on the output side of the resistors, and a total current of 4.75 amps will flow back into the positive terminal of the battery connected to point B.

No matter how many devices are connected in parallel, the voltage across each is equal to the voltage applied and the voltage across every resistor will be the same. Calculators make it easy to find total resistance for a parallel circuit. First enter the resistance value for R_1, then press the $1/x$ key, press the + key, enter the value of R_2, then press the $1/x$ key, press the + key, enter the value of R_3, press the $1/x$ key, and continue this process for all resistors in parallel. When the last resistor value has been entered and the $1/x$ key pressed, press the = key, then press the $1/x$ key. Voilà, we have the total resistance for the parallel circuit.

A.20 SUMMARY

Electricity has been around since the beginning of time, but only after discovering what it is could we begin to understand how to control it. Most of the discovery process occurred in the late 1700s and 1800s. The invention of a direct current electric source (the battery) by Count Alessandro Volta in 1800 allowed other inventors to accelerate their research work. This allowed us to get a clearer view of how electricity behaved and led to the use of electricity for long distance communication (telecommunications).

Electric current is the flow of electrons. The battery is a device containing chemical cells. The chemical action in a cell will produce an excess supply of electrons on one plate (making it negative) and will create a deficiency of electrons on another plate (making it positive). This difference in electron concentrations creates an external EMF, which will make electrons flow in one direction. This force is measured in volts and the current it causes to flow is measured in amperes. The opposition that a device, conductor, or insulator has to flow of current is measured in ohms. The relationship that exists between volts, amperes, and ohms is stated by Ohm's law ($E = I \times R$; $I = E/R$; and $R = E/I$). Voltage and current are directly related. Double the voltage and the current will double. Current and resistance are inversely related. Double the resistance and the current flow will be cut in half.

Conductors are elements or compounds that have loosely attached electrons in the outer shell of their atom. It does not take much EMF to force these electrons to move. Insulators are poor conductors. They are elements or compounds with electrons bound tightly to their atoms. It takes a high EMF to force these electrons to move.

APPENDIX REVIEW QUESTIONS

1. How is a DC voltage produced?
2. What is the difference between a cell and a battery?
3. What is a molecule?
4. What is an element?
5. What is an atom?
6. What is an electron?
7. What is electric current flow?
8. What is resistance?
9. What is voltage?
10. What determines the amount of electric current flow in a circuit?
11. How much voltage is developed by a carbon-zinc dry cell?
12. How much voltage is developed by a lead-acid wet cell?
13. Why is the physical size of a battery important?

14. If four 100-Ω resistors are connected in series, what is the total resistance they offer?
15. If four 100-Ω resistors are connected in parallel, what is their total resistance?
16. What determines the amount of voltage produced by a cell?
17. What is an insulator?
18. What is a conductor?
19. What is 1 amp?
20. What is 1 V?
21. What is Kirchhoff's voltage law?
22. What is Kirchhoff's current law?
23. What is Ohm's law?
24. In a series circuit containing a 100-Ω resistance and two 150-Ω resistances, if the current through the 100-Ω resistance is 0.25 amp and the voltage drop across the resistance is 25 V, is the current and voltage drop for other resistances in the series circuit the same or different? Explain.
25. What is an ion?

GLOSSARY

Acid A substance made of two or more elements. One of the elements is usually hydrogen or an atom that can accept electrons from a positive plate in a cell. The chemical reaction of an acid with a metal results in converting electrically neutral atoms into positive ions, negative ions, and electrons. When an acid is used in this manner, it is called an *electrolyte.*

Ampere The unit of electric current flow. One ampere is a flow of approximately 6250 trillion electrons per second (6.25×10^{18} electrons per second).

Ampere-Hours The rating for a cell or battery that indicates how much electrical storage it has. A 1600 ampere-hour battery will deliver 1600 amps for 1 hr, 1 amp for 1600 hr, 400 amps for 4 hr, 200 amps for 8 hr, and so on.

Atom An atom is the smallest component of an element having all the properties of the element. It consists of a number of protons, neutrons, and electrons such that the number of protons determines the element.

Battery A combination of two or more cells electrically connected to work together to produce electrical energy.

Cell A device that generates electricity by a chemical reaction.

Compound A substance or material composed of two or more elements.

Conductor An element or compound that readily conducts electricity because its atoms have loosely held electrons in their outermost shell. Most metals are good conductors. The six best conductors are silver, copper, aluminum, zinc, brass, and iron. Carbon is also a good conductor.

Direct Current (DC) The electric current flow is at a constant rate and flows continuously in one direction. The direction of electric current flow (outside the battery) is from the negative terminal of the battery, through a device attached to the battery, to the positive terminal of the battery. The amount of electric current flowing is measured in amperes.

Direct Current (DC) Voltage   The voltage produced by a chemical reaction. Devices that produce DC electricity via a chemical reaction are called *cells.* Several cells connected together electrically are called a *battery of cells* or simply a *battery.*

Dry Cell   A chemical cell that generates electricity via the chemical action between an electrolytic solution and two dissimilar metals. The electrolytic solution is in the form of a paste.

Electric Current   The orderly flow of electrons in one direction.

Electromotive Force (EMF)   The electron moving force, energy, or pressure that tries to force electrons to move. EMF is measured in volts.

Insulator   A poor conductor. An element or compound with electrons held so tightly to their atoms that it takes a high voltage to dislodge them.

Ions   Atoms that have an electric charge. An atom with a positive charge is a positive ion. A negatively charged atom is a negative ion.

Material   The substances of which a thing is composed. The term is applied to almost anything considered to be matter.

Matter   Anything that occupies space and has weight.

Molecule   The smallest physical unit of a compound that still retains the characteristics of the compound. It consists of the two or more atoms that make up the compound. For a material that is an element, a molecule is one single atom of the element.

Multimeter   A test device used to measure voltage, resistance, or current.

Ohmmeter   A test device used to measure the resistance of a component. An ohmmeter is usually one part of a multimeter.

Resistance   The amount of opposition to electric current flow that a device has.

Static Electric Charge   The development of an electric potential on a material. The electric potential does not move (is static). A static charge on a material is usually developed by friction between it and another material.

Voltmeter   A test device used to measure voltage. A voltmeter is usually one part of a multimeter.

Volt, Voltage   A measure of how much potential difference or electrical pressure exists between the two terminals of a battery (or some other device that generates electricity). This electrical pressure is called *electromotive force (EMF)* and is measured in volts.

Wet Cell   A chemical cell that generates electricity via chemical means between an electrolytic solution and two dissimilar metals. The electrolytic solution is a liquid. The battery in a SPC central exchange is composed of 24 wet cells.

Appendix B

Electric Circuits

KEY TERMS

Ammeter
American Wire Gauge (AWG)
Circuit
Circular Mil
Closed Circuit
Conductance
Decibel (dB)
Decibels Referenced to Noise
    (dBrn)

Electron Pump
Fixed Resistor
Load
Multimeter
Ohmmeter
Open Circuit
Resistor
Signal-to-Noise Ratio (S/N Ratio)

Switch
Tolerance
Variable Resistor
Varistor
Voltmeter
Watt

In Appendix A we covered direct current (DC) electricity, which can be generated by a chemical reaction. In Appendix C we will cover alternating current (AC) electricity, which is generated at a commercial power plant and then distributed to our homes and businesses. Before moving on to AC, we must gain a thorough understanding of how electricity is put to work. The principles we learn about putting DC to work will also serve us well with respect to AC circuits. DC circuits contain only a resistive component. AC circuits are affected by properties known as *inductance* and *capacitance,* which we will discuss in Appendix C. DC circuits are not affected by these properties. What we learn in this appendix about the relationship between voltage, current, and resistance in DC circuits serves as a building block for understanding their relationships in AC circuits.

B.1 WHAT IS A CIRCUIT?

A *circuit* is a circular journey. An electrical circuit is a complete electrical path between the terminals of a voltage supply. This complete path allows electrons to start their circular journey out of the voltage source, go through a conducting path, return to the voltage source, and then flow through the voltage source to the starting point, thereby completing the circular journey. In a DC circuit, electrons leave the negative terminal of a battery, flow through a conductor, through some device that uses the electric current to do some work, flow through a conductor back to the positive terminal of the battery, and then flow through the battery to the negative terminal. There will be just as much current entering the positive terminal of the battery as there was leaving its negative terminal. The voltage of the battery and the resistance in the circuit govern the amount of current flowing in the circuit. The battery can be thought of as an *electron pump.* It pulls electrons in on the positive end and pumps them out of the negative terminal.

The electric circuit must be complete for current to flow. As long as the circular path remains unbroken, it is referred to as a *closed circuit* and current flows until the battery is discharged. If the circuit is opened anywhere in the circular path, electron flow ceases immediately. With an *open circuit,* no current will flow. We control the opening and closing of a circuit by a *switch* inserted in the circuit, as we saw in Figure A-8. Closing the switch completes the circuit to allow current to flow through the lightbulb, which is the load of this simple circuit. We want electricity to accomplish some useful purpose and attach a device to an electric supply that will use the electricity to perform this purpose. Any device attached to use or convert electricity into some other form of energy is called the *load* of the circuit. Conductors are used to connect the load to the electric source. The conductors in most circuits consist of copper wire with insulation around the wire.

The load in a DC circuit is purely resistive. Remember that inductance and capacitance do not affect DC circuits, so that we will hold off discussing these until Appendix C. When we say the circuit is purely resistive, we mean that the only thing affecting current flow in the circuit is the voltage of the power supply (battery), and the resistance of the load and conductors connecting the load to the battery. As we learned in Appendix A, this resistance or opposition to current flow is measured in ohms. Every device and conductor has resistance to current flow.

B.2 RESISTORS

Every device has resistance that can be duplicated by a device called a *resistor.* Unlike devices designed for using electricity to perform some external function, the resistor merely converts electricity to heat. A resistor is a device used to purposely add additional resistance to a circuit. Its major function is to assist in controlling the amount of current that flows through a device or circuit; it uses up some of the electrical energy and leaves less for the remaining components in the circuit. Look at the circuit of Figure B-1. We have a device that needs to receive 10 V for proper operation. The device has a resistance of 50 Ω. Using Ohm's law ($I = V/R$), we find that the current flow will be 0.2 amp when 10 V is impressed across a resistance of 50 Ω.

If a battery that generates 13.2 V is used as the power supply for the Figure B-1 circuit and no other device is in the circuit, the full 13.2 V will appear across the device and 0.264 amp will flow in the circuit. This is about 33% more current than the device was designed for, and excessive current will destroy it. The excessive current through the device generates excessive heat, which causes it to burn up. We know from Appendix A that the total resistance of the circuit limits the amount of current that flows in it. We can increase the total resistance of the Figure B-1 circuit by adding a resistor to the circuit. But what value of resistor do we need? We must use a multistep approach to answer this question. We will use Ohm's and

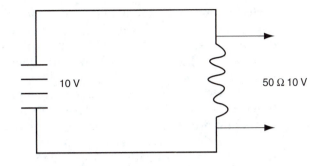

Kirchhoff's laws to determine the value of additional resistance needed. We will then use a resistor specifically designed to provide that amount of resistance.

Let's start our analysis by using Kirchhoff's voltage law. It states that the sum of all voltage drops in an enclosed loop or circuit equals the voltage applied to the circuit. In our case, we applied 13.2 V, and 13.2 V will be dropped across the load. The load must use all of the supplied voltage per Kirchhoff's voltage law. Since we do not want all of the 13.2 V to be used by the device, we must add an additional load to the circuit. We can use a resistor to add the additional load. If the supply voltage is 13.2 V and the device needs only 10 V, the additional load must be designed to use 3.2 V. This will satisfy Kirchhoff's voltage law. The sum of all voltage drops in the circuit will be 13.2 V (10 V across the device and 3.2 V across the resistor that was added).

Now that we know the amount of voltage the resistor must use, we can calculate the value of the resistor by using Ohm's law for resistance ($R = E/I$). We know from Appendix A that the amount of current flowing in a series circuit is the same everywhere in the circuit. We also know that the value of the current in this circuit must be 0.2 amp to provide 10 V to the device. Therefore, the current through the resistor will also be 0.2 amp. The voltage across the resistor must be 3.2 V. Since $R = E/I$, the resistance needed = 3.2/0.2 = 16 Ω. If we add a 16-Ω resistor in series with the device (see Figure B-2), the device will receive 10 V. We can

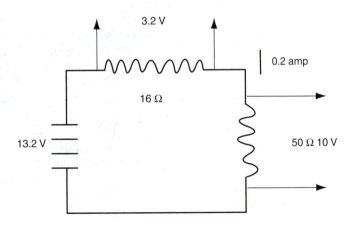

FIGURE B-2 13.2-V supply/
10-V device.

verify this by using Ohm's law for the total circuit. Total voltage dropped in the circuit (Et) equals the total current in the circuit (It) times the total resistance in the circuit (Rt). Thus, $Et = (It) \times (Rt)$. The total resistance in a series circuit is found by adding the value of each resistance. In our circuit, $Rt = 50 + 16 = 66$. In the circuit of Figure B-2; $Et = 0.2 \times 66 = 13.2$ V.

As seen in the previous example, we use resistors in a circuit to add additional resistance in the circuit. Resistors provide a circuit designer with a means of supplying proper voltages to devices in a circuit. We also use resistors as a substitute for devices so that we can study electricity in a laboratory. If we know a device will be attached to a battery supply using 22-gauge copper wire, we can substitute a resistor for the wire when designing the device in a lab. A 22-gauge wire has a diameter of approximately 0.025 in. The resistance of 22-gauge wire is approximately 16 Ω per 1000 ft. Suppose we wished to test a device to make sure it would work properly when it is 500 ft from a 13.2-V battery. The device needs 10 V to operate properly. We plan on using 500 ft of 22-gauge wire to connect the negative terminal of the battery to the device, and 500 ft of wire to connect the device to the positive terminal of the battery. In the lab, we replace the 1000 ft of wire with a 16-Ω resistor to achieve an electrically equivalent circuit. This allows us to test the device in the lab before installing it at the site.

Resistors come in a wide variety of forms. Some are *fixed resistors* and some are *variable resistors.* Some resistors are made using small-diameter wire (called *resistance wire*) and some are made from carbon graphite. Wire-wound resistors are ordinarily used when a large current will flow through them. Carbon graphite resistors are used in low-current applications. Wire-wound resistors are made by winding resistance wire around a porcelain core and attaching the ends of the wire to terminals at the ends of the core. Resistance wire has enamel coating to provide insulation. This insulation allows the wire to be wound many hundreds or thousands of turns on top of each other. The insulation ensures that each turn of wire is insulated from all other turns. Once the desired number of turns have been wound on the porcelain core and the ends of the wire attached to the end terminals, the entire assembly is encased in porcelain or enameled glass. Some wire-wound resistors are constructed with a movable slider that can move along the length of the core while making electrical contact with the wire wound around the core. This is a variable resistor. The resistance between the ends of the resistor will remain constant, but the resistance between the slider and one of the end terminals will depend on where the slider is positioned (see Figure B-3).

Carbon resistors are made of carbon graphite mixed with clay and adhesives. The carbon graphite is much like that found in a lead pencil, and it is mixed with varying amounts of clay to vary the resistance. A heavy-gauge wire is attached at each end of the carbon-clay mixture, and the assembly is encased in ceramic or plastic with a few inches of the heavy-gauge wire protruding from each end of the case (see Figure B-4). By varying the width and length of the mixture and the ratio of carbon to clay in the mixture, it is possible to construct resistors with many different values of resistance.

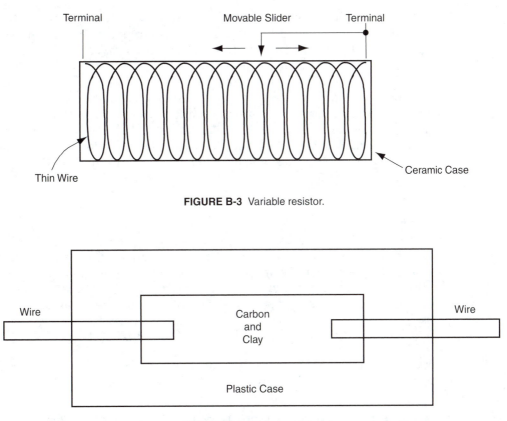

FIGURE B-3 Variable resistor.

FIGURE B-4 Carbon resistor.

B.3 POWER RATINGS FOR DEVICES AND RESISTORS

Any device that uses electricity has a power rating. Power is measured in *watts.* The wattage is found by multiplying the current through a device times the voltage across the device ($W = IE$). Devices that use electricity to perform some external function are designed to reduce the amount of heat they generate. Heat robs devices of power. But in the case of a resistor, we are purposely introducing a power loss to a circuit by having the resistor generate heat. In Figure B-2, the device has 10 V across it and the resistor has 3.2 V across it. Both have 0.2 amp of current flowing through them. We can use the power formula to determine the amount of power used by each:

$$W = IE$$

The power in watts consumed by the device = $0.2 \times 10 = 2$ W

The power in watts consumed by the resistor = $0.2 \times 3.2 = 0.64$ W

Devices and resistors are rated according to how much power they need or can handle. The device of Figure B-2 uses 2 W of power, and the resistor will use 0.64 W. The resistor uses electrical power and converts it into heat. So the resistor used in Figure B-2 must be capable of handling 0.64 W. In this case a resistor rated to handle 1 W would be used. If a 1/4-W resistor was used it would burn up and open the circuit, and the device would cease to function due to the open-circuit condition. The power formula can be rearranged by making substitutions in the formula. We know according to Ohm's law that $E = IR$. If we do not know the voltage across a device or resistor but know the current and resistance, we can substitute IR for E in the power formula. Substituting IR for E results in: $W = I \times I \times R$. This can be restated as:

$$W = I^2R$$

In our example for Figure B-2, the power used by the device = $0.2^2 \times 50 = 2$ W. The power dissipated by the resistor = $0.2^2 \times 16 = 0.64$ W. The formula can also be rearranged if we know the voltage and resistance but do not know the current by substituting for I in the formula. According to Ohm's law, $I = E/R$, and if we substitute E/R for I, we get power in watts = $E \times (E/R)$. This can be restated as:

$$W = E^2/R$$

In our example, the power used by the device = $10^2/50 = 100/50 = 2$ W. For the resistor: $W = 3.2^2/16 = 10.24/16 = 0.64$ W.

B.4 TOLERANCE RATINGS AND COLOR CODES FOR RESISTORS

When resistors are manufactured, they are manufactured to provide a certain resistance and to dissipate a certain amount of heat. The resistance value they provide is stated in ohms, and the heat-handling capability is stated in watts. The manufacturing process for a resistor may provide a resistor that is exactly the resistance we desire, but sometimes it does not. The manufacturer will tell us how closely it has manufactured a resistor to meet its stated value. If the value of a resistor is within 10% of the stated value, we say it has a *tolerance* of 10%. The actual value of a resistor may be 10% more or 10% less than the stated value. For a 1000-Ω resistor manufactured with 10% tolerance, the actual value of any given resistor may fall anywhere between the values of 900 Ω and 1100 Ω. If the resistor were manufactured with a tolerance of 5%, the actual value would be between 950 Ω and 1050 Ω. Most carbon resistors are made with tolerances of 5% or 10%, but they are also available in 1% or 20% tolerances.

The value of a carbon resistor is stated on the plastic case by means of a color code (see Table B-1). The plastic case will have four individual bands of color close to one end of the resistor. The resistor is held so that the bands of color are on the left end. The first band from the left end indicates the first digit of the resistance value. The second band represents the second digit of the resistance value. The third

TABLE B-1 Resistor Color Code

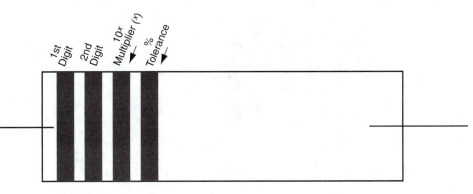

Color of Band

Black	0
Brown	1
Red	2
Orange	3
Yellow	4
Green	5
Blue	6
Violet	7
Gray	8
White	9

If first band is red = 2
If second band is orange = 3
If third band is green = $\times 10^5$

Value of Resistor = 23 x 10^5 = 2,300,000

Many people just use the color of the multiplier band as an indicator of how many zeros to add after the first two digits (but remember that black in the multiplier band means add 0 zeros (or add none), while brown means add 1 zero).

The fourth band indicates the range of values the resistor can have:
Gold = $\pm$ 5% of the stated value
Silver = $\pm$ 10% of the stated value
No band = $\pm$ 20% of the stated value

band represents the power of ten used as a multiplier. The fourth band represents the tolerance range. For a 1000-Ω resistor manufactured in a 10% tolerance environment (see Figure B-5), the color code from left to right should read: brown, black, red, and silver.

The first band is brown, informing us that the first digit is 1. The second band of black informs us that the second digit is 0. The first and second digits taken together equal 10. The third band informs us the power of ten to which the number we have (10 in our example) should be raised. If the third band is red, we should multiply by 10^2 (100). These three bands result in $10 \times 100 = 1000$. The fourth band

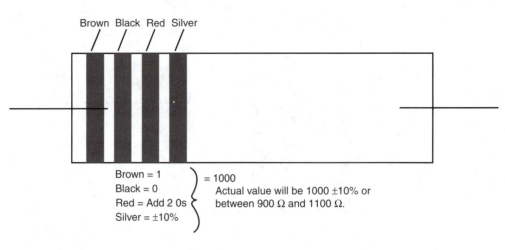

Brown Black Red Silver

Brown = 1
Black = 0   } = 1000
Red = Add 2 0s   } Actual value will be 1000 ±10% or
Silver = ±10%   } between 900 Ω and 1100 Ω.

FIGURE B-5 1000-Ω resistor.

informs us that this resistor will have ±10% tolerance. The actual value of the resistor will be between 900 Ω and 1100 Ω (1000 – 10% of 1000 Ω). Note that the third-column color could be used to inform us of the number of zeros to add to the first two digits. In our example, the color red stands for 2, and we simply add 2 zeros to 10 and get 1000. The only danger in this approach is the need to remember that black means 0; if the third color is black, we add 0 zeros or add no zeros. If the color code is black, brown, black, silver, the resistor is 10 Ω plus or minus 1 Ω. Black = 1, brown = 0; black means do not add any zeros. Black in the third band is 10^0, which is 1. The sequence black, brown, black results in $10 \times 1 = 10$.

B.5 MEASURING RESISTANCE

A test instrument called an ***ohmmeter*** measures resistance. This meter is often part of a multimeter. A ***multimeter*** can serve as a voltmeter, an ammeter, and an ohmmeter. When measuring the resistance of a device or resistor, it is important to make the measurement with no power being supplied to the circuit. Trying to use an ohmmeter to measure resistance with power applied may damage the meter. Using this technique will definitely provide an erroneous reading. The ohmmeter is designed to place a voltage across the resistor being measured and to measure the current that flows. The voltage used by portable meters comes from the 9-V dry cell used to power the meter. This voltage is placed in series with a resistor inside the meter case to limit current flow. Actually, the resistor placed in series with the battery supply can be one of several different resistors. The resistor to use is selected by a switch that allows the user to set the meter to measure a wide range of external resistors. The range selector usually states *X1, X100, X1K, X100K,* and *X1Meg.* The user will note the setting of the selector switch and multiply the meter reading by that factor.

Several ranges are provided to get the most accurate reading possible. The different resistors used for the various selector switch settings ensure that excessive current does not flow through the meter movement and destroy it. To measure a low value of external resistance, a high internal resistor is used. To measure a high external resistance, a low internal resistor is used. The voltage from the battery will cause current to flow when a resistor is attached to the external meter leads. Current will flow through the internal resistor selected, through the meter movement, and through the external resistor and will then return to the battery. The meter reads the amount of current flowing. Maximum current causes the meter to deflect completely to the right.

Maximum current occurs when the external resistance being measured is zero. Thus the right-hand side of the meter movement scale is zero Ω. When the circuit is open, no current flows and external resistance is maximum or in this case infinity. Thus, the left-hand side of the meter movement scale is infinity. The ohmmeter scale for a meter using an analog meter movement (as opposed to a digital readout on LEDs) is not linear. The scale has wider divisions at the right-hand side than at the left-hand side. The meter actually reads current, but the scale of the meter is in ohms. The scale was calibrated by measuring resistors of known value. When the meter movement indicated the amount of current flowing, the face of the scale at this point was labeled with the value of the known calibrating resistor (see Figure B-6).

Now you can see why resistance must be measured with no power supplied to the device being measured. If power is supplied, the device will have current flowing through it. An ohmmeter attached across the device will simply become a parallel load to the device. Ohmmeters have a low resistance and will provide a low-resistance parallel load that absorbs most of the current in the circuit. This current may be high enough to burn up the meter movement. In any event, the meter is indicating current flowing due to the power supplied to the circuit and not current flowing due to the battery supply of the meter. This does not allow us to measure the value of the resistor (see Figure B-7).

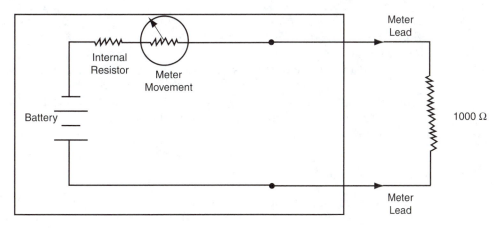

FIGURE B-6 Ohmmeter.

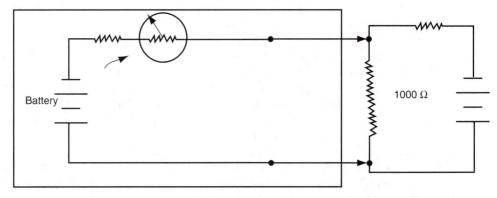

FIGURE B-7 Current flow in an active circuit will distort measurement and usually results in damage to the meter movement. Never measure resistance in a live circuit. Ohmmeter improperly connected.

B.6 MEASURING VOLTAGE

A test instrument called a *voltmeter* measures voltage. The meter movement used in analog meters will actually measure current flowing through the meter movement. Voltmeters measure voltage indirectly, much as ohmmeters measure resistance indirectly. The voltmeter leads are also attached the way the ohmmeter is attached to a device. Both meters are attached across the device or parallel to the device. The difference is that a voltmeter is attached across a device with power supplied to the device by the power supply of the circuit. The external circuit also powers the meter movement. When a multimeter is set to read voltage, an internal resistor will be placed in series with one test lead but no internal battery connected to the circuit. The source of power for the meter must be from the device being measured.

The meter will be in parallel with a device in the circuit. The meter movement itself has a very low resistance, but the internal resistor connected in series with it has a very high resistance. The meter movement is designed such that when 1 milliamp of current flows through it, the meter deflects the pointer of the meter to the extreme right-hand side of the scale. To prevent damage to the meter movement, the user should set the meter to read a high voltage and then switch to read lower voltages, until a reading occurs. A selector switch allows setting the meter to read high and low voltages by selecting low- or high-value resistors to be in series with the meter movement and meter lead. A voltmeter is calibrated by measuring a known voltage and marking this voltage on the scale at the point indicated by the meter pointer (see Figure B-8).

B.7 MEASURING CURRENT

As noted earlier, the ohmmeter and voltmeter are really ammeters in disguise. They both actually read current, but the faces of their scales are marked in ohms or volts respectively. They are both attached parallel to the device being measured. An *ammeter* attaches to a circuit by opening the circuit and then inserting an ammeter in the path

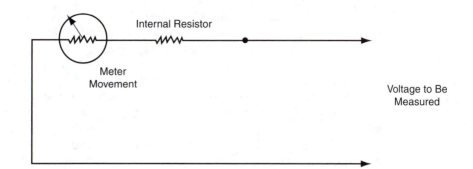

FIGURE B-8 Voltmeter.

to reestablish the circuit. An ammeter is placed in series with the circuit to measure current flowing through it. Although the meter leads are placed in series with the circuit, the meter movement itself is not. An ammeter has a very low resistance wire (or copper bus) attached internally between the terminals of the meter that external leads are connected to. This bus has a resistance of about 1/1000 of an ohm.

The meter movement has an internal resistor connected in series with it, and both of these are parallel to the low-resistance bus. The value of this resistor in series with the meter movement will depend on the setting of the selector switch. Because the meter movement and internal resistor are in parallel with the low-resistance copper bus, most circuit current will flow over the bus. A minuscule amount of current will flow through the meter movement. Meter movements are constructed such that when 1 milliamp of current flows through them, it causes a full-scale defection of the pointer to the right-hand side of the scale. The ammeter scale is calibrated by measuring currents of a known value in a lab environment (see Figure B-9).

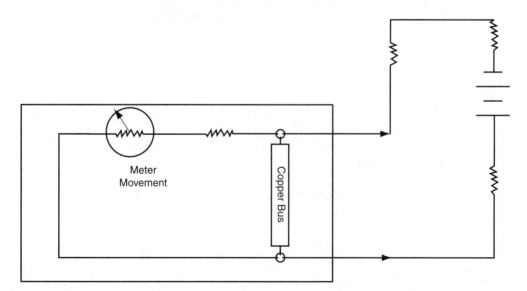

FIGURE B-9 Ammeter connected into circuit. A small amount of current flows through the meter movement.

Electric Circuits

The use of different values of internal resistors in series with the meter movement allows the user to vary the sensitivity of the meter by turning the selector switch. The user first selects the highest ammeter setting. This inserts the highest-value internal resistor in series with the meter movement. The deflection of the meter's needle takes a high current to make it deflect to the extreme right-hand position. Very little deflection occurs for a low current. If the needle of the meter only deflects slightly, the user increases the sensitivity of the meter by selecting a lower value of internal resistor. This increases current flowing through the meter movement, and the meter will be more sensitive in reading the external current. It now takes less current in the external circuit to cause a full deflection of the meter's needle.

The user continues to switch the selector to read lower currents until the meter can display the reading somewhere near the middle of the scale. This provides the most accurate reading for the current being measured. The use of a copper bus between the meter leads ensures that the meter itself has little effect on the circuit. If the meter had a high resistance and was placed in series with the circuit, it would impact the amount of current flowing. Adding additional resistance to the circuit would cause the amount of current in the circuit to decrease. We do not want the meter to affect the circuit being measured. Therefore, it must rob as little power as possible from the circuit in doing its job.

B.8 RESISTANCE VS. CONDUCTANCE

Resistance is measured in ohms. Every device, material, or element has a certain resistance or opposition to electric current flow. Materials with a very high resistance are called *insulators;* we use these materials to contain electron flow within a desired path. Materials with a very low resistance are called *conductors;* we use these materials to transport electrons from a power source to a device requiring electrical power. The measure of how good a conductor is at carrying current is **conductance.** The symbol for conductance is *G* and its unit of measure is the siemens (*S*). Conductance is the reciprocal of resistance, where resistance is stated in ohms:

$$G = 1/R$$

Every conductor has a certain amount of resistance. If we look at several elements that are all 1 ft long and have a cross-sectional area of 1 circular millimeter and compare their resistances or resistivity, we will find that the resistivity of various elements is as follows:

Element	Resistivity per Circular Mil/Foot in Ohms
Silver	9.9
Copper	10.7
Aluminum	17.0
Zinc	37.4

Brass	42.0
Nickel	47.0
Iron	70.0

If we know the resistivity of a material, we can calculate the resistance for any length and diameter of the material by using the formula $R = (pl)/a$ where resistance in ohms equals resistivity (p) times length (l) divided by area in circular mils. Note that area in *circular mils* or *cmils* is not the same as the algebraic area of a circle. The algebraic area of a circle is found by the formula πr^2, but circular mil is diameter squared (d^2). This is the same formula used to find the area of a square. The designation circular mils informs us that the circular area is being treated as a square area. It does not make any difference to us how the cross-sectional area of a conductor is determined because we are only using the measures for comparison purposes.

The use of circular mils greatly simplifies comparisons between conductors of different diameters. Suppose you take a circle that has a radius of 1 cm versus a circle that has a radius of 8 cm. Using the formula for area (πr^2), the areas are 3.14 cm^2 and 200.96 cm^2 respectively. The value 200.96 is 64 times greater than 3.14. A circle with a radius of 8 cm has 64 times the area of a circle with 1 cm. Now look at a square. A square with a length of 1 cm has an area of 1 cm^2. A square with a length of 8 cm has an area of 64 cm^2 or 64 times the area of a 1-cm^2 square. Thus, once the resistivity of an element has been defined in terms of a circular square area for a diameter of 1 mil, we can simply refer to that area as a circular mil to make comparisons to larger-diameter conductors simpler. In electricity the area of 1 circular mil is found by using the algebraic formula πr^2. More specifically, r is 0.5 mm and area $= \pi \times 0.5^2 = 3.1415 \times 0.25 = 3.1415/4 = 0.7854$ mm^2. This area is established as the area of a circular mil having a diameter of 1 mm, and the circular mil area is d^2 or 1 mm^2.

Now that we have defined a circular mil and its uses to compare areas, you should understand that a conductor four times as large in diameter as another conductor of the same material should have one-fourth the resistance of the smaller conductor. This is the same as stating the larger conductor is actually composed of four smaller conductors in parallel. If the resistance of a 1-cmil conductor 1 ft long is 9.9 Ω, then four of these connected in parallel provide a total resistance 2.48 Ω. Using the parallel resistance formula, $1/Rt = 1/R_1 + 1/R_2 + 1/R_3 + 1/R_4$ or $1/Rt = 1/9.9 + 1/9.9 + 1/9.9 + 1/9.9 = 4/9.9$. So if $1/Rt = 4/9.9$, then $Rt/1 = 9.9/4$ or $Rt = 2.48$ Ω. It stands to reason that if 1 circular mil of a material has a stated resistance of 9.9 Ω per foot, then a 10-ft long section has a resistance of ten 1-ft sections connected in series, or a total resistance of 99 Ω. The formula used to find the resistance of a given conductor is $R = pl/$circular mils.

Copper is used extensively as a conductor for electricity because of its low cost, high availability, and low resistivity. Table B-2 lists the resistance for various sizes of copper wire. The *American Wire Gauge (AWG)* lists the sizes of wire from four ought (0000) to 40. The larger the number of the gauge, the smaller the wire and the higher the resistance of the wire. If the gauge of a wire changes by 3, its cross-sectional area (and cmil) changes by a factor of 2. Thus, going from a 16-gauge

TABLE B-2 The American Wire Guage Table for Copper Wire

AWG	Diameter (mils)	Area (cmils)	Ω/1000 ft	Max. Amps
0000	460.0		0.0490	230
000	409.5		0.0618	200
00	364.8		0.780	175
0	324.9		0.0983	150
1	289.3		0.1240	130
2	257.6		0.1563	115
3	229.4		0.1970	100
4	204.3		0.2485	85
5	181.9		0.3133	75
6	162.0		0.3951	65
7	144.3		0.4982	55
8	128.5		0.6282	45
9	114.4		0.7921	40
10	101.9		0.9981	30
11	90.74		1.260	25
12	80.81	1.588	20	
13	71.96		2.003	17
14	64.08		2.525	15
15	57.07	3.184	10	
16	50.82		4.106	6
17	45.26		5.064	4
18	40.30		6.385	3
19	35.89		8.051	
20	31.96		10.15	
22	25.35		16.14	
24				
26	15.94		40.81	
30	10.03		103.21	
40	3.145		149.0	

wire to a 19-gauge wire (up by 3 in gauge size) cuts the cross-sectional area in half, and the 19-gauge wire has twice the resistance of a 16-gauge wire.

B.9 RESISTANCE OF WIRES CONNECTING TO A TELEPHONE

By referring to Table B-2, you can see that a 19-gauge wire 1000 ft long has a resistance of 8.051 Ω and a 16-gauge wire of the same length has a resistance of 4.016 Ω. Thus, a difference in gauge size of 3 results in a resistance change by a factor of 2. In telecommunications, the gauges of wire used are 26, 24, 22, and 19 gauge, with 26 gauge predominating. The cables that connect our telephone to the local central office switching system must not contain more than 1800 Ω of resistance when the local switching system is a computer-controlled switch. With older electromechanical switching systems, the resistance of wires connecting a telephone to a switch was limited to 1200 Ω.

Telephones introduced within the last ten years require 0.02 amp of current flow through the phone to provide adequate power for the transmitter and the touchtone dialing pad. Computer-controlled switching systems use a DC power supply of 52 V. All telephones attach to switching systems using one pair of wires. At the switching system this pair of wires connects to a device called a *line circuit*. The job of the line circuit is to interface a telephone to a switching system. The switching system connects the negative terminal of a 52-V battery through the line circuit to one of the wires connected to a telephone. This wire is called the *Ring*. The positive terminal of the 52-V battery is connected through the line circuit to the second wire connected to a telephone. This wire is called the *Tip*. The line circuit has a resistance of 200 Ω between the –52 V DC and one wire going to the telephone. The line circuit also has a resistance of 200 Ω between the positive terminal of the battery and the other wire going to the telephone (see Figure B-10). The positive terminal of the 52-V battery is connected to earth ground. Thus, the positive terminal is grounded and the negative terminal is –52 V with respect to ground.

Since a telephone needs at least 20 milliamps of current flowing through it to operate properly, the resistance of the wires attaching a telephone to a central switch must be selected to ensure this criterion is met. Notice in the circuit of Figure B-10 that the resistance of each wire is 900 Ω. The resistance of the telephone is 400 Ω and the resistance for each side of the line circuit is 200 Ω, for each connection to the wire pair. Ohm's law dictates that with 52 V applied to the circuit, 20 milliamps of current will flow through the telephone (and everywhere in the series circuit). $I = E/R = 52/(200 + 900 + 400 + 900 + 200) = 52/2600 = 0.02$ amp.

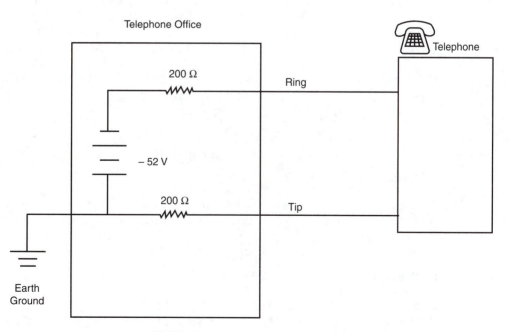

FIGURE B-10 Battery supply for telephone.

Now that we know 1800 Ω is the maximum resistance allowed in wires connecting a telephone to a local switch, we can select the gauge of wire needed to ensure that this value is not exceeded. We can look in Table B-2 and see that the resistance per 1000 ft of 26-gauge wire is 40.81 Ω. We need two wires to connect a telephone to the local switch. Electrical current will flow out over one wire from the switch to the telephone and will return to the switch over the second wire. Thus, the Ring wire, the telephone, and the Tip wire are electrically in series. The resistance for 2000 ft of wire is 81.62 Ω. We refer to the two wires as a loop because they form a loop in connecting a telephone to a central switch. The resistance per 1000-ft loop using 26-gauge wire is 81.62 Ω. If we divide 1800 by 81.62, we will find that the maximum allowable loop using 26-gauge wire is 22,000 ft or a little more than 4 mi. If the telephone is located more than 4 mi from the central switching system, a larger-diameter wire will be needed. The gauge of wire to use in the loop will be governed by the length of the loop needed to connect a telephone to a switch (the distance from the switch to the telephone).

Since we know that resistors can be substituted for actual components in a laboratory environment to perform tests or measurements, let's do that for a telephone. We stated earlier that telephones have a resistance of 400 Ω. This is referred to as the *nominal resistance value* of the telephone. Its actual resistance varies depending on the voltage applied to it. The resistance of the telephone cannot be measured accurately with an ohmmeter because the ohmmeter does not place a high voltage on the telephone. The ohmmeter will provide us with the resistance of a telephone under very low voltage conditions. We can calculate the resistance of a telephone in a live circuit indirectly by measuring voltage and current in a live circuit and using Ohm's law to find resistance. In the electrical circuit of Figure B-10, we are applying 52 V to the circuit and 0.020 amp flows through the circuit. The Ohm's law formula for voltage dropped across a device is $E = IR$. We can use this formula to calculate the various voltages dropped across each component of the circuit:

1. For one-half of the line relay in the central switch: $E = 0.02 \times 200 = 4$ V.
2. For one wire connecting the telephone to the switch: $E = 0.02 \times 900 = 18$ V.
3. For the telephone: $E = 0.02 \times 400 = 8$ V.
4. For the return wire telephone to switch: $E = 0.02 \times 900 = 18$ V.
5. For one-half of the line relay in the central switch: $E = 0.02 \times 200 = 4$ V.

Total of all voltage drops: 52 V.

Kirchhoff's voltage law is satisfied. The sum of the voltages dropped in the circuit equals the sum of the voltage applied to the circuit. We can also see that the voltage dropped across the telephone in this circuit is 8 V.

B.10 RESISTANCE OF A TELEPHONE VARIES

Now let's test for the resistance of a telephone using an 8-V power supply in the lab. Attach an ammeter in series with an 8-V supply and a telephone and then read the amount of current flowing through the telephone. Raise the voltage of the power supply to 20 V and take note of the amount of current flowing. We know that we supplied 8V to 20V to the telephone, and we can use the amount of current flowing and Ohm's law to calculate the resistance of the telephone: $R = E/I$. In most telephones, the resistance varies. This allows it to compensate for the length of the loop connecting the telephone to the switch.

We do not want less than 20 milliamps flowing in the telephone, but we also do not want more than 60 milliamps flowing in this circuit. The resistance of the telephone increases on shorter loops to limit current flow below 60 milliamps. Shorter loops will have less voltage loss across the wires connecting a telephone to switch, which results in a higher voltage at the telephone. For higher voltages at the telephone, we want to increase the resistance of the phone. The device in the telephone that does this is called a *varistor,* which is a variable resistor. Its resistance varies directly with the applied voltage. The varistor is in series with the main components of a telephone. When voltage increases, the resistance of the varistor increases. Thus, the varistor absorbs increases in voltage and keeps the voltage across the main working components of the telephone at a constant level.

B.11 POWER IN A CIRCUIT

Each device in a circuit uses power. Earlier, we stated that power in watts equals current in amperes times voltage applied ($W = IE$). Power is also the number of joules used per second ($W = j/s$). Recall from Appendix A that a volt is defined as the potential difference necessary to obtain 1 joule of work when 1 coulomb of charge flows. A joule is the amount of heat energy produced by a 1-Ω resistor when a current of 1 amp flows through it for 1 sec.

$$V = W/Q$$

Voltage = energy (W) in joules divided by charge in coulombs (Q). By rearranging the formula, we get $W = VQ$. A flow of 1 coulomb per second is 1 amp. Thus, $W = VI$ (W will be in joules per second when Q is in coulombs per second). Since 1 W is equal to 1 joule per second, W also can be stated as watts. The amount of power used by a device depends on the amount of current flowing through the device and the voltage across the device. In Appendix A, we found that the voltage dropped across a device equals the current through the device times the resistance of the device ($V = IR$). Thus, the higher the resistance of the device, the higher the voltage developed across it. Since $W = EI$, if I remains constant the higher the voltage is the higher W will be. Thus, if I remains constant, the higher the resistance of the device the higher E will be and the higher W will be.

Of course, if a higher-value resistor is used in a circuit with a specific non-changing voltage, the current in a circuit drops and tends to offset the change in resistance. For example, if a 10-V supply is placed across a 10-Ω resistor we get a current flow of 1 amp and the resistor will use 10 W of power. If the resistor is replaced by a 20-Ω resistor, we get 0.5 amp of current and use 5 W of power. Using higher-value resistors will result in less current flow and less wattage across the resistor. *But* this is not how electronic circuits are designed. These circuits are designed based on how much power a device needs. The devices are rated for a specific voltage and current value. Resistors are then placed in a circuit to use up some of the power supplied to the circuit in order to ensure that the proper voltage and current are delivered to the active device.

Suppose I have a device that requires a supply of 5 V and a current of 1 amp to operate properly. If I use a 5-V supply, I can simply attach the device directly across the supply, but what if I use a 10-V supply? I must calculate the value of resistor needed in series with the device to limit current flow to 1 amp and the voltage across the device to 5 V. We would find by Ohm's law that the value of resistor needed is 5 Ω. We would also calculate that 1 amp through a 5-Ω resistor requires that the resistor be a 5-W resistor ($W = I^2R$). The voltage supplied to every light in our house is 120 V. A 60-W lightbulb requires a current flowing through it of 0.5 amp: $W = IE$; therefore; $I = W/E = 60/120 = 0.5$ amp. A 120-W lightbulb would require 1 amp of current flowing through it. This means that the resistance of a 120-W lightbulb is half the resistance of a 60-W lightbulb. Again, it is the difference in resistance of various devices that determines the amount of power or energy they use. In a circuit, as each device uses energy, it causes the voltage in the circuit to drop and less voltage appears at the next component in a series circuit.

B.12 A VOLTAGE DIVIDER

Resistors are often used in an electrical circuit to limit the amount of voltage and current supplied to other parts of the circuit. A voltage divider is a circuit designed to provide a certain voltage to other parts of the circuit. Figure B-11 is a diagram for

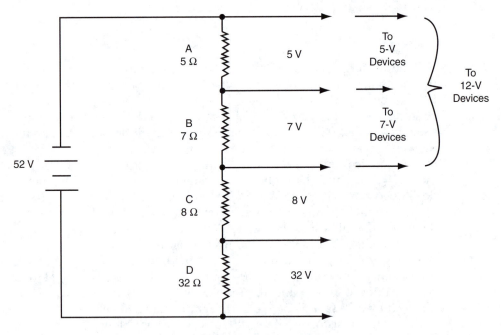

FIGURE B-11 Voltage divider.

a voltage divider. Suppose we have a source voltage of 52 V DC from a central of-fice battery and need to use that source to power an electronic circuit requiring +5 V DC and +12 V DC. We can use leads A and B of the voltage divider to provide these voltages. The voltage divider has a total resistance of 52 Ω and will have a current of 1 amp flowing through all resistors of the voltage divider. This will cause voltage drops of 5 V across resistor A, 7 V across resistor B, 8 V across resistor C, and 32 V across resistor D. The devices attached to points A, B, and C receive voltage from the divider but also form a parallel path with part of the divider. For that reason, the resistance of devices attached to a voltage divider must have a resistance that is at least ten times greater than the resistance of the voltage divider component. If this rule is not followed, the voltage developed across the voltage divider compo-nent will fall below its desired value.

Figure B-12 shows a voltage divider with high-resistance devices attached to the divider. These devices require small amounts of current, and the voltages de-veloped across the voltage divider remain stable and fairly constant. Figure B-13 indicates what happens if you try to provide voltage to a low-resistance device via a voltage divider. The device that receives power from a voltage divider is actually in parallel with the resistance of the voltage divider and forms a parallel circuit to the resistor of the voltage divider. Note what happens if a low-resistance device is powered from the 20-V tap on the voltage divider. With a device that has a resis-tance of 4 Ω connected to the 20-V tap on the voltage divider, the low resistance of the device will draw most current from the voltage divider and the voltage at the tap will be lowered to 4.9 V. This results in a current flow of 0.245 amp flowing through the voltage divider resistors A, B, and C. With this current flowing

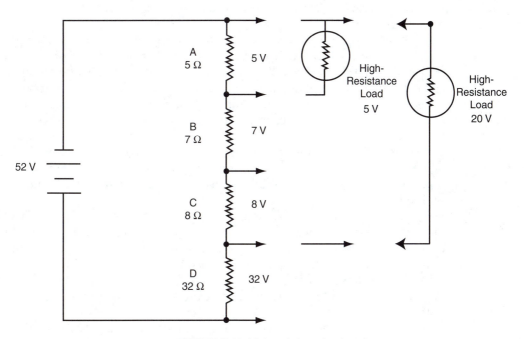

FIGURE B-12 High-resistance load.

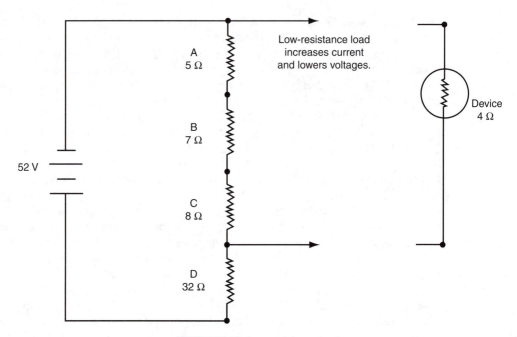

FIGURE B-13 Low-resistance load.

through resistors A and B, the voltage across A and B will be 1.2 V and 1.7 V respectively. A low-resistance load placed on a voltage divider lowers the voltages of a voltage divider.

To maintain the voltages desired across each resistor of a voltage divider, the resistors must have a much lower resistance than the total resistance of devices powered from the divider. All the devices attached at a particular tap are connected such that they are in parallel. The designer of a circuit ensures that only a few devices are powered from a particular tap. For the reasons discussed above, meters and other measuring devices must have a very high resistance between the point at which a measurement is taken and ground. This is done to prevent the measuring device from offering a low-resistance parallel path to circuit current. The high resistance of a meter is referred to as a *high impedance.* This high impedance prevents measuring devices from affecting a working circuit when we try to measure signals within the circuit.

You can see that the voltage developed across each resistor of a voltage divider is proportional to the value of the resistor as compared to the total resistance of the voltage divider. If we have a 100-volt supply impressed across four 25-Ω resistors, then 25 V is developed across each of the resistors. This can be stated in a formula as follows: V across resistor x = (value of resistor x divided by total resistance) times value of the voltage supplied to all resistors. In the example above, the voltage across $x = (25/100) \times 100 = 1/4$ of $100 = 25$ V. We can find the voltage at point A with respect to ground by using this formula: voltage across A = $(5/52) \times 52 = 5$ V.

This will be the voltage across A before adding a load to the voltage divider of Figure B-11.

The resistors used for a voltage divided must be high-wattage resistors. As we have seen in Figure B-11, we get very high current values through the resistors. This results in the need to dissipate high values of heat. The wattage of resistor $A = E^2/R = 5^2/5 = 25/5 = 5$ W. If less than a 5-W resistor is used, the resistor will burn open and the power supply voltage divider ceases to function. The 32-Ω resistor used for D must be $32^2/32 = 32$ W resistor.

B.13 KILOWATT-HOUR

The electric power company installs a meter on the side of our house to measure how much electrical energy (watts) we use. The meter measures the amount of electric current used. The power company uses three electrical leads to our house. One is a ground lead; the other two provide 120 V with respect to ground and 240 V with respect to each other. The meter measures the current provided by each 120-V lead. The meter provides readings in the number of kilowatt-hours used. Thus, if a 100-W lightbulb is left on for 10 hr, the meter will count that as 1 kilowatt-hour. If an electric furnace is running that has two 20-kW (kilowatt) heating elements, that system will use 40 kW per hour and the meter will count 40 kilowatt-hours for 1 hr of use at 40 kW. Electrical rates vary between companies, and most companies also use a graduated rate schedule. Initial kilowatt-hours are more expensive than kilowatt-hours used after the first 500 kilowatt-hours. Since the voltage being supplied to our house remains constant at 120 V on each hot lead, the kilowatt-hour meter is really measuring the amount of current we use, and we pay for the current used in the form of so much per kilowatt-hour.

B.14 DECIBELS

One watt of electricity is equal to one ampere of electricity flowing through one ohm of resistance. If the current is doubled, the power goes up by a factor of 4. The formula is as follows: power (in watts) = I^2R. Power can also be restated in the following way: $P = IE$. Remember that according to Ohm's law, $R = E/I$. If we substitute in the original formula ($P = I^2R$) for R, we get $P = I^2E/I$. This becomes $P = EI$. One watt = $1V \times 1$ amp. The current will be doubled through a stated resistance if the voltage is doubled. Thus, doubling the voltage doubles the current and $P = 2 \times 2 = 4$ W. The current was doubled and the power went up by a factor of 4. Power can also be stated as follows: $P = E^2/R$. In this case we use Ohm's law ($I = E/R$) to substitute for I in the equation ($P = EI$) to get $P = E(E/R) = E^2/R$. In telecommunications, we deal with extremely low power levels such as one-thousandth of a watt (1 mW). The milliwatt is the unit of measure for power in telecommunications circuits.

In telecommunications, we are often concerned with the comparison of one power level to another. The unit of measure used to compare two power levels is the *decibel (dB).* A decibel is not an absolute measure; it is relative. The decibel level indicates the relationship of one power level to another. The relevant formula for a decibel is: dB = 10 log P_2/P_1. If we have an output signal of 2 mW at an amplifier and the input to the amplifier is 1 mW, the relative power of output to input will tell us the gain of the amplifier in decibels:

dB = 10 log (power ratio)

dB = 10 log (output power / input power)

dB = 10 log (2mW/1mW)

dB = 10 log of 2

The log of 2 is 0.30103; therefore,

dB = 10 × 0.3 = 3

Suppose the output of an amplifier is 2000 W and the input is 1000 W. You can see that this provides a ratio of 2 and the gain of the amplifier is also 3 dB. The first amplifier increased the power by 1 mW, and the second amplifier had to increase the power by 1000 W to result in a 3-dB gain. That is why decibels are relative measures and not absolutes. A 3-dB gain in one circuit is quite different from a 3-dB gain in another, but they are similar in one respect: when comparing two power levels, if one power is twice another, the difference between the two is 3 dB. No matter what power levels we deal with, if the power is doubled, we have a 3-dB gain. Also notice what happens when we reverse our comparison. If we compare the ratio of input to output, the ratio is 1/2 or 0.5. The log of 0.5 is –0.30103.

dB = 10 log (power input/power output)

dB = 10 log (1/2)

dB = 10 × –0.30103

dB = –3

When the power is cut in half, it is –3 dB. When the power is doubled, it is +3 dB (see Figure B-14). Let's check a few more power levels. If the power ratio is 10, the log of 10 is 1 and we find that for any power change by a factor of 10, it is a 10-dB change. For example, a change in power level from 1 to 10 mW is a gain of 10 dB. A change from 10 W to 100 W is also a gain of 10 dB. A change of 10 dB means the power level changed by a factor of 10.

dB = 10 log (P_2/P_1)

dB = 10 log (10/1)

dB = 10 log 10

dB = 10 × 1

dB = 10

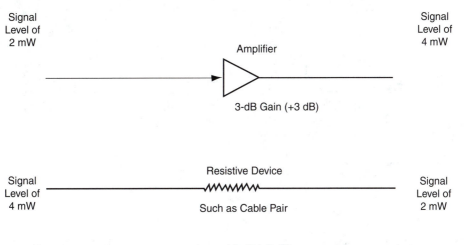

FIGURE B-14 Power gain or loss is stated in decibels.

If the power changes by a factor of 100, it is a 20-dB change, and a power ratio of 1000 is a 30-dB change, because the log of 100 is 2 and the log of 1000 is 3. Thus, a comparison of 1 W (1000 mW) to 10 mW, and 1 W to 1 mW, results in the following:

1 W to 10 mW	1 W to 1 mW
dB = 10 log (P_2/P_1)	dB = 10 log (P_2/P_1)
dB = 10 log (1000/10)	dB = 10 log (1000/1)
dB = 10 log 100	dB = 10 log 1000
dB = 10 × 2	dB = 10 × 3
dB = 20	dB = 30

Similarly, if the power ratios are turned around and we state that one power is a tenth of another power, the log of 1/10 (0.1) is –1 and 10 times –1 is –10. Thus, we state this as –10 dB. When one power is 1/100 of another, it is –20 dB, and the 1/1000 power ratio is –30 dB. We usually state overall circuit loss in a circuit in one of two ways. In terms of gain, we can say that the gain of the circuit is –10 dB and so on, or we can state it as a loss, indicating that the loss is 10 dB and so forth. The use of decibels is a convenient method of determining overall circuit gains and losses, because decibels are added together (Figure B-15).

In Figure B-15, the total gain or loss for the circuit is found by adding the decibel gain and loss of each section of the circuit: (–5, +3, –4, +4, –6, +3, –5) = –10-dB overall circuit loss. What does this mean? The overall loss of –10 dB means we receive 1/10 of the transmitted power level. The power level of 0 dBm is 1 mW.

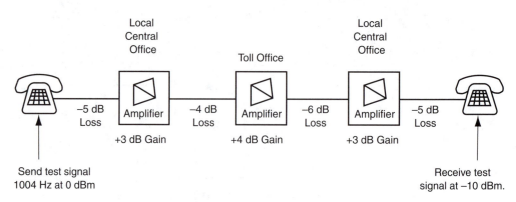

FIGURE B-15 Use of dB and dBm.

Therefore, the received power level is 0.1 mW. What if the overall loss had been –13 dB? We can come close to calculating the total power loss in our head by calculating the losses independently. We know that –10 dB = 1/10 of the original power and a –3-dB loss would be 1/2 the power it received. So if the overall loss was –13 dB, we would have 1/2 of 1/10, or 1/20 ($0.5 \times 0.1 = 1/20$). Thus, we would have approximately 1/20 of the original power from a transmitter available at the receiver, if the overall loss was 13 dB. For –10 dB, we have 1/10 of the original power. For –13 dB, we have 1/20 of the original power.

If we wish to provide a precise calculation, we can do so by rearranging the dB formula to find the power ratio:

$$dB = 10 \log (P_2/P_1) \text{ Therefore:}$$
$$dB/10 = \log (P_2/P_1)$$
$$\text{Antilog } (dB/10) = \text{antilog } (\log (P_2/P_1)$$
$$\text{Antilog } (dB/10) = P_2/P_1$$

Here is the formula to convert decibels to the power ratio:

$$\text{Antilog } (dB / 10) = P_2/P_1 \text{ (the power ratio)}$$

Substituting the values from Figure B-15 (that is, we want to find the power ratio of –10 dB), we arrive at the following:

$$\text{Antilog } (-10/10) = P_2/P_1$$
$$\text{Antilog } -1 = P_2/P_1$$

The antilog of –1 = 0.1; therefore, 0.1 = the ratio of P_2/P_1 or P_2 is 1/10 the power of P_1.

We can also find the overall power loss of a 13-dB loss (–13-dB gain) circuit. Since –10 dB = 1/10 (0.1), and –3 dB = 0.5, –13 dB = 0.1×0.5 or approximately 0.05, which is 1/20 (1/20 = 0.05). Thus, we can state that the –13-dB circuit results in an overall power loss such that 1/20 of the power at the transmitter reaches the receiver. I think this exercise shows you the advantage of using decibels. It is much easier to add decibels than to try calculating how much power is lost over each part of the circuit.

B.15 DECIBELS REFERENCED TO 1 mW (dB)

In Figure B-15, a test signal generator is attached to one end of the circuit and sends a test tone at 0 dBm. Where did the *m* in *dBm* come from? *dBm* means a comparison of the measured power level to a 1-mW power level. By comparing every signal to a power level of 1 mW, we achieve an absolute rather than a relative measure. In actuality, we are comparing all signals relative to one signal level, and that one signal level is 1 mW. The use of one level to compare all other signals against results in an absolute measure. When signals are measured, they are measured relative to a 1-mW signal and are stated as decibels referenced to 1 mW (dBm). But dBm is only used to state a measured signal or a transmitted signal from a test set. The gain (or loss) of each circuit path is stated in dB, not dBm.

Let's return to Figure B-15. If we were to measure the test signal where it enters the first central office, the measurement would be –5 dBm. The signal is sent by the test tone generator at 0 dBm. It travels over wire that has a loss of 5 dB. Thus, it will measure 5 dB down in level from the transmitted level of 0 dBm and will be measured as a –5-dBm signal level. What is the power level of a 0-dBm test tone? We can use our formula for decibels to determine the power level of 0 dBm.

$$0 \text{ dBm} = 10 \log (P_2/1 \text{ mW})$$

P_1 is stated as 1 mW because the reference signal (P_1) for dBm is 1 mW. Remember that the formula used to find our power ratio is the following: Antilog of dBm/10 = power ratio. If dBm/10 = 0, then the antilog of 0 equals the power ratio. The antilog of 0 is 1. Therefore, $P_2/P_1 = 1$. In dBm, P_1 is 1 mW. For P_2/P_1 to be 1, P_2 must also be 1 mW. Thus, 0 dBm = 1 mW. And it can be stated as:

$$\text{Power level of 0 dBm} = 1\text{mW}$$

If we measure all power levels relative to 1 mW, and a doubling of the power = 3 dB, then a measured power level of 3 dBm will be double the power at 0 dBm. Since the power of 0 dBm is 1 mW, the power at 3 dBm must be 2 mW. Another doubling of power would result in a reading of 6 dBm and would be 4 mW. An 8-mW power level would give a reading of 9 dBm. Let's double-check this:

$$\text{dBm} = 10 \log (P_2/1 \text{ mW})$$
$$\text{dBm} = 10 \log (8 \text{ mW} /1 \text{ mW})$$
$$\text{dBm} = 10 \log 8$$

The log of 8 is 0.90309; therefore,

$$\text{dBm} = 10 \times 0.90309$$
$$\text{dBm} = 9.0309$$

Now let's look at the overall circuit loss of 13 dB. If we transmit at 0 dBm and have a circuit loss of 13 dB, the signal will measure as a –13-dBm signal. What is the actual received power? Earlier, we stated that it would be about 1/20 of the transmitted level. If the transmitted level is 1 mW (0 dBm), the received level is about 1/20 of a milliwatt. We can double-check this assumption by using the following formula:

$$-13 \text{ dBm} = 10 \log (P_2/1 \text{ mW})$$
$$-13/10 = \log (P_2/1 \text{ mW})$$
$$-1.3 = \log (P_2/1 \text{ mW})$$
$$\text{Antilog } -1.3 = \text{antilog } (\log (P_2/1 \text{ mW}))$$
$$\text{Antilog } -1.3 = P_2/1 \text{ mW}$$
$$0.0501187 = P_2/1 \text{ mW}$$
$$1 \text{ mW} \times 0.0501187 = P_2$$
$$0.0501187 \text{ mW} = P_2 = \text{the power level at the receiver}$$

This is approximately 1/20 of a milliwatt and agrees with our earlier findings for Figure B-15.

When using decibels to describe the relationship in sound volume changes, a decibel is a change in power that the human ear can barely detect. A change in power level that sounds twice as loud is a change of 3 dB. If one sound is 10 times as loud as another, it is 10 dB higher, and so on. The starting point for measuring sound is one thousand-trillionth of a watt (0.000000000000001 W) or 0.001 pW. A 1000-W speaker produces a sound power of 1 mW (0.001 W or 1/1000 of a watt). Anything louder than this will probably break your eardrum. Levels between 100 and 1000 W will cause some damage to the eardrum. In telecommunications, we did not choose to use 0.001 pW as a starting reference point because most power levels used in telecommunications are in milliwatts. This is the reason 1 mW was chosen as the starting reference point. Sound decibels and electrical decibels do not have direct correlation because of different starting reference points. The ratios do have the same meaning in either case. A doubling of power is 3 dB, and so forth.

B.16 NOISE MEASUREMENTS

Decibels referenced to 1 mW provides a convenient measure for indicating the signal levels in telecommunications because most signal levels in the telecommunications system fall between +10 and –10 dBm. The level of noise present on a circuit should be well below the level of electrical voice signals. As a convenient starting reference point for noise measurements, –90 dBm was chosen. The power level of –90 dBm equals 1 picowatt (pW) of power. Let's double-check this with our formula for finding power levels referenced to 1 mW:

$$\text{dBm} = 10 \log (P_2/P_1)$$
$$\text{dBm}/10 = \log (P_2/P_1)$$

$$\text{Antilog (dBm/10)} = P_2/P_1$$
$$\text{Antilog (dBm/10)} \times P_1 = P_2$$

Since $P_1 = 1\text{mW}$,

$$\text{Antilog (dBm/10)} \times 1 \text{ mW} = P_2$$
$$\text{Antilog } (-90/10) \times 1 \text{ mW} = P_2$$
$$P_2 \text{ in milliwatts} = \text{antilog } (-90/10)$$
$$P_2 \text{ in milliwatts} = \text{antilog } -9 \text{ and the antilog of } -9 = 0.000000001$$
$$P_2 \text{ in milliwatts} = 0.00000001 \text{ mW}$$

This can be converted into watts by moving the decimal point three places to the left:

$$0.00000001 \text{ mW} = 0.000000000001 \text{ W}$$

Watts can be converted into picowatts by moving the decimal point 12 places to the right:

$$0.000000000001 \text{ W} = 1 \text{ pW}$$

The results of our formula inform us that a signal 90 dB below 0 dBm is in fact a power level of 1 pW. This signal is called *−90 dBm*. We choose −90 dBm as the reference starting point for measuring noise but give it a new name. We call this noise reference point *0 dBrn (0 dB referenced to noise).* Another factor enters into measuring noise in telecommunications. We only care about noise that will be passed by our telecommunications system. Since the telecommunications system is designed to only pass voice frequencies between 300 and 3400 Hz, we are only concerned with noise that falls between these two frequencies. To measure only the noise signals between 300 and 3400 Hz, a filter is placed in the transmission measuring test set. This filter is called a *C message filter.* This filter will only pass signals between 300 and 3400 Hz to the test set. A transmission-measuring test set is called a *TMS.*

The C message filter causes a loss in the signals it passes to the TMS. This loss is approximately 1.5 dB. Thus, if the actual noise on a circuit is −90 dBm (0 dBrn), the TMS will measure −91.5 dBm (−1.5 dBrn). To adjust for the 1.5 dB of loss in a C message filter, the TMS meter display is adjusted such that −91.5 dBm (−1.5 dBrn) will read 0 on a scale called *dBrnC0 (dB referenced to noise through a C message filter at a 0 level test point).* Thus, the scale is calibrated to compensate for the loss of the filter. The actual noise prior to going through the filter is −90 dBm, and the meter displays this as 0 dBrnC0. This scale is called a *C message weighted scale.* One point to note is that noise readings in dBrnC0 can be converted to dBm simply by subtracting 90 from the dBrnC0 number. For example, 0 dBrnC0 = −90 dBm, and 30 dBrnC0 = −60 dBm. A conversion from dBm scale to dBrnC0 scale can be done by adding 90 (0 dBm = 90 dBrnC0). An advantage of using the dBrnC0 scale is that all measurements are usually positive numbers. Noise measurements in dBrnC0 are determined by the following formula: 0 dBrnC0 = −90 dBm.

B.17 SIGNAL-TO-NOISE RATIO

The term *signal-to-noise ratio (S/N ratio)* is used to express in decibels how much higher in level a signal is to the noise on the circuit. Suppose that at some test point, we measure the voice signal a facility is carrying and measure the noise at the same point. We get the following readings: voice signal = –5 dBm and noise = 50 dBrnC0. The two signals can be compared if the noise signal is converted to dBm (dBm = dBrnC0 –90). Thus, dBm = 50 dBrnC0 –90 = –40 dBm. To compare the signal-to-noise ratio, the noise level (in dBm) is subtracted from the voice level (in dBm):

$$S/N \text{ ratio} = \text{signal-level dBm} - \text{noise-level dBm}$$
$$S/N \text{ ratio} = S \text{ dBm} - N \text{ dBm}$$

In our example,

$$S/N \text{ ratio} = -5 - (-40)$$
$$S/N \text{ ratio} = 35 \text{ dB}$$

In telecommunications, a S/N ratio of 30 dB or more is satisfactory for voice signals and with today's data communication technology is usually satisfactory for data as well. In fact, the high-frequency signals used in ADSL modem technology encounter losses as high as 80 dB over the local loop. This brings the signal level down to the level of noise, but the signal logic processor in the receiving ADSL modem is so smart that it can distinguish the difference between noise and the ADSL signal.

The calculation of the S/N ratio was stated above as the value of the signal stated in dBm—the value of noise stated in dBm. Many people refer to a S/N ratio as the ratio of power in the signal to power in the noise. Let's examine this premise. We stated earlier that a signal level of 0.001 W is equal to 0 dBm. Suppose we have a noise level of 0.000001 W. The S/N ratio would be 0.001/0.000001, which is ratio of 1000 to 1. Now let's convert the power levels to dBm. We already know that 0.001 W is equal to 0 dbm, so let's find the dBm value for 0.000001 W.

$$dBm = 10 \log (0.000001 / 0.001)$$
$$= 10 \log (0.001)$$

the log of 0.001 is –3; therefore,

$$dBm = 10 \times -3$$
$$dBm = -30$$
$$S/N = \text{signal level in dBm} - \text{noise level in dBm}$$
$$= 0 - (-30)$$
$$= 30 \text{ dB}$$

Remember from earlier discussions that 30 dB represents a P_2/P_1 power ratio of 1000 to 1.

This exercise verifies that the S/N ratio can be expressed in dB or as a power ratio. Use the appropriate formula for whatever measurements are given. It is important to note that the S/N ratio is more important than the actual level of the noise. The higher the power of a signal, the higher the level of noise it can withstand before the intelligence within the signal is unrecoverable by the receiver.

B.18 SUMMARY

Electricity can exist as static charges, as direct current, or as alternating current. It is difficult to use static electricity to power electrical devices. The invention of the battery and electric generators gave us devices that could serve as a continuous source of electricity. As the availability and development of these power sources became more pronounced, electric motors were invented to convert electrical power into many forms of mechanical power. Electric motors increased the productivity of manufacturing operations and by replacing manual labor made many tasks safer. The next major use of electricity was for lighting and heating. Switches were invented that allowed us to control the electrical devices. The birth of electrical circuits was accomplished by including a switch in the wires that connected an electrical power source to a motor, light, heater, or any other device that needs electricity as a power source.

An electrical circuit is any continuous loop between the terminals of an electrical power source where the loop consists of a conducting material, switch, and a device designed to use the electrical power. Electricity will flow from one terminal of the power source out over the conducting loop, through a switch, and through a device, and then return over the other side of the conducting loop to the other terminal of the power source. Every electrical power source has two terminals, and an electrical circuit must be connected to both terminals. When a battery is the power source, one terminal is marked negative (it has an excess supply of electrons) and the other is marked positive (it has a deficiency of electrons).

The difference in electron concentration between the negative and positive terminals of a battery is measured in volts. If a circuit is attached to the terminals and the switch in the circuit is turned on, electrons will flow through the circuit on their way from the negative to the positive terminal. The flow of electrons through the electrical device in the circuit will cause it to carry out the function it was designed for. In the case of a simple flashlight circuit, when the switch is turned on, the flow of electrons through the lightbulb (the electrical device of this circuit) will cause the filament of the lightbulb to heat up. When the filament gets hot, it gives off light.

Sometimes it is necessary to control the application of electricity through the use of devices called *resistors*. A resistor converts electricity into heat. This process uses electrical power. If the device used in an electrical circuit is designed to use less power than is supplied by a power source, a resistor can be inserted in series with the device to absorb some of the power. Resistors can also be used in a laboratory environment to replace other devices in a circuit. One thousand feet of 22-gauge wire has a resistance of 16.14 Ω. If we wish to test a device designed to work on a 5-mi loop of 22-gauge wire, we can use a resistor instead of 5 mi of wire.

Resistors can be used in a laboratory environment to provide us with an equivalent DC circuit for design and testing purposes. Resistors are color coded to provide a visual indicator of how much resistance they were designed and manufactured to provide. The designer of a circuit must also know the amount of power that will be consumed by a resistor in order to select a resistor with an adequate wattage rating. If a resistor with too low wattage is used in a circuit, it will burn open and the circuit will cease to function due to the open-circuit condition. A fuse is designed purposely to burn open when too much power exists in a circuit. The fuse burns open and stops the flow of electricity in the circuit, thereby protecting devices in the circuit from abnormally high currents.

We use an ohmmeter to check the resistance of various components that make up a circuit. The ohmmeter must be used when no electrical power is present in the circuit component being measured. After electrical power has been connected to a circuit, we cannot use an ohmmeter but can use a voltmeter to test for the value of voltage present across the power supply and in each component in the circuit. We could also insert an ammeter in series with any component to measure the amount of electric current flowing through the component. In addition, we could use a wattmeter to measure the amount of power consumed. A wattmeter consists of a device that measures both the voltage and current present in a circuit and displays the product of voltage times current as watts. The power company uses this type of meter to determine how much electricity a customer uses.

The power loss of a transmission facility is often stated in decibels. A loss of 3 dB means the power received will be one-half the power transmitted. A loss of 10 dB equates to receiving one-tenth of the power transmitted. The amount of noise induced onto a facility is not nearly as important as the signal-to-noise (S/N) ratio. A receiver can detect a signal in the presence of higher noise if the signal has a high power level. Noise power can overwhelm a low-powered signal. The S/N ratio should be 30 dB or greater. This means the power in the signal should be at least 1000 times greater than the power of the noise.

APPENDIX REVIEW QUESTIONS

1. What is a circuit?
2. What are resistors and what are they used for?
3. What device is used to measure voltage?
4. What device is used to measure current?
5. What device is used to measure resistance?
6. What device is used to measure power?
7. What is a decibel?
8. Why is the S/N ratio more important than the power of the noise?
9. What unit of measure is used for noise, and how does it relate to dBm?

10. What is the relationship between the gauge of a wire and its resistance?

11. What is the maximum distance at which a telephone can be located from a stored program control central exchange and still receive 20 milliamps of loop current, if 26-gauge wire is used?

12. Why must the resistance of a load be much higher than the resistance of the voltage divider resistors?

13. Since resistors used to construct a voltage divider must have a low resistance, they will have large values of current flow and must have a _____ wattage rating.

14. Circular mil is found by squaring the diameter of a wire. That is the formula for finding the area of a square. The formula for finding the area of a circle is $A = \pi r^2$ Is something wrong here?

15. What is Kirchhoff's voltage law?

GLOSSARY

Ammeter A test device used to measure current in a circuit. The ammeter is usually one part of a multimeter.

American Wire Gauge (AWG) A standard that lists wire conductors according to their diameters. The higher the AWG number, the smaller the diameter and cross-sectional area of the wire. The table was set up so that a change in the AWG number by 3 results in a change in wire cross-sectional area by a factor of 2. Going from a smaller number to a larger number such as from 16 to 19 means the 19-gauge wire is smaller than the 16-gauge wire and is half the size of the 16-gauge wire.

Circuit A circular journey. A circuit contains a conducting path between the terminals of a power source.

Circular Mil The area of a circle that has a diameter of 1 mil. The circular mil area is diameter squared. It is not the same area found by using the algebraic formula $A = \pi r^2$, but it does provide the same linear relationship and makes it easier to compare conductors of various diameters.

Closed Circuit A circuit in which the conducting path is unbroken.

Conductance A measure of how good a conductor is at carrying electric current. It is the reciprocal of resistance. The symbol for conductance is G, and it is measured in siemens. $G = 1/R$.

Decibel (dB) The ratio of one power level to another. Decibels state the relative comparisons of two levels of power. For example, 3 dB means the power level of one signal (P_2) is twice the power level of another signal (P_1). The formula is dB = 10 log (P_2/P_1).

Decibels Referenced to Noise (dBrn) The starting power level for measuring noise. 0 dBrn = –90 dBm and 0 dBrn = 1 pW.

Electron Pump A chemical cell or several cells connected to form a battery for the purpose of supplying (pumping) electrons to a circuit.

Fixed Resistors Resistors having one value of resistance.

Load   The active device in a circuit. It makes use of the electric current in the circuit to perform some useful function.

Multimeter   A test device used to measure voltage, resistance, or current.

Ohmmeter   A test device used to measure the resistance of a component. An ohmmeter is usually one part of a multimeter.

Open Circuit   A circuit where the conducting path has been broken open.

Resistors   Human-made devices that simulate a specific desired resistance. Resistors convert electrical energy to heat.

Signal-to-Noise Ratio (S/N Ratio)   The comparison of the power in the desired signal to the power in the noise signal. The S/N ratio should be 30 or more.

Switch   A device used to connect two parts of a circuit together so electric current flows when the switch is closed or ceases when the switch is open.

Tolerance   The rating of a resistor that indicates how closely the resistor will match its stated value. The color of the fourth band on a resistor states the manufacturing tolerance. A gold band is ±5% and a silver band is ±10%.

Variable Resistors   Resistors that contain a movable component so that we can vary the resistance desired from 0 Ω to some maximum value.

Varistor   A resistor that changes value according to the amount of voltage applied to it. The higher the applied voltage, the higher its resistance will be.

Voltmeter   A test device used to measure voltages. A voltmeter is usually one part of a multimeter.

Watts   A measure of the amount of power that can be handled or consumed by a device. $W = IE$

Appendix C

Alternating Current Electricity

KEY TERMS

Alternating Current (AC)
Alternator
Artificial Magnet
Bar Magnet
Capacitance
Capacitive Reactance
Capacitor
Counter EMF
Electromagnet
Flux Density

Flux Line
Generator
Impedance
Inductance
Inductive Reactance
Inductor
Left-Hand Rule
Magnetic Field
Magnetism
North Pole

Permanent Magnet
Power Supply
Reluctance
Residual Magnetism
Ringing Machine
Root Mean Square (RMS)
South Pole
Temporary Magnet
Tuned Circuit
Tuned Frequency

In Appendix A we covered direct current (DC) electricity, which can be generated by a chemical reaction within a cell. This chapter covers *alternating current (AC)* electricity, which is generated by mechanical means. DC current flows in one direction (negative to positive) because the polarity of the voltage supply does not change. The negative terminal of a battery is always negative and the positive terminal of a battery is always positive. Alternating current (AC) is a current that continually changes direction. The current continually changes direction because the voltage, which makes the current flow, is continually changing polarity. The most common way to produce AC voltage is by a mechanical device called a *generator* or *alternator.*

C.1 MAGNETS

The AC voltage *generator* produces an AC voltage by passing a wire through a *magnetic field.* Magnetic fields exist around iron, cobalt, and nickel when they have been magnetized. They also exist around natural magnets. A long time ago Greeks discovered that some rocks found near the city of Magnesia in Asia Minor had the power to attract and pick up small pieces of iron. This rock is composed of a type of iron ore called *magnetite,* and its power to attract other metals is called *magnetism.* These rocks are called *natural magnets.*

It was discovered that these natural magnets would interact with the magnetism of our planet when they were suspended on a string. One side of the rock would always point north. These magnets were used to construct compasses. It was also discovered that if a piece of iron was stroked with a natural magnet, the iron would become magnetized. Since the magnetism did not occur naturally, these were called *artificial magnets.* As we will learn later, artificial magnets can also be made by wrapping many turns of wire around a bar of iron and passing current through the wire. It has also been found that an artificial magnet made from an alloy of nickel and cobalt makes a stronger magnet than iron.

Today's magnets are strong and inexpensive. The magnetic surfaces used for secondary storage in a personal computer are made by embedding small iron or iron alloy particles in ceramic or plastic. Iron becomes magnetized more easily than other materials but also loses its magnetism easily. For this reason, magnets made of soft iron are called *temporary magnets.* Magnets made of steel alloys are much harder to magnetize, but they also hold their magnetism for a very long time and so are called *permanent magnets.* Steel is not a natural element. It is an alloy made of iron and carbon and usually some other metal such as nickel, chromium, or manganese, plus small amounts of phosphorous and sulfur.

The magnetic effect of magnets appears to be concentrated at the two ends of a magnet. The ends of a magnet are poles. The pole that points to the north if the magnet is suspended is called the *north pole.* The end of the magnet that points south is called the *south pole.* If a magnet is broken into many small pieces, each piece will be magnetized. The magnetic effect comes from electrons spinning around an atom. All atoms have a magnetic field, but the atoms in most materials are arranged randomly in the material. The random arrangement of atoms results in the north pole of each and every atom pointing in many different directions. This results in no external magnetic field. Materials that can be magnetized allow their atoms to be arranged so that many atoms have their north poles pointing in the same direction. By default, many south poles are also pointing in one direction. Whenever we stroke iron with another magnet, many atoms align their north poles in the same direction. With many atoms aligned in one direction, the iron now exhibits an external magnetic field.

C.2 MAGNETIC FIELDS

Magnetic fields are invisible, but we can observe the effects of these fields. A bar of steel that has been permanently magnetized is called a *bar magnet.* If a sheet of paper is place on top of a bar magnet and a small amount of iron filings distributed across the paper, we can see that the iron filings align themselves in a pattern similar to the one shown in Figure C-1. The arrangements of these iron filings provide a visible picture of the magnetic lines extending from the north pole to the south pole. These magnetic lines are referred to as *flux lines* or lines of flux.

The flux lines form a definite flux pattern. The number of flux lines per unit area is called the *flux density.* When the flux density is stated in terms of lines per square centimeter, the measure of density is the gauss. Figure C-2 provides a clearer picture of the path taken by the invisible magnetic lines of flux. The lines of flux

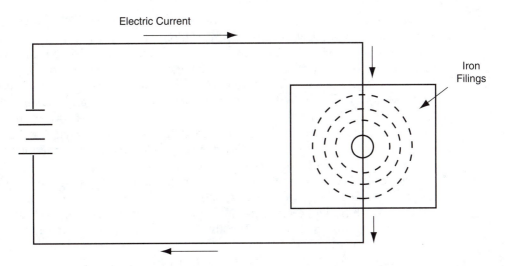

FIGURE C-1 Magnetic field surrounding a wire carrying electric current.

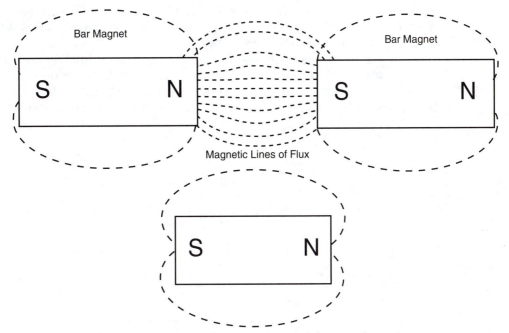

Flux density is greatest when south pole and north pole are close to one another.

FIGURE C-2 Magnetic lines of flux.

leave the north pole and enter the south pole. Because the lines are packed more tightly (denser) at the poles, the magnetic effect is stronger at the poles. Like poles attract and unlike poles repel. A north pole repels another north pole. A south pole repels another south pole, but attraction exists between a north and south pole.

These actions are due to magnetic fields around the magnets. There is no known insulator for magnetic lines of flux. They will pass through all materials.

Although magnetic lines pass through all materials, they pass through a few materials more easily than others. The measure of opposition to magnetic lines is *reluctance*. Materials with a low reluctance, such as iron, pass magnetic lines of flux easily. Materials with a high reluctance, such as wood, air, and plastic, have a high opposition to magnetic flux. We use this knowledge to concentrate magnetic flux where we want it, and to keep it out of areas where we do not want it. A steel or iron shield can be placed in front of a device we wish to protect from magnetism and the shield will absorb most magnetic lines of flux.

We can bend a bar magnet into the shape of a horseshoe to make a U-shaped magnet. Bending the bar into a U shape places the north and south poles closer together. The external magnetic lines of flux do not have to travel far as they go through air from the north to south pole. The air distance between the north and south pole of a horseshoe magnet is much shorter than for a bar magnet. The lines of force are packed tighter together making the magnetic field stronger (see Figure C-3).

Magnetic lines of flux encounter high opposition in air but low opposition in iron. Therefore, the magnetic flux is stronger in the iron than in the air surrounding the iron. The closer two magnets are placed, the stronger the force between them. The attraction and repulsion due to lines of flux is due to the fact that lines of flux cannot cross each other (see Figure C-4).

The source for electricity used in our homes and a business is an alternator at a local power plant. An alternator and a generator produce electricity by the interaction that occurs between a wire conductor and a magnetic field. When a wire moves past a magnet, electricity is produced in the wire. Generators move the wire

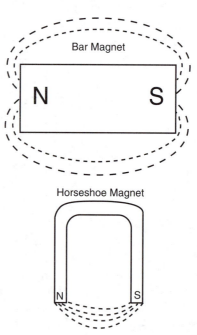

FIGURE C-3 Bar vs. horseshoe.

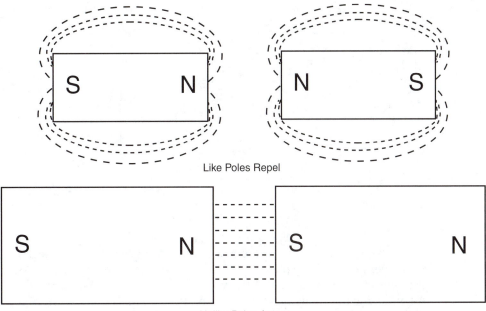

FIGURE C-4 Properties of magnetic lines of force.

through the magnetic field. Alternators also produce AC electricity, but in an *alternator,* the magnetic field is moving past the wire. An alternator spins the magnets. It does not matter which moves. If the relationship between a wire and a magnetic field is such that movement is occurring between the two, electricity is produced in the wire. If an ammeter is connected between the two ends of the wire, it registers current flow only while the magnetic field or wire is moving. If the magnet is placed near the wire and neither is moving no current will flow.

If you move the wire parallel to the face of a pole on a magnet, the ammeter will indicate maximum current flow when the wire is moving across the center of the pole. Minimum current will occur when the wire is moving at the edges of the face, and zero current when the wire is past the face and out of the lines of flux (see Figure C-5). If the wire is moved back across the face, you will notice that the ammeter indicates current is now flowing in the opposite direction. Thus, the direction in which the wire moves determines the direction of current flow. To provide a continuous motion to the wire, the wire could be moved back and forth in the magnetic field. The easiest way to move the wire in a magnetic field is to rotate the wire in a circle between two magnets. As the wire moves through the magnetic field for the first half of a revolution, current flows in one direction. As the wire moves through the field for the second half of the rotation, the direction of wire movement has changed by 180°. The direction of current flow also changes by 180°.

As the wire moves parallel to the lines of force, no lines of flux are cut by the moving wire and no electricity is produced. When the wire moves down through the magnetic field, current flows in the opposite direction. The EMF (amount of

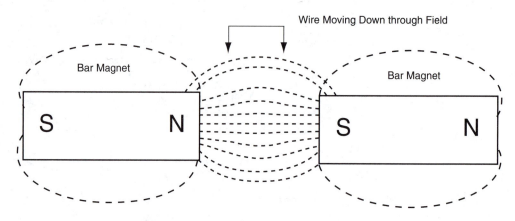

FIGURE C-5 Wire moving through a magnetic field.

volts) produced across the ends of a wire depends on the number of magnetic lines of flux being cut by the moving wire. The number of magnetic lines cut depends on flux density and the speed of the wire. The voltage can be increased by increasing the magnetic flux density, which is why the voltage is highest when the wire moves across the center of the face of a pole. At this time the wire is also closest to the pole. This is where the highest flux density occurs. Voltage could also be increased by moving the wire faster. The faster it moves, the more lines of flux the wire will cut for a given time period. When a moving wire cuts flux lines at the rate of 100 million (10^8) magnetic lines per second, 1 V is induced in the wire.

The amount of voltage produced can be increased by using more than one moving wire and connecting all the wires together so that the voltage generated in each wire adds to the voltages generated in the other wires. In a generator, this is done by winding the wire into a coil and rotating the coil of wire in the magnetic field (see Figure C-6). The wires are wound around a cylindrical iron core. The combination of loops of wire wound on an iron core is called an *armature.*

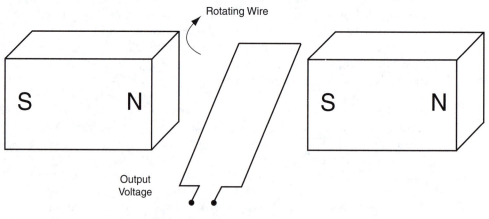

FIGURE C-6 Coil of wire in magnetic field.

C.3 ELECTROMAGNETS

Since increasing the strength of the magnetic field can increase voltage produced, research was done in this area. It was found that when a wire carries an electric current, a magnetic field surrounds the wire. The direction of the magnetic field depends on the direction of current flow. If a wire is held perpendicular with current flowing down the wire, the magnetic field has a counterclockwise direction. If current flowed up the wire from bottom to top, the magnetic field has a clockwise rotation. The direction of the magnetic field can be determined by holding the conductor in your left hand with your thumb pointing in the direction of electron flow. The fingers will point in the direction the lines of flux take. This is called the *left-hand rule* (see Figure C-7).

You can hold a compass next to a wire to prove that a magnetic field is produced around a wire carrying current. If no current is present, the compass points toward the earth's magnetic north pole no matter where the compass is held. When the wire is carrying current, the needle of the compass points in a direction perpendicular to the conductor at every point around the conductor. If the compass is moved around the conductor, the needle remains at a right angle to the conductor. If the current through the conductor is reversed, the compass needle will spin 180° and point in the opposite direction but still perpendicular to the conductor. You can verify the left-hand rule by observing the direction of current flow and the direction in which the compass needle points, and comparing these observations to the left-hand rule.

Winding the wire in a coil can increase the strength of the magnetic field. When wire is wound into a coil, to create a magnetic field, there must be insulation around the wire. The insulation ensures that the electric current will flow through the complete length of wire. If wire without insulation was used, the wires would be touching each other and current would not flow through the wire. The current would simply take a shortcut from the input to the output leads of the coil by flowing out the side of the wire to other wires it was touching. Without insulation, all wires are electrically connected together as one big, wide conductor.

Wrapping insulated wire into a coil concentrates the magnetic lines of flux into a small area. The strength of the magnetic field can be increased further by replacing the air inside the coil of wire with iron. Iron has a lower reluctance than air and concentrates the magnetic lines of flux to create a stronger magnetic field. By wrapping many turns of wire around an iron core and then passing electricity through the wire, we can create a magnet. When a soft iron core is used, the iron is magnetized as long as current flows through the wire around the iron but rapidly loses its magnetism when the current stops flowing. The magnetism developed depends on the flow of electricity flowing in the wire. This type of magnet is called

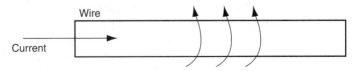

FIGURE C-7 Left-hand rule.

Magnetic Lines of Flux

an *electromagnet.* The strength of an electromagnet depends on the number of turns of wire in the coil, the amount of electric current flowing in the wire, and the composition of the core material.

For a typical electromagnet, wire is wrapped into many coils to increase the strength of the magnetic field. The direction of the magnetic field for a coil of wire is found by applying the left-hand rule. For this application of the rule, the left hand is wrapped around the coil of wires with the fingers pointing in the direction of electron flow. The thumb will then point to the north pole of the electromagnet.

Electromagnets are used in generators because they can develop very strong magnetic fields. A very small part of the generator or alternator output is fed through a rectifier to convert the AC voltage to DC voltage. That DC voltage is used to supply current through the electromagnets. How long an electromagnet will remain magnetized after electric current is turned off depends on the composition of the core material. As stated earlier, a core made from soft iron losses its magnetism rapidly when current ceases to flow in the wire. If we wish to retain magnetism for longer periods we use materials other than soft iron for the core. A hardened steel alloy composed of iron, carbon, aluminum, nickel, and cobalt is called *Alnico.* It retains magnetism for a long time and makes an excellent permanent magnet. The electromagnets in the generator and alternator are made using Alnico instead of soft iron because we do not want to lose all the magnetism when the generator or alternator is turned off.

The amount of magnetism that remains in an electromagnet when the current stops flowing is called **residual magnetism.** By using Alnico for the electromagnets of a generator, the residual magnetism is strong enough in the generator that it can start producing some voltage when the wire first starts spinning. As voltage is developed, current starts flowing in the electromagnets, making them stronger. This leads to higher voltage output and more current in the electromagnet, which leads to an even higher voltage output. This process continues until the generator is producing the voltage that it was designed to produce. At this point the voltage levels will stabilize, and current through the electromagnets also stabilizes.

C.4 AC VOLTAGE GENERATION

We will continue our discussion using a generator as an example. Keep in mind that our discussion would also hold true for an alternator. The only difference between the two is the spinning device. The stationary device is called a *stator* and the moving device is called a *rotor.* For a generator, the rotor contains the wire in which an EMF is induced and the stator is composed of electromagnets to generate the magnetic field. For an alternator the rotor contains the exciter (magnetic field) and the stator contains the wire in which an EMF is induced. It is connected to our load.

A simple generator consists of a device that has two large magnets facing each other. One magnet is a south pole and the other is a north pole. Lines of magnetic flux extend from one pole to the other. Between these two magnets is a coil of wire. When wire moves in a magnetic field a voltage is induced in the wire. The generator is turned by another device. When the generator turns or spins, it turns the coil of wire placed between the magnets. In a local power generating plant, steam is used to power a turbine and the turbine turns the rotor of an alternator. The steam is generated from heat produced by a coal fired furnace or by nuclear fusion. Since water is needed to produce steam, power plants are located on rivers. As the turbine turns the magnetic field inside an alternator, a voltage is induced in the wire.

A generator is constructed so that it spins wire coils inside of a magnetic field. As the wire spins in a circle, it moves down through the magnetic field and then returns back up through the field. On the return path, the motion of the wire is in an opposite direction and the voltage induced will also be opposite. The lines of magnetic flux are much stronger in the middle of the two poles and more voltage will be induced when the wire is in the center of the magnetic flux field than when the wire is moving in the outer regions of the flux. This is because the voltage polarities are reversed when the wire moves back up through the magnetic field relative to when the wire was moving down through the field.

The voltage across the wire is continually changing polarity, which will cause current in the external circuits attached to the output of a generator to continually change direction. The amount of current flow depends on the amount of voltage generated. The voltage varies depending on how many lines of magnetic flux are being cut by the moving wire. Voltage is maximum and current is maximum when the wire is moving in the center of the magnetic field. Voltage is zero and current is zero when the wire is moving in the outer region of the magnetic field where it is moving parallel to the lines of magnetic flux and does not cut across the lines of flux (see Figure C-8).

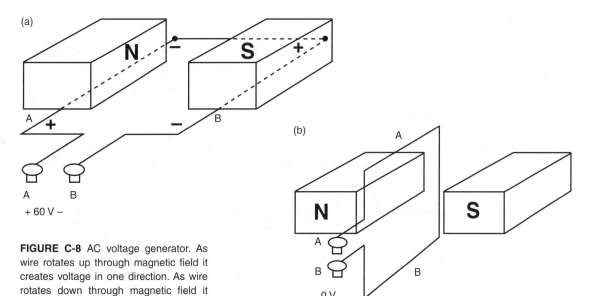

FIGURE C-8 AC voltage generator. As wire rotates up through magnetic field it creates voltage in one direction. As wire rotates down through magnetic field it creates voltage in the opposite direction. Thus, if voltage induced in wire A is 30 V and voltage in wire B is 30 V, the total voltage produced = 60 V.

When wire has rotated to a vertical position it is travelling parallel to the magnetic lines of flux between N and S. Since it cuts no magnetic flow, 0 V are produced.

(continued on next page)

Alternating Current Electricity

487

(c)

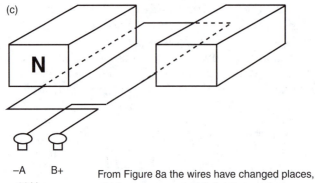

−A   B+
− 60 V +

From Figure 8a the wires have changed places,
which causes a reversal of voltage output.

(d)

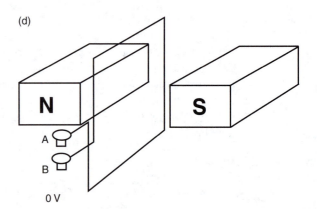

0 V

FIGURE C-8 (Continued)

Wires are outside of magnetic field and
moving parallel to it, therefore 0 V is produced.

Figure C-8 presents a simple generator consisting of two magnetic poles. Most generators contain more than one set of electromagnets. They contain many electromagnets located closer to each other. This creates a stronger magnetic field between magnets due to less air between the poles of the magnets. U-shaped electromagnets are located around the perimeter of the path taken by the spinning armature of the generator. The magnets are placed so that every other pole that the spinning wire will cross is a north pole and every other pole is a south pole (Figure C-9). Since several sets of north/south poles are used in a generator, the armature of the generator contains as many individual coils as there are sets of north/south poles.

Separate coils of wire are placed in strategically located slots on the armature such that one side of every individual coil is immediately opposite the center of a north pole at the same time. The other side of every coil is immediately opposite the center of a south pole. With this arrangement, all coils are generating exactly the same voltage at every point in time. When one winding has zero volts induced, all windings have zero volts induced. When one winding has maximum positive voltage induced, all windings have maximum positive voltage induced, and so on.

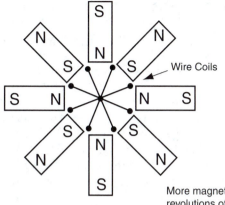

Wire coils are connected in
series to increase output voltage.

More magnetic poles allow speed or
revolutions of coils to be reduced but
still produce the same output voltage.

FIGURE C-9 Multiple poles and windings.

The voltages generated in each winding are in sync. The ends of all windings that are opposite a north pole at the same time are connected together and connected to one output terminal. The opposite end of these windings are connected to one output terminal. Thus, all windings are in parallel and each winding carries a share of the current drawn by a load attached to the generator. The use of parallel windings increases the current carrying capacity of an AC generator.

As stated earlier, commercial power plants use alternators. We used a generator for the discussion on how an AC voltage is generated, because the generator process is easier for me to explain. The generator process is also easier for most people to understand. The voltage developed by an alternator at a commercial power plant is between 14,000 V and 22,000 V depending on the particular design of the plant. If the power is to be transported for a long distance, the voltage is raised by using a step-up transformer at the power plant site. The voltage is stepped up or raised to 161,000 V. We will discuss transformers later. They simply transform one voltage level into another voltage level. Transporting power at high voltages reduces power loss in the transmission wire and more power is delivered to the end user.

Remember from our discussion of power in Chapter 2 that power is found by multiplying voltage times current and can also be stated as current squared times resistance. If the voltage across a device is 120 V and it needs 100 amp of current, the power needed and used by the device is 12,000 W. If we were to try transmitting that power to the device over a pair of wires that has 100 Ω of resistance, we would lose approximately 1,000,000 W (I^2R). To have 12,000 W available to the device, we would have to transmit at 1,012,000 W. The huge loss of power in the transmission medium (the wire) is unacceptable.

Instead of transmitting 100 ampere at 120 V, we can transmit the same power by transmitting at 161,000 V. 12,000 W of power at 161,000 V consist of 13.417 amps (161,000 V × 13.417 amps = 12,000 W). The power loss will be much lower in the wire with a lower current value. A current of 13.417 amps squared times 100 Ω of wire resistance results in a power loss of 18,000 W. This is significantly less power loss than the 1,000,000-W power loss we would have if we tried to transmit the power at 120 V. The wire used in a power distribution system is very thick and has a much lower resistance than 100 Ω. I selected a 100-Ω value for the transmission wire so that you could readily see the benefit of transmitting power at high voltage and low current. We step the voltage up at the power plant to reduce power loss in the transmission medium.

High-voltage transmission lines connect the power plant to a centralized local distribution point known as a substation. Step-down transformers are used at substations to reduce the voltage from 161,000 V to 13,200 V. Power is then distributed from the substation to residential and business customers at 13,200 V. Transformers are used at each customer location to step the voltage down to 220 V or 120 V as needed by the customer.

C.5 RMS VOLTAGE

AC signals, voltages, and currents have many instantaneous values across one cycle of signal. At times the value is 0, at other times they reach some maximum value, and for all other times they assume some value between 0 and maximum. We can compare signals to one another by stating the maximum value produced, the average value produced, or the root-mean-square value for the signal. We use the term *root mean square (rms),* to correlate alternating voltages or currents to direct voltages or currents. Using rms values allows AC signals, voltages, and currents to be evaluated in terms of how they compare to a DC signal, voltage, or current. A 120 V rms AC signal will do the same work as a 120 V DC signal. The formula for converting an AC signal to rms is:

RMS Formula

RMS = 0.707 × one-half the peak-to-peak signal

The power supplied to our homes has an AC voltage with a positive peak of +170 V and a negative peak of –170 V. The peak to peak voltage is 340 V rms = 0.707 × 1/2 (340) = 0.707 × 170 = 120 V. We state the voltage at our electrical outlets is 120 V. It is 120 V rms. When AC is stated in the rms values, we can use these values in Ohm's law formulas and in power formulas.

An AC voltage signal is voltage that continuously changes polarity. The voltage will start at zero volts and go to a maximum positive voltage. From a maximum positive voltage, the signal gradually decreases in voltage until it is zero again. The signal then goes to maximum negative voltage, and then back to zero. This trip from zero to maximum positive, back to zero, to maximum negative, and then back to zero volts is called a *cycle*. The number of cycles that an AC signal completes in one second is called the *frequency of the signal*. The AC voltage to our house is a 60 cycle-per-second (60 cps) signal. The signal goes from 0 V, to +170 V, to 0 V, to –170 V, and back to zero in 1/60 of a second. Thus the signal can complete 60 of these cycles in one second.

The number of cycles completed in one second is the frequency of the signal and can be stated as cycles per second (cps). Cycles per second are often called *Hertz (Hz)* as a tribute to German physicist Heinrich Hertz's discovery of electromagnetic radiation (the birth of radio waves) in 1886. In a simple generator that has only one north and south pole, the frequency of the signal is directly correlated to the speed of the coil of wire. If the coil of wire makes 60 revolutions per minute, the frequency of the output voltage is 60 Hz. The generator at a power plant contains many electromagnets that alternate in polarity around the circumference of the generator. As the wire spins past a north to south pole, current flows in one direction. As the wire spins from the south pole to the next north pole, the direction of current reverses. Thus, the frequency of the output voltage will be equal to the speed at which the coil of wire turns times the number of sets of north south poles. If the generator contains 20 north poles and 20 south poles (20 sets of poles), then an output frequency of 60 Hz is obtained by rotating the coil of wire at 3 revolutions per second.

An AC signal is a signal where the electric current continually reverses direction. In our electric generator example, the value of electric current and the direction of current depend on the value of and polarity of voltage generated. In Figure C-8, we are shown the voltage developed at four positions of rotor movement. As the rotor turns inside an alternator or generator, the alternator or generator will generate or develop varying levels of voltage. The voltage developed depends on the speed with which the rotor is turning, the strength of the magnetic field, and the angle of interaction between the magnetic field and the wires in which a voltage is being induced.

Figure C-8 shows only four of the numerous positions that are possible for a spinning rotor. As the rotor of a simple two-pole generator makes one complete turn or revolution we can describe the position of a particular point in a revolution as being anywhere between 0° and 360°. Figure C-8 depicts only the positions of the rotor when it is at 0°, 90°, 180°, 270°, and 360°.

Figure C-8a shows the voltage developed when the moving wire of a generator is exactly halfway between a north and south pole. At this position the wire is moving parallel to the lines of magnetic flux. The angle between the moving wire and the lines of flux is 0°. As a result, no voltage will be induced in the wire. We can graph this position as point A on the time line graph of Figure C-10.

Figure C-8b shows the voltage developed when the moving wire of a generator is exactly opposite the faces of the north and south poles. At this position, the moving wire is moving directly across the magnetic lines of flux and there is a heavy concentration of magnetic lines at the face of a pole. The angle of movement between the moving wire and the magnetic field is 90°. At this angle of movement, maximum voltage is induced in the wire. We can graph this position as point B on the time line graph of Figure C-10. If the time required to make a complete revolution is 1/60 of a second, and if we have made a quarter of a revolution (90°), then the time required to make this quarter revolution is 1/4 of 1/60 of a second or 1/240 of a second. Thus, point B occurs on our time line 1/240 of a second after point A occurs.

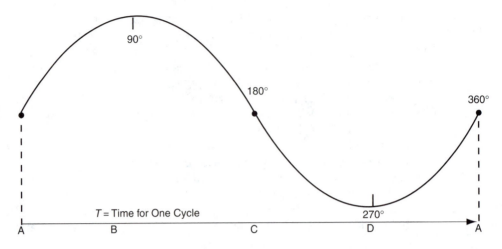

FIGURE C-10 Sine wave.

Figure C-8c shows the voltage developed when the moving wire is exactly halfway between a north and south pole. At this position the wire is moving parallel to the lines of magnetic flux. The angle between the moving wire and the lines of flux is 0°. As a result no voltage will be induced in the wire. We can graph this position as point C on the time line graph of Figure C-10. This point represents the voltage that is occurring when the rotor has made half of a revolution. Thus we plot this point on the time line 1/120 of a second after point A occurred and 1/240 of a second after point B occurred.

Figure C-8d shows the voltage developed when the moving wire of a generator is exactly opposite the faces of the north and south poles. At this position, the moving wire is moving directly across the magnetic lines of flux and there is a heavy concentration of magnetic lines at the face of a pole. The angle of movement between the moving wire and the magnetic field is 90°. At this angle of movement, maximum voltage is induced in the wire. Because the movement of the wire across the face is opposite that of C-8b, the polarity of voltage developed is also opposite the voltage developed at C-8b. We can graph this position as point D on the time line graph of Figure C-10. Thus, point D occurs on our time line 3/240 of a second after point A occurs, 2/240 of a second after point B, and 1/240 of a second after point C occurred.

After making one complete revolution (360°) we are back to the position shown in Figure C-8a. Figure C-8a shows the voltage developed when the moving wire of a generator is exactly halfway between a north and south pole. At this position the wire is moving parallel to the lines of magnetic flux. The angle between the moving wire and the lines of flux is 0°. As a result no voltage will be induced in the wire. We can graph this position as point E on the time line graph of Figure C-10. One revolution requires 1/60 of a second (4/240 of a second). Thus point E is placed on the time line graph 1/60 of a second after point A occurred.

The line developed on the graph of Figure C-10 is called a sine wave. A sine wave depicts a periodic oscillation, as a simple harmonic motion, having the same geometric representation as a sine function. In trigonometry, we learn that a sine is equal to the ratio of the side opposite a given angle to the hypotenuse (the side opposite the right angle in a right triangle). We use the sine law of trigonometry to determine the voltage at any point in the rotation of rotor. The formula for voltage developed at any point in the rotation is: $v = V$max sin θ (where v is the instantaneous value of voltage for any angle of rotation, Vmax is maximum voltage that is produced at 90° rotation and sin θ is the angle of rotation at a particular point in the rotation of the rotor).

Some typical sine values are listed in Table C-1.

Using the formula, we can determine the instantaneous voltage at any point along a sine wave because the induced voltage is proportional to the sine of the angle of rotation. The current flow will be directly proportional to the applied voltage and can also be determined by using a sine wave function. If the circuit attached to the voltage supply contains purely resistive components with no inductance or capacitance in the circuit, the current flow sine wave and voltage sine wave will be in-phase. This means maximum current occurs at maximum voltage and zero current occurs at zero voltage and so on.

Using the formula, we can determine the voltage developed in a moving wire or rotor for any point in their rotation. This formula is used to plot the value for many points of rotation. The values plotted are then joined together by drawing a line to connect the dots. The line drawn to connect the dots is known as a sine wave. The more points we calculate for a signal by using the sine formula, the more accurately the sine wave will represent the signal. A sine wave is nothing more than a graphical representation of the values of a signal over a period of time. You will note that if the last half of a sine wave is turned backwards, the graph becomes a circle. A sine wave simply extends the circular motion of the signal into a straight line graph.

TABLE C-1 Degree and Sine Value

Degrees of Rotation	Sine for That Degree	Degrees of Rotation	Sine for That Degree
		180	0
0	0	210	−0.5
30	0.5	225	−0.707
45	0.707	240	−0.866
60	0.866	255	−0.9659
75	0.9659	270	−1.0
90	1.0	285	−0.9659
105	0.9659	300	−0.866
120	0.866	315	−0.707
135	0.707	330	−0.5
150	0.5	360	0

C.6 FREQUENCY

As stated earlier, the frequency of signal is how often a signal repeats its pattern or cycle. If one cycle of a waveform can be completed in 1/60 of a second, then we can complete 60 Hz. If it takes 1/1000 of a second to complete one cycle, 1000 Hz will be produced. This can be stated in a formula as:

$F = 1/t$ (where F = frequency and t = the time for one cycle of the signal)

For example,

$$F = 1/(1/60) = 60 \text{ Hz}$$
$$F = 1/(1/1000) = 1000 \text{ Hz}$$

C.7 INDUCTIVE REACTANCE

AC signals do not behave like DC signals. With a DC signal, we are only concerned with voltage, current, and resistance. An AC signal entails two additional factors: *inductive reactance* and *capacitive reactance.* The combination of resistance, inductive reactance, and capacitive reactance is called *impedance.* All three components of impedance are measured in ohms; therefore, impedance is stated in ohms. Impedance is the opposition to flow of an alternating current. Inductive reactance exists in all conductors due to a property called *inductance,* which is present in all conductors. Inductance is measured in a unit called henry. Inductive reactance is very high when changes occur in a signal and is zero when no change is occurring. DC circuits have a steady current flow. There is no change in the flow of current and no inductive reactance is present in a DC circuit. However, an AC signal is constantly changing and inductive reactance comes into play in an AC circuit. The faster a change occurs, the higher the inductive reactance. How fast a signal changes depends on the frequency of the signal. The higher the frequency, the more times the signal changes, and the higher the inductive reactance.

Inductance is caused by the magnetic field surrounding a wire carrying electric current. As long as the current in the wire remains at a constant level, the magnetic field remains constant. When the current in a wire changes, the magnetic field changes. If the current in a wire changes direction, the magnetic field changes direction. When a wire is carrying an AC current, the magnetic field is continually changing. At zero current, there is no magnetic field. As current starts to increase in one direction, the magnetic field increases and expands outward from the wire. As current starts to decrease, the magnetic field starts contracting and collapsing inward. When current is zero, the magnetic field ceases to exist. As the current starts flowing in the opposite direction, a magnetic field also starts building in the opposite direction. As the current begins to increase, this magnetic field also increases and expands outward. As the current starts to decrease, the magnetic field collapses inward. When current is zero, the magnetic field ceases to exist.

As AC current flows in a wire, the magnetic field is continually expanding, contracting, and reversing polarity at the same frequency that the AC current changes. As the magnetic field is changing, it cuts the current carrying wire. We learned earlier that if a wire is in a moving magnetic field, a voltage is induced in the wire. As current in a wire begins to build in one direction, the magnetic field it creates induces a voltage in the wire which opposes the current flow in the wire. This voltage is called *counter EMF* because it is opposite in polarity to the voltage that is making the current flow in the wire. When current reaches a maximum level, the magnetic field stops expanding. With no changing current, the magnetic field stops moving and the counter EMF falls to zero. Now current begins to fall to lower levels. This causes the magnetic field to start collapsing. As it collapses, it cuts the current carrying wire in the opposite direction. This creates an EMF that aids or adds to the EMF making current flow in the wire.

The EMF being generated by the expanding and collapsing magnetic field is a voltage that opposes changes in the signal. As the original signal moves from zero current to maximum current, the counter EMF opposes that change. As the supply voltage making current flow in the wire starts to decrease and current starts falling, the EMF opposes that change and tries to keep current at its maximum level. As the current changes direction, the same process occurs. The induced EMF first opposes current flowing in the opposite direction but after current reaches a maximum in that direction, the induced EMF tries to keep it flowing in that direction. The EMF that is developed by the changing magnetic field is continually changing polarity and voltage level. The voltage developed lags behind the supply voltage changes by 90°.

An AC voltage goes from zero volts to a maximum positive volts, back to zero volts, then to a maximum negative volts before returning to zero volts. This is one cycle. One cycle is 360°. The time it takes to go from zero volts to maximum positive volts is one-quarter of a cycle (90°). When the voltage is maximum positive, maximum current flows in one direction. At this point counter EMF is zero because the magnetic field becomes stationary. As the supply voltage starts to decrease, the collapsing magnetic field creates an EMF of the same polarity of the supply voltage. As the supply voltage is decreasing toward zero, the induced EMF is building toward its maximum voltage. When the supply voltage has reached zero and changes polarity to start current flowing in the opposite direction, the magnetic field rapidly changes direction. At this point the supply voltage is 180° into its cycle. This change of magnetic field induces the highest counter EMF. When supply voltage is zero (at 180°), counter EMF is maximum (90°). This process continues with the counter EMF always 90° behind the supply voltage. This effect is what we measure as inductive reactance.

Inductive reactance is directly related to the frequency of the signal and the inductance in the circuit. The higher the inductance is of a component, the higher its inductive reactance. The inductance for a 24-gauge copper wire, one mile long, is approximately 0.5 mh. Since the local loop is composed of two wires (Tip and Ring), the inductance per loop mile is approximately 1 mh (0.001 henry). Inductive reactance is represented by the symbol XL. The formula for finding inductive reactance in a circuit is $XL = 2 \times 3.14159 \times F$ (frequency in Hz) $\times L$ (inductance in henries).

Inductive Reactance Formula
$$XL = 2 \times 3.14159 \times F \times L$$

Inductive reactance of a 100-Hz signal on a 24-gauge cable pair is:

$$XL = 2 \times 3.14159 \times 100 \times 0.001$$
$$XL = 0.628318 \ \Omega$$

Because a direct correlation exists between inductive reactance and frequency, a 500-Hz signal will have 5 times the inductive reactance of a 100-Hz signal and a 1000-Hz signal has 10 times the inductive reactance of a 100-Hz signal (6.28 Ω on a local loop). Note that the inductive reactance for a 5000-Hz signal on a local loop is as follows: $XL = 2 \times 3.14159 \times 5000 \times 0.001 = 31.4159 \ \Omega$. Since a 5000-Hz signal is equal to 50 times a 100-cps signal, the XL for a 5000-Hz signal should be 50 times the XL of a 100-Hz signal. Thus, $50 \times 100 = 5000$ and $50 \times 0.628318 = 31.4159 \ \Omega$.

C.8 TRANSFORMERS

The discussion thus far on inductance has been on the effect that it has on a wire carrying an AC signal. The wire carrying the AC signal will be affected by the magnetic field it creates. This is called *self-inductance*. The expanding and collapsing magnetic field created by the AC current in the wire creates an EMF that is self-induced. Any time a wire is in a moving magnetic field, an EMF is induced in the wire. The wire carrying an AC signal will be laying in its own moving magnetic field and is affected by its own magnetic field.

The magnetic field around a wire can be concentrated by winding the wire into many turns to form a coil of wire. The changes in the magnetic field of each loop comprising the coil affects all loops of the coil. This increases the self-induced EMF (inductance). The ability of a device, circuit, or coil to generate an EMF of self-inductance that *opposes changes in the current* flowing through it is measured as inductance in a unit called *henry*. Anything that increases the magnetic flux increases the inductance. Inductance is increased by winding the current carrying wire into a coil. Increasing the number of adjacent turns in a coil increases the inductance. Iron is a much better conductor of magnetic flux than air and using an iron core can increase the inductance of a coil.

When wire is wound into a coil, the induced EMF in each coil adds to the induced EMF of the other coils. If the induced EMF for a straight wire carrying an AC signal were 0.01 V at any point in time, the induced EMF would be 1 V if the wire were wound into a 100-turn coil. The changes in current for each turn creates a magnetic force field change affecting every turn of the coil. Wire size itself does not affect inductance directly, but indirectly it does. All wire made of the same material has the same inductance regardless of its gauge but the thinner a wire is, the more turns we can have in a given space and the number of turns does affect inductance.

If another wire is lying close to the current carrying wire, it will be in the changing magnetic field and it will also have an induced EMF. This effect is called *mutual induction*. A transformer is a device constructed so that one wire lies close to another wire. The wire, which has an AC voltage source connected to it, is called the *primary*. The changing current in the primary generates a changing magnetic field that cuts the second wire and generates an EMF in it. This second wire is called the *secondary*. To increase the induced EMF, the primary wire is usually wound several turns to form a primary coil. To increase its ability to pick up induction, the secondary wire is also wound several turns to form a secondary coil. When an AC signal is placed on the primary winding of a transformer, an AC signal is induced into the secondary winding. The primary winding and secondary winding are not physically connected to each other. They are electrically isolated from one another. The primary and secondary coils are magnetically coupled to each other. A transformer transfers electrical power from the primary to the secondary by means of a changing magnetic field.

How efficient a transformer is at transferring power from the primary to the secondary depends on the degree of magnetic coupling between the two. Maximum coupling occurs when all the magnetic flux lines generated at the primary cut the secondary. This degree of coupling is a coefficient of coupling of 1. To achieve as high a coefficient of coupling as possible, the primary and secondary coils are usually wound on a steel core. Both coils are wound on the same core. Many times a boxed- or circular-shaped core is used. Because steel has low reluctance, the magnetic lines are efficiently coupled to the secondary. Most transformers achieve 99.99% efficiency.

When a voltage is first applied to a conductor, the induced EMF equals the applied voltage and opposes the EMF of the supply. This is why instantaneous current does not occur, even in a DC circuit. When a DC voltage is first applied to a conductor, the voltage changes from zero to maximum. The current also tends to change from zero to maximum, but anytime current changes in a circuit, an EMF is developed to oppose that change. With a DC circuit, current gradually reaches its maximum value and remains steady. With a steady (nonchanging) current, induction ceases to occur and induced EMF is zero.

This is not the case for AC voltage and current. They are never steady. They are always changing values. When an AC voltage is first impressed on a conductor, it develops a counter EMF in the conductor equal to the applied voltage. With induced EMF equal to applied EMF, no current will flow. Suppose the AC voltage supply has a maximum 120 V rms, and we attach the supply to an inductor at that point in time when it is 120 V positive. The induced EMF will be 120 V negative. Initial current is zero. As the voltage decreases from 120 V positive to 0 V, the induced EMF slowly drops its opposition. Current starts flowing and rising current levels cause the induced EMF to rise from –120 V to 0 V.

When current reaches its maximum level, the induced EMF will be 0V At this point in time, supply voltage is 0 V and will start falling toward –120 V, current is maximum positive and will start falling toward 0 amp, and induced voltage is 0 V and will start rising toward +120 V. As the supply voltage moves toward –120 V, the counter EMF moves toward +120 EMF. The induced EMF is always out of phase with, but equal to, the supply voltage and lags the supply by 180°. The current in a purely inductive circuit lags the supply voltage by 90° and leads the induced voltage by 90°.

Since induced EMF is equal to supply EMF for the primary winding, if the secondary winding has the same number of turns as the primary, its induced EMF is equal to EMF induced in the primary. If we place twice as many turns in the secondary, the magnetic flux generated by the primary will cut twice as many conductors in the secondary. This creates twice the voltage in the secondary winding. So placing more windings on the secondary coil than are placed on the primary coil makes a step-up transformer. If you want a step-down transformer, put less turns on the secondary than on the primary. If the secondary transformer has half the turns of the primary, the voltage induced in the secondary is half that of the primary.

If you want to step the voltage up by a factor of 10, use 10 times as many turns on the secondary as you do the primary. This is called a *10:1 turns ratio.* If the power company sets the output of the power plant at 16,000 V, they can use a 10:1 transformer to step the voltage up to 160,000 V for transmission to a substation. The substation can use a 1:12 transformer and step the voltage down for distribution. At our home or business, a 1:120 transformer would be used to step the voltage down from 14,400 to 110 V/120 V. The exact voltage reaching our outlet will depend on the power losses that occur in the power distribution system, but it will be somewhere between 110 V and 120 V.

The reason we use AC for commercial power is because it is so easy to convert voltages by the use of a transformer to any voltage needed by the customer. We use transformers in telecommunications because of the high efficiency with which they can connect an AC signal from one circuit to another while maintaining electrical isolation between the circuits. As we will see later, the electrical signal created when a speech wave hits the transmitter of a telephone is a varying DC signal. Because it is a varying signal, the transformer treats it the same way it treats an AC signal.

C.9 CAPACITIVE REACTANCE

Capacitive reactance occurs in all AC circuits and is due to the *capacitance* existing in a circuit. Capacitance exists when two conducting materials are separated by an insulator and is measured in farads. Every cable pair used to connect a telephone to the central exchange has capacitance between the two conductors because the conductors are separated from each other by the insulation around each wire. Capacitance per loop mile of a 24-gauge cable pair is approximately 0.083 µf. Capacitive reactance *(XC)* is a measure of how the capacitance of a circuit is affecting the circuit. It is inversely related to the frequency of the AC signal and the value of the capacitance. The higher the signal or the higher the capacitance, the lower the capacitive reactance will be. The formula for finding the capacitive reactance of a circuit is as follows:

Capacitive Reactance Formula
$$XC = 1/(2 \times 3.14159 \times F \times C)$$

Capacitance in the formula is stated as farads. For a 1000-Hz signal on a mile of local loop, the capacitive reactance is: $XC = 1/(2 \times 3.14159 \times 1000 \times 0.000000083)$ = 1917.5 Ω.

Capacitance exists down the length of the cable pair and is said to be distributed along the cable pair. The distributed capacitance is 0.083 μf per mile. The capacitance for 4 mi of cable is four times the capacitance of 1 mi. Since capacitive reactance is inversely proportional to capacitance, the capacitive reactance for four miles of cable is one-fourth the capacitive reactance at one mile. Thus, the capacitive reactance for 4 mi is 1917/4 = 479 Ω at 1000 Hz. Since capacitive reactance is also inversely proportional to frequency, the capacitive reactance for 4 mi and a 5000-Hz signal is 479/5 = 96 Ω. Remember that the capacitive reactance exists between the pairs. This 96 Ω of capacitive reactance between the pairs will effectively short the signal across the pair, and very little of the signal placed on one end of the loop reaches the other end of the circuit.

We must do something to correct the problem caused by distributed capacitance. To correct this problem, inductance is purposely placed in the loop. Capacitance and inductance are affected in opposite directions by frequency. The higher a frequency in a circuit, the higher the inductive reactance will be and the lower the capacitive reactance. Because inductive reactance tends to offset capacitive reactance, we can use inductance to offset the effect of capacitance. A human-made device that is purposely designed to add inductance to a circuit is called an *inductor*. A human-made device that is purposely designed to add extra capacitance to a circuit is called a *capacitor*.

Because inductors and capacitors have opposite effects on an AC signal, capacitive reactance and inductive reactance offset each other in a circuit. If a capacitor and inductor are in series, the net reactance is found by subtracting the larger from the smaller. If XC is 400 Ω and XL is 200 Ω, the circuit has a net impedance of 200 Ω XC (400 XC – 200 XL = 200 XC). Likewise if a circuit has 500 XL and 250 XC, the net impedance is 250 XL (500 XL – 250 XC = 250 XL). From this math it can be seen that when XL = XC, the total impedance of the series LC circuit is 0 Ω. The formula for total impedance in a series circuit is

$$Z = \sqrt{R^2 + (XL - XC)^2}$$

Thus when XL = XC, the circuit impedance is due solely to the resistance in the circuit.

C.10 TUNED CIRCUITS

Because inductance and capacitance have opposite effects on an AC signal, we can design circuits to take advantage of this condition. When an AC signal is placed on an inductance, the inductive reactance *(XL)*, or opposition to current flow, becomes greater as the frequency of a signal goes higher. When an AC signal is placed on a capacitor, the capacitive reactance *(XC)*, or opposition to current flow diminishes as the frequency of a signal increases. At very high frequencies, inductors have so much opposition to current flow that very little current will flow. At very high frequencies, capacitors have very little opposition to current flow. The reverse is true for very low frequencies. At very low frequencies, inductors have very little XL and capacitors have very high XC. A DC circuit has no frequency; the frequency is zero.

The formula for XL is:

$$XL = 2 \times 3.14 \times F \times L$$

If F (frequency) $= 0$, then XL is 0 because anything multiplied by 0 is 0. The formula for XC is:

$$XC = 1/(2 \times 3.14 \times F \times C)$$

If $F = 0$, this formula becomes $1/0$. Anything divided by 0 is an extremely high number that cannot be defined. Thus, if the frequency is 0, the XC is so high that no current will flow. This is why a capacitor blocks the flow of current when placed in a DC circuit.

Let's look at the XL and XC at a particular frequency for a particular inductor and capacitor. We will use a frequency of 10,000 Hz, an inductor of 1 millihenry, and a capacitor of 253.3 nf. Let's calculate the XL and XC for these two components:

$XL = 2 \times 3.14159 \times F \times L$

$XL = 6.28318 \times 10,000 \times 0.001$ (inductance must be stated in henries)

$XL = 62.83\ \Omega$

$XC = 1/(2 \times 3.14159 \times F \times C)$

(Capacitance must be stated in farads.)

$XC = 1/(6.28318 \times 10,000 \times 0.0000002533)$

$XC = 1/0.015915308$

$XC = 62.83\ \Omega$

Notice how $XL = 62.83\ \Omega$ and $XC = 62.83\ \Omega$. I purposely selected an inductor and capacitor that would provide the same impedance to the flow of electricity when the frequency of the AC signal is 10,000 Hz. When $XL = XC$, we say the circuit is a **tuned circuit** and the frequency of the signal is called the **tuned frequency.** The total impedance of a circuit is composed of total resistance, total inductive reactance, and total capacitive reactance. Because inductive reactance and capacitive reactance have opposite effects on current, they tend to offset each other in a series circuit. The formula for total impedance in a series circuit is:

Total Impedance Formula
$$Z = \sqrt{r^2 + (XL - XC)^2}$$

If $XL = XC$, total impedance is equal to the resistive component only because when XL and XC are equal, $XL - XC = 0$. But what if they are not equal? Let's look at a frequency that is not the tuned frequency. Connect the capacitor and inductor in series and pass a 12,000-Hz signal through the circuit.

$$XL = 2 \times 3.14159 \times 12{,}000 \times 0.001$$

$$XL = 75.398224 \ \Omega$$

$$XC = 1/(2 \times 3.14159 \times 12{,}000 \times 0.0000002533)$$

$$XC = 52.36 \ \Omega$$

$$Z = \sqrt{R^2 + (XL - XC)^2}$$

$$Z = \sqrt{R^2 + (75.398 - 52.36)^2}$$

$$Z = \sqrt{R^2 + 23.038^2)}$$

You can see from our example that when XL does not equal XC, the circuit will have higher impedance. Also notice the same happens if XC is larger than XL. When XC is larger than XL, we get a negative number by subtracting XC from XL. When we square this negative number, we have a positive value. Thus, it does not matter whether XL or XC is larger. If they are not the same, the series circuit will have greater opposition (impedance) to current flow when they are not the same value than when they are the same value. Let's look at the circuit when an 8000-Hz signal is passed through it:

$$XL = 2 \times 3.14159 \times 8000 \times 0.001$$

$$XL = 50.26 \ \Omega$$

This agrees with the statement made earlier: the lower the frequency, the lower the XL. Now let's look at XC:

$$XC = 1/(2 \times 3.14159 \times 8{,}000 \times 0.0000002533)$$

$$XC = 78.54 \ \Omega$$

This also agrees with an earlier statement: the lower the frequency, the higher the XC. Because XL does not equal XC at 8000 Hz, the circuit has more impedance than it has at the tuned frequency of 10,000 Hz. There is only one frequency for a series circuit at which impedance will be its lowest. This tuned frequency (resonant frequency) is found by the following formula:

Tuned Circuit Formula to Find Resonant Frequency

$$Fr = 1/(2 \times 3.14159 \times \sqrt{LC})$$

This formula was arrived at by using our knowledge of the fact that at the tuned frequency for the tuned circuit $XL = XC$:

$$XL = 2 \times 3.14159 \times F \times L \text{ and } XC = 1/(2 \times 3.14159 \times F \times C)$$

Therefore, at the tuned frequency where XL must equal XC:

$$XL = XC$$

From this it follows that:

$$2 \times 3.14159 \times FL = 1/(2 \times 3.14159 \times FC)$$
$$(2 \times 3.14159 \times FL) \times (2 \times 3.14159 \times FC) = 1$$
$$(6.28318 \times FL) \times (6.28318 \times FC) = 1$$
$$FF \times 6.28318 \times 6.28318 \times LC = 1$$
$$FF = 1/(6.28318 \times 6.28318 \times LC)$$

Now take the square root of each side of the equation:

$$F = 1 \ / \ 6.28318\sqrt{LC}$$

This formula can be used to find the tuned frequency (called the *resonant frequency*) for any value of inductance and capacitance. In telecommunications, we use inductors and capacitors to tune devices to a particular frequency. This is what tuners in a radio do. The tuner has one device that is variable. In older radios the tuning knob was attached to a variable capacitor. Turning the knob changed the value of the capacitor. By changing the value of the capacitor, we changed the frequency the radio was tuned to.

Tuned circuits pass one particular frequency with maximum power, but the tuned circuit also passes frequencies close to the tuned frequency. The power level of other frequencies passed by the *LC* filter will be less than the power of the tuned frequency. If a frequency exits an *LC* filter with 70.7% of the voltage of the tuned frequency, it is considered to be passed by the filter. The passband of a filter is the difference between the highest frequency passed with 70.7% of the voltage of the tuned frequency and the lowest frequency passed with 70.7% of the voltage of the tuned frequency. The *Q* of the coil providing the inductance to the circuit determines the pass band. A high *Q* results in a narrow passband and a low *Q* results in a wide passband.

C.11 AC-TO-DC CONVERTERS

AC voltage can be converted into DC voltage by a device called a *battery charger*. Sometimes these devices are called a *power supply*. The power supply of a personal computer is plugged into a wall outlet that supplies 120 V RMS AC. The power supply converts the 120 V AC into a +5V, –5V, +12V, –12V, and whatever other DC voltages are needed by the components in the computer. The power supply and battery charger use devices called diodes to rectify (convert) AC to DC. The battery charger in a central office is designed to convert 240 V AC into 52V DC. It is large enough to keep the batteries charged and also supply DC voltage and current to the central office equipment.

Just as your car does not run off its battery (it runs off of the generator or alternator), a central office does not run off the batteries. The battery in a car is only used to supply power when the engine is not running and the alternator is not working. The battery in a central office is only used to supply power when the AC powered battery charger stops working. The 52-V DC supply not only supplies voltage for central office equipment, it also supplies power for the telephone. This supply must be free of noise. In addition to providing a standby emergency supply, batteries also act as a noise filter. If a battery charger were used to supply the energy by itself, the supply would be noisy. The battery attached across the output of a battery charger absorbs the noise and it does not reach our phone. In computer-controlled central offices, additional noise filters are placed throughout the office on the battery supply that is going to be sent out to telephones.

Just as there are devices to convert AC electricity to DC electricity, there are also devices to convert DC to AC. The ringing signal used to ring our telephones is generated by a solid state, transistorized device that converts 52 V DC to an AC ringing signal. This device is called a *ringing machine*. Today, most ringing machines have an output of 20 Hz, 90 V RMS.

C.12 SUMMARY

Commercial electric power is AC voltage produced by a generator. The generator spins a coil of wire in a magnetic field to produce an AC voltage. The voltage produced in the United States is 60 Hz 14,000 V rms AC. The electricity is distributed at high voltage to customer locations and then passed through a transformer to step the voltage down to 220 V or 120 V as needed. The voltage developed by an AC generator depends on the number of flux lines cut per second by a moving wire. This depends on flux density and the speed of the moving wire. The faster it moves the more lines of flux the wire will cut. When a moving wire cuts flux lines at the rate of 100 million (10^{18}) magnetic lines per second, one volt is induced in the wire.

Iron, nickel, and cobalt conduct magnetism better than other materials. These materials are used to concentrate the lines of magnetic flux into a small area thereby creating a high density of magnetic flux. Alnico is composed of aluminum, nickel, and cobalt. It makes an excellent strong, lightweight, permanent magnet. When a conductor is moved in a magnetic field, a current will flow in the wire due to induced EMF in the wire. When current flows in a wire, a magnetic field is built up around the wire. This magnetic field can be increased by winding the wire in a coil and can be increased even more by using iron instead of air for the core. This is how electromagnets are made. Electromagnets make the most powerful magnets, and they are used in many applications. The magnets inside a generator are electromagnets.

AC voltage and current are compared to their DC counterparts by a measure known as *root mean square (rms)*. The formula is: AC rms = 0.707 AC peak V. If an AC voltage has a peak of 1V, it can do the same amount of work that is done by 0.707 V DC. Also 1 amp of AC is equivalent to 0.707 amp DC. Two properties that affect the total impedance of an AC circuit and thereby affect AC current flow that do not affect DC current flow are inductance and capacitance.

Inductance is due to the magnetic field that surrounds a current carrying wire. With an AC current, the magnetic field is continually changing in magnitude and direction. This induces an EMF in the wire that is 90° out of phase with the supply voltage. At times the induced EMF opposes the supply, and at other times it aids the supply. Inductance is measured in henries and it causes opposition to current flow. This opposition is called inductive reactance *(XL)* and is measured in ohms. The faster an AC current changes (the higher its frequency), the higher the inductive reactance. This is stated by the formula for inductive reactance: $XL = 2\pi FL$.

Capacitance exists when an insulator separates two conducting materials. When an AC current is placed over a circuit with capacitance, the capacitance causes attenuation of the signal. This attenuation is opposition to current flow and is measured in ohms. This is called capacitive reactance *(XC)* and the formula is: $XC = 1/2\pi FC$.

Capacitive reactance and inductive reactance offset one another. We use this knowledge to construct tuned circuits. Each unique combination of a capacitor and inductor in a circuit will have only one frequency of AC where $XL = XC$. For a series circuit this is where opposition to AC current is lowest. For a parallel circuit, this is where opposition to AC current is highest. Tuned circuits are used to tune a receiver so that it accepts only one AC signal out of many signals present.

Many devices are available to convert AC voltage into DC voltage for use by electronic equipment such as personal computers, radios, and televisions. These devices are called power supplies. They plug into an AC outlet and are designed to provide DC voltages needed by the electronic equipment. A battery charger converts AC into the DC voltage needed to charge a battery.

APPENDIX REVIEW QUESTIONS

1. What type of electricity does the power company supply?
2. How is AC different from DC?
3. Why did the telephone engineers select DC as a power supply?
4. What is the value of expressing AC in RMS values?
5. What is a tuned ringer?
6. How are *XL* and *XC* related to one another at the tuned frequency?
7. Why do generators use many sets of magnetic poles instead of just one set?
8. What is the *XL* for a 1-mh inductor at 1 KHz?
9. What is the *XC* for a 1-pf capacitor at 1 MHz?
10. What is the tuned frequency for the above components?
11. What is the best material to use for a permanent magnet?
12. What is the best material to use for a temporary magnet?
13. What is reluctance? Which element has the lowest reluctance?
14. What determines whether the magnetic lines of flux flow in a counterclockwise or clockwise direction?
15. If a coil is standing vertically and the wires are wound such that current flows in a clockwise direction, where is its north pole?

GLOSSARY

Alternating Current (AC)   An electric current that reverses directions at regular intervals. The magnitude of the current varies from 0 amp to some maximum value, falls back to 0 amp, rises to a maximum value flowing in the opposite direction, falls back to 0, and then repeats this cycle over and over.

Alternator   A mechanical device for generating an AC voltage by spinning a magnetic field inside a coil of wire.

Artificial Magnet   A magnet that does not attain its magnetism naturally.

Bar Magnet   A bar of iron or steel that has been artificially magnetized.

Capacitance   The property of a device that allows it to store an electrical charge.

Capacitive Reactance   A measure that states the amount of opposition that capacitance provides to an AC current. It is found by the formula $XC = 1/(2\pi FC)$.

Capacitor   A human-made device used to store electrical energy. It is made by separating two conductors with an insulator. If extra capacitance needs to be added to an electrical circuit, the designer places a capacitor in the circuit.

Counter EMF   Voltage opposing the changes of current in an AC circuit.

Electromagnet   An artificial magnet made by wrapping a coil of wire around an iron or steel core and then sending an electric current through the coil of wire.

Flux Density   A measure of the amount of magnetic flux per unit of cross-sectional area.

Flux Lines   Invisible lines of magnetic flux that leave the north pole of a magnet, traveling through air or some other medium to enter the south pole of a magnet. They then flow through the magnet back to the north pole.

Generator   A mechanical device that generates an AC voltage in a wire by spinning the wire inside a magnetic field.

Impedance   The total opposition that a circuit or component has to AC current. This total opposition is composed of resistance, inductive reactance, and capacitive reactance.

Inductance   The property of a circuit or device by which a change in current induces an electromotive force.

Inductive Reactance   A measure of the amount of opposition that inductance offers to an AC current. It is found by the formula $XL = 2\pi FL$.

Inductor   A human-made device used to purposely add extra inductance to a circuit.

Left-Hand Rule   A rule used to determine direction of magnetic flux around a wire carrying an electric current. By wrapping your left hand around a wire such that the thumb points in the direction of current flow, the fingers around the wire point in the direction the magnetic flux is flowing. The left-hand rule is also used to determine the north pole of an electromagnet. By wrapping your left hand around a coil of wire such that the fingers point in the direction of current flow, the thumb points toward the north pole of the electromagnet.

Magnetic Field   The area in which magnetic lines of flux are flowing.

Magnetism   The properties of attraction possessed by magnets.

North Pole   The magnetic north pole that exists at one end of a magnet.

Permanent Magnets   A steel alloy that holds magnetic properties for a long time once it has been magnetized.

Power Supply   A device that converts AC voltage into DC voltages and then supplies these voltages to other devices.

Reluctance   The resistance to flow of magnetic flux.

Residual Magnetism   The magnetic properties left in an electromagnet after the current has been reduced to zero.

Ringing Machine   A device in a telephone central exchange that converts 52 volt DC into an AC voltage used to ring a telephone.

Root Mean Square (RMS)   The effective value of AC compared to DC. The rms formula to get the DC effective working power of an AC signal: rms = 0.707 AC.

South Pole   The magnetic south pole of a magnet.

Temporary Magnets   Human-made magnets that readily lose their magnetism when the external magnetizing force is removed.

Tuned Circuit   A circuit containing inductance and capacitance so that at one particular frequency of AC they are the same ($XL = XC$). The circuit is tuned by XL and XC to be most responsive to one particular frequency.

Tuned Frequency   The frequency of an AC signal that a tuned circuit is tuned to.

Appendix D

Answers to Odd-Numbered Chapter Review Questions

CHAPTER ONE

1. The telephone is faster than the telegraph and provides instant communication. Users of a telephone do not need to know the signaling protocols used; the user of a telegraph needs to know Morse code.

3. The name of the company that manufactured repeaters for Western Union was *Western Electric,* and the primary stockholder in that company was Elisha Gray.

5. The types of signals used in telecommunications are electrical signals, radio waves, and light waves.

7. We still have operators to handle special situations such as third-party billing and to provide customers with the option of using an operator if they prefer to.

9. SS7 is the system is used to transmit caller ID to the requester.

11. A keysystem is preferred over a PBX when a business has more than 24 central office lines (trunks) or more than 100 stations.

13. An ACD automatically routes inbound calls to the next available agent on a *most idle agent first* basis.

15. A keysystem is a voice communication system used by a small business that needs multiple central office lines and more than one telephone.

17. A local area network, or LAN, is used to connect PCs and file servers together so they can share files, printers, and programs. A LAN consists of PCs, printers, and file servers connected within one building or department.

19. A binary signal is a signal that can assume either of two discrete steps. A binary signal is most often a digital signal and uses two discrete voltage levels such that one voltage represents the digit 0 and the other represents the digit 1.

CHAPTER TWO

1. The telephone was invented by Alexander Graham Bell and patented in 1876.

3. The first Independent was formed in 1893.

5. The Communications Act of 1934 effectively eliminated the provisions of the Graham Act.

7. The Federal Communications Commission (FCC) regulates interstate telecommunications service.

9. Legislation was enacted by Congress in 1949 that extended coverage of the 1936 Rural Electrification Act to include coverage of rural telecommunications. The amended act provided low-cost government loans for the establishment of telecommunications in rural America. State legislation guaranteed a positive rate of return and franchised territory to carriers.

11. The antitrust lawsuit filed in 1949 was settled seven years later by the 1956 Consent Decree.

13. Privately owned equipment is called *customer-provided equipment (CPE)*.

15. The MCI ruling of 1969 by the FCC and settlement of the lawsuit brought against AT&T by MCI (the MCI Decision of 1976) led to deregulation of long distance service.

17. The Modified Final Judgment of 1984 forced AT&T to divest itself of the local Bell Operating Companies. Now that the Bell Companies are separate companies, they can no longer be forced by AT&T to buy solely from Western Electric; they buy from manufacturers providing equivalent equipment at the lowest price. Although the 1956 Consent decree stopped some of Western Electric's pricing schemes, it did not prohibit the Bell Companies from doing business with Western Electric. As long as Western Electric and local Bell Companies were part of AT&T, the Baby Bells would continue to use Western Electric as their favored supplier.

19. The toll revenue sharing arrangement was referred to as *separations and settlements.* The process was killed by the 1984 MFJ.

21. Common carriers providing local telephone service are called *local exchange carriers (LECs).*

23. The PSTN is composed of the networks of all common carriers, and services on the PSTN are available to the general public. A private network is privately owned and managed for the benefit of its owners.

25. The major provisions of the MFJ were:
    a. The breakup of the Bell System by forcing AT&T to divest itself of the local Bell Companies.
    b. Deregulation of the long distance services and requiring the new Bell Companies to provide all IEC customers with equal access to their IEC. This requirement led to the establishment of POPs.
    c. The establishment of LATAs and the ruling that calls within a LATA must be handled by a LEC while calls between LATAs must be handled by an IEC.

27. Any PIC can be overridden by using the Feature Group B access code for the IEC a caller wishes to use for the current call.

29. A point of presence (POP) office is the central office location that connects the circuits of the LEC to the circuits of the IEC.

31. The Telecommunications Reform Act of 1996 (Public Law 104-104) replaces the 1984 MFJ. The LECs are allowed to sell CPE; the IECs are allowed to provide local service. Any qualified common carrier can provide local service. If the Bell LECs do not hinder competition in the local services arena, they have been able to provide interLATA toll services after February 1997. The Telecommunications Reform Act of 1996 officially deregulates the local service market and also allows the Bell Companies to enter into the long distance market.

CHAPTER THREE

1. The three parts of a communication system are transmitter, medium, and receiver.

3. Half-duplex systems transmit messages in both directions over one medium; the transmitters on each end of the medium take turns sending messages. Full-duplex systems are capable of transmitting in both directions at the same time.

5. The carbon transmitter was invented by Thomas Edison.

7. AC power provides the power to ring a phone.

9. The first central exchange was installed in 1878, almost immediately after the first phones were installed.

11. The two wires are called *Tip* and *Ring*.

13. *Trunk* is the name given to a pair of wires connecting one exchange to another.

15. Calls were completed between lines appearing on different switchboards by the use of interposition jacks.

17. The American National Standards Institute (ANSI) provides telecommunications standards for North America.

19. AT&T's Bell Labs developed most standards that have become de facto standards for the telecommunications industry. Many of these standards are found in the *Local Switching Systems Generic Requirements (LSSGR)*.

21. The first automated switching system was invented in 1892 by Almon B. Strowger.

23. The major factor delaying the implementation of time division multiplexing was the lack of an underlying technology. It was invented before the development of integrated semiconductor chip circuits. It could not be implemented effectively using vacuum tube technology.

25. The most important advances have entailed improvements to switching and multiplexing technologies.

27. XBAR is cheaper, requires less maintenance, and can select alternative routes if a primary route is busy.

29. The number of telephone exchanges had grown so much in some area codes that it was necessary to add new area codes in order to have more exchange codes.

31. SXS cannot work on DTMF tones, and a converter is required between the linefinder and first selector to convert DTMF into rotary-type dial pulses.

33. A *stored program control (SPC)* switch.

35. The primary common control components include marker, register, and translator.

37. The primary purpose of a translator is to translate the dialed telephone number into an equipment address location.

CHAPTER FOUR

1. An application program that turns a PC into a telecommunications terminal is "TAPI."

3. Thomas Edison improved the carbon granule transmitter found in most telephones.

5. The standard single-line telephone set requires DC voltage to power the transmitter and DTMF pad.

7. Station equipment includes telephones, computer terminals, and modems.

9. A multiline phone allows the connection of up to four lines to one telephone set with pushbutton access to each line.

11. A hybrid keysystem requires dialing 9 to access a central office trunk.

13. The old 1A2 keysystem required 25-pair cables for each extension because all central office trunks had to be wired to each phone. There was no switching matrix. With the 1A2 keysystem, switching was done at the phone. The pushbuttons for each line connected the handset to the line for that button. An EKTS uses four-conductor station wire. Switching is done by a space division switching matrix in the KSU. One voice pair and one signaling pair are needed between the KSU and the phone. The signaling lead will tell the KSU which central office trunk it is to connect to the voice pair.

15. System features are programmed using the attendant's telephone.

17. The purpose of a hybrid circuit in the telephone is to convert the two-way, full-duplex, two-wire, local loop to the four-wire circuit of the telephone (transmitter and receiver).

19. When a DTMF button is depressed, two tones are sent out over the local loop.

21. DID is *direct inward dial.* This feature allows a person to call a CBX station directly without having to go through a switchboard attendant.

23. An IDF is an *intermediate distributing frame* used in a CBX environment to connect station cables to vertical cables. An MDF is used in a CBX environment to connect stations (via IDFs) to CBX line circuits.

25. The two predominant quick connect blocks are Type 66 and BIX.

27. The three major output devices for a PC are visual display unit, printer, and disk storage.

29. A UART interfaces the data bus of a PC to a serial port.

CHAPTER FIVE

1. Copper wire will continue to be the facility of choice for the local loop because it is already in place, it costs less than other media, and the telephone receives its power over the local loop. Wire is a medium that conducts electricity and can provide power to the phone from the central office.

3. The device that governs design of the local loop is the carbon transmitter found in most telephones. The older transmitter requires a minimum of 23 milliamps of current to work properly; newer transmitters require 20 milliamps of current to work properly.

5. The two wires of the local loop are called *Tip* and *Ring*.

7. The telephone receives its power from the central exchange battery via the local loop.

9. A telephone can be located about 7 1/2 mi from the exchange before loop treatment is needed.

11. Subscriber carrier is used to serve a group of customers when there is not enough cable from the exchange. When a new subdivision is built, it may be built in an area where cable pairs are in short supply. Subscriber carrier is a cheaper alternative to placing a new cable over a long distance.

13. Mutual capacitance is offset by placing load coils on the cable pairs.

15. The local loop is loaded if the loop will be more than 3 mi. The first load coil is placed 3000 ft from the office. The spacing for additional load coils is every 6000 ft.

17. *SONET* stands for *synchronous optical network*. This is the standard for fiber optic transmission facilities.

19. The major source of noise in the local loop is from deterioration of the plastic insulation on wires due to water entering the cable. The outer covering of the cable may be damaged by excavation activity or animals eating the covering. When the outer covering is damaged, moisture is sucked into the cable.

CHAPTER SIX

1. Multiplexing involves combining two or more signals and placing them over one transmission facility.

3. Frequency division multiplexing occurs when each signal has its own distinct frequency over the transmission medium. The signals can be combined for transmission and then separated at the distant end by using a receiver tuned to each frequency.

5. The purpose of a channel unit is to change a signal so that it may be multiplexed. The channel unit also contains a receiver designed to pull out only one signal from the multiplexed signal.

7. Amplitude modulation is a technique used by a channel unit in an FDM system to convert the incoming voice signal (0 to 4000 Hz) into a higher-frequency signal. The amplitude of the voice signal causes the amplitude of the carrier signal for the channel to vary in direct relationship to the

amplitude of the voice wave. Each channel unit has a different carrier frequency. This allows combining the channel units into a FDM signal.

9. There is a big difference between FM and FDM. FM is a modulation technique to prepare a signal so that it may be multiplexed with other signals using FDM; however, FM is rarely used. AM is used in most carrier systems to prepare signals for FDM. FDM combines many signals of different frequencies onto one medium. The signals can be AM or FM.

11. A T1 system carries 24 voice channels. A T3 system carries 672 voice channels.

13. PAM transmits the amplitude of a small portion of a signal. PCM takes the PAM signal, converts it to a digital code, and transmits the digital code that represents the signal.

15. The sampling rate of PCM in the PSTN is 8000 samples per second.

17. Since each amplitude sample taken can assume any value, the number of different levels is infinite. With an 8-bit coding technique, PCM can only code 256 signal levels. Quantization is the process of converting an input sample amplitude level to the nearest of the 256 levels that can be coded.

19. The DS0 signal rate is 64,000 bps. The DS1 signal rate is 1,544,000 bps.

21. There are 193 bits in each DS1 frame.

23. When a T1 system uses cable pairs for a transmission medium, repeaters must be placed in each central office and at 6000-ft intervals along the cable route.

25. OC-48 contains 32,256 channels.

27. The major advantages of SONET standards are that different vendors' equipment can work together and that lower-level signals can be stripped out of higher-level signals without having to demultiplex the entire signal.

CHAPTER SEVEN

1. The predominant method used to communicate over a short distance is to use electricity to represent voice or data signals.

3. The three components of a communication system are the transmitter, medium, and receiver.

5. An analog signal is a signal that continually varies in amplitude.

7. The characteristics of an analog signal that can be changed to convey information are its amplitude, frequency, or phase relationship to a reference signal.

9. The local loop must be twisted-pair copper wire because the telephone receives its electrical power from the battery at the central exchange and wire is needed to convey this electricity from the central exchange to the telephone.

11. Digital voice signals provide quieter communication because the decoder pays no attention to the amplitude variances that a particular signal may contain.

13. NRZ-L is the protocol used by the EIA-232 interface of a personal computer.

15. In NRZ-L the signal never returns to zero voltage, and the value during a bit time is a level discrete voltage. NRI also uses two discrete voltage levels for its signal, but the value of the voltage present does not represent 1s and 0s. Data is coded as the presence or absence of a transition from one voltage state to the other. It does not matter whether the transition is from the higher voltage state to the lower (such as from +3 V to –3 V) or from the lower voltage state to the higher (such as from –3 V to +3 V). If a transition occurs at the beginning of a bit time in either direction, it is coded as a 1. If no transition in the signal occurs at the beginning of a bit time, it is coded as a 0.

17. Manchester coding is used on local area networks (LANs) that use the *Ethernet* signaling protocol.

19. Twisted-pair copper wire can be used for 10-Mbps transmission on local area networks because the distance is short enough that the loss at high frequencies is much lower than it would be on a long cable pair.

CHAPTER EIGHT

1. A data communication system handles data signals that are binary digital electrical signals. A voice communication system handles voice signals that are analog electrical signals. Because today's voice communication systems convert the analog voice signal into a digital signal, the distinction between the two systems is diminishing.

3. The most common connection between a DTE and a DCE device is an EIA-232 interface.

5. The two primary types of DCE are a modem and a CSU/DSU.

7. Serial transmission involves sending bits one behind the other over a transmission medium.

9. Synchronous transmission is used between two high-speed modems. The computers attached to the modem may use asynchronous transmission between the computer and the modem. A PC to a modem is asynchronous, but between the modems, synchronous transmission is used.

11. Two high-speed modems use forward error correction, usually LAP-M with continuous ARQ.

13. LAP-M provides continuous ARQ.

15. Standard QAM achieves 16 different detectable events using 8 phases and 2 amplitudes.

17. A baud is the rate at which a signal changes. If a 1000-cycle signal changes back and forth between a 900-cycle signal and a 1000-cycle signal 3000 times a second, it is a 3000-baud signal.

19. A V.34 modem uses a baud rate of 3200. The signal changes 3200 times a second. Each signal represents 9 bits. Thus, $3200 \times 9 = 28,800$ bps.

21. A *PAD* is a *packet assembler/disassembler*. The PAD packages data into packets for transmission over a packet switching network. It also receives packets of data and depacketizes the data.

23. SPC offices use the Signaling System 7 (SS7) network to communicate with each other. Signals are sent in the form of word commands from one office to another.

25. The signaling used between the central office and the caller ID equipment at the customer's location is a 1200-Hz and 2200-Hz signal in asynchronous 1200-baud mode.

CHAPTER NINE

1. Data is transmitted over the PSTN without using ISDN by using a modem or data service unit.

3. Cable modems are approximately 60 times faster than ISDN.

5. The physical cable pair used by BRI is a digital subscriber line (DSL). The DSL circuit is one two-way digital circuit operating at 160 Kbps. The ISDN line circuit at the central exchange and the terminal device at the customer's location divide this 160-Kbps bit stream into four logical data channels (B1-, B2-, D- and M-channels).

7. The S/T interface uses AMI line coding.

9. The baud rate of 2B1Q is 80 Kbaud.

11. The 32,000 bits are part of the overhead and control sent over the S/T interface to the NT1. The NT1 uses this information to identify which device is active. The S/T bus uses 48 Kbps for control, performance monitoring, and timing. The 2B1Q protocol only needs 16 Kbps for framing and control.

13. SPID stands for *service profile identifier*. The SPID is in the database of the LEC central exchange and defines which services can be accessed by an ISDN line.

15. TE is ISDN-compliant terminal equipment, and a TA is an interface device used to connect a non-ISDN device such as a regular telephone to an NT1.

17. A LAPD frame contains the following fields: (1) FLAG field—the flag signals the beginning and end of a frame; the flag consists of the bit pattern 01111110; (2) ADDRESS field—contains the SAPI and terminal equipment identifier that is sending or is to receive the message; (3) CONTROL field—identifies the type of frame; (4) INFORMATION field—contains layer 2 and 3 messages; (5) FRAME CHECK SEQUENCE field—uses CRC to check for errors.

19. The central office knows what type of service or connection to provide on an ISDN call by the TEI and SAPI.

21. ADSL was originally developed as a vehicle to carry video signals over twisted-pair copper wire. This technology would allow the LECs to provide cable TV services in direct competition against cable TV companies.

23. The speed of ADSL is about the same as that of a cable modem.

25. DSL has the same speed in both directions. ADSL has a much faster download speed and a much slower upload speed.

CHAPTER TEN

1. A LAN is a data communication system allowing a number of independent devices such as computers and printers that are usually located within 500 m of each other to communicate directly with each other over a common physical medium. We use a LAN to connect computers that are physically close together (such as within one building, or on one floor of a building).

3. DOS-based LANs use hardware inserted into one of the expansion slots of a computer to send and receive data over the communication medium. The use of a network interface card provides a higher speed because the interface card accesses the parallel data bus of the computer. These DOS-based LANs include LANtastic, NetWare Lite, and Windows for Workgroups.

5. A peer-to-peer network is a network that allows PCs to work as a server or client.

7. A network interface card (NIC) is the hardware that is used to interface a PC to a LAN.

9. The two sublayers of the link layer in IEEE 802.3 are the MAC layer and the logical link control layer.

11. The session layer of the OSI model contains the API.

13. Cat-5 is category 5 cable. It has a tight twist to enable it to carry 100-Mbps signals over a distance that does not exceed 100 m.

15. The layers of the OSI model are the application, presentation, session, transport, network, data link, and physical layers.

CHAPTER ELEVEN

1. A WAN is a wide area network. It is used to connect LANs together that are separated by a distance greater than a half mile.

3. A PDN is a *packet (or public) data network* that is used to provide WAN facilities for PCs or LANs.

5. A LCN is a *logical channel number* assigned in a PDN to the terminals that connect to the PDN.

7. A SVC is a *switched virtual circuit*. The SVC is assigned a new LCN each time the terminal starts a new session over the PDN.

9. LAPF is the protocol used at the data link layer by frame relay.

11. A call setup packet is not required for a PVC because the LCN is permanently assigned to a PVC.

13. A FRAD is used to access the frame relay network.

15. ATM is the packet switched network designed to support SMDS.

CHAPTER TWELVE

1. Tim Berners-Lee developed the browser software.

3. TCP is a transport layer protocol.

5. IP is not a connection-oriented protocol.

7. The basic function of TCP is to turn the unreliable IP connectionless protocol into a reliable connection.

9. Port numbers are used to identify which application or protocol service a packet should be delivered to.

11. FTP is used for file transfers between computers.

CHAPTER THIRTEEN

1. PCS is a concept. The concept of PCS is to assign someone a personal telephone number (PTN). This PTN is stored in a database on the SS7 network. That database keeps track of where a person can be reached. When a call is placed for that person, the artificial intelligence network (AIN) of the SS7 determines where the call should be directed.

3. *MTS* was the first *mobile telephone system* installed (1946). These systems used radio frequencies between 35 and 45 MHz. Although *MTS* usually stands for "mobile telephone system," it also meant "manual telephone system." All calls had to be handled by an operator.

5. *Digital advanced mobile telephone system (DAMPS)* is a technique that places multiple calls over one radio frequency using code excited linear predictive coding (7.95 Kbps) and time division multiplexing (TDM).

7. *Code division multiple access (CDMA)* has been adopted as IS-95 Standard for PCS networks. CDMA can provide ten times the capacity of DAMPS. CDMA is not backward compatible with analog AMPS in the same way that D-AMPS is. CDMA cannot reuse DAMPS technology. To implement CDMA, the service provider must install a new network. CDMA is spread spectrum technology. Many conversations are multiplexed over one frequency. Each conversation is assigned a code. The receiver strips out conversations individually by using the code.

9. The BTS for a PCS 1900 system will consist of a small building to house the base station transmitters and receivers, and an antenna tower approximately 150 ft high. The antenna tower will contain nine antennas arranged in a triangular array. Each side of the triangle will contain one transmitting antenna and two receiving antennas.

11. The *home location registry (HLR)* contains information about the services the PCS subscriber is allowed to access. It also contains a unique authentication key and associated challenge/response generators. The AUC will verify that the customer is legitimate and allow the call to proceed.

13. A BTA (basic trading area) is a subset of an MTA.

15. Companies had to bid at auctions held by the FCC.

APPENDIX A

1. DC voltage is most often produced by a chemical reaction in a device called a *chemical cell* or *battery.*

3. A *molecule* is a material composed of two or more atoms of elements. A molecule is the smallest piece of matter that will still possess the characteristics of the larger mass. If the matter is broken down smaller than a molecule, it is broken down into atoms of the elements comprising the molecule.

5. An *atom* is the smallest piece of an element that still retains the characteristics of the element. An atom is composed of a nucleus containing positive charges called *protons* (and possible *neutrons,* which have no charge) and negative charges called *electrons* surrounding the nucleus.

7. *Electric current flow* is the orderly flow of electrons. For DC current flow, the electrons flow in one direction. For AC current flow, the electrons flow in one direction for a brief period of time, then flow in the opposite direction for a like amount of time, then reverse direction again. The flow of current continues to alternate between the two directions at twice the frequency of the signal.

9. *Voltage* is a force that tends to cause electric current flow. Voltage occurs when there is a difference in the number of electrons a material has compared to what it normally has.

11. A carbon-zinc dry cell develops 1.5 V per cell.

13. The physical size of a battery determines how long it can provide a voltage before it becomes discharged.

15. If four 100-Ω resistors are connected in parallel, their total resistance is 25 Ω.

17. An *insulator* is a material that has a high opposition to the flow of electric current.

19. One ampere is the flow of 6.25×10^{18} electrons flowing past a given point in 1 sec.

21. Kirchhoff's voltage law states that the sum of all the voltage drops around a single closed loop in a circuit is equal to the total source voltage in that loop. In other words, total voltage = $VR_1 + VR_2 + VR_3 + \ldots$.

23. Ohm's law states these relationships between voltage, current, and resistance in the formula: current in amperes = voltage/resistance ($I = V/R$).

25. An ion is an atom that has an electric charge. An atom with a positive charge is a positive ion. A negatively charged atom is a negative ion.

APPENDIX B

1. A circuit is a circular journey. An electrical circuit is a complete electrical path between the terminals of a voltage supply. This complete path allows electrons to start their circular journey out of the voltage source, go through a conducting path, return to the voltage source, and then flow through the voltage source to the starting point, thereby completing their circular journey. In a DC circuit, electrons leave the negative terminal of a battery, flow through a conductor, flow through some device that uses the electric current to do some work, flow through a conductor back to the positive terminal of the battery, and then flow through the battery to the negative terminal. There will be just as much current entering the positive terminal of the battery as there was leaving the negative terminal.

3. A device used to measure voltage is a voltmeter.

5. A device used to measure resistance is an ohmmeter.

7. A decibel (dB) is the unit of measure used to compare two power levels. It is not an absolute but a relative measure. In telecommunications, power losses are stated in decibels. The decibel level indicates the relationship of one power level to another. The relevant formula for a decibel is the following: $dB = 10 \log P_2/P_1$.

9. The measurement used for noise is dBrnC0, and –90 dBm = 0 dBrnC0.

11. The maximum distance at which a telephone can be located from a stored program control central exchange and still receive 20 milliamps of loop current, if 26-gauge wire is used, is approximately 4 mi.

13. Since resistors used to construct a voltage divider must have a low resistance, they will have large values of current flow and must have a high wattage rating.

15. Kirchhoff's voltage law states that the sum of the voltage drops around a closed loop equals the voltage applied to the loop.

APPENDIX C

1. The power company supplies AC electrical power to our homes and businesses.

3. The telephone engineers selected DC as a power supply because batteries were more readily available than AC power.

5. A tuned ringer is a ringer that has the inductance of the ringer coil and a capacitor in series with the coil acting as a tuned circuit. The ringer only works on one ringing signal.

7. Generators use many sets of magnetic poles instead of just one set to reduce the speed of rotation by the armature.

9. The XC for a 1-pf capacitor at 1 MHz is 159,154 Ω.

11. The best material to use for a permanent magnet is Alnico.

13. Reluctance is the opposition that a material offers to magnetic lines of flux. Iron has the lowest reluctance.

15. If a coil is standing vertically and the wires are wound such that current flows in a clockwise direction, the north pole is at the top of the coil.

Appendix E

Hypertext Markup Language (HTML)

Hypertext Markup Language (HTML) is used to develop web pages so they can be viewed by using a web browser. The web browser software is designed to interpret the HTML document and display its contents according to the HTML instructions embedded within the document. The HTML instructions are called tags. The first instruction (tag) of an HTML document is HTML. This tag indicates that the document is an HTML document.

Several web browsers are available. Most people use Netscape's Navigator or Microsoft's Internet Explorer. You enter the location of a web site that you wish to visit in the Uniform Resource Locator (URL) address field. In Netscape this field is called *Location*; in Internet Explorer, it is called *Address*. Type the name of the site or the actual dotted decimal address into the address or location box provided in the menu bar at the top of the window provided by the browser. After entering the URL, press the Enter key and the browser will activate your connection software and connect to the site on the Internet that you have specified in the address or location box.

When you purchase Internet access services from an Internet service provider (ISP), it provides you with some storage space on one of its disk drives. This space is provided so you can post your own web page on its web server. You create your web page on your computer using a regular text editor or a special HTML editor. You save this page to your hard drive. You use a File Transfer Program (FTP) to upload your web page from your hard drive to the storage location provided to you by the ISP. This location will usually be of the following form: www. **ISP name/~your name.**

Once you have Internet service via your ISP and have disk space on its server to store your web page, you need to decide whether you want to post a web page. If you decide to post a page for everyone on the Internet to see, you should spend some time designing the layout and content of your page. Many people post pictures of their family and pets on their web page. Some people even have active web page sites. They have a video camera running and connected via their PC and ISP server to provide steaming video pictures of their activities on their web site.

The design layout and content of your web site depends a great deal on what the site will be used for. For most of us a simple site will do. Usually, the simpler a site is, the better the Internet surfers will receive it. A site that is overloaded with graphics takes too much time to download, and when people encounter these sites, they usually exit quickly. Try to keep it simple. Do not try to provide too much content at your site. You often lose visitors that link out of your site to other sites, so put links where they are convenient to the surfer but where they force the visitor to stay a while before they can link out to other sites. Your web page will probably end up being multiple pages with links between the pages and graphics files.

Commercial web sites will require the talents of a seasoned web page developer. These HTML gurus will probably use a program such as Microsoft's FrontPage to help them provide polished web pages for their customers. Unless you have a couple hundred bucks or more that you can part with to buy a web publishing software package that will see limited use in developing your home page, you will probably want to develop your page using HTML tags. Most HTML tags have beginning and closing tags, and each of the tags is enclosed in brackets, such as a beginning tag <HTML> that indicates that the document is an HTML document and an ending tag</HTML> to indicate the end of the HTML document. The new paragraph tag<P> does not require an ending tag. Thus, </P> is not required and is seldom used. The following are the most common tags used in HTML:

HTML placed at the beginning of an HTML document

/HTML placed at the end of an HTML document

HEAD placed after the HTML tag and before the text of an HTML document to provide a heading for the document

/HEAD placed at the end of a heading to close the heading

TITLE placed at the beginning of the title information for the web page

/TITLE placed at the close of the title information

BODY placed at the beginning of the body of text that is your web page contents

/BODY placed at the end of your text for the page

Now let's use a text editor such as Notepad or DOS Editor to compose the following web page (substitute your own name where you see mine).

```
<HTML>
<HEAD>
<TITLE>Marion Cole's Web Page</TITLE>
</HEAD>
<BODY> This is my first web page </BODY>
</HTML>
```

Create a directory called homepage on your C drive and save the file we just created as MYPAGE.TXT. Also save the file as MYPAGE.HTML in the homepage directory. Click on the file you saved as MYPAGE.HTML, and the browser should automatically launch and open the file. Another alternative would be to open your browser and type the location and name of the file (C:\homepage\MYPAGE.HTML) in the address / location box of the browser. You should see the line of text displayed that reads "This is my first web page". Now let's add more tags to our page.

<P> New paragraph does not need a closing tag

<H1> <H2> <H3> <H4> <H5> and <H6> are heading tags that allow you to structure your page with various headings. Each of these has a different-size font to provide structure within a document. If you wish to change the size of font used by your text, you should use the FONTSIZE tag for that purpose and reserve the heading tags for headings. Each heading tag should be closed by its counterpart </H1> to </H6> as appropriate. It is important to remember the closing tag, or all of your page will be displayed as the HEAD chosen from that tag to the end of the document.

Use <FONT SIZE=1> to <FONT SIZE=6> to change the size of FONT. If a FONT SIZE has not been specified, the default font size = 3. Again remember to use a closing tag. The closing tag for all FONT SIZEs is simply </FONT>. You do not need to specify the size being closed. The </FONT> tag simply restores you to the default font size of 3.

<BOLD> Displays text in bold and needs to be closed with </BOLD>

<I> Displays text in italics and needs to be closed with </I>

<TT> Displays in monospaced typewriter font. Close with </TT>

<U> Underlines text. Close with </U>

<BIG> Bigger than surrounding text. Close with </BIG>

<SMALL> Smaller than surrounding text. Close with </SMALL>

<PRE> Placed in front of text that you wish to retain all formatting for. Browsers pay no attention to spaces, carriage returns, and other types of formatting. If you wish to retain the text as it is formatted, use the <PRE> tag before the text and the closing </PRE> after the text.

Now let's add to our first web page. Let's open up Notepad and open the file MYPAGE.TXT. We should have the following lines of text:

<HTML>
<HEAD>
<TITLE>Marion Cole's Web Page</TITLE>
</HEAD>
<BODY> This is my first web page </BODY>
</HTML>

Add the text that follows between <BOLD> This is my first Web Page <BOLD> and </BODY>.

<BOLD>This is my first Web Page</BOLD>

I am trying out different size fonts. <Bold><I>This is bold and italic</I></BOLD>

 The BR tag starts a new line.
<P> The P tag starts a new line with a space added between lines
<P> The P and BR tags do not need closing tags
<P><H1> This is heading 1; it requires a closing tag </H1>
<P><H2> This is heading 2; it requires a closing tag </H2>
<P><FONT SIZE=1> This is font size 1</Font>

How did everything look? If something looks wrong, check to make sure all closing tags are used and that they all include the forward slash (/).

</BODY>
</HTML>

Save the file as MYPAGE2.TXT and as MYPAGE2.HTML and then open MY-PAGE2.HTML with your browser. It usually pays to develop your web pages a few lines at a time and to verify the work done by opening the HTML file via your browser. This allows you to catch mistakes easily. Trying to debug a large HTML file may be difficult for a first-time user. Develop a little, then debug. Develop more and then debug again, and so on. Now we can try the preformatted text tag or the <PRE> tag. Type a document in Notepad using different-size fonts with paragraphs and spacing included, such as the text below.

<PRE>
This is Times New Roman Bold at font size 14.

I entered two carriage returns (Enter key) and changed to italics.

I entered two carriage returns and now will enter
ten spaces   How does this look?
</PRE>

Copy the above text and PRE from the Notepad document tags into the bottom of the body section of our MYPAGE2.TXT file and then save it as MYPAGE3.TXT and as MYTEXT3.HTML. Open MYPAGE3.HTML with your browser. You will notice that the browser keeps the carriage returns and spaces but pays no attention to the font selections. Everything is in FONT SIZE = 3. That's as it should be. The PRE tag simply keeps carriage returns and spaces in the document. If you need to place spaces in a page and do not use the PRE tag, you can use &#32 for each space desired. Wherever a space is required in a web page, simply type &#32. If you need four spaces, type &#32&#32&#32&#32. For double quote marks, you type &#34 wherever the quote marks should appear.

E.1 LISTS

HTML can provide several types of lists. The most common forms of lists are numbered (or ordered) lists, which are surrounded by the ordered list tags (<OL> and </OL>) and unnumbered lists, which are surrounded by the unnumbered list tags (<UL> and </UL>). The following is an example of an ordered list. Open MYPAGE3.TXT and save as MYPAGE4.TXT. Add the following lines of text provided from <P> to </OL> at the bottom of the page. Add them right after the closing </PRE>tag. Then save the file again as MYPAGE4.TXT and then save as MYPAGE4.HTML. Open the HTML file with your browser. Is the ordered list numbered?

<P>The line circuit in a SPC switching system performs seven functions:

<OL>
<LI>Battery feed
<LI>Overvoltage protection
<LI>Ringing signal
<LI>Signaling
<LI>Coding
<LI>Hybrid network
<LI>Test access
</OL>

The following is an example of an unnumbered list:
<P>The line circuit in a SPC switching system performs seven functions:

<UL>
<LI>Battery feed
<LI>Overvoltage protection
<LI>Ringing signal
<LI>Signaling
<LI>Coding
<LI>Hybrid network
<LI>Test access
</UL>

Open your MYPAGE4.TXT file and change the <OL> and </OL> tags to <UL> and </UL>. Save the file as MYPAGE5.TXT and then as MYPAGE5.HTML. Open the HTML file, and the list should now be in order but unnumbered. Bullets have replaced the numbers. Notice that in both cases, the <LI> tag precedes each item in a list. An ordered list is used when items must be done in a specific order, and an unordered list is used when the items in the list can be in any order. You can also form lists using <MENU> and </MENU> tags or <DIR> and </DIR> tags as the opening and closing tags, with each item in the list preceded by an <LI> tag. Try them out and see how the lists appear in a browser.

E.2 LINKS

A link tag <A> and </A> is used to link to a file on your hard drive or to another web site. You use the <A> tag with additional information about where you want to link to within the brackets of the tag. If you want to link to another file on your hard drive that is in the same directory, the link would be something like the following:

<A HREF="MYPAGE4.HTML">Go to MYPAGE4</A>

Notice that the file that we will link to is placed inside straight quote marks not curly marks. Also notice that we provide a name for the link in the space between the closing bracket of the <A . . . > tag and the opening bracket of the closing </A> tag. We could have started our link procedure by first providing a name for our link, surrounding the name of the link with <A> and </A> tags and then filling in the reference location information for the link as follows:

Go to MYPAGE4
<A>Go to MYPAGE4</A>
<A HREF="MYPAGE4.HTML">Go to MYPAGE4</A>

The same procedure is followed to link to a web page on the Internet. The HREF will specify the URL of the web page desired, such as linking to NASCAR's home page from your home page. Insert the text below into your MYPAGE5.TXT. Save the file as MYPAGE 6.TXT and MYPAGE6.HTML.

<P>Check out<A
HREF="http://**www.nascar.com**"> Stock Car Ricing at NASCAR</A>

Open a connection to the Internet via your ISP. After an Internet connection is established, open MYPAGE6.HTML and click on the link. Did you connect to NASCAR's home page? Notice when this link is displayed in your browser that only the text between the <A> and </A> tags is highlighted as the name of the link. Also notice that the http:// has to be included as part of the HREF or the browser will think the link is to a file on your hard drive.

When you link from one file to another, you must use a method that will remain effective after all files have been uploaded to your ISP server. If all files are in one directory, you will not have to specify the directory path to the linked file. The browser assumes that links to files are to files within the same directory. You can also link to a specific place within a web page by providing anchors for the link. This is done by using a NAME tag within the document you link to. Many people use this method to link to sections within one large web page. Each section is given a name such as <A NAME="Chapter One">Chapter One</A>.

Suppose you set up a large preformatted file as an HTML file and set up the Table of Contents as an unordered list with links to the NAME tag anchors. The title for Chapter One would be written as in the previous paragraph. The start of Chapter Two would be written as <A NAME= "Chapter Two">Chapter Two</A>.

The start of each chapter would be designated by the title of the chapter written in NAME tag format. The Table of Contents would then look like this:

```
<UL>
<LI><A HREF=#Chapter One>Chapter One</A>
<LI><A HREF=#Chapter Two>Chapter Two</A>
<LI><A HREF=#Chapter Three>Chapter Three</A>
<LI><A HREF=#Chapter Four>Chapter Four</A>
```

The Table of Contents would be completed as shown for each chapter and appendix contained in the web page. The user could then link quickly to a specific section within the file. Large files all use this technique. If you wish to link to a particular section in a large file from another file, you must also specify the name of the file that contains the section you wish to link to. If you wished to link to a bread pudding recipe in a file called *recipes,* you would provide an anchor in the recipes file at the start of the recipe for bread pudding as follows:

```
<A NAME="Bread Pudding">Bread Pudding</A>
```

Now you can provide a link from another web page in the same directory to the bread pudding recipe with the following link:

```
<A HREF="recipes.html#Bread Pudding">Bread Pudding Recipe</A>
```

E.3 LINKS TO GRAPHICS FILES

Once you have a graphics file that you would like to include in your web page, you simply place the file in the same directory as your web page and the link to it. Of course the graphics file does not have to be in the same directory. You could place the file anywhere and specify the path to the file in your link but why complicate matters? Keep it simple. Place all files for your home page in the same directory. Suppose you have a picture of the family saved as **family.gif.** You can place a link in your web page to display the picture by using the IMG tag.

```
<IMG SRC="family.gif">
```

The SRC indicates the source location of the image file and is a necessary part of the IMG tag. No closing tag is needed, but remember to use straight quote marks to enclose the location and/or name of the gif file. The same convention applied to the A HREF = also applies to the IMG SRC=. Some people also include the ALT attribute of the IMG tag when doing links to images in order to provide a name of the image to a browser that does not display gif or jpeg files. It would be in the following format:

```
<IMG SRC="family.gif" ALT="a picture of my family">
```

E.4 OTHER HTML TAGS

If you wish to separate text with horizontal lines, you can use the <HR> tag. It does not require a closing tag. Do not overdo the use of horizontal lines on your web page, but use them as necessary to separate sections. You can also use the ALIGN=LEFT, ALIGN=CENTER, or ALIGN=RIGHT to specify the alignment of text. You can use the BGCOLOR attribute of the BODY tag to establish the background color you desire on your web page. See the examples that follow:

```
<BODY BGCOLOR="#000000">
<BODY BGCOLOR="#FFFFFF">
<BODY BGCOLOR="#FF0000">
<BODY BGCOLOR=WHITE>
<BODY BGCOLOR=BLUE>
<BODY BGCOLOR="Backgrd.gif"
```

As seen in the previous examples, the color can be specified by using the red, green, blue values; 000000 is black and FFFFFF is white. The first two hexadecimal numbers specify the amount of red, the next two specify the amount of green, and the last two specify the amount of blue color. Thus, to get a background color of red you would specify heavy red, no green and no blue, or FF0000. Likewise, green would be 00FF00 and blue would be 0000FF. There are 16 colors on the Windows color palette and you can specify one of these 16 colors by using the name of the color rather than the RGB values. The 16 colors are black, white, gray, red, green, blue, maroon, purple, navy, teal, yellow, olive, lime, aqua, fuchsia, and silver. The background color can also be supplied from a gif file if desired.

Colors can also be used as a BODY tag attribute to identify the color of text and links, as shown in the following example:

```
<BODY BGCOLOR=BLUE  TEXT= WHITE  LINK = PURPLE>
```

E.5 FRAMES

You can design your web page using frames. A frame can be designed to display several web pages at one time, with each page sitting inside a frame within the display. You can create a frame by using the FRAMSET tag. This tag cannot be used in conjunction with a BODY tag. The FRAMSET tag replaces and is used instead of the BODY tag. The FRAMSET tag must contain the attributes that you wish to assign to a frame.

The COLS and ROWS attributes define the vertical and horizontal sizes of each frame. <FRAMSET COLS = (10%, 50%,*)> will separate the screen into three frames. The first will be 10% of the viewing area; the second is 50% of the viewing area and the third is what remains (40%). You can also define size as the number of pixels instead of percentage of screen. The ROWS attribute splits the screen into horizontal frames. <FRAMSET ROWS=(50%, 50%)> will provide two frames of equal size, with the first frame sitting on top of the second frame. The FRAMSET tag defines the sizes of frames but will not display anything until the contents of each frame are specified with the FRAME tag.

The FRAME tag tells the browser which file is to be loaded into a specific frame. <FRAME SRC=(index1.html)> instructs the browser to load the index1.html file into the FRAME specified. The home page controlling document is index.html. This document is automatically loaded if no file is specified. You will find that the source code spells out how the frames are to be set up.

If you use <FRAMSET COLS=165, *> and <FRAMSET ROWS=80, *> to establish the first two frames on the left side of the screen. The first frame is 165 pixels wide and 80 pixels tall. The second frame is also 165 pixels wide and uses the remaining screen after 80 pixels at the top of the screen. Thus, you have two frames 165 pixels wide stacked one on top of the other. If we stopped here, the right-hand side of the screen would contain nothing. We define the third frame as being 100% of the area available and stick the index1.html into this area with the tags <FRAMSET ROWS= 100%> and <FRAME SRC=index1.html>. Of course, you must also remember to use the closing </FRAMSET> tags where appropriate.

E.5 DESIGNING A WEB PAGE

Using the information supplied in this section, try constructing a web page for your use. You can always find help on HTML at Netscape's home site, or you can simply use a search engine to locate helpful tips on HTML. Jump in and try your hand at it. I learned much of what I know about HTML from Laura Lemay's *Teach Yourself Web Publishing with HTML 3.2 in 14 Days*. (Indianapolis, IN: Sams Net, 1996). If you need to know more than the basics I have presented, I highly recommend this book.

Appendix F

Summary of 1996 Telecommunications Reform Act

The Library of Congress

Summary of Senate Bill 652

S.652

PUBLIC LAW: 104-104 , (became law 02/08/96)

SPONSOR: Sen Pressler, (introduced 03/30/95)

TABLE OF CONTENTS:

Telecommunications Act of 1996—Title I: Telecommunication Services—Subtitle A: Telecommunications Services—Amends the Communications Act of 1934 (the Act) to establish a general duty of telecommunications carriers (carriers): (1) to interconnect directly or indirectly with the facilities and equipment of other carriers; and (2) not to install network features, functions, or capabilities that do not comply with specified guidelines and standards.

Sets forth the obligations of local exchange carriers (LECs), including the duty: (1) not to prohibit resale of their services; (2) to provide number portability; (3) to provide dialing parity; (4) to afford access to poles, ducts, conduits, and rights-of-way consistent with pole attachment provisions of the Act; and (5) to reestablish reciprocal compensation arrangements for the transport and termination of telecommunications.

Imposes additional obligations on incumbent LECs (incumbent LEC requirements), including the duty to: (1) negotiate in good faith the terms and conditions of agreements; (2) provide interconnection at any technically feasible point of the same quality they provide to themselves, on just, reasonable, and nondiscriminatory terms and conditions; (3) provide access to network elements on an unbundled basis; (4) offer resale of their telecommunications services at wholesale rates; (5) provide reasonable public notice of changes to their networks; and (6) provide physical collocation, or virtual collocation if physical collocation is impractical.

Directs the Federal Communications Commission (FCC) to complete, within six months, all actions necessary to establish regulations to implement such requirements. States that nothing precludes the enforcement of State regulations that are consistent with those requirements.

Requires the FCC to create or designate one or more impartial entities to administer telecommunications numbering and to make such numbers available on an equitable basis. Directs that the cost of numbering administration and number portability be borne by all carriers on a competitively neutral basis.

Exempts a rural telephone company from incumbent LEC requirements until such company has received a bona fide request from interconnection, services, or network elements and the State commission determines that such request is not unduly economically burdensome, is technically feasible, and is consistent with universal service provisions, except the public interest determination. Sets forth provisions regarding: (1) State termination of the exemption and the establishment of an implementation schedule; and (2) limits on the exemption.

Authorizes an LEC with fewer than two percent of the subscriber lines installed in the aggregate nationwide to petition for a suspension or modification of specified requirements for the telephone exchange service facilities specified in the petition. Directs the State commission to grant such petition to the extent that it is necessary to avoid significant adverse economic impacts on users of telecommunications services or to avoid imposing an undue economic burden or a technically infeasible requirement, where such suspension or modification is in the public interest.

Provides for the continued enforcement of exchange access and interconnection requirements. Authorizes an incumbent LEC to voluntarily negotiate and enter into a binding agreement with a requesting carrier without meeting incumbent LEC requirements. Directs that such agreement: (1) include a detailed schedule of itemized charges for interconnection and each service or network element included in the agreement; and (2) be submitted to the State commission. Permits any party negotiating such an agreement to ask a State commission to participate in the negotiation and to mediate any differences arising in the course of the negotiation.

Authorizes the carrier or any other party to the negotiation, from the 135th through the 160th day after the date on which an incumbent LEC receives a request for negotiation, to petition a State commission to arbitrate any open issues. Sets forth provisions regarding the duty of the petitioner, opportunity to respond, action by the State commission, refusal to negotiate, standards for arbitration, and pricing standards. Requires any interconnection agreement adopted by negotiation or arbitration to be submitted for approval to the State commission. Sets forth provisions regarding grounds for rejection, preservation of authority by the State commission, the schedule for decision, failure of the State commission to act, and review of State commission actions.

Authorizes a Bell operating company (BOC) to prepare and file with a State commission a statement of the terms and conditions that such company generally offers within that State to comply with incumbent LEC requirements and applicable regulations and standards. Sets forth provisions regarding State commission review, the schedule for review, and authority to continue review. Specifies that submission or approval of the statement shall not relieve a BOC of its duty to negotiate the terms and conditions of an agreement regarding interconnection.

Sets forth provisions regarding: (1) consolidation of State proceedings; (2) a required filing by the State commission; and (3) availability of any interconnection, service, or network element provided under an approved agreement to which the LEC is a party to any other requesting carrier on the same terms and conditions as those provided in the agreement.

Preempts any State and local statutes, regulations, or requirements that prohibit or have the effect of prohibiting any entity from providing interstate or intrastate telecommunications services. Preserves a State's authority to impose, on a competitively neutral basis and consistent with universal service provisions, requirements necessary to preserve and advance universal service, protect the public safety and welfare, ensure the continued quality of telecommunications services, and safeguard the rights of consumers.

Authorizes a State, without violating the prohibition on barriers to entry, to require a competitor seeking to provide service in a rural market to meet the requirements for designation as an eligible carrier. Makes this provision inapplicable to: (1) a service area served by a rural telephone company that has obtained an exemption, suspension, or modification that effectively prevents a competitor from meeting such requirements; and (2) a provider of commercial mobile services.

Requires: (1) the FCC to institute and refer to a Federal-State Joint Board a proceeding to recommend changes to any of its regulations to implement specified requirements, including the definition of the services that are supported by Federal universal service support mechanisms and a specific timetable for completion of such recommendations; (2) one member of the Board to be a State-appointed utility consumer advocate nominated by a national organization of State utility consumer advocates; and (3) the Board, after notice and opportunity for public comment, to make its recommendations to the FCC within nine months.

Directs the Board and the FCC to base policies for the preservation and advancement of universal service on: (1) availability of quality services at just, reasonable, and affordable rates; (2) access to advanced telecommunications and information services to all regions of the nation; (3) access and costs in rural and high cost areas that are reasonably comparable to that provided in urban areas; (4) equitable and nondiscriminatory contribution by all telecommunications services providers; (5) specific and predictable support mechanisms; (6) access to advanced telecommunications services for schools, health care, and libraries; and (7) such other principles as the Board and the FCC determine are in the public interest.

Defines "universal service" as an evolving level of telecommunications services that the FCC shall establish periodically, taking into account advances in telecommunications and information technologies and services.

Requires all carriers providing interstate telecommunications services to contribute to the preservation and advancement of universal service. Authorizes the FCC to exempt a carrier or class of carriers if their contribution would be "de minimis."

Provides that only designated eligible carriers shall be eligible to receive specific Federal universal service support.

Grants States authority to adopt regulations not inconsistent with the FCC's rules. Requires all providers of intrastate telecommunications to contribute to universal service within a State in an equitable and nondiscriminatory manner, as determined by the State. Permits a State to adopt additional requirements with respect to universal service in that State as long as such requirements do not rely upon or burden Federal universal service support mechanisms.

Directs: (1) the FCC, within six months, to adopt rules to require that the rates charged by providers of interexchange telecommunications services to subscribers in rural and high cost areas shall be no higher than those charged by each such provider to its subscribers in urban areas; and (2) such rules to require that a provider of interstate interexchange telecommunications services provide such services to its subscribers in each State at rates no higher than those charged to its subscribers in any other State.

Requires a carrier, upon receiving a bona fide request, to provide telecommunications services: (1) which are necessary for the provision of health care services in a State, including instruction relating to such services, to any public or nonprofit health care provider that serves persons who reside in rural areas in that State at rates that are reasonably comparable to those charged for similar services in urban areas in that State; and (2) for educational purposes included in the definition of universal service for elementary and secondary schools and libraries at rates that are less than the amounts charged for similar services to other parties, as necessary to ensure affordable access to and use of such services. Permits a carrier providing such service to have an amount equal to the amount of the discount treated as an offset to its obligation to contribute to the mechanisms, or receive reimbursement utilizing the support mechanisms, to preserve and advance universal service.

Directs the FCC to establish competitively neutral rules to: (1) enhance access to advanced telecommunications and information services for all public and nonprofit elementary and secondary school classrooms, health care providers, and libraries; and (2) define the circumstances under which a carrier may be required to connect its network to such public institutional telecommunications users.

Specifies that: (1) telecommunications services and network capacity provided to health care providers, schools, and libraries may not be resold or transferred for monetary gain; and (2) for-profit businesses, elementary and secondary schools with endowments of more than $50 million, and libraries that are not eligible to participate in State-based plans for funds under the Library Services and Construction Act are ineligible to receive discounted rates.

Requires the FCC and the States to ensure that universal service is available at rates that are just, reasonable, and affordable.

Prohibits a carrier from using services that are not competitive to subsidize those that are subject to competition. Requires the FCC, with respect to interstate services, and the States, for intrastate services, to establish any necessary cost allocation rules, accounting safeguards, and guidelines to ensure that services included in the definition of universal service bear no more than a reasonable share of the joint and common costs of facilities used to provide those services.

Requires that: (1) if readily achievable, manufacturers of telecommunications and customer premises equipment ensure that equipment is designed, developed, and fabricated to be, and providers of telecommunications services ensure that service is, accessible and usable by individuals with disabilities; and (2) whenever such requirements are not readily achievable, such a manufacturer or provider shall ensure that the equipment or service is compatible with existing peripheral devices or specialized customer premises equipment commonly used by such individuals to achieve access, if readily achievable.

Directs the Architectural and Transportation Barriers Compliance Board to develop guidelines for accessibility of telecommunications and customer premises equipment in conjunction with the FCC and to review and update the guidelines periodically.

Requires the FCC to establish procedures for its oversight of coordinated network planning by carriers and other providers of telecommunications service for the effective and efficient interconnection of public telecommunications networks used to provide such service. Authorizes the FCC to participate in the development by industry standards-setting organizations of public telecommunications network interconnectivity standards that promote access to public telecommunications networks used to provide service, network capabilities and services by individuals with disabilities, and information services by subscribers of rural telephone companies.

Directs the FCC to: (1) complete a proceeding for the purpose of identifying and eliminating market entry barriers for entrepreneurs and other small businesses in the provision and ownership of telecommunications and information services, or in the provision of parts or services to providers of such services; (2) seek to promote the policies and purposes of the Act favoring diversity of media voices, vigorous economic competition, technological advancement, and promotion of the public interest; and (3) periodically review and report to the Congress on any regulations prescribed to eliminate such barriers and the statutory barriers that it recommends be eliminated, consistent with the public interest.

Prohibits a carrier from submitting or executing a change in a subscriber's selection of a provider of telephone exchange service or telephone toll service except in accordance with such verification procedures as the FCC shall prescribe. Makes any carrier that violates such procedures and collects charges for such a service from a subscriber liable to the carrier previously selected by the subscriber in an amount equal to all charges paid by such subscriber after such violation.

Directs the FCC to prescribe regulations that require incumbent LECs to share network facilities, technology, and information with qualifying carriers where the qualifying carrier requests such sharing for the purpose of providing telecommunications services or access to information services in areas where the carrier is designated as an essential carrier. Establishes the terms and conditions of such regulations. Requires LECs sharing infrastructure to provide information to sharing parties about deployment of services and equipment, including software.

Prohibits any LEC subject to interconnection requirements under this Act from: (1) subsidizing its telemessaging service directly or indirectly from its telephone exchange service or its exchange access; and (2) preferring or discriminating in favor of its telemessaging service operations in its provision of telecommunications services. Directs the FCC to establish procedures or regulations thereunder for the expedited receipt and review of complaints alleging violations that result in material financial harm to providers of telemessaging services.

(Sec. 102) Specifies that a common carrier designated as an "eligible telecommunications carrier" shall: (1) be eligible to receive universal service support; and (2) throughout the service area for which the designation is received, offer the services that are supported by Federal universal service support mechanisms either using its own facilities or a combination of its own facilities and resale of another carrier's services, and advertise the availability of such services and the charges thereof using media of general distribution.

Requires a State commission to designate such a carrier for the service area. Authorizes (in the case of an area served by a rural telephone company) or requires (in the case of all other areas) the State commission to designate more than one common carrier as an eligible carrier for a service area designated by the State commission, as long as each additional requesting carrier meets the requirements of this section and such designation is in the public interest.

Sets forth provisions regarding: (1) designation of eligible carriers for unserved areas; and (2) relinquishment of universal service (in areas served by more than one eligible carrier).

(Sec. 103) Amends the Public Utility Holding Company Act of 1935 (PUHCA) to allow registered holding companies to diversify into telecommunications, information, and related services and products where the Securities and Exchange Commission (SEC) determines that a registered holding company is providing telecommunications, information, and other related services through a single purpose subsidiary, designated an "exempt telecommunications company" (ETC). Requires prior State approval before any utility that is associated with a registered holding company may sell to an ETC any asset in the retail rates of that utility as of December 19, 1995.

Specifies that the ownership of ETCs by registered holding companies shall not be subject to prior approval or other restriction by the SEC, but the relationship between an ETC and a registered holding company shall remain subject to SEC jurisdiction, with exceptions.

Requires any registered holding company or subsidiary thereof that acquires or holds the securities, or an interest in the business, of an ETC to file with the SEC such information as the SEC may prescribe concerning: (1) investments and activities by the registered holding company, or any subsidiary thereof, with respect to ETCs; and (2) any activities of an ETC within the holding company system that are reasonably likely to have a material impact on the financial or operational condition of the holding company system.

Prohibits public utility companies from assuming the liabilities of an ETC and from pledging or mortgaging the assets of a utility for the benefit of an ETC.

Sets forth provisions regarding: (1) protection against abusive affiliate transactions; and (2) non-preemption of rate authority.

Prohibits reciprocal arrangements to avoid the provisions of this section among companies that are not affiliates or associate companies of each other.

Authorizes State commissions to: (1) examine the books and records of the ETC and any public utility company, associate company, or affiliate in the registered holding company system as they relate to the activities of the ETC; and (2) order an audit of a public utility company that is an associate of an ETC.

(Sec. 104) Amends the Act to specify that a purpose of the Act is to make available service to all the people of the United States without discrimination on the basis of race, color, religion, national origin, or sex.

Subtitle B: Special Provisions Concerning Bell Operating Companies—Requires a BOC to obtain FCC authorization prior to offering "interLATA" (i.e., long-distance; "LATA" means "local access and transport area") service within its region unless those services are previously authorized or incidental to the provision of another service, in which case interLATA service may be offered after the date of this Act's enactment. Permits a BOC to offer out-of-region services immediately after such date.

Sets forth requirements for a BOC's provision of interLATA services originating in an in-region State, including: (1) the presence of a facilities-based competitor or competitors (but the presence of a competitor offering exchange access, telephone exchange service offered exclusively through the resale of the BOC's telephone exchange service, and cellular service does not meet such requirement); or (2) the failure of a facilities-based competitor to request access or interconnection.

Establishes specific interconnection requirements, including a competitive checklist that a BOC must satisfy as part of its entry test (e.g., interconnection in accordance with specified requirements, nondiscriminatory access to 911 services, and reciprocal compensation arrangements).

Sets forth administrative provisions regarding applications for BOC entry. Authorizes the Attorney General to provide to the FCC an evaluation of an application using any standard the Attorney General deems appropriate. Sets forth provisions regarding FCC determinations, limits on FCC actions, publication of determinations, and enforcement of conditions required for approval. Directs the FCC to establish procedures for the review of complaints concerning failures by BOCs to meet such conditions.

Prohibits joint marketing of local services obtained from the BOC and long distance service within a State by carriers with more than five percent of the nation's presubscribed access line for three years after the date of enactment, or until a BOC is authorized to offer interLATA services within that State, whichever is earlier.

Requires any BOC authorized to offer interLATA services to provide intraLATA toll dialing parity coincident with its exercise of that interLATA authority. Bars States from ordering a BOC to implement toll dialing parity prior to its entry into interLATA service. Provides that any single-LATA State or any State that has issued an order by December 19, 1995, requiring a BOC to implement intraLATA toll dialing parity is grandfathered under this Act, with the prohibition against "non-grandfathered" States expiring three years after this Act's enactment date.

Sets forth "incidental" interLATA activities that the BOCs are permitted to provide upon the date of enactment.

Prohibits a BOC (including any affiliate) which is an LEC from providing specified services (including manufacturing activities, origination of interLATA telecommunications services other than incidental interLATA services, out-of-region services, or previously authorized activities and interLATA information services other than electronic publishing and alarm monitoring services) unless it does so through an entity that is separate from any entities that provide telephone exchange service.

Delineates structural and transactional requirements that apply to the separate subsidiary, including operating independently from the BOC, maintaining separate books and records, having separate officers, not obtaining credit under any arrangement that would permit a creditor upon default to have recourse to the BOC's assets, and conducting transactions with the BOC on an arm's length basis.

Sets forth provisions regarding: (1) non-discrimination safeguards; (2) biennial audit requirements; (3) sunset of provisions of this section; and (4) joint marketing.

Permits a BOC to: (1) engage in manufacturing after the FCC authorizes the company to provide interLATA services in any in-region State; (2) collaborate with a manufacturer of customer premises or telecommunications equipment during the design and development of hardware or software; and (3) engage in research activities relating to manufacturing and enter into royalty agreements with manufacturers of telecommunications equipment.

Requires each BOC to maintain and file with the FCC information on protocols and technical requirements for connection with and use of its telephone exchange service facilities.

Sets forth provisions regarding: (1) manufacturing limitations for standard-setting organizations; (2) alternate dispute resolution; (3) BOC equipment procurement and sales; and (4) FCC enforcement authority.

Prohibits a BOC or any affiliate from engaging in the provision of electronic publishing that is disseminated by means of such BOC's or any of its affiliates' basic telephone service, but allows a separated affiliate or electronic publishing joint venture (EPJV) operated in accordance with this section to engage in electronic publishing.

Requires a separated affiliate or EPJV to be operated independently from the BOC and to maintain separate books and records. Prohibits the affiliate from incurring debt in a manner that would permit a creditor upon default to have recourse to the BOC's assets. Sets forth provisions governing the manner in which transactions by the affiliate must be carried out (to ensure that they are fully auditable) and governing the valuation of assets transferred to the affiliate (to prevent cross subsidies). Prohibits the affiliate and the BOC from having corporate officers or property in common.

Prohibits the separate affiliate or EPJV from marketing the name, trademarks, or service marks of an existing BOC except for those that are owned by the entity that owns or controls the BOC.

Prohibits a BOC from engaging in joint marketing of any promotion, marketing, sales, or advertising with its affiliate, except that a BOC may: (1) provide inbound telemarketing or referral services related to the provision of electronic publishing if the BOC provides the same service on the same terms, conditions, and prices to non-affiliates as to its affiliates; (2) engage in non-discriminatory teaming or business arrangements; and (3) participate in EPJVs, provided that the BOC or affiliate has not more than a 50 percent (or, for small publishers, 80 percent) direct or indirect equity interest in the publishing joint venture.

Requires a BOC that enters the electronic publishing business through a separated affiliate or EPJV to provide network access and interconnection to electronic publishers at just and reasonable rates that are not higher on a per-unit basis than those charged to any other electronic publisher or any separated affiliate engaged in electronic publishing.

Entitles a person claiming a violation of this section to file a complaint with the FCC or to bring suit as provided in the Act.

Prohibits a BOC or affiliate thereof from engaging in the provision of alarm monitoring services before five years after the date of this Act's enactment, except for such services by a BOC that was engaged in providing such services as of November 30, 1995, directly or through an affiliate (but such BOC may not acquire an equity interest in or obtain financial control of any unaffiliated alarm monitoring services entities from November 30, 1995, until five years after the enactment date).

Provides that an incumbent LEC engaged in the provision of alarm monitoring services shall: (1) provide nonaffiliated entities, upon reasonable request, with the network services it provides to its own alarm monitoring operations on non-discriminatory terms and conditions; and (2) not subsidize its alarm monitoring services directly or indirectly from telephone exchange service operations. Requires the FCC to establish procedures for the receipt and review of complaints concerning violations of such provision or the regulations thereunder that result in material financial harm to a provider of alarm monitoring service. Bars an LEC from recording or using in any fashion the occurrence or contents of calls received by providers of alarm monitoring services for purposes of marketing such services on behalf of such LEC or any other entity.

Directs the FCC to adopt rules that eliminate discrimination between BOC and independent payphones and subsidies or cost recovery for BOC payphones from regulated interstate or intrastate exchange or exchange access revenue. Authorizes the FCC, if it determines that it is in the public interest, to allow the BOC's to have the same rights as independent payphone providers in negotiating with the interLATA carriers for their payphones. Grants the location provider the ultimate decision-making authority in determining interLATA services in connection with the choice of payphone providers.

Title II: Broadcast Services—Requires the FCC, if it determines that it will issue additional licenses for advanced TV services, to: (1) limit the initial eligibility for such licenses to persons who are licensed to operate a TV broadcast station, who hold a permit to construct such a station, or both; and (2) adopt regulations that allow such licensees or permittees to offer such ancillary or supplementary services on designated frequencies as may be consistent with the public interest, convenience, and necessity. Provides for the: (1) recovery for FCC reallocation or reassignment of the original or additional license of a person licensed to operate a TV broadcast station; and (2) charging and collection of fees from licensees by the FCC for the authorized use of designated frequencies. Requires a report from the FCC to the Congress on the implementation of this provision. Requires the FCC, within ten years after the first issuance of additional licenses, to conduct an evaluation of the advanced TV services program.

(Sec. 202) Directs the FCC to modify its multiple ownership rules to eliminate its limitation on the number of radio stations which may be owned or controlled nationally. Limits the number of radio stations an entity may own, operate, or control in a local market, with an exception when the FCC determines that such ownership, operation, or control will increase the number of radio broadcast stations in operation. Directs the FCC to: (1) eliminate its limitation on the number of TV stations which may be owned or controlled nationally; (2) increase to 35 percent the national audience reach limitations for TV stations; and (3) conduct a rulemaking proceeding to determine whether its rules restricting ownership of more than one TV station in a local market should be retained, modified, or eliminated.

Directs the FCC to extend its waiver policy with respect to its one-to-a-market ownership rules to any of the top 50 markets. Directs the FCC to permit a TV station to affiliate with an entity that maintains two or more networks unless such networks are composed of: (1) two or more of the four existing networks (ABC, CBS, NBC, FOX); or (2) any of the four existing networks and one of the two emerging networks (WBTN, UPN). Directs the FCC to: (1) permit an entity to own or control a network of broadcast stations and a cable system; and (2) revise ownership regulations if necessary to ensure carriage, channel positioning, and nondiscriminatory treatment of nonaffiliated broadcast stations by a cable system. Requires the FCC to revise all such rules biennially. Repeals current restrictions on broadcast-cable crossownership under the Communications Act.

(Sec. 203) Provides an eight-year license term for both TV and radio broadcast licenses.

(Sec. 204) Revises provisions regarding renewal procedures for the operation of TV broadcast stations. Includes standards for both renewal and denial of an application. Requires each renewal applicant to attach to such application a summary of comments and suggestions from the public regarding violent programming. Makes such amendment effective with respect to applications filed after May 1, 1995.

(Sec. 205) Extends to direct broadcast services current protections against signal piracy. Empowers the FCC with exclusive jurisdiction to regulate direct-to-home satellite services.

(Sec. 206) Provides that any ship documented under U.S. laws operating under the Global Maritime Distress and Safety System provisions of the Safety of Life at Sea Convention shall not be required to be equipped with a radio telegraphy station operated by one or more radio officers or operators.

(Sec. 207) Directs the FCC to promulgate regulations to prohibit restrictions that impair a viewer's ability to receive video programming services through devices designed for over-the-air reception of TV broadcast signals, multichannel multipoint distribution service, or direct broadcast satellite services.

Title III: Cable Services—Revises the definitions of "cable service" and "cable system" for purposes of the Act. Directs the FCC to: (1) review any complaint submitted by a franchising authority after the date of enactment of this Act concerning an increase in rates for cable programming services; and (2) issue a final order within 90 days, unless the parties agree to extend the review period. Terminates such review authority for cable programming services provided after March 31, 1999. Makes such provision inapplicable with respect to: (1) operators providing video programming services in areas subject to effective competition (as defined); or (2) any video programming offered on a per channel or per program basis. Exempts from certain cable rate regulation provisions small cable operators (serving fewer than one percent of all cable subscribers in the United States, serving no more than 50,000 subscribers, and not affiliated with any entity whose gross annual revenues exceed $250 million). Revises provisions with respect to cable TV market determinations, requiring an expedited decisionmaking process. Prohibits any State or franchising authority from restricting in any way a cable system's use of any type of subscriber equipment or transmission technology. Sets forth provisions with respect to: (1) cable equipment compatibility; and (2) subscriber notice (allowing any reasonable means at the cable operator's discretion). Repeals anti-trafficking restriction provisions of the Act. Directs the FCC to allow cable operators to aggregate equipment costs into broad categories, regardless of the function levels of such equipment within such categories. Provides for the treatment of prior-year losses of a cable system.

(Sec. 302) Subjects common carriers providing video programming to subscribers using radio communications to the requirements of title III and to the ownership and joint venture restrictions set forth in the following paragraph, but not to other requirements of title VI of the Act. States that such carriers providing such programming on a common carrier basis shall be subject to such requirements and restrictions, but not to other requirements of title VI. Allows such carrier to elect to provide such programming by means of an open video system, stating that such a provider need not make capacity available on a nondiscriminatory basis to any other person for the provision of cable service directly to subscribers.

Prohibits any LEC or affiliate from purchasing or otherwise acquiring more than a ten percent financial interest, or any management interest, in any LEC providing telephone exchange service within such cable operator's franchise area. Prohibits an LEC and a local cable operator from entering into a joint venture to provide video programming directly to subscribers or to provide telecommunications services within such market. Provides exceptions, including exceptions for joint ventures in rural areas, joint use of transmission facilities in limited circumstances, acquisitions made in competitive markets, exempt cable systems (cable systems serving less than 17,000 subscribers, with other restrictions), and small cable systems located in nonurban areas. Authorizes the FCC to waive such financial interest or joint venture restrictions in cases of undue economic distress, economic viability, anticompetitive effects of such restrictions, or when the local franchising authority approves such waiver.

Authorizes an LEC to provide cable service to its subscribers through an open video system that complies with this section. Outlines, with respect to the provision of such service through such system, provisions concerning: (1) certificates of compliance; (2) dispute resolution; (3) FCC regulations; (4) consumer access; (5) reduced regulatory burdens for such systems; and (6) FCC implementation of appropriate rules and regulations within six months after the enactment of this Act. States that an operator of an open video system may be subject to the payment of fees based on gross revenues in lieu of cable TV franchising fees.

(Sec. 303) Sets forth provisions regarding preemption of franchising authority regulation of telecommunications services. Prohibits a franchising authority from ordering a cable operator to discontinue the provision of a telecommunications service or a cable system to the extent it is used to provide a telecommunications service by reason of the failure of the cable operator to obtain a franchise or franchise renewal for the provision of such service. Prohibits a franchising authority from requiring a cable operator to provide any telecommunications service or facilities, other than institutional networks, as a condition of the initial grant of a franchise, franchise renewal or franchise transfer.

(Sec. 304) Directs the FCC to adopt regulations to ensure the commercial availability of convertor boxes, interactive equipment, and related equipment used to access multichannel video programming (MVP) from manufacturers, retailers, or other vendors not affiliated with any MVP distributor. Ensures the continued system security of MVP services. Provides FCC waiver authority with respect to provisions adopted under this section.

(Sec. 305) Directs the FCC, within 180 days after the enactment of this Act, to complete an inquiry to ascertain the level at which video programming is closed captioned. Provides closed captioning accountability criteria and requires a schedule of deadlines for the provision of such service. Provides exemptions from such requirements in cases of economic burden, inconsistency with current contracts, or undue burden of a significant difficulty or expense (with specified factors). Directs the FCC to: (1) commence an inquiry to examine the use of video descriptions on video programming in order to ensure the accessibility of such programming to persons with visual impairments; and (2) report to the Congress on its findings.

Title IV: Regulatory Reform—Directs the FCC to forbear from applying any regulation or provision of the Act to a telecommunications carrier or service if it determines that: (1) enforcement is not necessary to ensure that charges, practices, and classifications are just and reasonable and not discriminatory; (2) enforcement is not necessary for the protection of consumers; and (3) forbearance is consistent with the public interest. Directs the FCC to consider whether such forbearance will promote competitive market conditions. Allows any carrier to petition for such forbearance, requiring an FCC ruling within one year of such petition. Prohibits State enforcement of a regulation or provision after FCC-granted forbearance.

(Sec. 402) Directs the FCC, in every even-numbered year beginning with 1998, to: (1) review all regulations issued under the Act that apply to the operations or activities of a provider of telecommunications services; and (2) determine whether such regulation is no longer necessary in the public interest. Requires the FCC to repeal or modify any regulation so determined. Provides procedures for streamlining such repeals or modifications.

(Sec. 403) Eliminates or reduces specified FCC regulations, functions, and authority with respect to: (1) amateur radio examination procedures; (2) the designation of inspection entities; (3) instructional TV fixed service processing; (4) the setting of depreciation rates; (5) the use of independent auditors; (6) the delegation to private laboratories of equipment testing and certification; (7) the uniformity of license modifications; (8) jurisdiction over Government-owned ship radio stations; (9) the operation of domestic ship and aircraft radios without licenses; (10) fixed microwave service licensing; (11) foreign directors; (12) limitations on silent station authorizations; (13) construction permit requirements; (14) inspections of broadcast station equipment and apparatus; and (15) inspections by entities other than the FCC.

Title V: Obscenity and Violence—Subtitle A: Obscene, Harassing, and Wrongful Utilization of Telecommunication Facilities - Communications Decency Act of 1996— Revises provisions of the Communications Act prohibiting obscene or harassing telephone calls and conversation to apply to obscene or harassing use of a telecommunications facility and communication. Increases the penalties for violations. Prohibits using a telecommunications device to: (1) make or initiate any communication which is obscene, lewd, lascivious, filthy, or indecent with intent to annoy, abuse, threaten, or harass another person; (2) make or make available obscene communication; (3) make or make available an indecent communication to minors.

Provides that no person shall be held to have violated such prohibition solely for providing access or connection to a telecommunications facility, system, or network not under such person's control. Provides employers with a defense for actions by employees unless the employee's conduct is within the scope of employment and is known, authorized, or ratified by the employer. Establishes as a defense to prohibited communications that a person has taken, in good faith, reasonable, effective, and appropriate actions to prevent access by minors or has restricted access by requiring use of a verified credit card, debit account, or adult access code or personal identification number.

(Sec. 504) Requires cable operators, upon request, to fully scramble or block programming to which the subscriber does not subscribe.

(Sec. 505) Requires a multichannel videoprogramming distributor: (1) to fully scramble or block sexually explicit adult programming so that nonsubscribers do not receive it; and (2) until it complies with such requirement, to not provide such programming during the hours of the day when a significant number of children are likely to view it.

(Sec. 506) Allows cable operators to refuse to transmit any public access or leased access program which contains obscenity, indecency, or nudity.

(Sec. 507) Amends the Federal criminal code to specify that current obscenity statutes prohibit using a computer to import or transport in interstate or foreign commerce, for sale or distribution, obscene material, including material designed, adapted, or intended for producing abortion or for any indecent or immoral use.

(Sec. 508) Prohibits using any facility or means of interstate or foreign commerce to persuade, induce, entice, or coerce a minor to engage in prostitution or any sexual act for which any person may be criminally prosecuted.

(Sec. 509) Provides that no provider or user of an interactive computer service shall be held liable for any voluntary action taken to restrict access to, or to enable information content providers to restrict access to, material that the user or provider considers to be objectionable, whether or not such material is constitutionally protected.

Subtitle B: Violence—Directs the FCC, if it determines that video programming distributors have not, within one year, voluntarily established rules for rating programming that contains sexual, violent, or other indecent material about which parents should be informed before it is displayed to children and voluntarily agreed to broadcast signals that contain such ratings, to: (1) establish an advisory committee to recommend guidelines and procedures for rating such programming; (2) prescribe such guidelines and procedures; and (3) prescribe rules requiring programming distributors to transmit such rating to permit parents to block inappropriate programming. Directs the FCC, not less than two years after enactment of this Act, to require apparatus designed to receive TV signals that are shipped in interstate commerce or manufactured in the United States and that have a picture screen of 13 inches or greater (measured diagonally) to be equipped with a feature designed to enable viewers to block display of all programs with a common rating. Authorizes the FCC to allow apparatus manufacturers to comply with such requirement using alternative technology that meets certain standards of cost, effectiveness, and ease of use.

(Sec. 552) Encourages broadcast television, cable, satellite, syndication, and other video programming distributors to establish a technology fund to encourage electronics equipment manufacturers to facilitate the development of technology which would empower parents to block programming deemed inappropriate for children and to encourage availability of such technology to low income parents.

Subtitle C: Judicial Review—Provides for the expedited review of any civil action challenging the constitutionality of this title by a district court of three judges and by direct appeal to the Supreme Court.

Title VI: Effect on Other Laws—Provides that any conduct or activity that was, before the enactment of this Act, subject to any restriction or obligation imposed by the AT&T Consent Decree, the GTE Consent Decree, or the McCaw Consent Decree shall, after enactment of this Act, be subject to the restrictions and obligations imposed by the Communications Act as amended by this Act.

Provides that nothing in this Act shall be construed to modify, impair, or supersede: (1) the applicability of the antitrust laws; or (2) any State or local law pertaining to taxation, except with respect to fees for open video systems. Repeals a provision of the Communications Act permitting the FCC to render a proposed merger of competing local telephone companies exempt from any Act of Congress making the transaction unlawful.

(Sec. 602) Exempts any provider of direct-to-home satellite service from the collection or remittance of any local tax or fee on such service.

Title VII: Miscellaneous Provisions—Prohibits a party calling a toll-free telephone number from being assessed a charge by virtue of being asked to connect or otherwise transfer to a pay-per-call service. Prohibits the calling party from being charged for information conveyed during a call to a toll-free (800) number unless the calling party: (1) has a written agreement specifying the material terms and conditions under which the information is offered and which includes the rate at which charges are assessed and certain identifying information; or (2) is charged for the information only after the information provider includes an introductory disclosure message regarding the charge, rate, and means of billing for the call and the calling party is charged by means of a credit, prepaid, debit, charge, or calling card. Outlines provisions concerning: (1) billing arrangements; (2) required use of a personal identification number by the subscriber to obtain access to the information provided; (3) exceptions to the written agreement requirement; and (4) termination of service if a telecommunications carrier reasonably determines that a complaint against an information provider is valid.

Amends the Telephone Disclosure and Dispute Resolution Act to authorize the FCC to extend the definition of "pay-per-call services" under such Act to other services that the FCC determines are susceptible to the unfair and deceptive billing practices addressed by such Act.

(Sec. 702) Makes it the duty of every telecommunications carrier to protect the confidentiality of proprietary information of other carriers, equipment manufacturers, and customers. Permits a carrier that receives proprietary information from another carrier or a customer for purposes of providing any telecommunications service to use such information only for such purpose. Directs a carrier to disclose customer proprietary network information upon the customer's request. Permits a carrier to use, disclose, or permit access to aggregate customer information for other purposes. Requires a carrier that provides telephone exchange service to provide subscriber list information to any person upon request for the purpose of publishing directories in any format.

(Sec. 703) Directs the FCC to prescribe regulations to: (1) govern the charges for pole attachments used by telecommunications carriers to provide telecommunications services, when the parties fail to resolve a dispute over such charges; and (2) ensure that utilities charge just, reasonable, and nondiscriminatory rates for the pole attachments. Requires a utility to apportion the cost of providing space on a pole based on the number of attaching entities. Requires any increase in the rates for pole attachments to be phased in over a five-year period. Requires a utility to provide a cable television system or any telecommunications carrier with nondiscriminatory access to any pole or right-of-way owned by it. Allows a utility company providing electric service to deny a cable television system or telecommunications carrier access to such poles when there is insufficient capacity and for reasons of safety, reliability, and generally applicable engineering purposes. Requires utilities that engage in the provision of telecommunications services or cable services to impute to its costs of providing such service an equal amount to the pole attachment rate for which such company would be liable. Requires utilities to provide written notification to attaching entities of any plans to modify or alter its poles or other rights-of-way. Requires any attaching entity that modifies its own attachments to bear a proportionate share of the costs of such modifications. Prevents a utility from imposing the cost of rearrangements to other attaching entities if done solely for the benefit of the utility.

(Sec. 704) Preserves State or local authority over decisions regarding the placement, construction, and modification of personal wireless service facilities, but prohibits State or local regulation thereof from: (1) unreasonably discriminating among providers of functionally equivalent services; or (2) prohibiting the provision of personal wireless services. Requires State or local action on requests regarding such facilities to occur within a reasonable time, with denials of requests to be in writing and supported by substantial evidence in a written record. Prohibits State or local regulation of such facilities on the basis of environmental effects of radio frequency emissions to the extent such facilities comply with FCC regulations. Provides for expedited judicial review and petitions of the FCC for relief from adverse State or local actions.

Directs the President to prescribe procedures by which Federal agencies may make available property and rights-of-way for the placement of new telecommunications services that are dependent upon the utilization of Federal spectrum rights.

(Sec. 705) Prohibits a commercial mobile services provider from being required to provide equal access to common carriers for the provision of telephone toll services. Directs the FCC, if it determines that subscribers to such services are denied access to the provider of telephone toll services of the subscribers' choice, contrary to the public interest, to prescribe regulations to afford subscribers unblocked access to the provider of telephone toll services of the subscribers' choice through the use of a carrier identification code assigned to such provider or other mechanism. Provides that such regulations shall not apply to mobile satellite services unless the FCC finds it to be in the public interest.

(Sec. 706) Requires the FCC and each State telecommunications commission to encourage the deployment of advanced telecommunications capability to all Americans by utilizing price cap regulation, regulatory forbearance, measures that promote competition, or other regulating methods that remove barriers to infrastructure investment. Requires the FCC to regularly initiate a notice of inquiry concerning such availability and, if it determines it to be necessary, to take action to accelerate deployment of such capability by removing barriers to infrastructure investment and by promoting competition in the telecommunications market.

(Sec. 707) Establishes the Telecommunications Development Fund as a corporate body in the District of Columbia to promote access to capital for small businesses in order to enhance competition in the telecommunications industry, to stimulate new technology development, to promote employment and training, and to support universal service. Directs the Fund to: (1) make loans, investments, or other extensions of credit and provide financial advice to eligible small businesses; and (2) prepare research, studies, or financial analyses.

(Sec. 708) Recognizes the National Education Technology Funding Corporation as a nonprofit corporation independent of the Federal Government and operating under the laws of the District of Columbia. Authorizes the Corporation to receive discretionary grants, contracts, gifts, contributions, or technical assistance from any Federal department or agency.

Requires audits of the Corporation by independent certified public accountants. Provides reporting and recordkeeping requirements. Requires the accessibility of Corporation books for audit and examination. Directs the Corporation to report annually to the President and the Congress on operations and activities of the previous fiscal year. Requires Corporation members to be available to testify before the Congress concerning such operations and activities.

(Sec. 709) Directs the Assistant Secretary of Commerce for Communications and Information to report annually to specified congressional committees concerning the activities of the Joint Working Group on Telemedicine, together with any findings in the studies and demonstrations on telemedicine funded by the Public Health Service or other Federal agencies. Specifies that such reports shall examine questions related to patient safety, the efficacy and quality of the services provided, and other legal, medical, and economic issues related to the utilization of advanced telecommunications services for medical purposes.

(Sec. 710) Authorizes appropriations.

Appendix G

History of Switching and Multiplexing

KEY TERMS

Bandwidth
Channel Unit
Electronic Mail (e-mail)

Register (Dial Register)
Space Division Switching
Switchboard

Teletype
Translator

This appendix is provided for those who would like to understand how the telecommunications technologies have evolved. The technologies discussions below have become history. Most people have enough to learn during the course of a term without a digression into material that discusses old technologies. Therefore in this edition, I have moved discussions on older technologies out of the main text of the book and placed it here. Those who have the time and desire to learn about older technologies may read on at their leisure.

At the turn of the century, all calls were handled by operators using equipment referred to as a switchboard. Operators answered calls with the word "Central," and the location of the operator switchboard was referred to as the central exchange. It was located in the center of the local telecommunications system. Operators could connect any telephone to any other telephone in town as each telephone was wired to jacks on the switchboard. The switchboard was the first central office switching system, and switching was done manually by an operator. As the operator function of connecting a calling party to a called party was automated, the switchboard evolved to an automated step-by-step exchange and then to a crossbar exchange. These systems have all but disappeared as computer-controlled switching systems have replaced them. The first sections below explain SXS and X-Bar switching systems.

The first multiplexing systems used analog techniques (AM and FDM) to multiplex trunk circuits between towns onto one physical facility. The first systems could multiplex 12 circuits onto two pair of wires (one pair for transmit and one pair for receive). These systems evolved to the SSB-SC FDM system discussed below. The most predominant of these systems was referred to as "L-Carrier." Initially, L-carrier used coaxial cable as a transport medium and later used microwave as a transport medium. WDM and TDM over fiber have replaced these L-carrier systems.

G.1 EVOLUTION OF TELECOMMUNICATIONS

Telecommunications began with the telegraph in 1837. In 1876, Elisha Gray and Alexander Graham Bell filed papers with the patent office for an invention called the *telephone*. Gray filed a disclosure notification, and Bell's father-in-law filed a patent application for Bell. The Supreme Court ruled in a split decision that Bell is to be recognized as the inventor of the telephone. Western Union made many improvements to the telegraph system over the years, but nothing could beat the direct person-to-person contact that the telephone offered.

When telephones were first introduced, the telephone office operated much like the telegraph office when someone wished to place a call to other towns. Only the telephone company and wealthy people had telephones. To expand its business beyond the wealthy, the telephone company accepted messages from walk-up customers at their local telephone office. It charged so much to deliver the messages to other towns that had a telephone office. It would call the distant telephone office and read the message over the telephone to an operator in that exchange. The message was given to a runner (messenger) who would take the message to the intended recipient. The telephone company employed many young boys as messengers.

As the price of telephone service came down to a point where more people could afford it, telephones began to replace the telegraph as the preferred service for telecommunications within a town or city. Once Bell implemented a long distance network under American Telephone and Telegraph (AT&T), the telephone began to replace the telegraph for long distance communication. Western Union still maintained a strong position in telecommunications as the preferred system for data communication.

Western Union continued to enhance its ability to carry data and established a network for **teletype** machines. These machines did not use Morse code, instead utilizing a five-state code known for its inventor as the *Baudot code*. The teletype machine consisted of an electromechanical typewriter, a paper-punch machine, and a tape reader. As an operator typed a message at the teletype, it would punch a five-level code in a strip of narrow paper. This strip of narrow paper is known as *paper tape* or *ticker tape*.

The teletype could be used to transmit electrical signals as the operator was typing a message, or it could simply be used to punch the code in a paper tape. Most of the time the machine would be off line while the tape was being punched. Once the paper tape had holes punched into it according to the Baudot code, it would be taken from the punch machine and fed into a paper tape reader that was also attached to the teletype. The teletype would be connected online to a distant teletype machine via Western Union's Teletype Network or AT&T's Teletype Network. Once the transmitting teletype was connected to the receiving teletype, the transmitting machine would start feeding the paper tape through the tape reader and the punched holes on the tape were converted into various electrical signals sent out to the receiving machine. The receiving machine could be set to generate a typed message on the typewriter, punch a paper tape on the tape puncher, or both.

The teletype networks were used by all major news services to transmit information to newspaper, radio, and television newsrooms across the country. The

Associated Press Network was a network that connected to teletypes at all major news organizations around the country. When a message was transmitted out onto the network, it reached all teletypes at one time. News was broadcast to all teletypes attached to the AP Network.

Businesses also used the teletypes, but they did not want to broadcast their data to all teletypes. They used a point-to-point network to transmit data from one location to another. They connected directly to the one and only teletype they wished to send information to. When a business had many locations or wanted to send information to a supplier or customer, it would need many different connections. To provide teletype users with the capability of reaching any other teletype, relay centers were established.

These relay centers were used for store and forward of messages between machines that could not connect directly together. A business would not want to pay for a direct connection to a machine it does not send messages to on a regular basis. Businesses could send messages to other businesses via a relay center. The center would receive messages from one location on a teletype in the center that was connected to the originating location. The teletype in the relay center would generate a punched paper tape from the received electrical signals. Personnel in the teletype relay center would take the paper tape to another machine connected via the teletype network to the desired recipient's teletype machine. They would feed the punched paper tape through a tape reader. The tape reader would convert the punched code into various electrical signals sent to the receiving teletype.

Relay centers were replaced by dial-up connections. Teletypes were connected to regular telephone lines, and the teletypes were equipped with auto-answer capabilities. A teletype user would dial the telephone number associated with another teletype, and when it answered the call, a connection between the two teletypes was established. The user could now type a message or run a coded paper tape through the tape reader, and the receiving machine received the message.

The store-and-forward technique of the old teletype relay center is similar to the process used with today's *electronic mail (e-mail)*. A person types an e-mail message and sends it to their mail server. Their mail server will forward the message to a mail server that services the intended recipient. That mail server will hold the message until the recipient goes online and requests delivery of stored e-mails. The more things change, the more they remain the same.

G.2 OPERATOR SWITCHBOARD

When the operator connected two lines using a patch cord, there was no indication when the call was completed. To alleviate this condition an operator *switchboard* was invented. The switchboard consisted of patch cords, patch panels, and switches (see Figure G-1). The switches are similar to a light switch but have more contacts. A light switch has one set of contacts. When the switch is *off* the contacts are *open*. When the switch is thrown *on*, the contacts are *closed*. Operation of the light switch connects the AC power line to a light socket. The switches used on the switchboard had more than one set of contacts. Some contacts opened and others closed when the switch was thrown on.

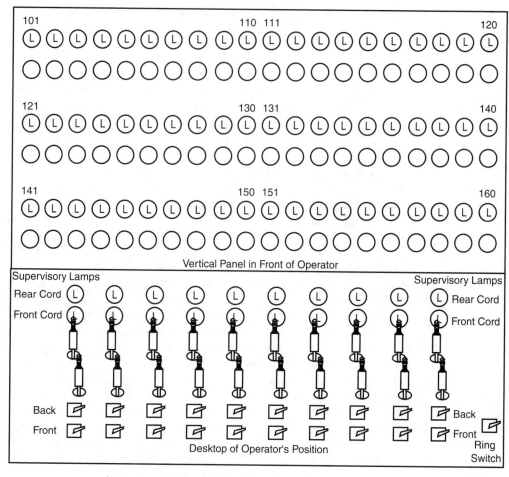

FIGURE G-1 Manual switchboard.

The patch cords used on the switchboard were made with a plug on one end only. The other end was wired to one of the switches on the switchboard. The patch cords were arranged in pairs. One cord was placed in front of the other; both cords sat directly behind the switch they were wired to. Two patch cords (a pair of patch cords) were wired to one switch. A pair of patch cords was used to handle one call. The switchboard was equipped with 20 patch cords (10 pairs of patch cords) wired to 10 switches. The patch cords and switches were placed in a desktop, and the patch panel containing jacks connecting to phone lines were placed in a vertical position, behind the desktop, in front of the operator. The operator's phone was replaced by a headset, and a steady ringing source was available at the switchboard position. The operator's headset was connected through a switch called the *monitor switch* to the contacts on the patch cord switches. Every set of patch cords was equipped with a monitor switch in addition to the patch cord switch. Each of the 10 monitor switches was placed directly in front of each patch cord switch.

The monitor switch allowed the operator the option of connecting both the transmitter and receiver of the headset to a patch cord or connecting only the receiver to the patch cord. The monitor key made it possible for the operator to talk to either person on the patch cords and to monitor a call without being heard. Operators would monitor a call once every few minutes—long enough to determine if the call was over. On determining that it was over, the operator would take down the connection by removing the patch cords from the jacks.

The patch cord switch was a multiposition switch. In one position the front patch cord of a pair was connected to the rear patch cord. The switch was placed in this position to connect the calling party to the called party. The switch could be placed in a second position to allow the operator to talk with the calling party without the called party hearing the operator's conversation. The switch could also be placed in a third position allowing the operator to talk in private with the called party. When the patch cord switch was in either the second or third position, the operator could operate a ring key on the position. The operation of the ring key would open the headset from the call and attach a 90-V AC ringing signal to the patch cord. This allowed the operator to ring over either patch cord when placing a call or to ring someone if the person hung up before the other party was through with the call.

An operator could handle 10 different calls at one time. If more than 10 calls needed to be handled at one time, a second operator switchboard position was added and all incoming jacks were cabled to both switchboards. The operator switchboard had a total of 21 switches and 20 cords. The 20 cords were arranged in 10 pairs, and 2 switches were associated with each pair. The 2 switches were an operator talk/monitor switch and a patch cord switch to either separate or tie the pair of cords to each other. The switchboard had one switch to connect either the operator's headset or the ringing signal to the patch cord keys. This switchboard was known as a *manual switchboard* because the operator manually switched incoming calls to their calling destination (see Figures G-1 and G-2).

G.3 STEP-BY-STEP OR STROWGER AUTOMATED SWITCHING

Strowger invented an automated switching device in 1892. His switching system is called the step-by-step (SXS) or *Strowger switch*. However, his device was not accepted for use by the telephone companies until around 1920. Between 1878 and 1920, telephone design changed to eliminate the local battery supply at the phone and the hand-crank generator. The central exchange design was changed to include a 50-V DC battery supply (in place of the earlier 24-V battery), a line relay for each line, and a 90-V AC ringing supply. The battery used in the Strowger central office consisted of 23 active cells (50 V). The battery used in today's computer-controlled switches consist of 24 active cells (52 V). The batteries in a central office are continuously being charged by a battery charger. The battery charger converts the AC power supplied by the electric company to an output of 50 (or 52) V DC, which is connected to the central office 50- (or 52-) V battery. Each telephone is powered from this 50- or 52-V DC supply via the wire pair from the central exchange to the telephone.

FIGURE G-2 The early manual telephone switchboard was installed in New York City in 1888. (Courtesy of AT&T Bell Laboratories)

Each wire pair leaving a central office is referred to as a *line* if it connects to a telephone. The 50- (or 52-) V supply passes through a relay in the central exchange before being placed on the wire pair. This relay is called a *line relay*. There is one line relay for every line coming into the central exchange. In a manual switchboard system, these relays were located outside the operator switchboard and were wired between incoming phone lines and the switchboard. Strowger wired the line relays to a mechanical switching system instead of the operator switchboard. The line relays were connected to a device called a linefinder. The purpose of a *linefinder* was to automatically find a line that was requesting service.

A telephone requested service when the telephone receiver was taken off hook. Once the linefinder found the line requesting service, it connected this line to the input of a first selector. The first selector would provide dial tone and collect the first dialed digit. On receiving the first dialed digit, the first selector connected the originating line to another selector, which would collect the next dialed digit. As each digit was dialed, the selector that a line was attached to would take one step for each value of the digit. If a 5 was dialed, the selector switch would take

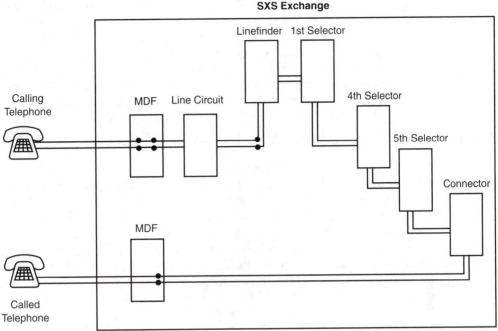

FIGURE G-3 Step-by-step exchange.

five steps. The Strowger SXS system was called a *direct-controlled switching system* because each switching stage (that is, selector) was under direct control of the originating telephone's dial.

Since all new switching systems employ SPC and a digital switching matrix, SXS systems have become history in all but a few rural areas. Most college and university engineering curriculums limit the discussion to the latest computer-controlled switching systems. This book supports that approach by focusing on the latest switching designs. The present chapter provides some detail on crossbar (XBAR) switching, because XBAR is the system on which computer-controlled switching is designed. For reference purposes, Figure G-3 presents a diagram of a local intraoffice call through the various switching stages of a Strowger SXS switching system.

G.4 STEP-BY-STEP SPACE DIVISION SWITCHING

As mentioned, the automated switching system invented by Strowger is referred to as step-by-step (SXS) or Strowger switching in honor of its inventor. In this type of switching system, each switching stage connects a pair of input wires to a pair of wires going to the next switching stage and finally through a connector switch to the pair of wires going to the line being called. The wires connected for a particular call remain connected for that call, and only one conversation takes place over these wires. Each call established through a step-by-step office uses separate wiring paths through the office for each call. The use of separate and individual wiring paths for each conversation path is referred to as *space division switching* technology.

A pair of wires connecting two devices to each other is referred to as a circuit. If the two devices (such as two phones) are connected directly, the circuit is known as a *dedicated point-to-point circuit*. When circuits are connected to a switch so that any one of the circuits can be connected by the switch to any other circuit, the switch is known as a *circuit switch*. Circuit switching is the type of switching used in the PSTN.

G.5 DISADVANTAGES OF DIRECT CONTROL STEP-BY-STEP SWITCHES

SXS switches have several drawbacks. They are incapable of offering many of the options customers want today, such as three-way calling, call waiting, caller ID, and the ability to select a long distance carrier of choice. Another problem with SXS switching is the need to have dial pulses. Stepping switches need electrical current flow to operate. Turning the flow of current off and on in a controlled manner using a rotary dial controls the stepping switch (see Figure G-4). But telephones that we call touchtone telephones do not turn the flow of electrical current off and on to transmit a digit. These telephones transmit different tone combinations to represent each digit. When touchtone telephones were introduced, they could not be used on a SXS switching system. A device was invented to store the tones generated by a touchtone phone, and this device converted the received tones into rotary dial pulses. These devices were placed between the linefinders and the first selectors. Once these devices had been added to a linefinder group, the lines in that group could have touchtone phones.

Because the telephone companies incurred additional expense in adding touchtone to dial pulse converters between all linefinders and first selectors, they charged an extra fee ranging from $0.75 to $1.25 each month for a touchtone phone. SXS switching has been replaced in all metropolitan areas with computer-controlled switching systems but is still found in small communities. Telephone

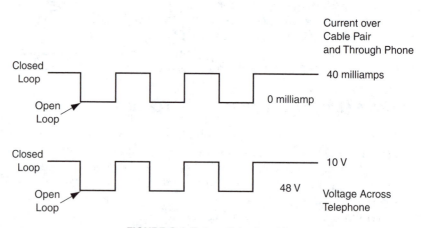

FIGURE G-4 Rotary dial pulse string.

companies in large cities no longer incur extra expense in providing touchtone service, and the additional charge for this service should no longer exist.

Figure G-4 illustrates dial pulses. These pulses can be depicted as either pulses of current or pulses of voltage. The top diagram shows that when a telephone is off hook, a certain amount of current flows in the circuit to the central office. The amount of current depends on the length of the local loop. In our example, 40 milliamps of current is flowing when the telephone is off hook. To dial the number 3 using a rotary dial, the caller inserts a finger in hole 3 of the dial and pulls the dial around until the finger hits a finger stop. The finger is removed from the dial and the dial spins automatically back to its normal at-rest position. As the dial returns to normal, the loop is opened three times. Each time it is opened current flow will cease (0 milliamp). After opening the loop three times, the dial will be back at normal and the loop remains closed with a steady current flow of 40 milliamps in the local loop and central office equipment. The equipment will recognize each absence of current flow as one pulse of a digit. The equipment counts the number of open-circuit conditions encountered and stores the count in a register as the dialed digit.

Figure G-4 also shows the dial pulses as voltage conditions. Each time the dial opens the circuit, a voltmeter would read the 48 V of the central office battery placed on the loop. When the dial contacts close the loop, current will flow and voltage drops occur across each component in the circuit (that is, the central office equipment, the local loop, and the telephone). The voltage drop read across the telephone will be between 8 and 13 V (depending on the length of the local loop). In Figure G-4, we are reading the voltage across the telephone, and in our example the voltage changes from 10 V (closed circuit) to 48 V (open circuit) for each open-circuit condition—that is, each pulse.

G.6 CROSSBAR SWITCHING

In 1938, the Bell telephone companies begin the deployment of a new switching technology called the *crossbar switch (XBAR)*. This switch was improved over the next ten years, and in 1948 the no. 5 crossbar switch was introduced. This switch consisted basically of line link frames, trunk link frames, and common control equipment. Common control equipment encompasses a group of equipment used to establish call setup. After the call is set up, the equipment is released and is available to set up other calls. Instead of using linefinders, the crossbar used linescanners to determine when a line connected to a line link frame has gone off hook. The scanner would send a signal to a device called the *marker*. The marker would assign another common control component called a *register* to the line. A set of crosspoints was made in the line link frame connecting the line to the register. This set of crosspoints was operated by the marker. The register is a storage device. In the XBAR switch, the register consisted of electromechanical devices called *relays* to store dialed digits. The register received all dialed digits from the telephone and stored them for later use (see Figures G-5 and G-6).

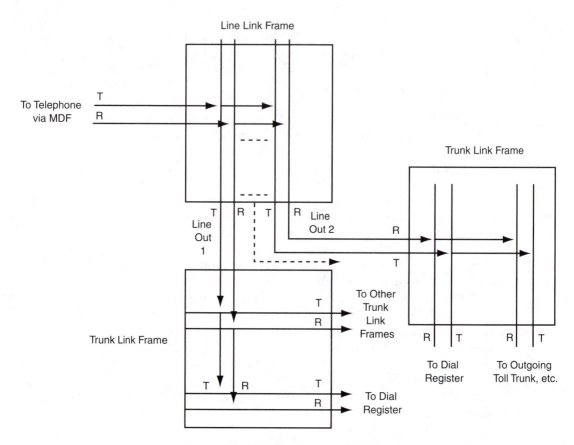

FIGURE G-5 Attaching a register in XBAR.

On completion of dialing, the register would request that the marker assign another common control component (translator) to the register. The *translator* looked at the dialed digits stored in the register and translated the dialed number into a series of switching instructions for the marker. The marker used the instructions from the translator to operate the necessary crosspoints in line link and trunk link frames to connect the calling line to the called line. Because each conversation path occupied its own wire path through the XBAR switching system, it is referred to as a *space division switching technology.*

The register, marker, and translator were all part of the common equipment used by all calls. Once they determined which crosspoints were to be activated in the switching matrix, they issued instructions to the switching matrix to operate the needed crosspoints. The crosspoints stayed operated for the duration of the call, but the common equipment was free to set up additional calls. The marker controlled all operations in the switch. Using pooled common equipment for switching control significantly reduced the size of electromechanical switching devices. This

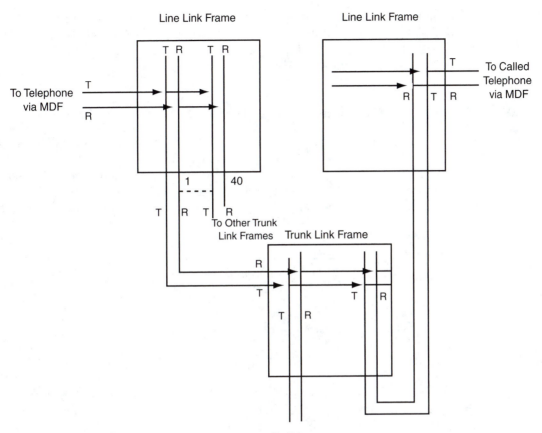

FIGURE G-6 Local call in XBAR.

meant less maintenance was required in a XBAR office compared to a SXS office, and XBAR equipment cost less than SXS.

G.7 DUAL-TONE MULTIFREQUENCY DIAL REGISTERS

When touchtone telephones were introduced, the registers of the XBAR switch were replaced with registers that could recognize both dial pulses and touch-tones. Included in the new registers were tuned circuits. There was a tuned circuit for each of the touchtones a telephone could send. When a button was pushed on the telephone, two tones were transmitted from the telephone to the register. The tuned circuits would determine what two tones were received and convert the tones into a dialed digit for storage in the register. If the telephone was a rotary dial phone, a stepping relay would count the dial pulses of each digit dialed for storage in the register. Thus it did not matter what type of telephone was on a line, because the register was capable of storing digits received from either type of phone.

G.8 DUAL-TONE MULTIFREQUENCY DIAL OR TOUCHTONE DIAL PAD

Telephones manufactured today are made with a dial that use tones for dialing signals. This dial is called a *dual-tone multifrequency dial or DTMF dial.* The dial is a keypad containing 12 or 16 buttons (see accompanying table). The buttons are arranged into four rows and three or four columns. A separate tone is assigned to each row and column. When any button is depressed, it sends out a unique set of tones to the central exchange.

	1209 Hz	1336 Hz	1477 Hz	1633 Hz
697 Hz	1	2	3	A
770 Hz	4	5	6	B
852 Hz	7	8	9	C
941 Hz	*	0	#	D

Depressing button 1 will cause a 697-cycle (Hertz) tone and a 1209-cycle tone to be sent to the central exchange. The receiver in the register at the central exchange decodes the different combinations of tones received into the equivalent dialed digit. The dial register stores the dialed digit in a storage device. The storage device for older electromechanical switches, such as SXS or XBAR, was relays. The storage device for SPC switching systems is RAM. The DTMF pad used on a telephone has 12 buttons: *, #, and 1 through 0. It would have been possible to use a single tone for each button pushed rather than the dual-tone design. This would require the ability to generate and detect 12 tones (16 if you include *A, B, C, D* buttons). The dual-tone design requires fewer tone detectors in the dial register of the central office than a single-tone design would. The use of fewer tones also reduces the sensitivity requirement of the receiver. A dual-tone receiver is better than a single-tone design in its ability to recognize tones distorted by abnormal conditions on a local loop that cause noise on the cable pair.

All push-button telephones are not DTMF telephones. Some push-button telephones do not send DTMF tones; they convert the button pushed to rotary dial pulses. You can recognize this type of push-button telephone because you can hear the clicks of the rotary dial in the phone receiver. You can also tell when a push-button phone is not a DTMF phone since this type of telephone will not work with voice messaging systems. It will work with the central office switching system because the dial register in the central office is designed to work with either a rotary or a DTMF dial. Voice messaging systems must receive DTMF tones to inform the system of what action it should take.

Rotary dial signaling pulses cannot pass through a central exchange. DTMF tones will pass through the exchange to the messaging system attached to the called number. Because a dial register in a central office can recognize both rotary dial pulses and DTMF tones, both a rotary dial telephone (called a *500 set*) and a touch-tone phone (called a *2500 set*) will work with any central office exchange. DTMF tones must be present for at least 40 milliseconds (ms) in order for the register at the central office to recognize the tones. A pause of 60 ms is required between digits. The total time for transmission of one digit is 100 ms; therefore it is possible to transmit 10 digits in 1 sec. DTMF is faster than rotary dial, which takes an average of $1^1/_2$ sec per digit or 15 sec for 10 digits.

G.9 STORAGE OF DIALED DIGITS IN A REGISTER

In a common control switching system, such as XBAR and SPC, the dialed digits are received by a device called a *dial register* or simply the *register*. The register recognizes a rotary dialed digit by counting the number of open-circuit conditions (pulses) sent by the dial (Figure G-4). The pulses are counted by a stepping relay, which steps once for each open circuit (pulse). The register also contains tuned circuits so it can detect the different tones sent by a touchtone keypad. In the dial register of a XBAR switching system, the output of the tuned circuits for a DTMF call (or the output of the stepping relay for a rotary dial) was wired through an additional stepping switch, called the *selector switch*, to 40 storage relays (see Figure G-7).

The selector switch would step once at the end of each dialed digit and select the storage location for the dialed digit. It could take ten steps to ten different levels. Each level of the selector contained four storage relays. Initially the selector switch would be stepping on level 1 and would connect four signal leads from the stepping relay/tuned circuits to four relays labeled W_1, X_1, Y_1, and Z_1 (storage location 1). The stepping relay and tuned circuits would cause an output to occur on the four signal leads *(W, X, Y, Z)* that would be different for each dialed or touchtone digit. This caused a different combination of relays to be operated for each digit sent by the telephone.

If the register recognized the digit as a 1, it operated the W and X relay. If it recognized the digit as a 2, the W and Y relays were operated. If the dialed digit or touchtone digit was a 3, the W and Z relays were operated. For a digit 4, the X and Y relays were operated, and so on.

As stated above, the additional selector switch initially connected the output of the tone detector circuits and stepping relay over signaling leads W, X, Y, and Z to relays W_1, X_1, Y_1, and Z_1. After the first digit was detected and stored, the selector switch stepped to level 2 for storage of the second dialed digit. The W, X, Y, and Z signaling leads were then connected to relays W_2, X_2, Y_2, and Z_2 for storage of the second digit sent by the telephone. After each digit was received, the selecting switch stepped to the next set of four storage relays. The selecting switch could take

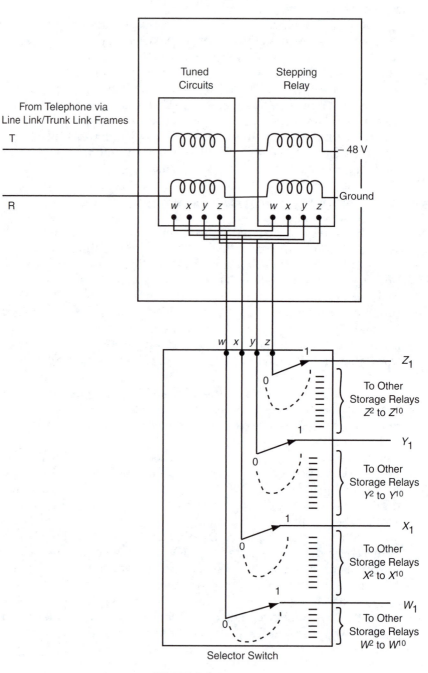

FIGURE G-7 Dial register.

ten steps. This allowed it to direct the storage of dialed digits to ten different storage locations for storage of up to ten dialed digits. The tenth storage location used relays W_{10}, X_{10}, Y_{10}, and Z_{10}.

Dialed Digit	W (*) Relay	X (*) Relay	Y (*) Relay	Z (*) Relay
1	OPERATED	OPERATED		
2	OPERATED		OPERATED	
3	OPERATED			OPERATED
4		OPERATED	OPERATED	
5		OPERATED		OPERATED
6			OPERATED	OPERATED
7	OPERATED			
8		OPERATED		
9			OPERATED	
0				OPERATED

G.10 XBAR COMMON CONTROL: THE MARKER AND TRANSLATOR

After receiving one digit, the register asked the marker to connect the translator to it. The marker was the brains of a XBAR switching system. The translator looked at the first dialed digit to see if it was a 1 or 0. If the first digit was a 1, the translator would tell the marker to connect the calling line to the *Direct Dial* Automated Long *Distance (DDD)* switching system. The DDD machine then received all additional dialed digits and determined how the call should be handled. If the first digit received by the register was a 0, the translator told the marker to connect the calling line to an idle trunk terminating on the operator's position. If the first digit received by the register was neither a 1 nor a 0, the translator informed the register to collect two additional digits and return after the third digit was received.

After receiving three digits, the register would ask the marker to connect it to the translator. The translator would look at the three digits to see if they matched the *NNX* exchange code assigned to this exchange. If they matched, the translator would direct the register to collect four more digits and come back after they had been collected. When the register came back and presented all seven dialed digits to the translator, the translator converted the dialed number into the equipment location of the terminating line circuit and instructed the marker to connect the originating line circuit to the called (or terminating) line circuit.

If the first three digits of the dialed number did not match the central office exchange code of the originating office, the translator instructed the marker to connect the calling line to an idle trunk circuit going to the desired central office. When the first three digits of a dialed telephone number did not match the exchange code of the originating office, the call had to be destined for a nearby central office. If the call was destined for an office many miles away, the caller would have dialed a 1 as the first digit and would have been connected to a toll exchange. The fact that the call remained attached to the register in the local class 5 exchange means the call was a local call but not a call within the same exchange. These types of calls are termed *EAS calls.* As mentioned earlier, EAS calls are not charged toll charges. The local service rate is raised by tariff filings as additional EAS points are added to a local exchange. The LEC calls the EAS points "free" calling areas, but they are not free. Calls to EAS points are paid for by way of higher local service rates. Local exchange rates are based on how many telephones can be called within an exchange. As the number of accessible phones rises, so do the rates.

G.11 BANDWIDTH AND SINGLE-SIDEBAND SUPPRESSED CARRIER

The **bandwidth** of an AM signal is the difference between the highest frequency and the lowest frequency. In the example used, the bandwidth is 8000 Hz (68,000 – 60,000). The bandwidth of an AM signal is twice the modulating frequency. When a low-frequency signal is present as an input to the mixer, the amplitude of both sideband signals changes as the frequency of the input signal changes. We say the sidebands have been modulated (changed) by the input signal. Since both sidebands have been modulated, they both contain the frequency changes of the modulating signal. We only need to demodulate one of these sidebands to recapture the intelligence of the modulating signal. Since only one sideband is needed to demodulate the modulating signal, some systems will transmit only one sideband with the carrier frequency. These systems are called *single-sideband (SSB)* systems. Systems that do not transmit the main carrier frequency, but transmit the sideband only, are called *single-sideband suppressed carrier (SSB-SC)* systems.

G.12 CCITT STANDARDS FOR FREQUENCY DIVISION MULTIPLEXING

Telecommunications standards are set by several organizations. The predominant organizations are the American National Standards Institute (ANSI), which establishes standards for North America, and the Consultative Committee on International Telegraphy and Telecommunications (CCITT) organization within the International Telecommunications Union (ITU), which creates worldwide standards. Figure G-8 provides a description of the SSB-SC system specifications per CCITT Recommendation G.232.

G.13 GROUPS AND SUPERGROUPS

The basic building block for any multiplexing system is the **channel unit.** One channel unit is needed for each input voice signal. Each channel unit is designed with a 3200-Hz bandwidth to pass the voice frequency signals between 200 and 3400 Hz (Figure G-8). In the SSB-SC system specifications per CCITT Recommendation G.232, channel 1 will modulate a 108-KHz carrier signal. By using only the lower sideband, channel 1 occupies frequencies 104 to 108 KHz. Channel 2 will modulate a 104-KHz channel and occupy the 100 to 104-KHz band. Each channel will have a 4-KHz bandwidth assigned to provide guardbands on each side of the

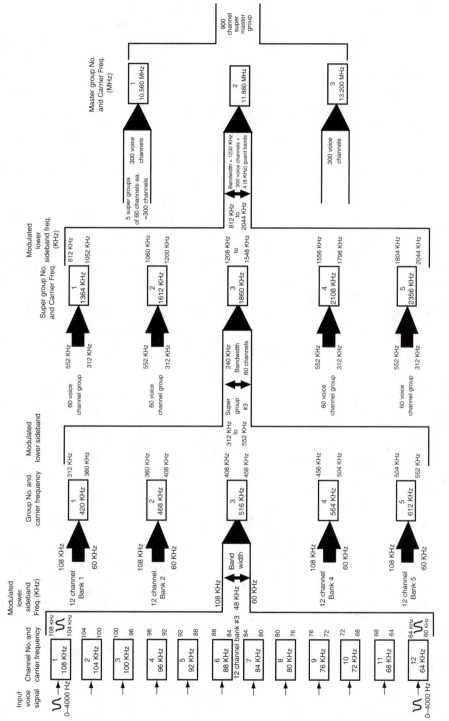

FIGURE G-8 CCITT SSB-SC system layout.

voice signal. Channels 3 through 12 each occupy the next lower 4-KHz band. Channel 12 modulates a 64-KHz carrier and occupies the 60- to 64-KHz band. The total bandwidth for a 12-channel group is from 108 KHz (the top frequency of the lower sideband for channel 1) down to 60-KHz (the bottom of the lower sideband of channel 12). This is a bandwidth of 48 KHz (108 − 60 = 48). The bandwidth for 12 channels needs to be 48 KHz (12 channels × 4 KHz per sideband = 48 KHz). Twelve channels grouped together as above form a group. Five groups of 12 channels each are combined into one 60-channel group called a *supergroup*.

To build a supergroup, we must change the frequencies of the groups before they can be combined (multiplexed together). The frequency band of each group is from 60 to 108 KHz. *Each* of the five groups in a supergroup consists of 12 channels multiplexed together, as stated above. The five groups are identical. To change the frequency of each group, we do a second modulation step. The first group of 12 channels is input to a second modulator, where they modulate a 420-KHz carrier. The second group of twelve channels modulate a 468-KHz carrier. The third group will modulate a 516-KHz carrier, the fourth group will modulate a 564-KHz carrier, and the fifth group will modulate a 612-KHz carrier. Only the lower sidebands from these modulators are used to create the supergroup of 60 channels.

Earlier we stated that a lower sideband is the difference between two signals when they are mixed together. The frequencies of the signals in a group are between 60 and 108 KHz. The carrier frequency for the first group of a supergroup is 420 KHz. When these two signals are mixed, the difference between 420 and 108 is 312 KHz, and the difference between 420 and 60 is 360 KHz. Thus, the lower sideband for group 1 of a supergroup is 312 to 360 KHz. The lower sidebands for all five groups of a supergroup are 312–360 KHz, 360–408 KHz, 408–456 KHz, 456–504 KHz, and 504–552 KHz. Each of these sidebands is 48 KHz wide to accommodate the bandwidth of 12 channels (12 × 4 KHz = 48 KHz). The standard group has a bandwidth of 48 KHz (60–108 KHz). Since a supergroup is composed of five 12-channel groups for a total of 60 channels, it should be 240 KHz wide (60 channels × 4 KHz per channel = 240 KHz). The bandwidth of a supergroup meets this criterion and is 240 KHz (552 KHz − 312 KHz = 240 KHz).

G.14 MASTERGROUP

A third modulation process can be added to the FDM process to create a basic mastergroup. The basic mastergroup contains five supergroups. Since each supergroup is 60 channels, a mastergroup contains 300 channels. It is formed by the same processes used above to create a supergroup from five groups. Frequencies 1364 KHz, 1612 KHz, 1860 KHz, 2108 KHz, and 2356 KHz are the carrier frequencies used for each of the supergroups within the mastergroup. The band of frequencies in a supergroup is from 312 to 552 KHz. When the first supergroup is mixed with the 1364-KHz carrier signal, the lower sideband will be 812 to 1052 KHz (1364 − 552 = 812 KHz and 1364 − 312 = 1052 KHz). Since each supergroup occupies a bandwidth of 240 KHz, the lower sideband of the first supergroup is also 240 KHz wide (1052 − 812 = 240). An 8-KHz guardband is placed between supergroups. The lower sidebands for all five supergroups in a mastergroup are 812–1052 KHz, 1060–1300 KHz,

1308–1548 KHz, 1556–1796 KHz, and 1804–2044 KHz. The total bandwidth for a mastergroup is 1232 KHz (2044 – 812 KHz = 1232 KHz) or (supergroup bandwidth 240 KHz × 5 + (4)8 KHz guardbands) = 1232 KHz.

A fourth mixer can be added to an FDM system to create supermastergroups. A supermastergroup can be made by combining three mastergroups using carrier frequencies 10,560 KHz, 11,880 KHz, and 13,200 KHz. Each mastergroup contains 300 channels; therefore, a supermastergroup contains 900 channels. Because there are a total of 15 supergroups in a supermastergroup, the system is often referred to as a *15-supergroup system.*

G.15 L CARRIER

The Bell System used SSB-SC systems called *L carrier* in its long distance network prior to the widespread acceptance of digital carrier. However, L carrier contains only 10 supergroups instead of the 15 of CCITT Recommendation G.232. L carrier uses ten supergroup carrier frequencies to create a mastergroup and does not create supermastergroups. Ten supergroups of 60 channels each results in a total of 600 channels per L carrier system. The system is called an *L600 system.* This system utilizes frequencies slightly different from those used by the CCITT standard for creating supergroups and mastergroups (see Figure G-9). AT&T also uses a U600 system. This system has the same design as an L600 system but uses a different set of frequencies for the ten supergroup carrier frequencies. The high bandwidth requirements (1.232 MHz) of the L600 and U600 systems require the use of coaxial cable or microwave radio as their transmission facility. AT&T continued to improve on FDM and developed an L5E system capable of multiplexing 13,200 voice circuits over one coaxial wire.

The L600 system used one mastergroup. The newer L5 system uses six mastergroups, each containing 600 voice circuits. Six mastergroups provides 3600 channels or voice circuits. These mastergroups were modulated to a higher order called *jumbo groups.* Each jumbo group contains 3600 circuits. The L5 system uses three jumbo groups for a total of 10,800 circuits. AT&T also has systems called *L5E,* which can multiplex 13,200 trunk circuits (voice circuits) over onto one pair (one transmit and one receive coaxial wire) of coaxial cable with amplifiers placed every mile. As mentioned earlier, coaxial cable is not the only medium used to carry FDM. Many FDM systems will use microwave technology to carry the L carrier systems, because microwave is cheaper. However, coaxial cable is quieter and does not require unsightly towers; it is used on many long distance routes.

G.16 FREQUENCY DIVISION MULTIPLEXING INPUT SIGNAL LEVELS

If the level of an input signal to an AM multiplier's channel unit is too high, it will cause crosstalk and noise. The input level to the channel is controlled by purposely introducing additional loss to the input signal. A resistive network called a *pad* is placed between the input signal and the channel input. The resistive network is adjusted so that the input signal level to the channel unit is –16 dBm. The output of the signal at the channel unit of the receiving multiplexer is +7 dBm. The FDM system provides a +23-dB gain for each channel signal.

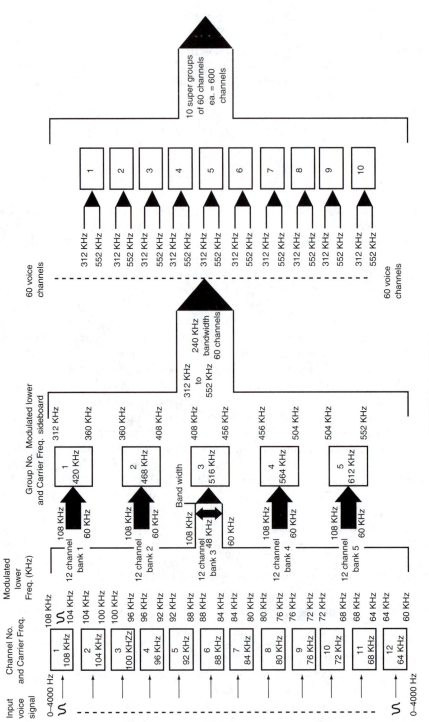

FIGURE G-9 SSB-SC system.

G.17 FREQUENCY MODULATION

AM signals are highly susceptible to noise caused by other electromagnetic signals. Another multiplexing technique—called *frequency modulation (FM)*—uses the same process used by FM radio broadcast stations. FM signals are less susceptible to noise than AM signals are. FM radio stations use radio transmitters that generate carrier frequencies between 88 MHz (88,000,000) and 108 MHz (108,000,000). In FM modulation, the input signal to a mixer causes the *frequency of the carrier signal* to change in step with the *amplitude of the input signal*. For example, at a particular amplitude, for a given input signal, the frequency of a 100-MHz carrier may change by 20 KHz in each direction (100,020,000 Hz and 99,980,000 Hz). *How often the carrier signal changes back and forth between these two frequencies is determined by the frequency of the input signal.* If the input signal has a frequency of 1000 Hz, the signal will change back and forth (between 100,020,000 and 99,980,000 Hz) 1000 times in one second. The *bandwidth* in FM is determined by the *amplitude* of the input signal, and in this example the bandwidth is 40 KHz. The bandwidth of any signal is the difference between the highest and lowest frequency of the signal. The bandwidth of the signal above is (100,020,000 – 99,980,000) or 40 KHz. A detector in the receiver detects how often the carrier frequency changes. In this example, the frequency of the carrier signal is being changed 1000 times each second. This represents the frequency of the original signal on the input of the transmitter. This demodulated signal is then fed to the speaker of the radio for conversion to sound waves of the same frequency (see Figure G-10).

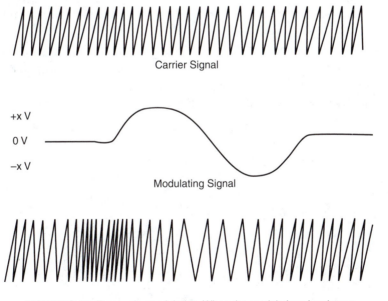

FIGURE G-10 Frequency modulation. When the modulating signal goes positive, the frequency of the carrier signal increases. When the modulating signal swings negative, the frequency of the carrier signal decreases.

When using FM in telecommunications, the channel units are wired to cable pairs and inputs just like the AM channel units were. The channel units of the FDM simply contain FM technology instead of AM technology. The use of FM channel units was limited. Most FDM systems use SSB-SC AM technology. Remember that *FDM means multiplexing* and does not refer to the FM modulation technique. A modulation technique such as AM or FM is needed to prepare signals for multiplexing by converting voice input signals to different higher-frequency signals. FDM is the actual multiplexing technique to combine these different frequency signals.

GLOSSARY

Bandwidth The width of a signal, which is determined by subtracting the highest frequency of the signal from its lowest frequency.

Channel Unit Performs the function of interfacing the multiplexer to a specific type of input signal. The channel unit makes changes to the input signal so it can be combined (multiplexed) with other signals for transmission over the transmission medium.

Electronic Mail (e-mail) A software application that allows a PC user to compose and send messages to another PC attached to the same network or any other network that the sender can access.

Register (Dial Register) The device in a common control switching system that receives dialed digits.

Space Division Switching A form of switching (SXS and XBAR) where each conversation occupies its own distinct and separate wire path through the switching system.

Switchboard The manual switchboard was used by operators to establish calls between two telephones. The manual switchboard was replaced by an automated switch.

Teletype A device used to transmit and receive data over a teletype network. This was the forerunner of today's data networks.

Translator In a stored program control switching system, a database that converts dialed digits into switching instructions.

Glossary

Acid A substance made of two or more elements. One of the elements is usually hydrogen or an atom that can accept electrons from a positive plate in a cell. The chemical reaction of an acid with a metal results in converting electrically neutral atoms into positive ions, negative ions, and electrons. When an acid is used in this manner, it is called an electrolyte.

ACK A positive acknowledgment (ACK) that the data was good. The receiving modem sends an ACK to the transmitting modem to indicate that the data is being received okay.

Address Bus The physical circuit path used (by a PC or other computer) to notify other components where data on the data bus should go. It controls the sending and receiving of data to and from the appropriate hardware device or memory location.

Address Resolution Protocol (ARP) The protocol used to resolve IP addresses to hardware addresses (MAC addresses). When the IP protocol is used, computers are referred to by higher-level applications using an IP address. In order for two computers on the same network to communicate with each other, they must use MAC addresses. When an IP packet is on the network containing the destination computer, ARP converts the IP address into the MAC address.

Advanced Mobile Phone System (AMPS) Also called *cellular radio*. The serving area for AMPS is broken up into cells. Low-powered transmitters are used for AMPS. Radio frequencies can be reused in nonadjacent cells. Cellular radio uses frequencies from 825 to 845 MHz and between 870 and 890 MHz.

Alerters A ring detector circuit that results in a chirping sound from a speaker.

Alternating Current (AC) An electric current that reverses directions at regular intervals. The magnitude of the current varies from 0 amp to some maximum value, falls back to 0 amp, rises to a maximum value flowing in the opposite direction, falls back to 0, and then repeats this cycle over and over.

Alternator A mechanical device for generating an AC voltage by spinning a magnetic field inside a coil of wire.

American Standard Code for Information Interchange (ASCII) The use of a seven-level binary code to represent the letters of the alphabet and the numerals 0 to 9.

American Wire Gauge (AWG) A standard that lists wire conductors according to their diameters. The higher the AWG number, the smaller the diameter and cross-sectional area of the wire. The table was set up so that a change in the AWG number by 3 results in a change in wire cross-sectional area by a factor of 2. Going from a smaller number to a larger number such as from 16 to 19 means the 19-gauge wire is smaller than the 16-gauge wire and is half the size of the I6-gauge wire.

Ammeter A test device used to measure current in a circuit. The ammeter is usually one part of a multimeter.

Ampere The unit of electric current flow. One ampere is a flow of approximately 6250 trillion electrons per second (6.25×10^{18} electrons per second).

Ampere-Hours The rating for a cell or battery that indicates how much electrical storage it has. A 1600 ampere-hour battery will deliver 1600 amps for 1 hr, 1 amp for 1600 hr, 400 amps for 4 hr, 200 amps for 8 hr, and so on.

Amplitude Modulation (AM) Used by AM broadcast stations. In amplitude modulation, the amplitude of the carrier frequency coming out of the mixer will vary according to the changing frequency of the input voice signal.

Analog Signal A signal that is analogous (similar) to a voice signal. The signal continuously varies in amplitude and frequency; it can assume an infinite number of values for amplitude, frequency, current, and voltage.

Application Layer The seventh layer of the OSI model. It does not include end-user application programs but is the layer that provides the support for these applications.

Applications Program A program developed for a special application such as word processing, project management, data management, and so on

Applications Protocol Interface (API) An interpreter for an application. The API resides at the session layer of the OSI model. The API permits applications to run on different network operating systems.

Architecture The way in which components of a switching system are physically and logically organized and how these components are connected together. Each manufacturer uses a proprietary design to interconnect the various components which make up a switching system. Thus, the architecture for CBXs will be different for each manufacturer and will even be different between the different models of switches sold by the same manufacturer. It is the difference in architecture which makes the models different.

Artificial Magnet A magnet that does not attain its magnetism naturally.

Asymmetric Digital Subscriber Line (ADSL) A digital subscriber line technology that can be used to deliver T1 speeds over the local loop. An ADSL has a high bit rate in one direction and a low bit rate in the opposite direction.

Asynchronous Transfer Mode (ATM) A transport and switching technology based on fixed-size packets called *cells*. The fixed-size cells allow the switching systems to process the packets very fast. Fixed-size cells also prevent one user from monopolizing the medium for extended periods of time.

Asynchronous Transmission   Also called *start-and-stop transmission*. The transmitting device sends a start bit prior to each character and sends a stop bit after each character. The receiving device will synchronize from the received start bits. Thus, synchronization occurs at the beginning of each character. Data is sent between two devices as a serial bit stream.

Atom   An atom is the smallest component of an element having all the properties of the element. It consists of a number of protons, neutrons, and electrons such that the number of protons determines the element.

Automatic Call Distributor (ACD)   A switching system used by telemarketing and/or customer service centers to connect incoming calls to agents. Calls are assigned to agents according to which agent has been least busy. ACDs can also serve as outbound systems, and with predictive dialing software, they can automatically dial telephone numbers and connect the called party to an agent.

Automatic Retransmission Request (ARQ)   The method of error detection used between high-speed modems. ARQ is either discrete ARQ or continuous ARQ. The receiving device returns a positive acknowledgment (ACK) when it receives a good block of data, and returns a negative acknowledgment (NAK) when the block of data received contains an error.

Average Busy Hour Peg Count (ABHPC)   An average of the number of calls placed over a group of circuits during the hours where most calls occur for several days.

Average Hold Time (AHT)   The average time for each call placed through a switching system or over a group of trunk circuits.

Bandwidth   The width of a signal, which is determined by subtracting the highest frequency of the signal from its lowest frequency.

Bar Magnet   A bar of iron or steel that has been artificially magnetized.

Baseband   The predominant method of data transmission on a LAN. The medium carries only one signal at a time and does not use FDM.

Base Station Controller (BSC)   Connects to all the base transceiver station (BTS) sites serving a BTA. The BTS provides an interface between the mobile phone and the BSC. The BTS is controlled by the BSC. Communication is constantly taking place between the BSC and the BTS. The BSC tells the BTS which radio channel to use for communication with a particular mobile station.

Base Transceiver Station (BTS)   The BTS site for a PCS 1900 system will consist of a small building to house the base station transmitters and receivers and an antenna tower approximately 150 ft high. The antenna tower will contain 9 antennas arranged in a triangular array. Each side of the triangle will contain one transmitting antenna and two receiving antennas.

Basic Rate Interface (BRI)   The basic building block of ISDN. A BRI consists of two 64-Kbps bearer channels and one 16-Kbps delta channel (2B+D). The delta (D) channel is used for signaling and sets up calls for the B-channels. The B-channels are used to carry customer voice or data.

Basic Trading Area (BTA)   Each of the 51 major trading areas (MTAs) was subdivided into a BTA. There are 492 BTAs. PCS licenses were sold for each BTA.

Battery   A combination of two or more cells electrically connected to work together to produce electrical energy.

Baud   The number of times a signal changes its state. If the amplitude of a signal changes 2400 times a second, the signal changes states 2400 times per second or at 2400 baud. If a signal changes back and forth between a frequency of 1800 Hz and 2200 Hz 3200 times per second, the baud rate is 3200. If a signal changes phase 2400 times a second, the baud rate is 2400 baud. Many people confuse baud rate and bit rate. Even some terminal emulation programs state that they will provide an option to change baud rates from 9200 to 19,200. This is incorrect. They should state that you can change the bit rate from 9200 to 19,200 bps. The selection of baud rate is done automatically by modems.

B-Channel   The channel that carries the customer's information. This information may be voice, data, or video.

Bell Operating Companies (BOCs)   Often referred to after deregulation as the *Baby Bells*. These were the local Bell Telephone companies. Before the 1984 Modified Final Judgment, 23 BOCs existed as subsidiaries of AT&T.

Binary Signal   A signal that assumes one of two discrete states. Also see **Digital Signal.**

Bit-Oriented Protocol   Includes HDLC and its derivatives such as LAPB. These do not use ASCII characters for control information but use each bit or group of bits for specific functions.

Blocked Calls   Calls which receive a fast busy signal or a recording stating all circuits are busy. When there are insufficient circuits to handle the volume of calls being placed through a switching system, or over trunk circuits to a particular location, some calls will be blocked by the system.

Blocking   A term used to describe a situation where a call cannot be completed because all circuit paths are in use.

Branch Feeder Cables   Cables that connect distribution cables to main feeder cables.

Bridge   A layer 2 device. It uses the MAC address to determine whether to pass a frame from one side of the bridge to the other. Most bridges join similar LANs. They can be used on Ethernet to break a larger LAN into smaller segments, thereby creating smaller collision domains. A special bridge known as a *translating bridge* can be used to join an Ethernet LAN to a token-ring LAN.

Bridged Ringer   A term indicating that the ringer is wired between the Tip and Ring leads. Bridged ringers were used on private lines as well as on two- and four-party lines.

Broadband   A transmission medium using FDM to allow the medium to carry multiple signals at the same time. Broadband requires special electronics and is seldom used for LANs.

Broadband ISDN (B-ISDN)   Provides speeds faster than 100 Mbps and can be used to establish wide area networks. Broadband ISDN is asynchronous transfer mode transmission over fiber optic cables.

Busy Hour   The hour of a day where the most telephone usage occurs.

Capacitance   The property of a device that allows it to store an electrical charge.

Capacitive Reactance   A measure that states the amount of opposition that capacitance provides to an AC current. It is found by the formula $XC = 1/(2\pi FC)$.

Capacitor   A human-made device used to store electrical energy. It is made by separating two conductors with an insulator. If extra capacitance needs to be added to an electrical circuit, the designer places a capacitor in the circuit.

Carrier Sense Multiple Access/Collision Detection (CSMA/CD)   The type of access methodology used by Ethernet. IEEE 802.3 specifies that the NIC will monitor the medium to determine if signals are present before allowing the NIC to transmit onto the medium.

Carrier Serving Area (CSA)   The creation of DS0 capabilities in a remote area of the exchange by using subscriber carrier to establish a remote wire center.

Carterphone Decision of 1968   A ruling by the Federal Communications Commission forcing AT&T to allow the attachment of a Carterphone to telephone lines at the customer's residence. This ruling was the beginning of deregulation of station equipment.

Cat-5 Cable   Category 5 cable. This cable contains tightly twisted wire pairs with about three twists per inch. Cat-5 cable is data grade cable.

Cell   A device that generates electricity by a chemical reaction.

Cellular Radio Also called *advanced mobile phone system (AMPS)*   The serving area for AMPS is broken up into cells. Low-powered transmitters are used for AMPS. Radio frequencies can be reused in nonadjacent cells. Cellular radio uses frequencies from 825 to 845 MHz and from 870 to 890 MHz.

Central Office   The central wire center or central exchange. All telephones in a small geographic area are wired to a central exchange, which serves all telephones in that area. Central offices are used in the PSTN to connect the calling party to the called party or to another switching system that can connect through the PSTN to the called party.

Centrex   A service offered by local exchange carriers as a direct replacement for a PBX. The Centrex service uses a special software program and the line circuits and switching equipment of the local class 5 switching office to provide Centrex.

Channel Service Unit (CSU)/Data Service Unit (DSU)   Also called *customer service unit/data service unit.* The CSU/DSU is a DCE device used to interface a computer to a digital leased line. It is a digital-to-digital interface. It can connect a low-speed digital device to a high-speed digital highway.

Channel Unit   Performs the function of interfacing the multiplexer to a specific type of input signal. The channel unit makes changes to the input signal so it can be combined (multiplexed) with other signals for transmission over the transmission medium.

Circuit   A circular journey. A circuit contains a conducting path between the terminals of a power source.

Circuit Switching   The process of connecting one circuit to another.

Circular Mil   The area of a circle that has a diameter of 1 mil. The circular mil area is diameter squared. It is not the same area found by using the algebraic formula $A = \pi r^2$, but it does provide the same linear relationship and makes it easier to compare conductors of various diameters.

Class 5 Exchange   Also called the *end office.* The lowest-level switch in the PSTN hierarchy; the local exchange.

Client   The workstation in a client/server LAN.

Client/Server An environment for a LAN where the PCs serve as either a workstation (client) or a server.

Closed Circuit A circuit in which the conducting path is unbroken.

Code Division Multiple Access (CDMA) Has been adopted as IS-95 standard for PCS networks. IS-95 Q-CDMA was codified as a standard in 1993. CDMA can provide 10 times the capacity of DAMPS. CDMA is not backward compatible with analog AMPS in the same way that D-AMPS is. CDMA cannot reuse DAMPS technology. To implement CDMA, the service provider must install a new network. CDMA is spread spectrum technology. Many conversations are multiplexed over one frequency. Each conversation is assigned a code. The receiver strips out conversations individually by using the code.

Code Excited Linear Predictive Coding (CELP) Also known as *vector-sum excited linear predictive (VSELP) coding*. This speech coding algorithm is EIA standard IS-54. This is the speech coding technique recommended for TDMA cellular radio systems. The bit rate is 7.95 Kbps. VSELP is also available in a 16-Kbps chip. AT&T makes a digital signal processor using this technology; it is called a *DSP-1616*.

Codec Coder/decoder is the hardware device that converts an analog signal into a digital signal and vice versa.

Collision Domain A section of a LAN medium where collisions will be detected.

Committed Information Rate (CIR) The minimum bandwidth or speed guaranteed to a user for the transfer of data across the network.

Common Carrier A company that offers telecommunications services to the general public as part of the public switched telephone network.

Communication A process that allows information to pass between a sender (transmitter) and a receiver over some medium.

Communications Act of 1934 Legislation passed by Congress and signed into law by the president in 1934 to regulate interstate telecommunications and radio broadcasts. This law created the Federal Communications Commission to administer the act.

Compound A substance or material composed of two or more elements.

Computer Inquiry Any of a series of rulings by the Federal Communications Commission. Issued Computer Inquiry I in 1971; stated that the Federal Communications Commission would not regulate computer services.

Computer Inquiry I A 1971 ruling by the FCC stating that it would not regulate computer services and networks.

Computer Inquiry II Issued in 1981. Mandated that station equipment was to be deregulated and could not be provided by a local exchange carrier; this equipment must be owned by the customer or some nontelephone company entity.

Computer Inquiry III In 1986, detailed the extent to which AT&T and the Bell Operating Companies could compete in the nonregulated enhanced services arena.

Computer Telephony Integrated (CTI) The integration of personal computer technology and telecommunications technology to provide PBX and/or ACD services via a personal computer system.

Computer Telephony Integrated (CTI) Systems Switching systems implemented using a PC with special applications software and special circuit cards that can interface to telephones and central office lines. CTI-based systems are available to serve as a PBX or an automatic call distributor (ACD).

Computerized Branch Exchange (CBX)
Standalone PBX switching systems implemented via stored program control switching systems. Almost all PBXs in service today are CBXs or CTI-based systems.

Concentration   The use of fewer outlets than inlets.

Conductance   A measure of how good a conductor is at carrying electric current. It is the reciprocal of resistance. The symbol for conductance is G and it is measured in Siemens. $G = 1/R$.

Conductor   An element or compound that readily conducts electricity because its atoms have loosely held electrons in their outermost shell. Most metals are good conductors. The six best conductors are silver, copper, aluminum, zinc, brass, and iron. Carbon is also a good conductor.

Connectionless Protocol   Also referred to as a *datagram service*. Rather than establish an end-to-end connection through a network, each frame of data contains the destination address that routers can use to route the packet through the network.

Connection-Oriented Network   A virtual circuit is established for the transfer of data. The virtual circuit provides a connection over which subsequent packets can be delivered.

Continuous ARQ   Also known as *sliding window ARQ*. Continuous ARQ eliminates the need for a transmitting device to wait for ACKs after each block of data. The device continuously transmits blocks of data and sends a block number with each block. The receiving device checks the blocks for errors and continuously returns positive ACKs to the transmitter. If an error in data occurs, the receiving device returns a NAK along with the block number affected. On receipt of a NAK, the transmitting device will retransmit the bad block of data.

Counter EMF   Voltage opposing the changes of current in an AC circuit.

Crosstalk   An undesirable condition where a circuit is picking up signals being carried by another circuit. For example, a customer hears other conversations or noise on a private line. This is usually caused by a breakdown in the insulation between wires of two different circuits.

Customer-Provided Equipment (CPE)   Also called *customer-premise equipment*. This is station equipment, on a customer's premises, that according to Computer Inquiry II must be provided by someone other than the local exchange carrier. The 1996 Telecommunications Reform Act overrides Computer Inquiry II and allows LECs to reenter the CPE market.

Cyclic Redundancy Check (CRC)   A form of error checking used between modems. A transmitting modem treats the block of data transmitted as representing a large binary number. This number is divided by a 17-bit divisor (CRC-16) or a 33-bit divisor (CRC-32). The remainder is attached in the trailer behind the block of data. The receiving modem performs the same division on the block of data received and compares its calculated remainder to the remainder sent in the trailer.

Data   Raw facts, characters, numbers, and so on that have little or no meaning in themselves. When data is processed, it becomes information. In telecommunications when we speak of data, we are referring to information represented by digital codes.

Data Bus   The physical circuit path inside a computer that data travels over to get from one location to another.

Data Communication   The transfer of data over telecommunications facilities.

Data Communications Equipment (DCE)   Also called *data circuit termination equipment*. A device that interfaces data terminal equipment (DTE) to the PSTN. A modem is a DCE used to interface a DTE to an analog line circuit on the PSTN. A CSU/DSU is a DCE device used to interface a DTE device to a leased digital line in the PSTN.

Data Link Connection Identifier (DLCI)   The logical channel assigned between a FRAD and the frame relay network.

Data Link Layer   The second layer of the OSI model. It controls how data is to be placed on the medium. It controls the size of frames transmitted on the LAN.

Data Switching Exchange (DSE)   Also known as *packet switching exchange*. These are the switches of a packet switched network. The DSE routes packets of information based on the packet address information found in each packet header.

Data Terminal Equipment (DTE)   A DTE device is used to transmit and receive data in the form of digital signals. The personal computer is the most common form of DTE.

D-Channel   The channel that carries signaling and control information. The D-channel is used to set up calls for the B-channels.

Decibel (dB)   The ratio of one power level to another. Decibels state the relative comparisons of two levels of power. For example, 3dB means the power level of one signal (P_2) is twice the power level of another signal (P_1). The formula is dB = 10 log (P_2/P_1)

Decibels Referenced to Noise (dBrn)   The starting power level for measuring noise. 0dBrn = –90 dBm and 0 dBrn = 1 pW.

Decimonic Ringing   A ringing system used by Independent telephone companies. The system could send out five different ringing signals (10-, 20-, 30-, 40-, and 50-cycle signals).

Dense Wave Division Multiplexer   A device that places many voice or data circuits over one glass fiber by splitting the infrared light spectrum into several different frequencies and then attaching a time division multiplexer to each of these frequencies.

Dense Wave Division Multiplexing (DWDM)   Technology that splits the light spectrum into many different frequencies. A time division multiplexer is then assigned to each of the different light wave frequencies. DWDM is used to multiplex multiple DC-48 (or DC-192) systems over one fiber pair.

Deregulation   The change of the telecommunications industry from a regulated monopoly to a competitive nonregulated market. One of the major provisions of the 1984 Modified Final Judgment was to deregulate long distance services.

Differential Manchester   A digital coding technique. Differential Manchester coding contains two transitions per bit time. The midbit transition is used only for receiver synchronization, and the absence or presence of a transition at the beginning of a bit time is used for the coding/encoding of data. The absence of a transition at the beginning of a bit time is used to represent the digit 1. The presence of a transition is used to represent the digit 0. Differential Manchester is both biphase and differential and is used for *token-ring* LANs.

Digital Access Cross-connect (DAC)   A device used to connect fiber circuits or cable circuits to each other. The cables or fibers connect to ports on the DAC. The ports contain a time slot interchange (TSI). The TSI at each port connects via TDM links to all other ports. The TSIs are programmed to connect any time slot (channel) of one port to any time slot of any port.

Digital Advanced Mobile Phone System (DAMPS)   A technique that places multiple calls over one radio frequency using pulse code modulation (PCM) and time division multiplexing (TDM). The PCM signal is not the standard 64,000-bps PCM signal found in the PSTN but is a 16,000 bps signal. DAMPS was introduced in 1992 for the existing AMPS and is backward compatible with analog AMPS. The TDM feature allows DAMPS to carry four times as many calls as AMPS.

Digital Signal   A signal that can assume one of several discrete states. Most digital signals take on one of two discrete states. The prefix *bi* means two, and a two-state signal is also called a *binary signal.* Each state is represented by a different electrical signal. Two different electrical states can be used to represent the two binary digits (1 and 0). Because this two-state signal is used to represent 1 or 0, it is also called a *digital signal.*

Digital Signal Processor (DSP)   A very large-scale integrated circuit chip that controls the sending and receipt of electronic signals. DSPs are used to perform data transfer functions in regular modems, ADSL modems, ISDN Modems, ISDN devices, PCs, remote access servers, and so on.

Digital Subscriber Line Circuit (DSL)   Usually refers to a local-loop cable pair that is handling a digital signal using 2B1Q line coding.

Direct Current (DC)   The electric current flow is at a constant rate and flows continuously in one direction. The direction of electric current flow (outside the battery) is from the negative terminal of the

battery, through a device attached to the battery, to the positive terminal of the battery. The amount of electric current flowing is measured in amperes.

Direct Current (DC) Voltage   The voltage produced by a chemical reaction. Devices that produce DC electricity via a chemical reaction are called *cells.* Several cells connected together electrically are called a *battery of cells* or simply a *battery.*

Direct Inward Dial (DID) Trunks   Trunk circuits at the class 5 central exchange that allow someone on the PSTN to dial directly into a particular station on a CBX.

Discrete ARQ   Also called *stop-and-wait ARQ.* An error control protocol that requires an acknowledgment from the receiver after each block of data sent. The transmitting modem sends a block of data and then waits for an acknowledgment before sending the next block of data.

Distribution Cable   A cable that is fed by a branch feeder cable from the central exchange and that connects to telephones via drop wires. The distribution cable is usually a small cable of 25 to 400 pairs.

Divestiture   A major provision of the 1984 Modified Final Judgment. This provision forced AT&T to divest itself of the 23 Bell Operating Companies.

Domain Name   The top level in the Internet naming conventions. There are 13 domain names: **.edu, .com, .gov, .int, .mil, .org, .net, .web, .arts, .info, .nom., .rec, and .store**

DOS-Based LANs   LANs that do not require a NOS.

Drop Wire (Drop Cable)   A pair of wires connecting the telephone to the distribution cable. One end of the drop wire connects to the protector on the side of a house or business, and the other connects to a terminal device on the distribution cable.

Dry Cell   A chemical cell that generates electricity via the chemical action between an electrolytic solution and two dissimilar metals. The electrolytic solution is in the form of a paste.

Dual Simultaneous Voice and Data (DSVD)   Allows a device to transmit voice and data at the same time over the same media or logical communications channel.

Dual-Tone Multifrequency (DTMF) Dial   The dial on a 2500 set that sends out a combination of two tones when a digit on the keypad is depressed.

EIA-232 Interface   Also called *RS-232*. The most common interface standard for data communication. CCITT standard V.24 is the same as EIA-232. EIA-232 defines the voltage levels needed for the various signal leads of EIA-232.

Electric Current   The orderly flow of electrons in one direction.

Electromagnet   An artificial magnet made by wrapping a coil of wire around an iron or steel core and then sending an electric current through the coil of wire.

Electromotive Force (EMF)   The electron moving force, energy, or pressure that tries to force electrons to move. EMF is measured in volts.

Electron Pump   A chemical cell or several cells connected to form a battery for the purpose of supplying (pumping) electrons to a circuit.

Electronic Data Interexchange (EDI)   A network that allows companies, their vendors, and customers to exchange data.

EDI uses a data network such as an *X.25* network.

Electronic Mail (e-mail)   A software application that allows a PC user to compose and send messages to another PC attached to the same network or any other network that the sender can access.

Equal Access   A key provision of the 1984 Modified Final Judgment. This provision forced the Bell Operating Companies to provide 1+ access to toll for all interexchange carriers.

E-Rate Fund   Established by a provision of the 1996 Telecommunications Reform Act; intended to provide schools (K–12) with funds to buy computer network services and to pay for internet access services. Financed by fees levied on the common carriers by the FCC.

Ethernet   A protocol used on Local Area Networks (LANs) to control a user's access to the LAN medium. A LAN technology using a bus or star topology and access via CSMA/CD with Frames that limit user data to 1500 bytes.

Exchange Boundary   Pertains to the limits of the serving area. The outermost customers being served by a particular exchange establish the boundary of the exchange.

Exchange Code   The first three digits of a seven-digit telephone number.

Extended Area Service (EAS) Tandem   A class 4 switching system used in a metropolitan area to connect class 5 switching systems together on EAS calls. Extends the rate base of a local exchange so that calls to neighboring exchanges are not toll calls but local calls.

Facility   A term typically used to refer to the transmission medium.

Feature Group   A series of services established by the Bell Operating Companies to provide access to interexchange carriers (IECs). *Feature Group A* gave customers access to IECs by having them dial a seven- or ten-digit telephone number plus a personal identification number. *Feature Group B* provided access via a seven-digit code of the form 950-10XX. This access code was later changed to 10XXX. *Feature Group D* is the 1+ access to IECs that was mandated by the 1984 Modified Final Judgment.

Feature Phone   A telephone containing electronics that allow it to provide features. When additional features have been added to the basic single-line telephone such as last number redial and speed dialing lists, the phone is called a feature phone.

Federal Communications Commission (FCC) The federal agency created by the Communications Act of 1934 to oversee enforcement of the act by regulating interstate telecommunications and broadcast communication.

Feeder Cable   Includes various types. Cables leaving the central exchange are called *main feeder cables.* The main feeder cables connect to branch feeder cables. The branch feeder cables are used to connect the main feeder cables to distribution cables.

File Transfer Protocol (FTP)   Used to request files from a file server on the Internet or any other network. FTP establishes two TCP/IP connections for a file transfer. Port 21 is used as a control port and Port 20 is used for the transfer of data.

Fixed Resistors   Resistors having one value of resistance.

Flag   A special pattern of bits used to flag the beginning and/or end of a frame. For the X.25 frame the flag bit pattern is 01111110.

Flow Control   Controlling the flow of data from one device to another usually via hardware flow control (RTS/CTS) or by software flow control (XON/XOFF). A modem contains a memory buffer to allow it to compress data before transmitting and to allow it to convert asynchronous data to synchronous data. It must be able to stop the transmitting PC when this memory buffer approaches a near-full condition. Some modems contain a very large memory buffer, which negates the need for flow control.

Flux Density   A measure of the amount of magnetic flux per unit of cross-sectional area.

Flux Lines   Invisible lines of magnetic flux that leave the north pole of a magnet, traveling through air or some other medium to enter the south pole of a magnet. They then flow through the magnet back to the north pole.

Foreign Exchange (FX)   A term used to refer to all class 5 exchanges that cannot connect to your particular telephone on a local call basis. If I have a telephone in Kansas City, then St. Louis, Chicago, New York, and so on are foreign exchanges to me.

Frame   The term used for a packet formed at the data link layer for transmission out onto the transmission medium connecting to the network.

Frame Relay   A fast packet network that uses a separate logical channel for call setup and control and does end-to-end error correction. It is a two-layer protocol designed to run on PRI facilities.

Frame Relay Assembler/Disassembler (FRAD)   The device used to access the frame relay network.

Frequency Division Multiplexing (FDM)
The process of converting each speech path to different frequency signals and then combining the different frequencies so they may be sent over one transmission medium.

Full-Duplex Transmission If a transmission system allows signals to be transmitted in both directions at the same time, the system is called a *full-duplex transmission system.*

Generator A mechanical device that generates an AC voltage in a wire by spinning the wire inside a magnetic field.

Global System for Mobile Communications (GSM) 1900 A standard for mobile phone service based on Time Division Multiple Access (TDMA).

Graham Act of 1921 Legislation passed by Congress and signed by the president in 1921 to exempt telecommunications from antitrust legislation.

Grounded Ringer A term indicating that the ringer is wired between the Tip lead and ground or between the Ring lead and ground. Grounded Ringing was used on eight- and ten-party lines.

Group Ware An application that allows users on a LAN to share information with each other via e-mail and scheduling applications.

Half-Duplex Transmission A type of transmission where transmitters on each end of a medium take turns sending over the same medium.

Harmonic Ringing A ringing system used by Independent telephone companies. The system could send out five different ringing signals (16 2/3-, 25-, 33 2/3-, 50-, and 66 2/3-cycle signals).

Header The term used for the addressing and control information placed in front of

user data when a packet is formed or placed in front of the Information Field when a Frame is formed.

Hookswitch Often called a *switch-hook.* This is the device in the telephone that closes an electrical path between the central office and the telephone, by closing contacts together, when the receiver is lifted out of its cradle.

Hybrid Keysystem A keysystem designed to handle more than 24 central office lines and 40 telephone sets. Central office lines are accessed by dialing 9 for access.

Hybrid Network A network that consists of a transformer and an impedance matching circuit. The transformer performs a two-to four-wire conversion and vice versa. In a telephone, the hybrid network connects the two-wire local loop to the four-wire transmitter/receiver.

Hypertext Markup Language (HTML) A structured use of special words called tags utilized in turn by a browser as instructions on how to display a web page. A software language used to create documents, such as home pages, for use on the Internet.

Impedance The total opposition that a circuit or component has to AC current. This total opposition is composed of resistance, inductive reactance, and capacitive reactance.

Improved Mobile Telephone System (IMTS) Introduced around 1964. The major benefit of IMTS was that it used several different frequencies and could support many different conversations at the same time. It was also connected to the local class 5 office instead of the toll office. IMTS was connected to regular telephone numbers and line circuits. An IMTS phone was assigned a regular PSTN telephone number. Anyone could reach an IMTS phone by dialing the PSTN number assigned to the mobile phone. This eliminated the need for operators to handle mobile phone calls.

Inductance   The property of a circuit or device by which a change in current induces an electromotive force.

Inductive reactance   A measure of the amount of opposition that inductance offers to an AC current. It is found by the formula $XL = 2\pi FL$.

Inductor   A human-made device used to purposely add extra inductance to a circuit.

Industry Standard Association (ISA) Bus   Originally a 60-pin bus that transferred data 8 bits at a time and had 20 address leads. Thirty-six additional pins were added to another socket referred to as the 16-bit bus. This allowed the transfer of data 16 bits at a time and provided an additional 4 address leads.

Information Service Providers   The term used for companies like Prodigy and America Online. Any company that contracts with customers to provide them information is an information service provider. Most of these companies provide information to their customers over the PDN.

Insulator   A poor conductor. An element or compound with electrons held so tightly to their atoms that it takes a high voltage to dislodge them.

Integrated Services Digital Network (ISDN)   The use of digital line circuits to provide end-to-end digital service. The *basic rate interface (BRI)* provides the user with two DS0 channels. The *primary rate interface (PRI)* provides the user with 23 DS0 channels.

Interactive-Voice-Response (IVR)   An automated system that responds to user input. The user input is often done via the touch-tone keypad of a telephone, but some systems accept voice input. On receiving

input from a user, the IVR system will respond with information or request additional input.

Interexchange Carriers (IECs or IXCs)   Common carriers that provide long distance telephone service. Major IECs are AT&T, MCI, Sprint, LDI, and so on.

International Data Number (IDN)   A number assignment technique used to assign terminal numbers to the DTE (with PAD software) connected to the PDN. The numbers are of the form 31264298997678. The addresses are usually 12 to 15 digits long.

Internet   A network that connects Internet service providers (ISPs) together. Some companies will serve as their own ISP and will connect their local network to the Internet via a router or server, but most people access the Internet via a local ISP or commercial online information access provider such as Prodigy or America Online (AOL). Anyone who can access the Internet can access any computer that is attached to the Internet.

Internet Protocol (IP)   A network layer protocol used by many networks in general and by the Internet in particular. It provides the routing functions necessary to route data packets across a WAN and is also used on LANs. IP provides for a connectorless, unreliable transfer of data.

Internet Service Providers (ISPs)   Companies that provide individual users and companies with access to the Internet.

Interoffice Calls   Calls completed between telephones served by two different central exchanges.

Intraoffice Calls   Calls completed between telephones served by the same central exchange.

Ions   Atoms that have an electric charge. An atom with a positive charge is a positive ion. A negatively charged atom is a negative ion.

IP Address   The network layer address. It contains both the network and station identity. The IP address for IP version 4 is 4 bytes long. To provide many more addresses, IP version 6 has been released. It has a 16-byte field.

IP Datagram   A packet of data that contains source and destination addresses. Encapsulated within the IP datagram are the TCP or UDP header information and the actual user data from an applications program.

Jumper Wire   A short length of twisted-pair wire used to connect wire pairs of two different cables. A jumper wire is a very small gauge of wire (usually 24 or 26 gauge). Except for its small size and much longer length, a jumper wire is similar to the jumper cable used to start a car with a dead battery.

Keysystem   A telephone system used by small businesses to allow several central office lines to terminate at each telephone attached to the system. Any telephone can access all the lines to the telephone office.

Kingsbury Commitment of 1913   An agreement signed by the vice president of AT&T in 1913. This agreement stated that AT&T would allow other telephone companies access to its network, would sell its Western Union stock, and would not buy any more Independent telephone companies without getting permission from the government.

Left-Hand Rule   A rule used to determine direction of magnetic flux around a wire carrying an electric current. By wrapping your left hand around a wire such that the thumb points in the direction of current flow, the fingers around the wire point in the direction the magnetic flux is flowing. The left-hand rule is also used to determine

the north pole of an electromagnet. By wrapping your left hand around a coil of wire such that the fingers point in the direction of current flow, the thumb points toward the north pole of the electromagnet.

Line   The circuit or cable pair which connects a telephone to the local central office switching system.

Line Circuit   The hardware device in a local telephone office (switching exchange) that a telephone connects to. Every telephone must connect to its own line circuit if the phone is on a private line.

Line Relay   An electromechanical device attached to a telephone line at the central exchange. When a subscriber took the handset of the telephone off hook, electric current flowing in the relay caused it to operate and signal the operator.

Link Access Procedure Balanced (LAPB)   An HDLC-based protocol used on point-to-point links. It is a full-duplex protocol that allows either end of the link to initiate actions.

Link Access Procedure on the D-Channel (LAPD)   The software protocol used at the data link layer (layer 2) of the D-channel. LAPD uses Carrier Sense Multiple Access/with Collision Resolution (CSMA/CR) to allow multiple TEs on the S/T bus. LAPD is a multipoint protocol.

Load   The active device in a circuit. It makes use of the electric current in the circuit to perform some useful function.

Load Coils   The devices that introduce additional inductance into a circuit. Load coils are added to cable pairs over 18,000 ft to improve the ability of the pair to carry voice signals.

Loading   Purposely adding inductance to a cable pair over 18,000 ft long to offset the mutual capacitance of the cable pair.

Local Access Transport Area (LATA)   A key provision of the 1984 Modified Final Judgment (MFJ). This provision established 184 geographic regions that conform to the Standard Metropolitan Statistical Index used by marketing organizations. The LATA was defined as an area within which calls must be carried by a local exchange carrier (LEC) and could not be carried by an interexchange carrier (IEC or IXC). These calls within the LATA are called *intra-LATA calls*. The MFJ further stated that calls between LATAs *(inter-LATA calls)* must be carried by an IEC, not by a LEC. The 1996 Telecommunications Reform Act replaced the 1984 MFJ and opened the door to competition in this area. It allows either type of call to be carried by either a LEC or an IEC.

Local Area Networks (LANs)   A network that exists in one building to connect the computers within that building to one another so they can share information, files, programs, printers, and other resources.

Local Exchange   The switching system that telephones are connected to.

Local Exchange Carrier (LEC)   The provider of local telephone services. Prior to the 1996 Telecommunications Reform Act, the LEC was your local telephone company. With the passage of that act, the local telephone company is now called the *incumbent local exchange carrier (ILEC)* and its competitors are called *competitive local exchange carriers (CLECs)*.

Local Loop   A term used to describe the facilities that connect a telephone to the central exchange. These facilities usually consist of a twisted pair of wires.

Logical Channel Number (LCN)   A number used by the DTE (PAD) and DCE at the PSE to identify the originator and destination of the packet containing the LCN.

Loop Extender   A hardware device. Can be connected between the line circuit of the switch, and the cable pair serving a remote area of the exchange, to extend the range of a central exchange. A loop extender adds a voltage boost to the circuit. By doubling the voltage applied to a line, the distance served can be doubled.

Loop Treatment   A term indicating that a local-loop cable pair has been specially conditioned. Typically this term describes the addition of extra equipment to a local-loop cable pair. To extend the range of a central exchange, a device called a *loop extender* can be connected between the line circuit of the switch and the cable pair serving a remote area of the exchange. A loop extender adds a voltage boost to the circuit. By doubling the voltage applied to a line, the distance served can be doubled. Another device, called a *voice frequency repeater (VFR)*, is also added to long local loops. The VFR will amplify voice signals in both directions to compensate for the extra decibel loss of long loops.

Magnetic Field   The area in which magnetic lines of flux are flowing.

Magnetism   The properties of attraction possessed by magnets.

Magneto   A device that generates an AC voltage by turning a coil of wire inside a magnetic field. The old hand-crank telephone had a magneto attached to the hand crank.

Main Distributing Frame (MDF)   In a central exchange, the place where all cables that connect to the switch are terminated. Outside plant cables that connect to telephones and to other central offices are also terminated at the MDF. Short lengths of a twisted-wire pair (called a *jumper wire*) are used to connect wires from the switch to wires in the outside plant cable. In a PBX environment, the MDF is where all cables from the switch, IDFs, and cables from the central exchange are terminated and jumpered together.

Main Feeder Cable   Cable that leaves the central exchange and connects to branch feeder cables.

Major Trading Area (MTA)   The United States is divided into 51 areas for mobile telephone licensing purposes. Each of these segments is called a *major trading area (MTA)*. Each MTA was subdivided into a basic trading area (BTA). There are 492 BTAs.

Manchester Coding   Used on local area networks (LANs) that employ the Ethernet signaling protocol. For Manchester coding, a transition occurs at the middle of each bit time. A high-to-low transition represents the digit 1, and a low-to-high transition represents the digit 0. The midbit transition is also used to provide receiver synchronization.

Material   The substances of which a thing is composed. The term is applied to almost anything considered to be matter.

Matter   Anything that occupies space and has weight.

MCI Decision of 1976   A ruling by the federal court in MCI's favor against AT&T; it allowed MCI and other common carriers to handle long distance services for the general public.

MCI Ruling of 1969   The ruling in 1969, by the Federal Communications Commission, that forced AT&T to allow private-line customers of Microwave Communications Inc. (MCI) to use local telephone lines for access to MCI's private-line network.

Media Access Control (MAC)   The lower layer of the data link layer. It controls access to and from the medium of the LAN.

Medium   The device used to transport information or a message between the transmitter and receiver.

Metropolitan Area Network (MAN)   A wide area network that spans a geographic area the size of a city.

Microwave Radio   Utilizes signals in the gigahertz band of frequencies. These signals must take a line-of-sight path, and due to the earth's curvature, microwave relay stations must be placed about 30 miles apart.

Mixer (Modulator)   An electronic device that has two inputs. One input is a low-frequency signal that contains intelligence. The other signal is a pure sine wave at a high frequency. This high-frequency signal is called the *carrier signal* because it carries our intelligent signal after modulation. The mixer combines the two signals so that we end up with a high-frequency signal that has the intelligence of the low-frequency signal imposed on it.

Mobile Telephone Serving Office (MTSO)   The central controller and switching system for cellular radio. The MTSO can be a stand-alone switch owned by a private cellular company, or it can be integrated into a local switch if the LEC is the cellular service provider.

Mobile Telephone System (MTS)   Mobile telephone service began in 1946 and was called *MTS*. These systems used the radio frequencies between 35 and 45 MHz. Although *MTS* usually stands for "mobile telephone system," it also meant "manual telephone system." All calls had to be handled by an operator.

Modem   A DCE device that interfaces the digital signal from a DTE device to the analog local loop and line circuit of the PSTN, by converting the digital signals into modulations of an analog signal.

Modified Final Judgment (MFJ) The 1984 agreement—reached on August 24, 1982, between AT&T and the Department of Justice—to settle an antitrust suit brought against AT&T by the Justice Department, on November 20, 1974. This judgment modified the 1956 Final Judgment. The major provisions of the 1984 MFJ were deregulation of long distance services and divestiture of the Bell Operating Companies by AT&T.

Modified Long Route Design (MLRD) Outside plant design that involves adding loop treatment to cable pairs serving remote areas of an exchange. Local loops longer than 24,000 ft are designed using MLRD.

Modulo 8 Indicates that the sliding window allows 7 frames to be outstanding without acknowledgment.

Modulo 128 Indicates that the sliding window allows 127 frames to be outstanding without acknowledgment.

Molecule The smallest physical unit of a compound that still retains the characteristics of the compound. It consists of the two or more atoms that make up the compound. For a material that is an element, a molecule is one single atom of the element.

Monopoly A form of market where one company is the sole provider of goods and/or services. There is no competition. Market demand does not set the price of goods or services. In a nonregulated monopoly, price is set by the monopoly service provider. In a regulated monopoly, price is set by the government agency charged with regulating the monopoly service provider.

Moves Adds and Changes (MACs) The changes made in the database for a SPC switch when a telephone is moved, added, or has its features or number changed.

Multimeter A test device used to measure voltage, resistance, or current.

Multiplexer A device that can combine many different signals or calls (data or voice channels) so they can be placed over one facility. A multiplexer also contains a demultiplexer so that it can demultiplex a received multichannel signal into separate voice channels.

Multiplexing A technique used on circuits between two points so that many circuits can be placed over one transmission medium. Many individual signals (voice or data) are combined and sent over this medium.

Multipurpose Internet Mail Extension (MIME) A protocol that extends the capabilities of Simple Mail Transfer Protocol (SMTP) to allow the transfer of binary files such as executable programs and graphics.

NAK A negative acknowledgment transmitted by the receiving modem when it detects that an error has occurred in a block of transmitted data. If a NAK is received, the modem retransmits the last block of data from its memory.

Narrowband ISDN Designates BRI and PRI, which use synchronous transmission.

National Science Foundation Network (NSFNET) Network that started out as a 56-Kbps X.25 packet-based wide area network to connect computers of several universities together for research work. The network has grown in size and speed and has become the network backbone for the Internet. This backbone is currently a 155-Mbps (OC-3) fiber optic facility.

Network An overused term. In telecommunications, the word network has many meanings, usually determined by the context. The term *public switched telephone network (PSTN)* refers to the interconnection of switching systems in the PSTN. *Switching network* refers to the component inside a switching system that switches one circuit to another circuit. The *network in a telephone* refers to the hybrid network, which performs a two-wire to four-wire conversion process.

Network Interface Card (NIC)   Designed to work with a specific technology and designed to plug into an ISA or PCI slot in a PC. The NIC contains a CPU, programming, and memory that allow it to control the flow of information between the NIC and the LAN without assistance from the main CPU of the host PC.

Network Layer   The third layer of the OSI model. The network layer allows for the establishment, maintenance, and control of sessions between computers not on the same physical network.

Network Operating Systems (NOS)   Designed specifically to allow PCs to share resources via a LAN. The NOS occupies the network, transport, session, presentation, and application layers of the OSI model and includes the APIs and software drivers necessary to interface the NOS to the OS and to user applications.

Network Termination 1 (NT1)   An interface device provided by a customer to terminate an ISDN line to the customer's premises. The NT1 interfaces the customer's ISDN equipment to the digital subscriber line of the LEC.

Network Termination 2 (NT2)   An interface device that interfaces devices to a PRI NT1. An NT2 is usually used in a PBX environment. NT2s will only be found in a PRI environment.

Network Termination 12 (NT12)   A device combining an NT1 and an NT2.

Nonreturn to Zero (NRZ)   Protocol that does not use the 0 V as a signal level but that uses a voltage level such as +3 V and –3 V (or some other two-state voltage such as +10 V and –10 V, + 15 V and – 15 V, and so on). The standard NRZ protocol is also referred to as *nonreturn to zero-level* (NRZ-L) to differentiate it from NRI. In NRZ-L the signal never returns to zero voltage, and the value during a bit time is a level discrete voltage. Typically, a negative voltage such as –5 V is used to represent a binary digit 1 and a positive voltage such as a +5 V is used to represent a binary digit 0. NRZ-L is the protocol used by the EIA-232 interface of a PC. The EIA-232 interface is used to connect the PC to devices such as an external modem or serial printer.

Nonreturn to Zero Invert (NRI)   Uses two discrete voltage levels for its signal, but the value of the voltage present does not represent 1s and 0s. Data is coded as the presence or absence of a transition from one voltage state to the other. It does not matter whether the transition is from the higher voltage state to the lower (such as from +3 V to –3 V) or from the lower voltage state to the higher (such as from –3 V to +3 V). If a transition occurs at the beginning of a bit time in either direction, it is coded as a 1. If no transition in the signal occurs at the beginning of a bit time, it is coded as a 0.

North Pole   The magnetic north pole that exists at one end of a magnet.

Null Modem   No modem. A null-modem cable is used to connect two PCs via their serial ports, when they are connected directly without using a modem.

Numbering Plan Area (NPA)   An area represented by an area code.

Ohmmeter   A test device used to measure the resistance of a component. An ohmmeter is usually one part of a multimeter.

Open Circuit   A circuit where the conducting path has been broken open.

Operating Systems   Software programs that control the basic input/output operations of a computer.

OSI Model   Developed by the International Standards Organization and used as a template to separate defined functions or protocols into seven separate layers. It is more of a conceptual than a practical framework.

Outside Plant   Cables, telephone poles, pedestals, and anything that is part of the telecommunications infrastructure and not inside a building.

Outside Plant Engineer   The person charged with properly designing the facilities that connect telephones to central exchanges and the facilities that connect central exchanges together (outside plant designer).

Overlay Programs   The term used to describe maintenance and administrative programs that do not permanently reside in memory but are contained on a secondary storage device. These overlay programs are only loaded into memory as needed and when loaded overwrite the last overlay program contained in memory.

Packet Assembler/Disassembler (PAD)   The software in the user's DTE that forms the user's data into packets according to the X.25 protocol and then delivers or receives these packets to or from the PDN. The PAD can also be a hardware device (with software) that allows for dialup connections to the PDN.

Packet Networks   Networks that use STDM techniques to transport data packets from many users over one physical circuit.

Packet (or Public) Data Network (PDN)   A public network that connects computers together and transports data between them. It is composed of circuits leased from IECs and LECs. The most common PDNs are Telenet, now owned by Sprint and called *SprintNet*, and Tyment, now owned by MCI and called *Concert.*

Packet Switching Exchanges (PSEs)   Nodes in the PDN used to route packets through the network.

Parallel Transmission   The use of several transmission leads to allow the simultaneous transmission of several bits at one time. A parallel data bus with 8 leads can process data 8 bits at a time. Or 64 leads on the data bus allows information to be processed 64 bits at a time. Parallel transmission is much faster than serial transmission but requires many more transmission leads.

Peer-to-Peer   A LAN arrangement where each PC can act as either a server or a client.

Peg Count (PC)   A term used to indicate how many calls are handled by a circuit, a group of circuits, or a switching system. This term is a hold-over from electromechanical traffic registers. When the relay of the register operated it was said to peg the meter. Each peg of a meter made it turn a numbered dial one step.

Peripheral Component Interconnect (PCI) Bus   Developed by Intel to provide for faster transfer of data between peripherals and the motherboard.

Peripheral Equipment   Equipment which exists on the periphery of a switching system matrix. Peripheral equipment interfaces external devices to the computer of a SPC switch to provide capabilities and features to these devices. Most peripheral equipment consists of line circuits to interface a telephone to the switch and trunk circuits to interface inter-office circuits to a switch. Also included in this category are dial registers, modems, and voice mail.

Permanent Magnet   A steel alloy that holds magnetic properties for a long time once it has been magnetized.

Permanent Virtual Circuit (PVC)   A circuit that has been permanently assigned a path through the data network by virtue of a permanently assigned logical channel number.

Personal Communication System (PCS)   Not so much a technology as a concept. The concept of PCS is to assign someone a personal telephone number (PTN). This PTN is stored in a database on the SS7 network. That database keeps track of where a person can be reached. When a call is placed for that person, the artificial intelligence network (AIN) of the SS7 determines where the call should be directed.

Personal Communication System 1900 (PCS 1900)   PCS is provided by radio frequencies in the 1900-MHz range. PCS 1900 is the latest evolution in mobile communication. PCS 1900 can be provided by the time division multiple access (TDMA) technology of DAMPS or can use code division multiple access (CDMA) technology.

Physical Layer   The lowest layer of the seven-layer OSI model. It contains the hardware necessary to make physical connection to a LAN medium and to transmit electrical signals onto the medium.

Plastic Insulated Cable (PIC)   Cable that contains wires electrically isolated from

each other by plastic insulation around each wire.

Point of Presence (POP)   The point at which the local exchange carrier and interexchange carrier facilities meet each other.

Port   The interface point between a switching system matrix and external devices. Each switch will be equipped with a certain number of ports to handle the number of devices that will be attached to the system. If a line circuit card is inserted in the port, it becomes a line port. If a trunk circuit card is inserted into a port it becomes a trunk port. If a dial register card is inserted in a port, it becomes a dial register port. Line circuits usually use one port per circuit. Some trunk circuits, dial registers, and special circuits may use two or more ports for each circuit.

Power Supply   A device that converts AC voltage into DC voltages and then supplies these voltages to other devices.

Preferred Interexchange Carrier (PIC)   Determined as follows: A customer tells the local exchange carrier who it wants to use as its long distance service provider. The LEC makes an entry in the database of the switching system serving this customer. That database entry will inform the switch which IEC should be used on long distance calls placed by that customer. With the 1996 Telecommunications Reform act, customers must be able to choose a carrier for toll calls within a LATA as well as the same or a separate carrier for long distance service.

Primary Rate Interface (PRI)   Consists of 23 64-Kbps B-channels and one 64-Kbps D-channel.

Private Automatic Branch Exchange (PABX)   A switching system that resides in a business location, is owned by the business (or third-party supplier). It connects telephones within the business to each other when the originater dials an extension number or connects the originater to the local central office switch when a "9" is dialed.

Private Branch Exchange (PBX) The name given to manual switchboards in a private business location. A PBX may also be a small SPC switching system used by large businesses. All PBXs sold today are computerized branch exchanges (CBXs) or CTI-based systems.

Private-Line Services Services provided by a common carrier to a private organization to help that organization establish its own private network. A private network cannot be accessed or utilized by the public. A private network is exactly that—it is private and can only be accessed and used by the private organization.

Proprietary Telephone A telephone designed by its manufacturer to only work with certain keysystems or PABXs made by that same manufacturer.

Protocol The rules of communication. Each protocol defines a formal procedure for how data is to be transmitted and received using that protocol.

Public Data Network (PDN) The packet data network accessible to the general public for the transmission of packet data.

Public Switched Telephone (or Telecommunications) Network (PSTN) A network of switching systems connected together to allow anyone to call any telephone located in the United States. A public network is accessible to any members of the general public. The PSTN is composed of many nodes (switching systems) and many transmission links connecting these nodes. Stations are the end points of the network and connect to the PSTN via their local exchange.

Public Utilities Commission (PUC) The state government agency that regulates telecommunications within the state.

Pulse Amplitude Modulation (PAM) The process used to take samples of analog voice signals so they can be multiplexed using TDM. The samples appear as pulses in the TDM signal.

Pulse Code Modulation (PCM) The industry standard method used to convert an analog signal into a digital 64,000-bps signal. PCM takes a PAM signal and converts each sample (pulse) into an 8-bit code.

Quadrature Amplitude Modulation (QAM) The transmission of 4 bits per baud. A V.29 modem transmits at 2400 baud with 4 bits per baud to achieve 9600-bps transmission. QAM uses 2 different amplitudes for each of 8 different phases of a 1700-Hz signal to achieve the 16 detectable events necessary to code and decode 4 bits at a time.

Radio Waves Electromagnetic waves. When these waves cut across a wire (antenna), they induce a voltage signal in the wire.

Receiver The device responsible for decoding or converting received information into an intelligible message.

Recommended Standard 232 (RS-232) The most common interface standard for data communication. CCITT standard V.24 is the same as RS-232. RS-232 defines the voltage levels needed for the various signal leads of RS-232.

Reference Points The points at which various devices are connected in an ISDN circuit. The U interface serves as a demarc. It is where the cable pair from the LEC attaches to an NT1. The S/T reference is the interface where a TE1 or TA connects to a NT1. The R reference is the interface where a non-ISDN device attaches to a TA. The V reference is the interface point between the LE and ET in the central exchange.

Register (Dial Register)   The device in a common control switching system that receives dialed digits.

Regulated Monopoly   A sole provider of goods or services; regulated by a government agency. A regulated monopoly must seek approval from the government agency for anything it wishes to do. The government agency sets the price that a monopoly can charge for its goods and services.

Relay Center   A center that relays information or data from one customer to another. Today the term designates telecommunications centers that relay information between people who are hard of hearing and those who are not.

Reluctance   The resistance to flow of magnetic flux.

Repeater   A hardware device that regenerates digital signals.

Request for Comments (RFC)   A procedure used to get feedback on recommendations that someone is making for data networks in general but for the Internet in particular. The RFCs will contain information on Standards as they are developed and can be referred to in order to gain an understanding of how a protocol such as IP or TCP works.

Residual Magnetism   The magnetic properties left in an electromagnet after the current has been reduced to zero.

Resistance   The amount of opposition to electric current flow that a device has.

Resistance Design   Designing a local loop, which uses twisted-pair copper wire, so

that the resistance of the wire serving a telephone does not exceed the resistance design limitations of the central exchange line circuit. Each switching system manufacturer will state how much resistance its switch is designed to support.

Resistors   Human-made devices that simulate a specific desired resistance. Resistors convert electrical energy to heat.

Revenue Sharing   The sharing of toll revenue between the Independent telephone companies and AT&T prior to the 1984 MFJ.

Revised Resistance Design (RRD)   A design approach in which the outside plant engineer uses the length of the local loop, as well as the design limitations of the central exchange, to determine the local-loop design criteria. Loops up to 18,000 ft are designed to use nonloaded cable pairs with a maximum loop resistance of 1300 Ω. Loops between 18,000 and 24,000 ft are designed to use loaded cable pairs with a maximum loop resistance of 1500 Ω. Loops longer than 24,000 ft are designed according to modified long route design (MLRD).

Ring Detector   An electronic solid-state transistorized device designed to detect the presence of a ringing signal (a 90-V AC signal).

Ringer   An electromechanical device (relay) that vibrated when a 90-V AC signal was received by the telephone. The vibrating device would strike metal gongs to create a ringing sound.

Ringing Machine   A device in a telephone central exchange that converts 52 V DC into an AC voltage used to ring a telephone.

RJ-11 Jacks   Regular telephone jacks. The jack that a telephone plugs into is an RJ-11 jack. These jacks contain four wires, though only two are used for a telephone.

RJ-45 Connectors   Telephone jacks that contain eight wires. They are used for data devices.

Roaming   The mobile telephone of AMPS or DAMPS is continuously reporting its location to the closest cell site. As you drive across country and pass from one service provider to another, the central controller of the MTSO owned by other service providers will report your location over the SS7 network to your home base.

Root Mean Square (RMS)   The effective value of AC compared to DC. The RMS formula to get the DC effective working power of an AC signal: RMS = 0.707 AC.

Rural Electrification Administration (REA)   Department of the federal government established during the Depression under President Franklin D. Roosevelt by the Rural Electrification Act of 1936. This "act" put people to work bringing electricity to rural America. The "act" was amended in 1949 to bring telephone service to rural America. This department was renamed the Rural Utility Services Department.

Selective ARQ   With selective ARQ, only frames that have errors are retransmitted. When a transmitting modem is using selective ARQ and receives a NAK, only the frame for which the NAK was received is retransmitted.

Serial Transmission   The transmission of bits, one behind the other, over one transmission medium.

Server   A PC or minicomputer containing resources or access to resources that can be shared via a LAN with clients (workstations) on the LAN. A print server is a PC that controls the access of clients to the various printers attached to a LAN. A file server contains files and applications software that can be shared with clients.

Service Profile Identifier (SPID)   Contained in a database maintained by the LEC of ISDN line equipment and the services available to each line. This database contains SPIDs. When a terminal is activated, the SPID is used to determine which services a terminal can have access to.

Session Layer   The fifth layer of the OSI model. This layer is responsible for maintaining and terminating sessions between user applications programs. In a client/server environment, the session layer ensures applications can cross between different operating systems on the client and server.

Sidebands   The two additional signals generated by mixing two signals. The upper sideband represents the sum of the carrier and modulating frequencies. The lower sideband will have a frequency that is the difference between the carrier frequency and the modulating frequency. Both sidebands contain the intelligence of the modulating signal.

Sidetone   In a telephone set, some of the transmitted signal is purposely coupled by the hybrid network to the receiver so you can hear yourself in the ear covered by the telephone receiver. This signal is called sidetone.

Signal-to-Noise Ratio (S/N Ratio)   The comparison of the power in the desired signal to the power in the noise signal. The S/N ratio should be 30 or more.

Signaling System 7 (SS7)   A network of switches and databases used to establish and control voice paths in the PSTN. SS7 is a data network that connects switching systems together and allows them to exchange information.

Simple Mail Transfer Protocol (SMTP)   The protocol used for sending e-mail over the Internet. SMTP is identified in the TCP header as port 25.

Simplex Transmission   Transmission of signals in one direction only is called *simplex communication*.

Sine Wave   A graphic representation of an AC signal along a time line.

Software-Designed Network (SDN)   A service sold by IECs to serve as a replacement for a business's private-line network. SDN allows you to replace WATS, FX trunks, and tie lines with connections via the IEC's SDN.

South Pole   The magnetic south pole of a magnet.

Space Division Multiplexing   A process in which multiple communications are placed over many different wire pairs inside one cable. Each communication channel (voice or data) occupies its own space (occupies its own set of wires).

Space Division Switching   A form of switching (SXS and XBAR) where each conversation occupies its own distinct and separate wire path through the switching system.

Specialized Common Carrier Decision of 1971   An extension of the 1969 MCI Decision by the FCC. This ruling allowed any common carrier to handle private-line networks.

Static Electric Charge   The development of an electric potential on a material. The electric potential does not move (is static). A static charge on a material is usually developed by friction between it and another material.

Station Equipment   The largest segment of station equipment is the telephone, but the term *station equipment* has been broadened to include anything a customer attaches to a telephone line. The most common piece of station equipment is the telephone. A modem, CSU/DSU, personal computer, and keysystem are all referred to as station equipment.

Statistical Time Division Multiplexer   A hardware device that multiplexes many input channels of data or voice onto a common TDM medium between two STDM multiplexers but only places data into a TDM slot when the input has data to send. This makes the STDM multiplexer more efficient than a synchronous time division multiplexer because time slots are not reserved for each and every input channel. Since the STDM multiplexer must be smart enough to inform the distant STDM multiplexer which output channel should receive the data, these multiplexers are also referred to as *intelligent multiplexers*.

Statistical Time Division Multiplexing (STDM)   A process of time division multiplexing where the time slots on the TDM medium between STDM multiplexers are dynamically allocated. Any input to the STDM multiplexer can be assigned to any time slot on the TDM medium. Time slots on the TDM medium are assigned to input channels that have data ready to send over the TDM medium to a distant STDM multiplexer. To accomplish dynamic allocation of any input to any time slot, the TDM multiplexer places a header on the data that each input channel has ready to transmit. This header informs the receiving STDM multiplexer which output channel is to receive the data.

Straight-Line Ringer   A ringer that will operate on any ringing signal. This type of ringer is used on single-party private lines.

Subscriber Line Carrier-96 (SLC-96)   SLC-96 uses TDM to multiplex 96 lines over two cable pairs. Multiple SLC-96 systems are often multiplexed onto a fiber facility.

Superimposed Ringer   The type of ringer used by Bell on four-party lines. The ringer included a diode. Ringing signals were superimposed on top of a DC voltage. The diode only allowed the signal with the correct polarity of DC for that station to pass to the ringer.

Switch   A device used to connect two parts of a circuit together so electrical current flows when the switch is closed or ceases when the switch is open.

Switchboard   The manual switchboard was used by operators to establish calls between two telephones. The manual switchboard was replaced by an automated switch.

Switched Virtual Circuit (SVC)   A circuit that is set up by a call setup packet. The call setup packet at the beginning of a data session establishes a virtual circuit for the subsequent transfer of data. When the data session ends, a call clear packet is issued to release the LCN assignments.

Synchronous Optical Network (SONET)   A network of fiber media connected together for the transport of digital signals using light waves.

Synchronous Time Division Multiplexer   A hardware device that places many signals over one medium using time division multiplexing. A T1 system is a time division multiplexer.

Synchronous Time Division Multiplexing   A multiplexer that has static (permanent) assignments of time slots on the TDM highway between two multiplexers. Each time slot is permanently mapped to an input and is reserved only for that input regardless of whether the input channel has information to send.

Synchronous Transmission   The transmission of data as blocks of bytes. Synchronization of the receiver occurs from a special bit pattern called a *sync signal* placed in front of the block of data information.

Synchronous Transport Level 1 (STS-1)   The basic building block signal level for SONET.

Tags   HTML tags. These tags are used by a browser as instructions on how a web page should be displayed.

Tandem Exchange   Class 1, 2, 3, and 4 exchanges that are part of the PSTN.

Tariff   The document filed by a common carrier that defines in detail any service proposed by the carrier and the charge proposed for the service. For interstate service, the tariff is filed with the FCC. For intrastate services, the tariff is filed with the PUC.

Telecommunications Application Program Interface (TAPI)   A special application program that allows the PC to act as a telecommunications device.

Telecommunications   The transmission of voice, data, or video over a distance using electrical, electromagnetic, or light signals. The communication process occurs via the public switched telephone network (PSTN) or privately owned networks.

Telecommunications Reform Act of 1996   Effectively replaces the 1984 MFJ. This act is designed to accelerate competition for providing services in the local and long distance markets. It eliminates the franchised local territory concept. The LEC is no longer the only service provider in town. Other common carriers can come into an incumbent local exchange carrier (ILEC) territory and provide local telephone service. These new local service providers are called *competitive local exchange carriers (CLECs)*. Prior to this act, the older incumbent LECs (ILECs) were restricted to providing local services only. Under this act, if the ILEC can prove that it has not hindered competition in its local exchange territories, it can now offer long distance services.

Teletype   A device used to transmit and receive data over a teletype network. This was the forerunner of today's data networks.

Temporary Magnets   Human-made magnets that readily lose their magnetism when the external magnetizing force is removed.

Terminal Adapter (TA)   A device that interfaces a non-ISDN device to a NT1 in BRI and to a NT2 in a PRI environment.

Terminal Equipment 1 (TE1)   Equipment designed to ISDN standards and for direct interface to an NT1.

Terminal Equipment 2 (TE2)   Equipment that is non-ISDN compliant. TE2 equipment must be attached to a terminal adapter.

Terminal Equipment Identifier (TEI)   Specific number assigned to each TE1 and TA. The TEI is usually assigned automatically the first time the terminal device is used. LAPD will invoke a query of the SPID database to obtain the TEI. The TEI is used by the LE and ET to determine what service to provide for a B-channel.

Time Division Multiple Access (TDMA)   TDMA technology is used by DAMPS. With TDMA, each conversation occupies a time slot on a common transmission medium. In DAMPS, each conversation occupies a time slot on a particular radio frequency.

Time Division Multiplexing (TDM)   The process of converting each speech path into samples, then combining the different samples by transmitting each sample at a different time, so they may be sent over one transmission medium.

Time Division Switching   A form of switching (stored program control or SPC) where each conversation occupies its own distinct and separate time slot on a common wire path through the switching system.

Token   The access methodology used by token-ring (IEEE 802.5) technology. A special bit pattern called a *token* is passed from one station to the next, and only the station with the token can access the medium of the LAN and transmit data onto the medium.

Token-Ring   A LAN configuration where data passes through each station attached to the LAN. The stations are wired in a Ring configuration by wiring all stations to a medium access unit (MAU).

Tolerance   The rating of a resistor that states how closely the resistor will match its stated value. The color of the fourth band on a resistor states the manufacturing tolerance. A gold band is + or –5% and a silver band is + or –10%.

Traffic   A term used for the volume of calls placed over a circuit, group of circuits, or a switching system. If a system is handling 10,000 calls an hour we say the traffic is 10,000 calls an hour.

Trailer   A trailer is used in synchronous communication; it is data placed behind the block of information transmitted. The trailer contains parity checking information and the address of the sender.

Translator   In a stored program control switching system, a database that converts dialed digits into switching instructions.

Transmit   This header informs the receiving STDM multiplexer which output

Transmission Control Protocol (TCP)   A transport layer protocol used in conjunction with IP at the network layer to form a TCP/IP protocol. TCP turns the unreliable connection of IP into a reliable connection by performing error detection and correction. The TCP protocol is identified inside an IP packet by a Protocol field number of hexadecimal 6.

Transmitter   The device responsible for sending information or a message in a form that the receiver and medium can handle. The device in a telephone that converts the air pressure of a voice signal into an electrical signal that represents the voice.

Transport Layer   The fourth layer of the OSI model, it provides for end-to-end reliability of network connections by providing flow control with error-detection and recovery mechanisms.

Trellis Coded Modulation (TCM)   TCM is a forward error-correction technique. An extra bit is added to the bits of data transmitted to help the receiver decode the data more reliably. QAM transmits 4 bits at a time. TCM added a 5th bit to the 4-bit code for error-correction purposes.

Tuned Circuit   A circuit containing inductance and capacitance so that at one particular frequency of AC they are the same $(XL = XC)$. The circuit is tuned by XL and XC to be most responsive to one particular frequency.

Tuned Frequency   The frequency of an AC signal that a tuned circuit is tuned to.

Tuned Ringer   A ringer that will operate on only one ringing signal. These ringers were used on party lines. They were tuned to only ring on the signal assigned to that party.

2500 Telephone Set   The standard single-line telephone, which contains a DTMF dial pad.

2B1Q   The coding technique used for a digital subscriber line. This coding technique allows 2 bits to be transmitted at one time by using four distinct voltage levels. Each voltage level can represent a particular 2-bit code.

Uniform Resource Locator (URL)   The identifier used to designate where an application or resource resides on the Internet. The URL consists of six parts: Service, Domain Name, Port Number, Pathname, Filename, and Variable. Prot Number, FIlename, and Variable are optional fields. The URL is typed into browsers' Address or Netsite field when using **http.** The computer then sends a request via the Internet to port 80 on the destination host for an HTML file located on that host.

Universal Asynchronous Receiver/Transmitter (UART)   A specially designed DSP chip that interfaces a parallel data bus to a serial data lead and vice versa.

UnPBX   A PBX system that is CTI based, and thus there is no stand-alone CBX system. The system is within the PC that serves to replace a stand-alone CBX.

User Datagram Protocol (UDP)   A transport layer protocol used with IP when error detection and correction are not needed. When data is being transmitted over a highly reliable error-free transmission medium, error detection and correction are not needed and the use of UDP will provide for a faster transfer of data than the TCP protocol. The UDP protocol is identified inside an IP packet by a Protocol field number of hexadecimal 11.

Variable Resistors   Resistors that contain a moveable component so that we can vary the resistance desired from 0 ohms to some maximum value.

Varistor   A resistor that changes value according to the amount of voltage applied to it. The higher the applied voltage, the higher its resistance will be.

Virtual Circuit   The term used to describe a circuit in a data network where many users share the same physical circuit in a TDM arrangement.

Voice Communications The transmission of voice signals over telecommunication facilities.

Voice Frequency Repeater (VFR) A device that will amplify voice signals in both directions to compensate for the extra decibel loss of long loops.

Voice-Over IP (VOIP) The transmission of voice using Internet protocol-based packets that contain digitized voice signals.

Volt, Voltage A measure of how much potential difference or electrical pressure exists between the two terminals of a battery (or some other device that generates electricity). This electrical pressure is called *electromotive force (EMF)* and is measured in volts.

Voltmeter A test device used to measure voltage. A voltmeter is usually one part of a multimeter.

Watts A measure of the amount of power that can be handled or consumed by a device. $W = I \times R$

Wave Division Multiplexer The hardware device attached to each end of a fiber pair that multiplexes 16 OC-48 systems onto the fiber pair.

Wave Division Multiplexing (WDM) Technology that splits the light spectrum into many different frequencies. A time division multiplexer is then assigned to each of the different light wave frequencies. Many TDM signals are multiplexed together so they can be placed on one fiber medium.

Wet Cell A chemical cell that generates electricity via chemical means between an electrolytic solution and two dissimilar metals. The electrolytic solution is a liquid. The battery in a SPC central exchange is composed of 24 wet cells.

Well-Known Port Numbers Port numbers that have been assigned by the Internet Assigned Numbers Authority to each application using TCP or UDP. These port numbers exist between the upper tow levels of the TCP/IP and UDP/IP protocol model. They identify which application should receive data from the transport layer.

Wide Area Network A data network that spans a large geographic area from across town to across the country. A WAN is used to connect LANs or computers in different locations together.

Wide Area Toll Service (WATS) A service sold by LECs and IECs. The LECs traditionally have sold intrastate WATS and the IECs have sold interstate WATS. With the 1996 Telecommunications Reform Act, LECs and IECs can sell both services. A customer pays so much a month or so much per minute of billing (or both) to be able to place long distance calls within a particular geographic region. The larger the region a business desires access to, the higher the cost will be.

Wiring Hub A device used to connect computers to a LAN when the medium used by the LAN is twisted-pair copper wire.

Word In a PC, the number of adjacent bits that can be manipulated or processed. This depends on the number of leads comprising the data bus, which is in turn dependent on the number of registers attached to the data bus. A word can be 8 bits or 1 byte in an 8088 microprocessor environment. A word can be 32 bits or 4 bytes in a 80486-based PC, or it can be 64 bits (8 bytes) in a Pentium-based PC.

X.25 The interface standard to a packet data network.

Zero-Slot LANs LANs that do not require a NIC and thus require no slot on the ISA or PCI bus. PCs are connected via the serial or parallel port.

Index

Major trading area (MTA), 397
Management by objective (MBO), 409
Manchester, 244
Mann-Elkins Act of 1910, 27
Map, area code, 88–89
Mark bit, 263
Marker, 553, 559–560
Mark parity, 264
Mastergroup, 562–563
Mate colors, plastic insulated cable, 158
Materials, 421
Matter, 420
MCI. *See* Microwave Communications Inc. (MCI)
Media access control (MAC), 328
    MAC address, 337
Medium, 64
    LAN, 332–333
    for long distance networks, 179–181
    twisted copper wire as, 65
Medium access unit (MAU), 324
Mergers, 51–52
Message length word, 278
Message type word, 278
Method of procedure (MOP), 408
Metric numbering system, 19–20
Metropolitan area network (MAN), 16, 308, 321
MF signaling, 101
Microwave Communications Inc. (MCI)
    decision of 1976, 32
    ruling of 1969, 32
Microwave radio, 4
Mixer, 194, 195
Mobile telephone serving office (MTSO), 394
Mobile telephone switching office (MTSO), 394–395
Mobile telephone system (MTS)
    AMPS, 392–395
    beginnings of, 390–391
    IMTS, 392
    managing, 411
    See also Personal communication system (PCS)
Modem, 3, 253
    See also Data communication equipment (DCE)
Modified Final Judgment (MFJ), 35
Modified long route design (MLRD), 162–163
Modulating signal, 198
Modulation, 193
    of carrier, baud rate and, 268–274
Modulator, 194, 195
Modulator/demodulator, 255–256

Modulo 8, 359
Modulo 128, 360
Molecules, 420
Monitor lights, 76
Monitor switch, 548
Monopoly, 26
Monopoly service providers, 26
    regulation of, 26–27
Morse code, 3–4
Multiline telephones, 122, 124
Multimeter, 440, 454
Multiple appearance directory number (MADN), 143
Multiple subscriber line carrier systems, 155
Multiplexer, 155
    intelligent and asynchronous, 215
Multiplexing, 9, 176, 192–218
    advances in technology, 78
    dense wave division (DWDM), 182
    history of, 545
    reasons to use, 195
    space division, 193
    See also Frequency division multiplexing (FDM); Time division multiplexing (TDM); Wave Division Multiplexing (WDM)
Multipurpose Internet Mail Extension (MIME), 384
Mutual capacitance, 170
Mutual induction, 496

N

Narrowband ISDN, 308
National Information Infrastructure (NII), 18
National Science Foundation Network (NSFNET), 370
Natural magnets, 479
Negative ion, 422
Negative terminal, 426
Network, 26
    AIN, 390
    balancing, 112
    connection-oriented, 356
    data, 411
    digital, converting PSTN to, 239–242, 288–290
    engineers, 411
    flat, 44–45
    hybrid, 112, 197–198
    ISDN. *See* Integrated Services Digital Network (ISDN)
    local area. *See* Local area network (LAN)

Positive terminal, 426
Power loss, vs. bandwidth, 246–248
Power-loss design, 173–175
Power ratings, for devices and resistors, 451–452
Power supply, 502
    local, at each telephone, 66–70
Predictive pulse code modulation, 238
Preferred interexchange carrier (PIC), 38
Primary center, 41
Primary colors, plastic insulated cable, 158
Primary rate interface (PRI), 292
Primary rate ISDN, and broadband ISDN, 308–310
Primary wire, 496
Private automated branch exchange (PABX), 77, 128
Private branch exchange (PBX), 12–13, 76, 127
    proprietary telephones, 122
    purpose and functions, 137–138
    system features, 138
Private line, 288
Private-line network, 346
Private-line services, 32
Private network, managing, 136–137
Proprietary telephone, 122
Protocol, 327, 335
    ARP, 378–380
    bit-oriented, 361
    communication signals and, 223–224
    connectionless, 363
    FTP, 382–384
    SMTP, 373
    TCP/IP, 370–374
    UDP, 372
    X.25. See X.25 protocol
Protons, 421–422
Public data network. See Packet data network (PDN)
Public switched telephone network (PSTN), 2–3, 26
    circuit switching via tandem exchanges in, 85
    converting to digital network, 239–242, 288–290
    data and, 256–258
    extended area service in, 83–85
    old analog, and data, 287–288
    toll calls over, 80–83
Public Utilities Commission (PUC), 28–29
Pulse amplitude modulation (PAM), 293
    TDM using, 203–205

Pulse code modulation (PCM), 207–208, 238
    adaptive differential (ADPCM), 238
    predictive, 238
    TDM using, 205–206

Q

Quadbit, 269
Quadrature amplitude modulation (QAM), 271
Quadrature phase modulation, 268–269
Quality of service (QoS), 363
Quantization, 207
Quantizing, 238
Quantizing noise, 207, 238
Quaternary symbols, 293
Quats, 293

R

Radio waves, 4
Range extenders, 162
Rates
    higher, as result of deregulation, 33–34
    reduced IEC access, 48–50
Reactance
    capacitive, 494, 498–499
    inductive, 494–496
Ready access terminals, 176
Receiver, 111
Recommended Standard 232 (RS-232), 253
Reference point, 296
Regional Bell Operating Companies (RBOCs), 36
Regional center, 42
Register, 553
    storage of dialed digits in, 99, 557–559
Regulated monopolies, 26
Relative density, 435
Relative phase modulation, 268–269
Relay centers
    for hearing impaired, 7
    and speed of delivery, 5–7
Remote access server (RAS), 270
Remote bridges, 329
Repeater, 327
Request for comments (RFC), 371
Residual magnetism, 486
Resistance, 432, 437–438
    measuring, 454–455
    parallel, 442–443
    relationship with current and voltage, 438–439
    series, in circuit, 440–442

Telecommunications-based information technology (TBIT), 414
Telecommunications management
  interpersonal skills for, 406–407
  leading and directing, 409
  making decisions, 414
  managing growth, 413–414
  role in planning, 407–408
  and staffing, 408–409
  technological component, 412–413
Telecommunications Reform Act of 1996, 45
  and ISPs, 54–55
  1997 and 1998 events regarding, 48
  summary of, 528–544
Telegraph, 3–4
Telephone
  connecting to central exchange, 229–232
  electronic feature, and answering machines, 121
  hands-free vs. speakerphone, 122–123
  introduction of central power and line equipment to, 74–76
  invention of, 5, 65, 546
  local power supply at, 66–70
  multiline, 122, 124
  proprietary, 122
  resistance of wires connecting to, 460–462
  varied resistance of, 462–463
  See also Mobile telephone systems
Telephone, integrated
  and data terminal, 123
  services digital network, 123–124
Telephone companies, independent, 27
Telephone numbering, 74
Teletype, 546
Telnet, 385
Temporary magnets, 480
10 Base-2, 326
Terminal adapter (TA), 296–297
Terminal equipment
  connections to NT1, 301–302
  and S/T bus, 297–298
Terminal equipment identifiers (TEI), 299–300
Terminal equipment 1 (TE1), 301
Terminal equipment 2 (TE2), 301
ThinNet, 326
Ticker tape, 546
Tie lines, 130
Time division multiple access (TDMA), 397
Time division multiplexing (TDM), 78, 92–93, 193, 199, 293, 344

circuit in, 93–94
European 30 + 2 system, 214
higher levels of, 215
STDM, 214–215
switching, 94–97
using PAM signals, 203–205
using PCM signals, 205–206
Time division switching, 100
Tip lead, 71, 232
Token ring, 324
  LAN, 246
Tolerance ratings, 452
Toll calls, over PSTN, 80–83
Toll office, 81
Toll revenue, dividing prior to 1984, 35
T1 carrier system, 243, 211–214
Total element long-run incremental costs (TELRIC), 48, 50
Touchtone dial pad, 556–557
  See also Dual-tone multifrequency (DTMF) dial
Trailer, 275
Transceiver, 65, 111, 325
Transducers, 227
Transformers, 496–498
Translating bridges, 329, 336
Translator, 559–560
Transmission
  asynchronous vs. synchronous, 261, 263–264
  full-duplex, 64–65
  half-duplex, 64–65
  serial vs. parallel, 261
  simplex, 64–65
  start-and-stop, 263
Transmit and receive paths, 195
Transmitter, 64, 110
  carbon, 65
  carbon granule, 111–112
  UART, 146, 208, 259
Transmitter design criteria, 161–162
Transport layer, 329
T reference point, 297
Trellis coded modulation (TCM), 271–272
Trunk circuit, 90
Trunks, interoffice, 73
Trunk side access, 37
T3 carrier system, 181
Tuned circuits, 499–502
Tuned frequency, 500
Tuned ringer, 114
2500 telephone set, 115